MW00635794

Mechanisms and Machines: Kinematics, Dynamics, and Synthesis

Mechanisms and Machines: Kinematics, Dynamics, and Synthesis

Michael M. Stanišić
University of Notre Dame

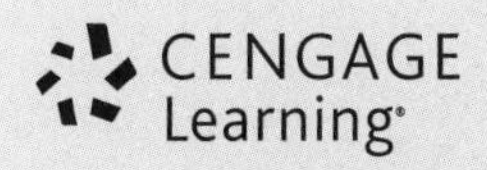

Australia • Brazil • Japan • Korea • Mexico • Singapore • Spain • United Kingdom • United States

Mechanisms and Machines: Kinematics, Dynamics, and Synthesis
Michael M. Stanišić

Publisher: Timothy Anderson

Development Editor: Eavan Cully

Senior Editorial Assistant: Tanya Altieri

Content Project Manager: D. Jean Buttrom

Production Director: Sharon Smith

Media Assistant: Ashley Kaupert

Intellectual Property Director: Julie Geagan-Chevez

Analyst: Christine Myaskovsky

Project Manager: Amber Hosea

Text and Image Researcher: Kristiina Paul

Senior Manufacturing Planner: Doug Wilke

Copyeditor: Connie Day

Proofreader: Pat Daly

Indexer: Shelly Gerger-Knechtl

Compositor: Integra Software Services

Senior Art Director: Michelle Kunkler

Cover and Internal Designer: Liz Harasymczuk

Cover Image: © Oleksiy Maksymenko/Getty Image

Library of Congress Control Number:2013954202

ISBN-13: 978-1-133-94391-4
ISBN-10: 1-133-94391-8

Cengage Learning
200 First Stamford Place, Suite 400
Stamford, CT 06902
USA

Cengage Learning is a leading provider of customized learning solutions with office locations around the globe, including Singapore, the United Kingdom, Australia, Mexico, Brazil, and Japan. Locate your local office at: **international.cengage.com/region**.

Cengage Learning products are represented in Canada by Nelson Education, Ltd.

For your course and learning solutions, visit **www.cengage.com/engineering**.

Purchase any of our products at your local college store or at our preferred online store **www.cengagebrain.com**.

Printed in United States of America
1 2 3 4 5 6 7 18 17 16 15 14

*To **L**auren, **E**mily, and **O**livia.*

Contents

Preface

This text is an introduction to the analysis and synthesis of mechanisms and machines, with an emphasis on the first. The intended audience is undergraduates who are studying mechanical engineering but the audience could also include students enrolled in multidisciplinary programs such as mechatronics or biomechanics.

The Vector Loop Method and Kinematic Coefficients

My main motivation for writing this book is to introduce the vector loop method and kinematic coefficients into an introductory Mechanisms and Machines course and to present this subject matter entirely in terms of them. Over the years, I have found that using this method provides students with systematic and easily computerized procedures for determining the kinematic and dynamic properties of machines and mechanism.

According to Hall[1], kinematic coefficients were first introduced by Eksergian[2] who formulated the kinematic and dynamic equations of motion of a machine in terms of "velocity ratios and their derivatives with respect to the fundamental coordinate of the mechanism . . ."

These velocity ratios and their derivatives are the kinematic coefficients. I suspect the method did not catch on then due to its computational intensity. At that time, digital computing was non-existent and analytical results were primarily obtained graphically. The legacy of that continues as most texts today spend considerable time presenting graphical methods, such as velocity and acceleration polygons and instantaneous centers.

Modrey[3] applied kinematic coefficients (he referred to them as "influence coefficients") to the velocity and acceleration analysis of planar mechanisms in paradoxical configurations, configurations that did not amend themselves to standard methods of analysis. Benedict and Tesar[4] applied kinematic coefficients to the dynamic equation of motion of a machine.

In the early 1970s, when the digital computer was becoming a tool available to all engineers, making large amounts of computation commonplace, Hall introduced the vector loop method and kinematic coefficients into the undergraduate curriculum at Purdue University. There were no textbooks at that time which focused on the vector loop method, much like today. A few years prior to becoming Emeritus, Hall published a packet of course notes.[5] [6] Hall's "Notes . . ." was a dense introduction to the vector loop method and kinematic coefficients and their application to velocity and acceleration

[1] A.S. Hall Jr., *Mechanism and Machine Theory*, vol. 27, no. 3, p. 367, 1992

[2] R. Eksergian, *J. Franklin Inst.*, vol. 209, no. 1, January 1930 to vol. 201, no. 5, May 1931.

[3] J. Modrey, *ASME J. of Applied Mechanics*, vol. 81, pp. 184-188, 1957.

[4] C.E. Benedict and D. Tesar, *J. of Mechanisms*, vol. 6 pp. 383-403, 1971.

[5] A.S. Hall Jr., *Notes on Mechanism Analysis*, Balt Publishers, Lafayette, IN, 1981.

[6] A.S. Hall Jr., *Notes on Mechanism Analysis*, Waveland Press, Prospect Heights, IL, 1986.

analysis and to the inverse and forward dynamics problems. The forward dynamics problem and dynamic simulation of a machine became tenable through the use of kinematic coefficients. Hall can be credited with bringing this method into the undergraduate curriculum, disseminating this powerful tool to thousands of engineers.

This book is a major expansion of Hall's "Notes..." The majority of examples and problems are new here. Hall introduced the rolling contact equation for modeling rolling contacts in mechanisms. Here, the rolling contact equation is extended to include geartrains and transmissions so that rolling contacts in both mechanisms and geartrains are modeled by the same rolling contact equation. Kinematic coefficients have been extended here to describe limit positions, time ratio, dead positions, transmission angle, mechanical advantage and to develop Freudenstein's equation, arguably the most powerful synthesis tool available to a mechanism designer.

Advantages of the Vector Loop Method

When studying the kinematics and dynamics of mechanisms and machines, the vector loop method and the kinematic coefficients offer several benefits:

- The method is purely algebraic. Geometric interpretations of kinematic properties such as limit positions and dead positions follow from interpretation of the algebraic expressions for the kinematic coefficients. Graphical methods and force analysis are not necessary to develop these properties of a mechanism. The algebraic results are also well suited to computerization. For example, the Jacobian matrix which is inverted to solve the position problem using Newton's method is the same Jacobian which needs to be inverted to solve for the kinematic coefficients. (Singularities of this Jacobian define the dead positions.) After solving the position problem there is no need to recompute the Jacobian since it is known from the last iteration in Newton's method. I believe young people today are more inclined to algebraic approaches rather than graphical approaches, and the vector loop method is purely algebraic.

- All kinematic and dynamic analysis is performed in a fixed frame of reference. In acceleration analysis the Coriolis terms arise naturally. The student no longer needs to recognize situations where the Coriolis terms must be included. Coriolis terms will appear during the differentiation required to find the second-order kinematic coefficients. All that is needed is the ability to apply the chain rule and the product rule for differentiation. Also, the equations that use kinematic coefficients to compute velocity and acceleration, either linear or rotational, have the same form.

- The most important feature of the vector loop method is that the dynamic simulation of a machine becomes a straightforward process. The same kinematic coefficients that describe the kinematic properties of velocity and acceleration appear seamlessly in the machine's equation of motion. This equation of motion is numerically integrated to yield the dynamic simulation. Development of the equation of motion, known as the Power equation, is systematic with the vector loop method.

- The Power equation also leads to a generic description of mechanical advantage as the ratio of the load force (or torque) to the driving force (or torque). The procedure for finding an algebraic expression of a mechanism's mechanical advantage is systematic and does not involve free body diagrams. It also leads to another way to describe limit positions (infinite mechanical advantage) and dead position (zero mechanical advantage).

The vector loops are the basis of any mechanism's or machine's motion. They are the fundamental equations which completely define the machine's kinematics and dynamics. *The subject of mechanism and machine kinematics and dynamics is unified by the vector loop method* and for this reason I believe this book teaches the subject in the most condensed and straightforward manner.

I have found in teaching undergraduate courses on mechanisms and machines that most textbooks contain much more information than can be covered in one or even two semesters. Existing texts are excellent references but are too large to be appropriate for introductory courses. Of the nine chapters in this book, seven chapters (1-5 and 7-8) can be covered in a fifteen week semester.

Objectives

There are two primary objectives of this book. The first objective is to teach students a modern method of mechanism and machine analysis, the vector loop method. The method predicts the kinematic performance of a mechanism, but most importantly, it makes the dynamic simulation of a machine tenable. This is through the machine's differential equation of motion, known as the Power equation. This nonlinear differential equation of motion is solved using a first-order Euler's method. The numerical solution allows a machine designer to select the motor or driver for a machine so that the machine achieves a desired time based response. This response may include things such as time to steady state operation or the flywheel size required to reduce steady state speed fluctuation to an acceptable level.

The second objective of the book is to give an introduction to mechanism synthesis. Of the myriad of synthesis problems that exist, we consider the two most basic, namely function generation and rigid-body guidance. In a single semester it is not possible to proceed further. Students interested in other types and more advanced forms of mechanism synthesis will need to proceed after the semester ends using the abundance of reference texts and archival literature that exists.

Organization

The books begins with a basic introduction in **Chapter 1** which teaches students how to visually communicate the geometry of a mechanism or machine through skeleton diagrams and then determine the gross motion capabilities of the mechanism or machine. Vectors later associated to the vector loop method are used for this purpose. **Chapter 2** introduces position analysis through the vector loop method. Newton's method is taught to solve the nonlinear position equations. **Chapter 3** introduces rolling contacts. In addition to mechanisms, automotive transmissions, and an automotive differential are analyzed.

Chapter 4 introduces kinematic coefficients. They are used for velocity and acceleration analysis. The physical meaning of the kinematic coefficients is given and they are used to algebraically predict the limit positions and dead positions of planar mechanisms. From these follow concepts such as time ratio and transmission angle.

Chapter 5 reviews the inverse dynamics problem. Kinematic coefficients are used to compute the accelerations of mass centers and angular accelerations of links. There are three purposes for studying the inverse dynamics problem. The first is to show students how solving this problem is a tool for selecting the driver or motor for a machine. For this a numerical example of a motor selection is presented. The second purpose of chapter 5 is to introduce students to the three dimensional aspects in the force analysis of planar mechanisms and machines. The third purpose is to prepare students for **Chapter 6** where joint friction is studied. There students see that the linear inverse dynamics problem becomes nonlinear when friction is considered. The method of successive iterations is used to solve these nonlinear equations.

Chapter 7 uses kinematic coefficients to develop the Power equation, the nonlinear differential equation of motion of a machine. This is the culmination of the vector loop method. A first-order Euler's method is taught to numerically integrate the equation, producing a dynamic simulation of the machine's motion, solving the forward dynamics problem. Continuing with the example in chapter 6, a dynamic simulation of the machine with its selected motor is developed. This verifies the motor selection and also shows students how a flywheel can be added to the machine to smooth its steady state operation. This example appears in both chapters 6 and 7 and illustrates the continuity between the inverse and forward dynamics problems. Kinematic coefficients and the Power equation are also used in chapter 7 to mathematically describe a mechanism's mechanical advantage.

Chapter 8 begins discussion of mechanism synthesis. A vector loop is used to develop Freudenstein's equation and a systemic procedure for synthesizing four bar mechanisms which develop desired third-order kinematic coefficients is presented. This generates mechanisms that develop third-order Taylor's series approximations of the desired function generation.

Chapter 9 presents the three position problem of a rigid body. This is the simplest possible guidance problem. It is beyond the scope of a fifteen week semester to consider the four and five position problem along with the issue of sequencing.

This course can be taught beginning with either analysis or synthesis, depending on the instructor's preference. Many mechanical engineering programs offer technical electives that cover mechanism synthesis in detail, in a semester long course. In that case the analysis portion of this book would be more important so as to prepare them for analyzing the designs they will develop in that follow on technical elective. Chapters 8 and 9 may then be neglected.

My personal preference is that synthesis is introduced first, in which case chapters 1, 8, and 9 are covered first. Chapters 8 and 9 are independent of the other chapters. The students naturally question how they might determine how well their designs are performing and how can they optimize amongst the many solutions they develop. This justifies the analysis that is covered by the remaining chapters, which should be taught in the order they are presented as chapter 3 depends on chapter 2, and so on.

Problems and Exercises

This book contains three types of problems: Exercises, Programming Problems, and Design Problems. In light of the computational orientation of the vector loop method there are relatively few numerical results.

Exercises are intended as homework problems which develop a student's ability to model a mechanism and set up the correct equations for use in a computer program. Many examples and exercises develop flowcharts which show the logic of a computer program that uses the developed equations.

To keep the student grounded, there are Programming Problems that implement the modeling equations and generate numerical results and plots. These predict the mechanism's gross motion capabilities, force and torque transmitting capabilities, or time-based response of the machine.

Design problems are open-ended. A complete survey of possible solutions to these open-ended design problems also requires a computer program.

Prerequisites

Students using this book should have taken the following courses, which are typically taken before or during the sophomore year:

1. Analytical Geometry
2. Calculus
3. Ordinary Differential Equations
4. Linear Algebra
5. Computer Programming (Matlab®, FORTRAN, C++, or any other scientific language)
6. Statics
7. Dynamics

MindTap Online Course and Reader

In addition to the print version, this textbook is now also available online through MindTap, a personalized learning program. Students who purchase the MindTap version will have access to the books MindTap Reader and will be able to complete homework and assessment material online, through their desktop, laptop, or iPad. If your class is using a Learning Management System (such as Blackboard, Moodle, or Angel) for tracking course content, assignments, and grading, you can seamlessly access the MindTap suite of content and assessments for this course.

In MindTap, instructors can:

- Personalize the Learning Path to match the course syllabus by rearranging content, hiding sections, or appending original material to the textbook content
- Connect a Learning Management System portal to the online course and Reader
- Customize online assessments and assignments

- Track student progress and comprehension with the Progress app
- Promote student engagement through interactivity and exercises
- Additionally, students can listen to the text through ReadSpeaker, take notes and highlight content for easy reference, and check their understanding of the material.

Acknowledgments

What a person knows or creates is a result of a life's experience, which is filled by input from other individuals, who in many cases were either teachers or role models. I would like to acknowledge three individuals without whom I would never have written this book. They, and those before them, have partial ownership of this book.

Professor Emeritus Allen Strickland Hall Jr. of the School of Mechanical Engineering at Purdue University was the finest of mentors and teachers. He was a patient man of few, but meaningful words. He was more of a guide than a teacher. Professor James A. Euler of the Mechanical Engineering Department of the University of Tennessee was a Ph.D. student at Purdue University when he led me at the age of nine, in the disassembly and rebuilding of a lawn mower engine. At that point I had disassembled many toys and small mechanical devices, or destroyed them as my parents said, generally failing to put them back together. This was the first machine I had encountered, and I began to wonder about the mechanical movements and their relative timing and coordination. Finally my father, Professor Emeritus Milomir Mirko Stanišić of the School of Aeronautics and Astronautics at Purdue University. He was an outstanding mathematical physicist and the finest of role models. He is continuously missed.

There are also many individuals who helped me in preparation of this book. Dr. Craig Goehler, of the Mechanical Engineering Department of Valparaiso University provided solutions to many of the homework problems. Ms. Mary E. Tribble reorganized many of the chapters and and corrected the errors of my Chicago-English. Mr. Conor Hawes checked the homework solutions. Dr. Zhuming Bi of Indiana University Purdue University Fort Wayne, Dr. Daejong Kim of University of Texas at Arlington, and Dr. D. Dane Quinn of The University of Akron, provided valuable input during the review process. Finally, I thank the Staff of Cengage Learning for their support and constructive criticisms.

Michael Milo Stanišić
University of Notre Dame

CHAPTER 1

Introduction

This text is an introduction to the kinematics and dynamics of mechanisms and machinery. Mechanisms are mechanical devices that consist of a system of interconnected rigid links. The purpose of a mechanism is to create a desired relative motion of the links. The dimensions of the links and the types of connections between the links dictate their relative motions.

Kinematics is the branch of mechanics that studies aspects of rigid body motions that are independent of both force and time. The subject matter is purely geometric. With regard to mechanisms, kinematics is the study of how the dimensions of links, and the connections between links, affect the relative motion of the links. There are two types of problems in mechanism kinematics. The first is the *kinematic analysis* problem. In this problem the mechanism is specified, and the relative motion of the links is determined. Chapters 2, 3, and 4 discuss kinematic analysis. In this text, the vector loop method is used for kinematic analysis. The second problem is the *kinematic synthesis* problem. In this problem the desired relative motion of the links is specified, and the mechanism that produces the desired motion is determined. Kinematic synthesis is a very broad field. The subject could not be fully examined in a one-semester course. Chapters 8 and 9 present two very basic methods of kinematic synthesis. The synthesis methods presented here are independent of the vector loop method and may be studied at any point. Both types of kinematics problems are equally important, and they go hand in hand. The synthesis problem typically yields many potential solutions. Analyzing the solutions determines the optimum design.

Dynamics is the branch of mechanics that studies the relationship between the forces and torques acting on a rigid body and its time-based motion. When a mechanism is used to transmit forces, torques, and power, it is referred to as a machine. With regard to machines, dynamics is the study of the relationship between the time-based motion of the machine's links and the forces and torques that are applied to the links. There are two types of dynamics problems. The first is the *inverse dynamics* problem. In the inverse dynamics problem the time-based motion of the machine is specified, and the driving forces or torques corresponding to this motion are determined. The second is the *forward dynamics* problem. In the forward dynamics problem the driving forces and torques are specified, and the resulting time-based motion of the machine is to be determined. These two problems also go hand in hand. For example, the inverse dynamics problem would be solved to size and select the motor needed to drive a machine under steady state conditions; then, using the selected motor, the forward dynamics problem would be solved to predict the time response of the machine.

In the remainder of this chapter we will look at the types of joints that exist between links in a planar mechanism. We will then learn how to make skeleton diagrams of planar mechanisms and how to compute the number of degrees of freedom (dof) in a planar mechanism.

1.1 JOINTS

We begin our study of mechanism kinematics by considering what types of connections, a.k.a. joints, exist between the links in a planar mechanism. The set of possible joints is very limited. First we need to make some definitions.

DEFINITIONS:

Link *A rigid body.*

Links are numbered, and each is referred to by its number.

Planar Motion *A motion in which all points belonging to a link move in a plane known as the plane of motion, while simultaneously the link is free to rotate about an axis perpendicular to the plane of motion.*

Figure 1.1 shows an abstraction of two links. Link 1 is a fixed link. The small dark blue patch attached to 1 indicates that it is fixed. Link 2 is undergoing an unconstrained planar motion relative to 1. Let us consider this motion. Attach a frame of reference to 1, whose X-Y axes define the plane of motion. We choose a reference point Q and a reference line ℓ on 2.

We define the location of Q relative to the origin of X-Y with the vector $\bar{r}$, which has X and Y components r_x and r_y,

$$\bar{r} = \begin{bmatrix} r_x \\ r_y \end{bmatrix}.$$

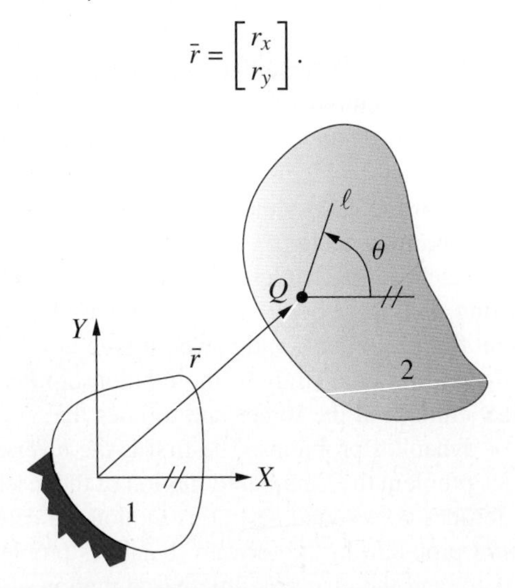

FIGURE 1.1 A link undergoing planar motion

We define the orientation of 2 relative to X-Y with the angle θ, measured from the positive direction of the X axis to ℓ. Since the motion of 2 relative to 1 is unconstrained, the two components of $\bar{r}$ and the angle θ are independent of each other. When $\bar{r}$ and θ are defined, the position of 2 relative to the X-Y frame (i.e., link 1) is known. Thus we make the following statement:

An unconstrained link undergoing planar motion has three degrees of freedom. This is because three parameters—r_x, r_y, and θ—must be specified in order to define the position of link 2. We also observe that without any joint between them, the motion of 2 relative to 1 has three degrees of freedom. In the upcoming discussion we will introduce various types of joints between links 1 and 2 and deduce how the joint has constrained their relative motion by eliminating degrees of freedom.

DEFINITIONS:

Joint *A permanent contact (connection) between two links.*

The words *joint*, *contact*, and *connection* can be used interchangeably. Although it may appear that a joint can come apart—for example, the joint between 3 and 4 in Figure 1.19—it is assumed that the contact is permanent and unbreakable.

Joint Variable(s) *The relative motion(s) between two connected bodies that occur(s) at the joint.*

Independent Joint Variable(s) *The joint variables associated with a particular type of joint that are independent, i.e., do not influence one another.*

There are two types of joints in planar mechanisms, P_1 joints and P_2 joints.

P_1 Joint *A joint between two links that constrains their relative motion by eliminating two degrees of freedom, thus allowing for one degree of freedom of relative motion.*

P_1 joints have one independent joint variable.

1.1.1 P_1 Joints

There are three types of P_1 joints: pin joints, sliding joints, and rolling joints.

The Pin Joint

The left side of Figure 1.2 shows links 1 and 2 with a pin joint between them. The right side shows the X-Y frame attached to 1 and the point of reference on 2, point Q. We choose to place Q at the center of the pin joint on 2, so r_x and r_y are fixed and these two degrees of freedom no longer exist. There is one remaining degree of freedom, θ. Knowing θ defines the position of 2 relative to 1.

Rotation θ is the independent joint variable of a pin joint.

The pin joint has eliminated two degrees of freedom from the motion of 2 relative to 1, allowing for a one degree of freedom motion of 2 relative to 1. This makes the pin joint a P_1 joint.

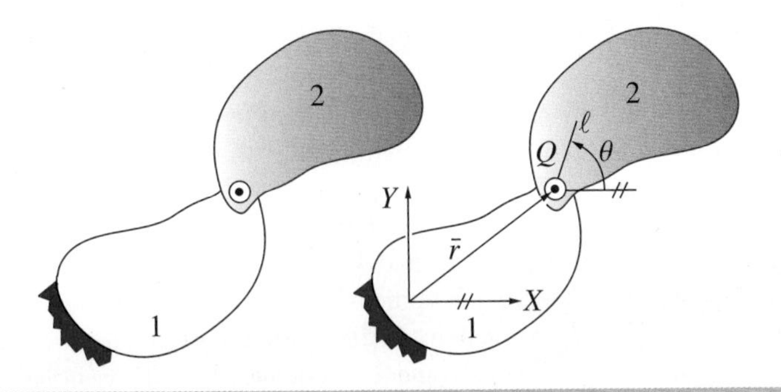

FIGURE 1.2 Pin joint

The Multiple Pin Joint

The left side of Figure 1.3 shows link 1, which is pinned to two links, 2 and 3. Clearly there are two pin joints there. Moving to the right in the figure, the two pin joints come closer so that distance d between them is reducing. There are still two pin joints. Finally, on the right hand side, d has gone to zero, so the two pin joints are coincident and appear to be one pin joint. However, that is not the case. There are still two pin joints there; they are just coincident. There is a simple rule that one can remember:

> The number of pin joints is one less than the number of links joined at that pin joint.

A mechanism is typically represented in a skeletal form, in what is known as a skeleton diagram. A skeleton diagram is a simplified drawing of a mechanism or machine that shows only the dimensions that affect its kinematics. We will look more into skeleton diagrams in Section 1.2. In a skeleton diagram, the three links pinned together on the right side of Figure 1.3 would have looked like the left side of Figure 1.4. The little black box indicates that link 1 is solid across the pin joint. The image on the right side

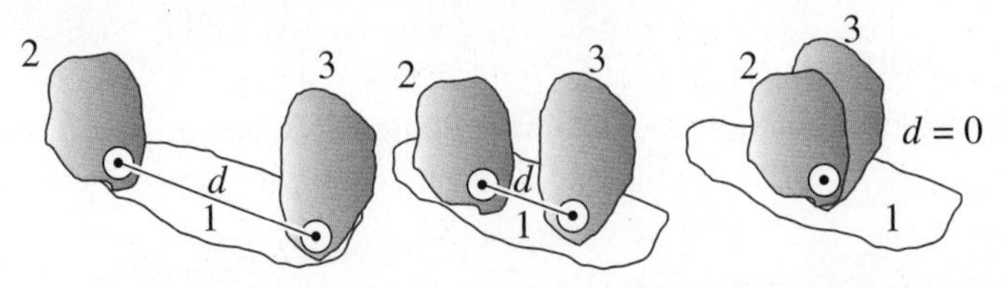

FIGURE 1.3 Multiple pin joint

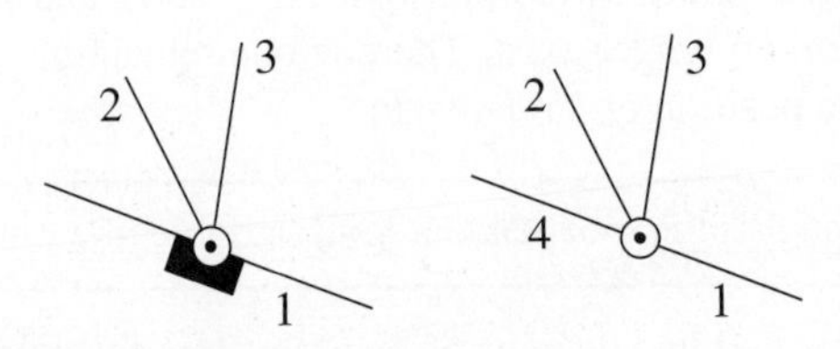

FIGURE 1.4 Skeleton diagram representations of a multiple pin joint

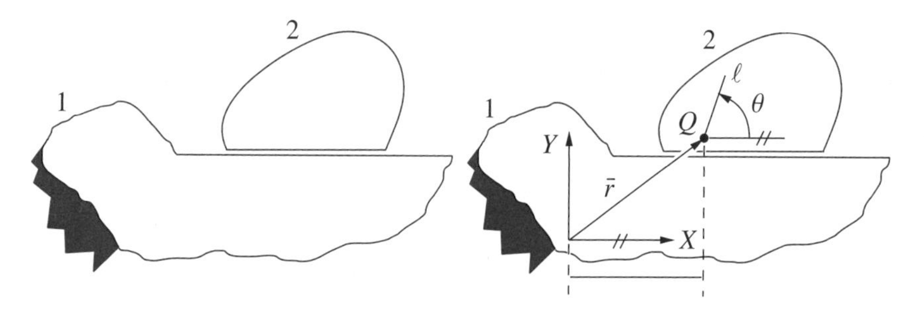

FIGURE 1.5 Sliding joint

of Figure 1.4, which does not have the black box, represents four links joined together at the pin joint, so there would be three pin joints at that location.

The Sliding Joint

Figure 1.5 shows links 1 and 2 with a sliding joint between them, the reference point Q, and the vector $\bar{r}$ and angle θ that define the position of 2 relative to 1. The values of r_y and θ are constant, so these two degrees of freedom are eliminated. The remaining degree of freedom is r_x. Knowing r_x defines the position of 2 relative to 1.

> Displacement r_x is the independent joint variable of a sliding joint.

The sliding joint has eliminated two degrees of freedom from the motion of 2 relative to 1, allowing for a one-degree-of-freedom motion of 2 relative to 1. So the sliding joint is a P_1 joint.

Figure 1.6 shows four examples of how a sliding joint between 1 and 2 may be represented in a skeleton diagram. The dimensions of the slider block, 2, are unimportant and arbitrary. The only significant feature of a sliding joint is the direction of sliding. The top left representation is most common. In some skeleton diagrams the slider block will be drawn as being encased in the mating body, as shown on the top right. In this representation you might think there are two parallel sliding joints, one at the top and the other at the bottom of 2. You should consider this to be only one sliding contact, because the second parallel sliding contact is redundant. If the two sliding contacts in

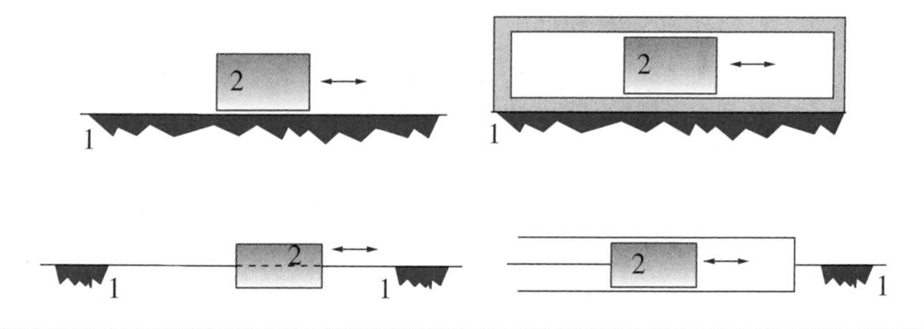

FIGURE 1.6 Skeleton diagram representation of a sliding joint

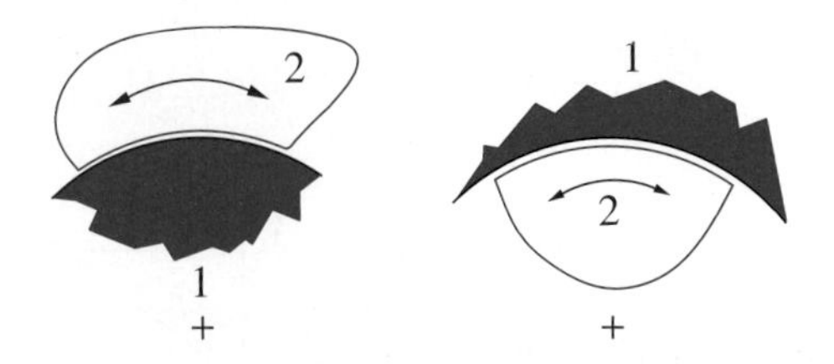

FIGURE 1.7 Circular sliding joints

that image were not parallel, then they would each count as a sliding contact. The bottom left depicts a block sliding along a shaft. The bottom right depicts a piston sliding in a cylinder, such as you see in the skeleton diagram of the front loader in Figure 1.16.

The Circular Slider

Figure 1.7 depicts circular sliding joints. Circular sliders in fact are oversized pin joints. The "+" indicates the center of the pin joint. As in a pin joint, the rotation indicated in Figure 1.7 is the joint variable of a circular slider. Like the revolute joint and straight sliding joint, the circular slider allows one degree of freedom in the motion of 2 relative to 1, eliminates two degrees of freedom, and is a P_1 joint.

Pin joints and sliding joints are line contacts. The bodies connected by a pin joint in Figures 1.2, 1.3, 1.4, and 1.7 contact along a line that is a circle (or part of a circle in 1.7). The bodies connected by a sliding joint in Figures 1.5, 1.6, and 1.7 also contact along a line that is straight in 1.5 and 1.6 and part of a circle in 1.7. As you will see, the remaining joints are all point contacts. This makes the pin joint and the sliding joint readily identifiable as P_1 joints, and we can state a simple rule:

> Any joint that involves a line contact between the connected bodies is a P_1 joint.

The Rolling Joint

Figure 1.8 shows a rolling joint between links 1 and 2. It is simplest to consider the case when 1 is flat and 2 is circular, although our discussion applies to a rolling joint between two links of any shape. Assign the reference point Q at the center of 2. The figure shows the vector $\bar{r}$ and angle θ that define the position of 2 relative to 1.

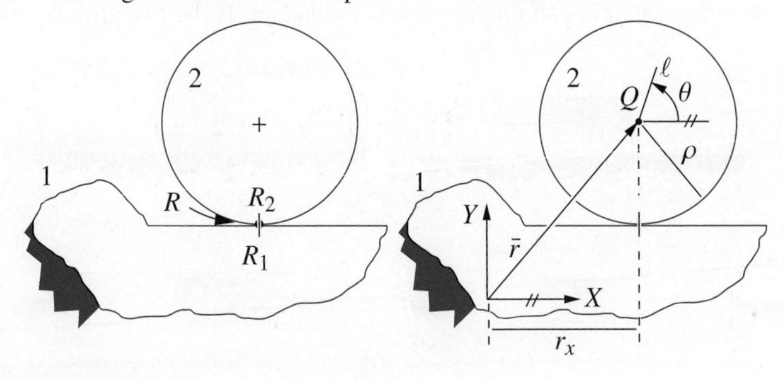

FIGURE 1.8 Rolling joint

The two bodies contact at point R, shown on the left. At point R there exist a pair of instantaneously coincident points R_1 and R_2 belonging to links 1 and 2, respectively. A hash mark is drawn at the point of contact to indicate that this is a rolling joint. The hash mark is drawn in the direction of the common normal to the two bodies at the point of contact. Point R and the hash mark are the location of a no-slip condition between the two rolling links. Let $\bar{v}_{r_1}$ and $\bar{v}_{r_2}$ represent the velocity of points R_1 and R_2 respectively. The no-slip condition means that the relative velocity between R_1 and R_2 is zero, i.e., $\bar{v}_{r_1/r_2} = \bar{0}$ and thus $\bar{v}_{r_1} = \bar{v}_{r_2}$. The no-slip condition also means there is no rubbing between the two bodies at the point of contact.

In the rolling joint, r_y is constant while r_x and θ are changing. r_x and θ are the joint variables of a rolling joint, but they are not independent. Their changes, Δr_x and $\Delta\theta$, are related to each other by the no-slip condition, namely

$$\Delta r_x = -\rho \Delta\theta. \tag{1.1}$$

The negative sign appears because a positive (counterclockwise) $\Delta\theta$ causes Q to roll to the left, and this corresponds to a $-\Delta r_x$. In this equation either $\Delta\theta$ or Δr_x is independent, but not both. So there is only one degree of freedom.

> Rotation θ, *or* displacement r_x, is the independent joint variable of a rolling joint.

We conclude that the rolling joint allows one degree of freedom for the motion of 2 relative to 1 and eliminates two degrees of freedom. Thus the rolling joint is a P_1 joint. For purposes of graphical communication, *a rolling joint in a skeleton diagram will always be indicated by a hash mark at the point of contact*, as shown in the figure. In many instances both links in a rolling joint are circular, as shown in Figure 1.9. Rolling joints are typically achieved by gears, and the two circular shapes in Figure 1.9 are the *pitch circles* of those gears.

We have considered all the known P_1 joints—the pin joint, the sliding joint, and the rolling joint—and we have seen that P_1 joints allow for one degree of freedom in the relative motion of the two connected links. Thus the following statement is true.

> All P_1 joints eliminate two degrees of freedom.

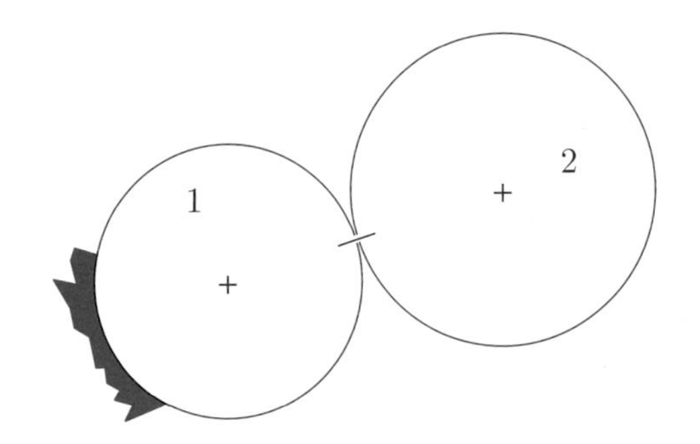

FIGURE 1.9 Rolling joint between circular shapes

1.1.2 P_2 Joints

DEFINITION:

P_2 Joint *A joint between two links that constrains their relative motion by eliminating one degree of freedom, thus allowing for two degrees of freedom of relative motion.*

P_2 joints have two independent joint variables. The only known P_2 joint is the slipping joint.

The Slipping Joint

Figure 1.10 shows a slipping joint between links 1 and 2. In a slipping joint, links 1 and 2 contact at a point, as in the rolling joint. Unlike the rolling joint, in a slipping joint there is rubbing at the point of contact and the no-slip condition, Equation (1.1) does not apply. In a skeleton diagram the only distinction between a rolling joint and a slipping joint is the hash mark. The hash mark, as in Figure 1.8, is used to indicate that there *is* a no-slip condition at the point of contact, and this corresponds to a rolling joint. *There is no hash mark at the point of contact in a slipping joint.* Without Equation (1.1), r_x and θ are independent of each other, so the slipping contact allows two degrees of freedom, and eliminates one degree of freedom, in the motion of 2 relative to 1.

Rotation θ *and* displacement r_x are the independent joint variables of a slipping joint.

The slipping joint is a P_2 joint, and the following statement is true of all P_2 joints.

A P_2 joint eliminates one degree of freedom.

Figure 1.11 shows several possible depictions of a slipping joint in a skeleton diagram. The version of the slipping joint shown at the bottom right of Figure 1.11 is referred to as a "pin in a slot" joint. In this case, when the two surfaces of 1 that contact the circular portion of 2 are parallel, there is only one P_2 joint and not two. The second P_2 joint would be redundant. If the two surfaces of 1 that contact 2 were not parallel, they would count as two P_2 joints.

To visually communicate the geometry and structure of a mechanism that are important to its kinematics, we use *skeleton diagrams*. The next section introduces you

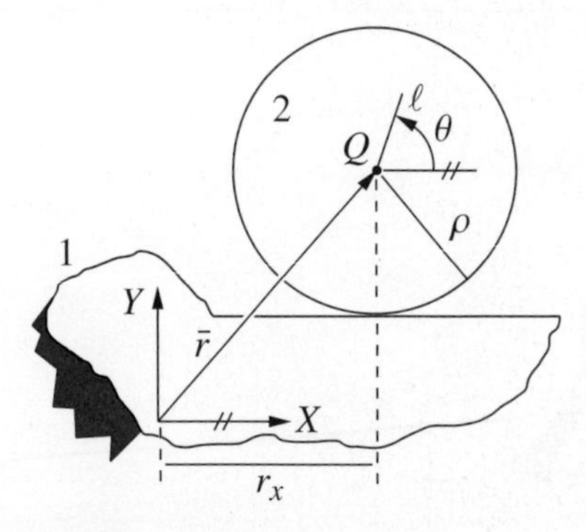

FIGURE 1.10 Slipping joint

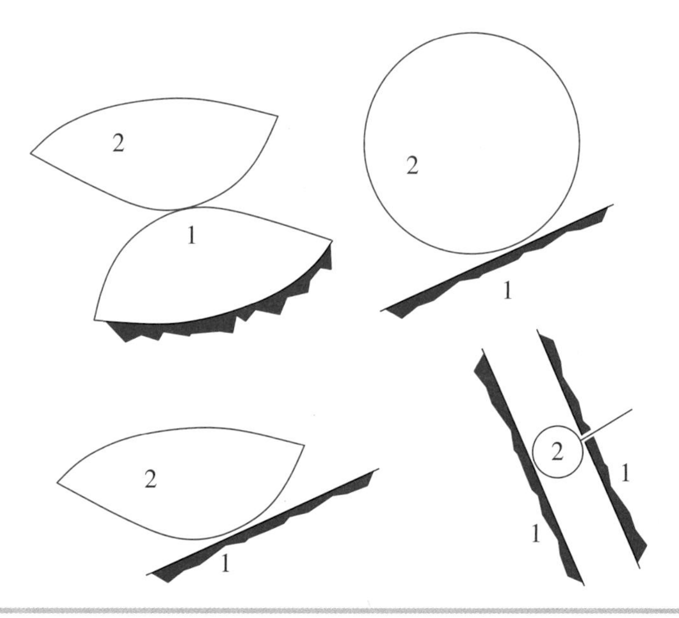

FIGURE 1.11 Skeleton diagram representations of slipping joints

to skeleton diagrams and also shows you several examples of mechanisms and machines that we see around us in our everyday lives.

1.2 SKELETON DIAGRAMS

A skeleton diagram is a simplified drawing of a mechanism or machine that shows only the dimensions that affect its kinematics. Figure 1.12 shows a connecting rod with its attached piston, from an internal combustion engine. The connecting rod and piston both have many geometric features, mostly associated with issues of strength and the size of the bearing at each joint. These features are kinematically unimportant.

The *only* geometric feature of either the connecting rod or the piston that is important to the kinematics of the mechanism containing them is the distance, l, between the

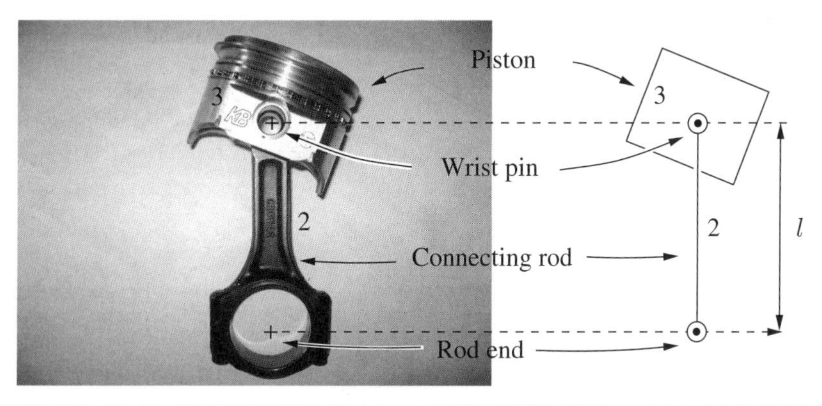

FIGURE 1.12 A connecting rod and its skeletal representation

centers of the connecting rod's bearings. (These bearings are commonly known as the wrist pin bearing and the rod end bearing.) In a skeleton diagram of the engine's mechanism, this connecting rod would be drawn as a stick whose length is l, as on the right hand side of Figure 1.12. The piston would be shown as a square or rectangle whose dimensions are not specified, because they are kinematically unimportant. The bearings would be pin joints. Let us consider some examples of mechanisms and machines that are around us in our daily lives and examine their skeleton diagrams.

1.2.1 Examples of Skeleton Diagrams

The top left of Figure 1.13 shows a six link mechanism that guides the hood of an automobile. The mechanism consists of six links, where link 1 is the body of the car (considered as the fixed link) and link 4 is the hood. The mechanism has a lot of features that are not important to the relative motion of the links. On the right side a skeleton diagram of the mechanism is overlayed on the actual mechanism, and at the bottom the skeleton diagram alone is shown. In the actual mechanism links 3 and 5 are curved. In the skeleton diagram link 3 is shown as a stick whose length is the dimension of link 3. Link 5 is shown as a bent stick, where the distance between the joints and the angle of the bend are important dimensions. The length of the actual hood to which link 4 is attached is unimportant. What is significant with regard to link 4 is the distance between the two joints labeled points A and B. Referring to the skeleton diagram, you see it contains two "loops." One loop consists of links 1-2-6-5. You can imagine standing on 1, walking to 2, then to 6, then to 5, and finally back to 1, forming a closed loop. The other loop consists of 1-2-3-4-5. In the same way, you can go from 1 to 2 to 3 to 4 to 5 and back to 1. There is a third loop, 2-3-4-5-6; however, it exists as a consequence of the previous two loops. In other words, this third loop is not an independent loop. If either of the previous two loops is removed, this third loop no longer exists. In Chapter 2 you will need to be able to identify the number of independent loops in a mechanism so that you can develop the correct system of equations to model its motion.

The purpose of the mechanism is to guide the hood as it opens and closes. This is known as *rigid body guidance*. Chapter 9 will introduce a method by which you can synthesize (design) a mechanism so that one of its links is guided in a desired manner.

In this mechanism the pin joints are rivets. These are called rivet joints. Rivet joints are fairly common in mechanisms whose links are stamped or are tubular. The spring is not included in the skeleton diagram because it has no effect on the kinematics of the motion. One may choose to include it for aesthetic purposes or to convey the geometry of its attachment points. The spring will affect the forces corresponding to the motion, so its characteristics and attachment points are important to the force analysis and dynamics but not to the kinematics.

Figure 1.14 is another mechanism used to guide an automobile hood. It is a well-known mechanism called the four bar mechanism. It consists of a single loop, 1-2-3-4. The hood would be attached to link 3, and link 1 would be attached to the body of the automobile. As links 2 and 4 rotate clockwise, the hood opens, and as 2 and 4 rotate counterclockwise, the hood closes. The figure shows a simple animation. The mechanism also uses rivet joints. What is interesting about this device is that as the hood approaches its closed position, links 2 and 4 become nearly parallel. You will see in

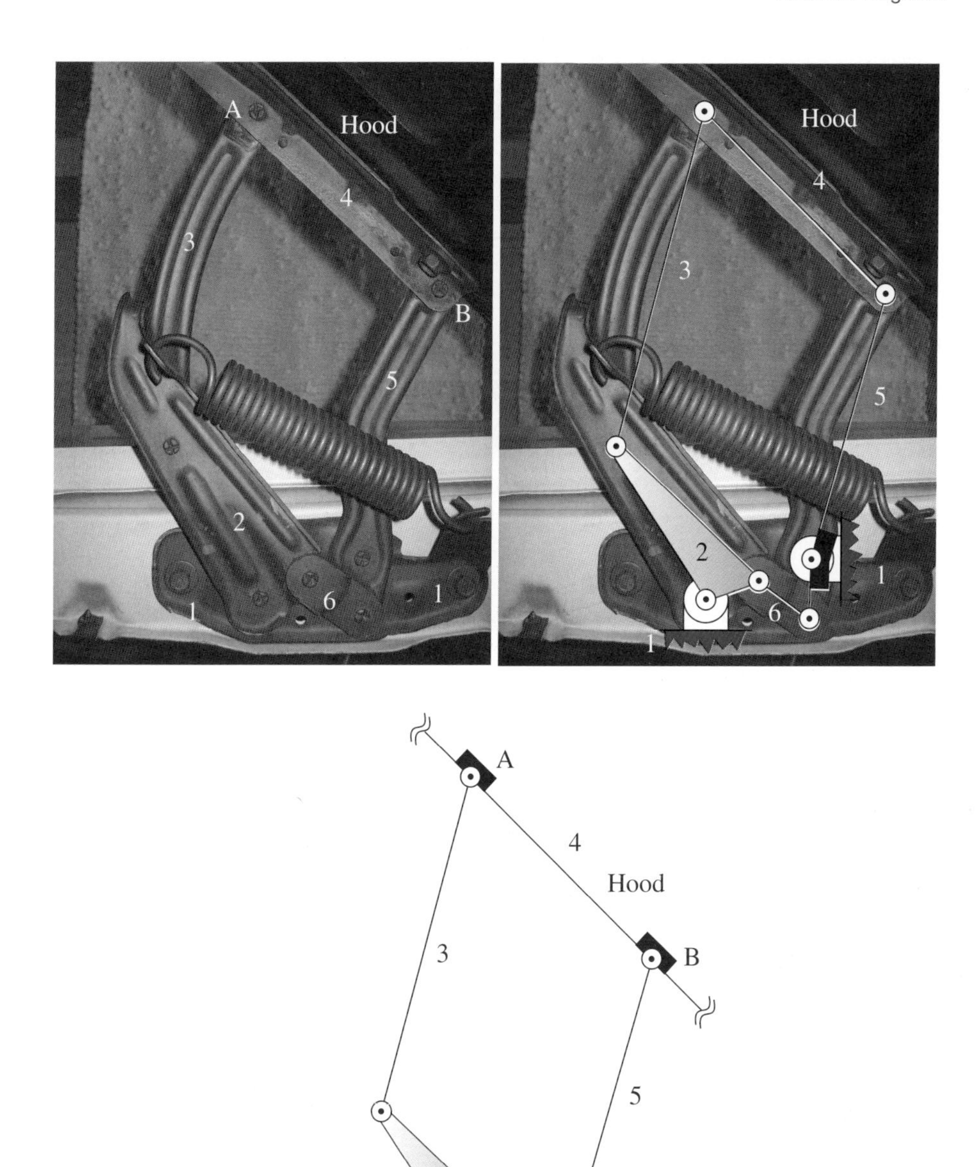

FIGURE 1.13 An automobile hood mechanism

the programming problem of Section 4.8 that when 2 and 4 are parallel, link 3 instantaneously stops rotating. This allows the hood to develop a uniform pressure along the seal between the hood and the body of the automobile.

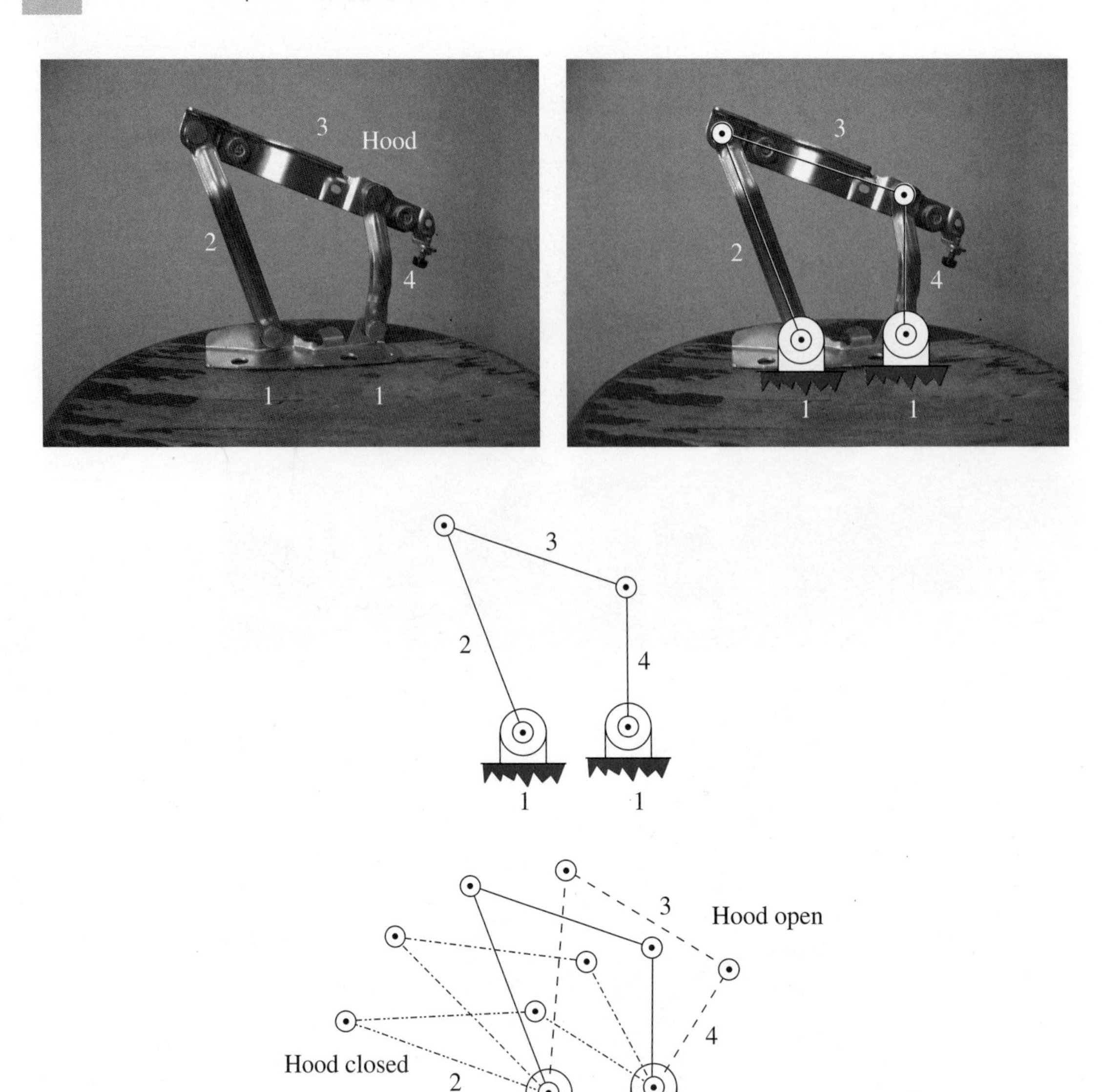

FIGURE 1.14 Another hood mechanism from an automobile

Figure 1.15 shows a door damper mechanism. It is also a four bar mechanism consisting of a single loop, 1-2-3-4. Link 1, the fixed link, includes the door frame and the housing of the large damper. The door, link 2, rotates relative to the frame via the door hinge. Link 4 rotates relative to 1 via a shaft that is a connected to the damper. Link 3 is connected to links 2 and 4 by pins at its ends. Beneath the image of the door is a skeleton diagram of the mechanism, viewing down onto the door from the ceiling. This would be a view onto the plane of motion. Such a door damper is an example of a

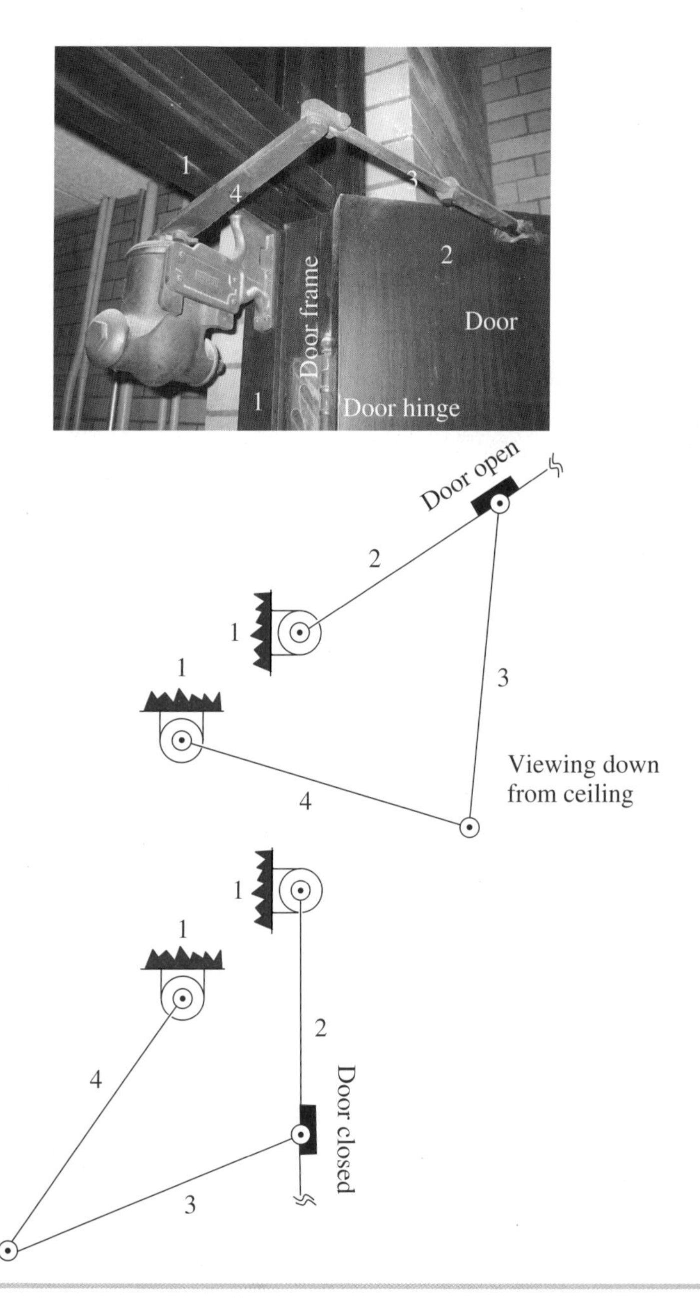

FIGURE 1.15 A door damper mechanism

© Cengage Learning.

mechanism that develops a desired relationship between the rotation of the door and the rotation of link 4 (which is resisted by the damper mechanism). This is known as *function generation*. Chapter 8 will introduce a method by which you can synthesize a mechanism to develop a desired function generation. Notice that as the door is closed, links 3 and 4 approach an overlapped condition. You will see in the programming problem of Section 4.8 that because of this geometric condition, when link 2 (the door) is being pulled shut the rotational rate of link 4 relative to 1 increases significantly, which makes the damper more effective in gently closing the door.

Figure 1.16 shows a small front loader. The front loader uses short steel rods (called pins) for its pin joints. The pins are kept from falling out of their holes by shaft clips or

FIGURE 1.16 A small front loader

cotter pins. The front loader consists of three independent closed loops: 1-2-3-4, 4-5-6-8, and 4-8-7-9. It uses two hydraulic cylinders to actuate the system. Each hydraulic cylinder consists of two links that slide on each other: links 2 and 3, and links 5 and 6. Fluid pressure creates the sliding action. The skeleton diagram is overlayed on the machine in Figure 1.16. Figure 1.17 shows the skeleton diagram alone.

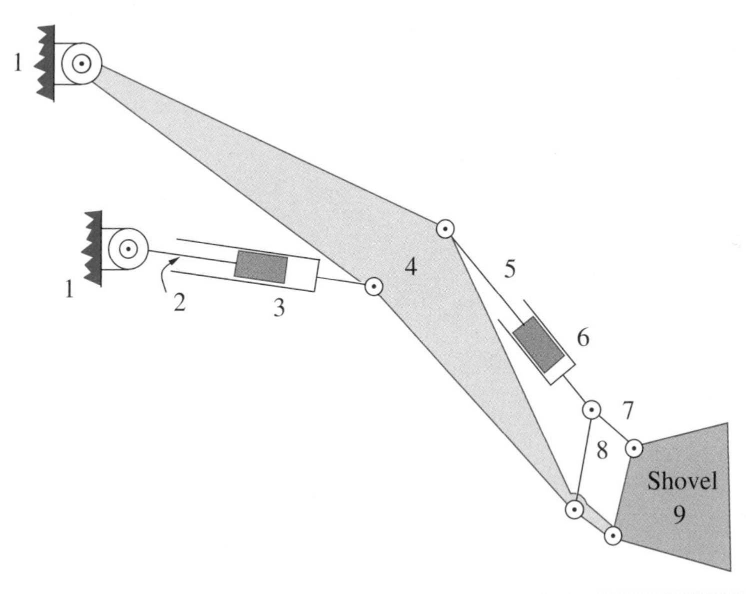

FIGURE 1.17 Small front loader and skeleton diagram

Figure 1.18 shows a folding chair lying on its side. Consider the seat back, link 1, to be fixed. This three link, single loop mechanism contains a "pin in a slot" joint. The pin that slips in the slot is actually a portion of the tube used to form link 2. The slot is a bent piece of sheet metal attached to the bottom of the seat. This mechanism also uses rivet joints for its pin joints.

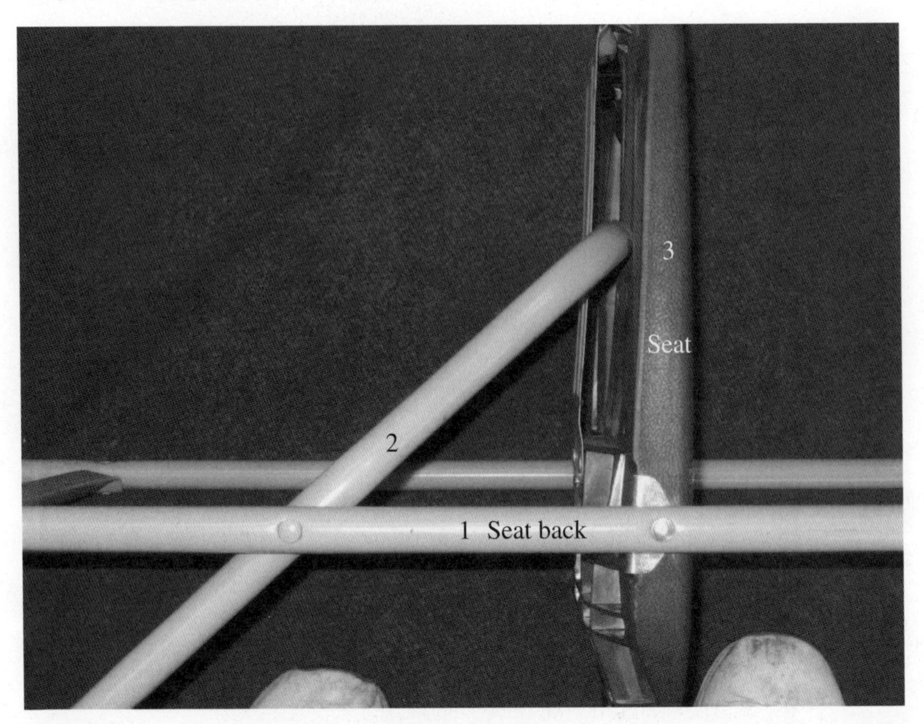

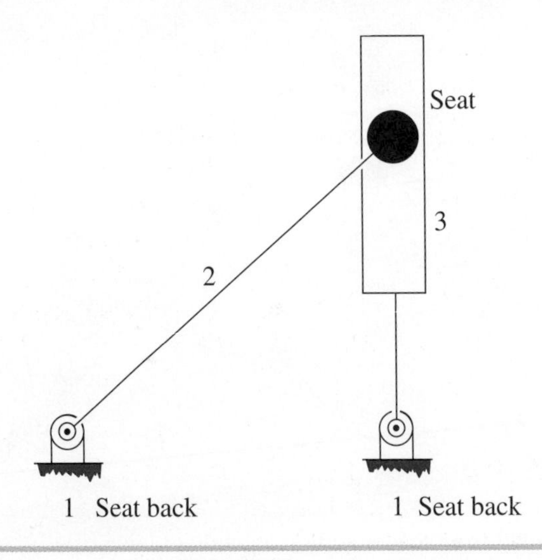

FIGURE 1.18 A folding chair lying on its side

Figure 1.19 shows an entry door to a racquetball court. It is a four link, single loop mechanism. The mechanism uses shoulder screws for the pin joints between links 1 and 2, and between links 2 and 3, and the door hinge is the pin joint between links 4 and 1. The slider is a small block of steel that slides in a track on the top of the door.

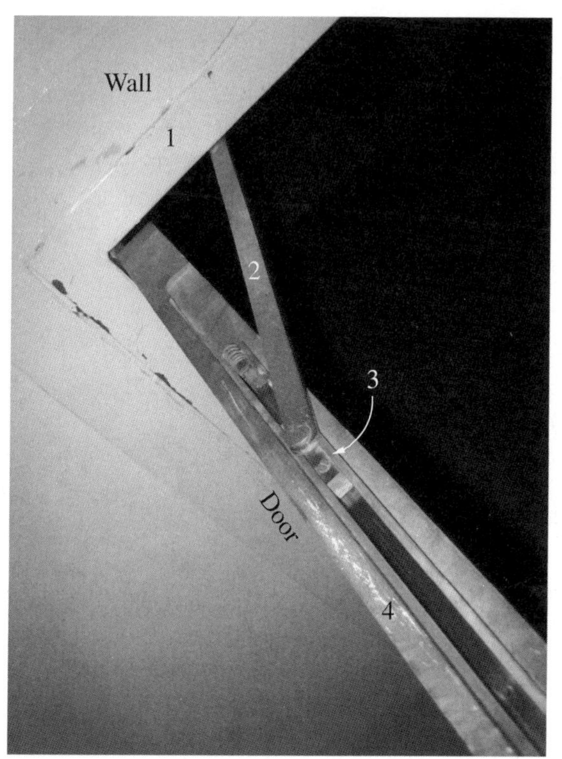

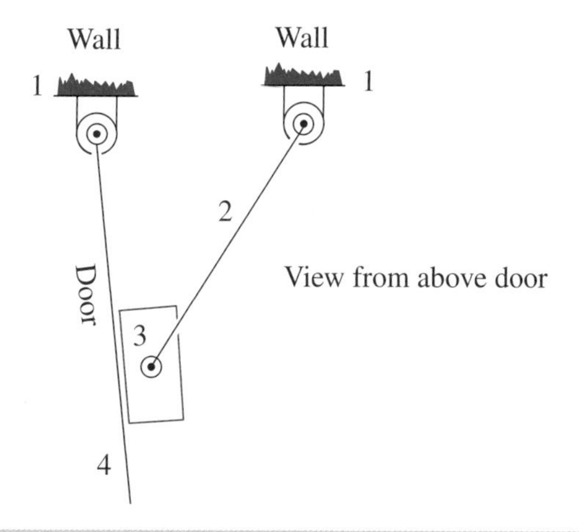

FIGURE 1.19 The top of an entry door to a racquetball court

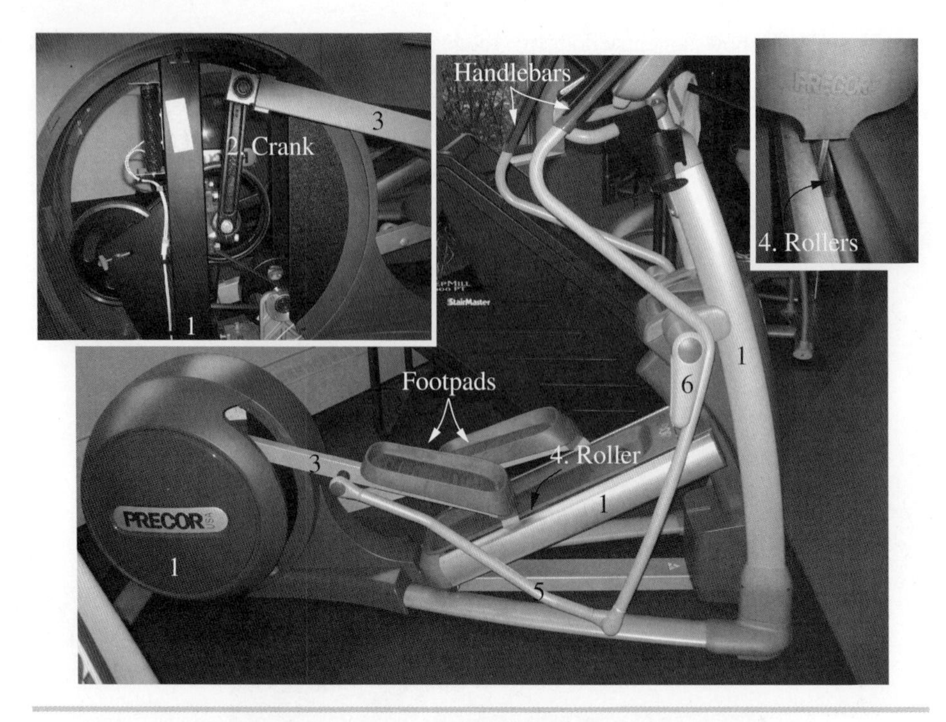

FIGURE 1.20 The Precor EFX 576i exercise machine

There is a small stiff spring at the end of the track that is compressed if the door is swung open too far.

Figure 1.20 shows an exercise machine. To use the machine a person would step onto the footpads (which are a part of link 3) and grab hold of the handlebars (which are an extension of link 6). The person would be facing to the right. The person's hands and feet are moved together, and the person is running in place. The trajectory of points on the feet is approximately elliptical, so this is called an elliptical machine. When running on pavement the foot impacts the ground with each step. In the elliptical machine the foot never completely disengages the footpads, so foot impact is greatly reduced.

The elliptical machine consists of two mechanisms that are kinematically identical. Consider the one that is on the operator's right. The machine consists of six links, which have been numbered. Link 2 is hidden under a cover, which has been removed in the top left image. Link 3 has a roller, link 4, attached to its right end. The roller can be seen in the top right image. The roller travels along a groove in link 1, the fixed link.

A crank is a link that is connected to ground with a pin joint and is capable of making continuous rotations. Link 2 is a crank. The two mechanisms are 180° out of phase because the crank for the mechanism on the operator's left lags the crank for the mechanism on the operator's right by 180°, so the left and right handlebars are swinging in opposite directions. Figure 1.21 shows the machine with its skeleton diagram overlayed on the machine and, below that, the skeleton diagram alone. Except for the contact between 4 and 1, the machine consists of only pin joints. The contact between 4 and 1 is a point contact, and the point contact has a hash mark drawn through it to indicate that it is a rolling joint and not a slipping joint.

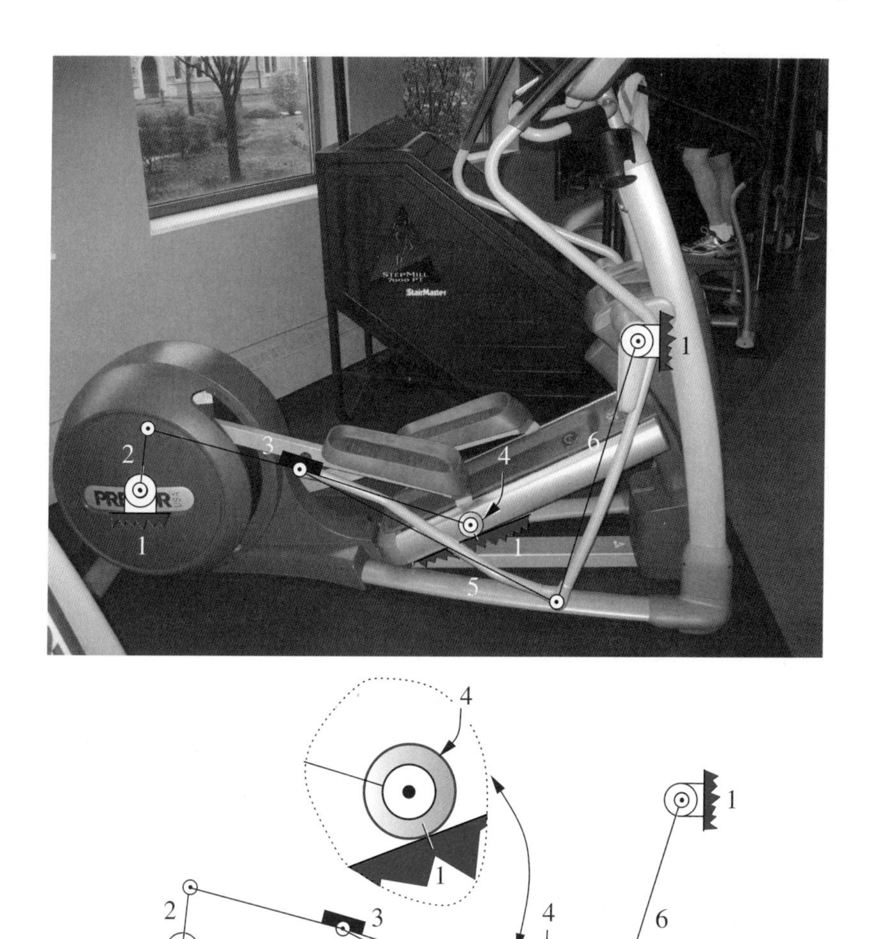

FIGURE 1.21 Skeleton diagram of the Precor EFX 576i

Figure 1.22 shows another exercise machine. The operator is facing to the left. It also consists of two identical mechanisms, one each on the person's left and right sides. Consider the mechanism on the person's left. The mechanism consists of seven links that are numbered. The side cover has been removed so you can see that link 2 is a crank that is part of a pulley/flywheel. It is not uncommon for cranks to be parts of flywheels. Flywheels add rotational inertia to a machine and smooth out its operation. A person stands on the footpad on the right end of link 6 and holds the handlebars, which are an extension of link 7. The mechanisms are again 180° out of phase, because again the crank for the mechanism on the operator's left lags the crank for the mechanism on the operator's right by 180°. In Section 1.4 we will see that this machine has properties that clearly distinguish it from that in Figure 1.20. Figure 1.23 shows the machine with an

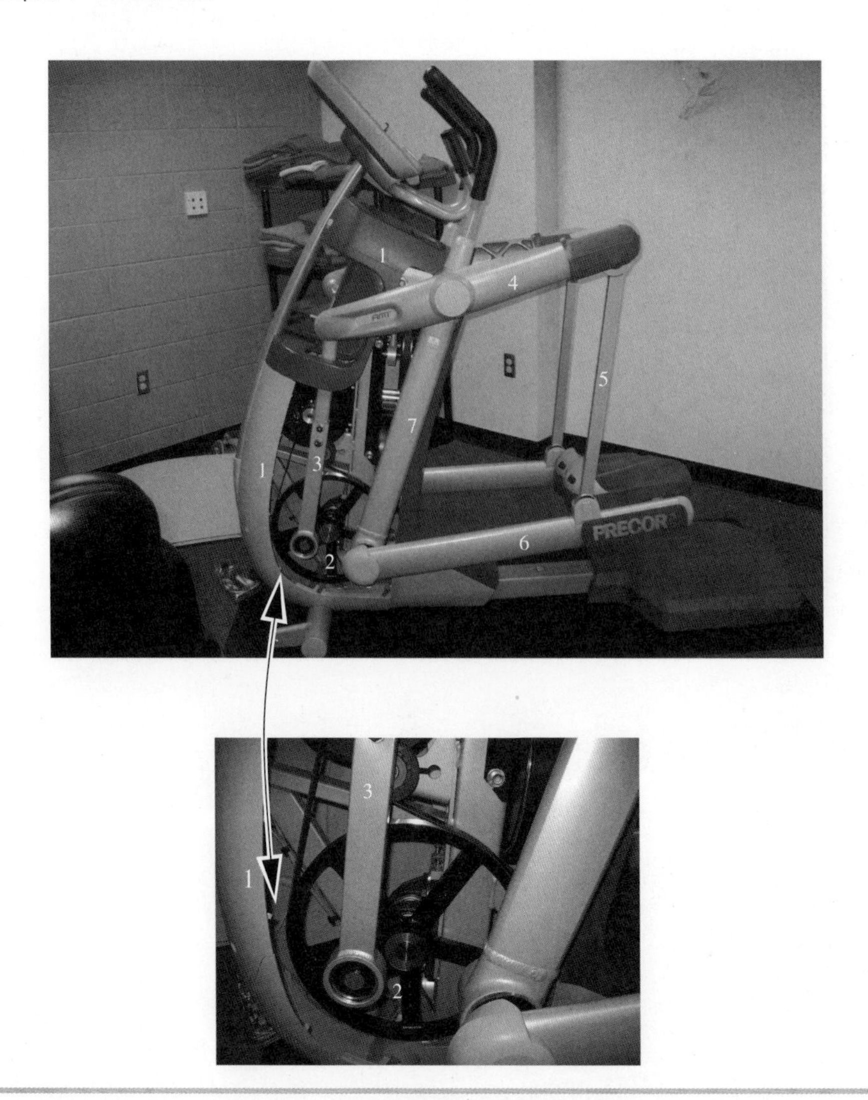

FIGURE 1.22 The Precor AMT 100i exercise machine

overlayed skeleton diagram and, below that, the skeleton diagram alone. The mechanism consists of only pin joints.

The top left of Figure 1.24 shows a planetary gear train. It consists of five bodies. They are generally referred to as the ring gear (link 4), sun gear (link 2), planet gear (of which there are three, all numbered as link 3), and arm (link 5). The arm is a circular plate that carries the center of the planet gears. The wooden base is the fixed link, 1. Two of the three planet gears are redundant. Only one is necessary and only one is counted. Planetary gear trains typically involve redundant planet gears so that the moving centers of mass of the planet gears will counterbalance each other. On the top right is the mechanism with its skeleton diagram overlayed. The diagram shows only one of the three planet gears. The gears are represented by their pitch circles. Typically gear teeth have

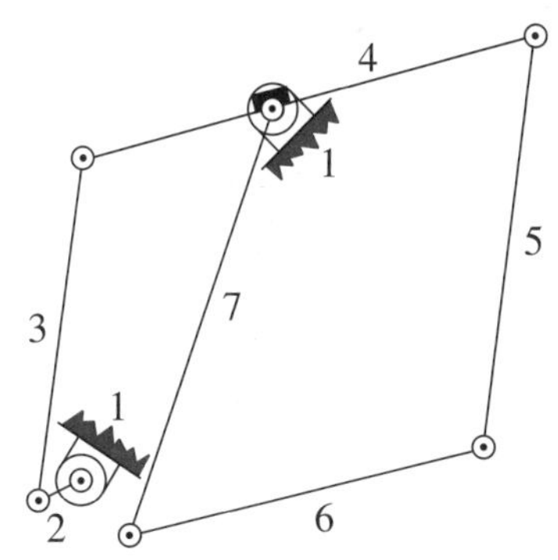

FIGURE 1.23 Skeleton diagram of the Precor AMT 100i

involute tooth profiles. When such gears are in mesh, their relative motion is identical to that of rolling pitch circles, which is why they are drawn as circles in the skeleton diagram. The bottom of the figure shows the skeleton diagram alone. Because of the arm it is not possible to see the hash mark at the point of contact between 2 and 3, so it is indicated by words. Links 2, 5, and 1 are joined together by a pin joint at the center. In accordance with the discussion in Section 1.3 this counts as 2 pin joints.

Figure 1.25 shows a pair of pliers that are capable of clamping onto a workpiece. The lower handle, link 1, is considered fixed. Links 1, 2, and 3 are connected by pin joints that are rivets. The joint between 4 and 1 is more complex. On the right end of 1 there is a screw. Consider this to be a part of 1 whose location is adjustable. The right side of the image shows a view down into 1 wherein the contact between 1 and 4 is found. The lower part of the image shows this connection disassembled. We see that the

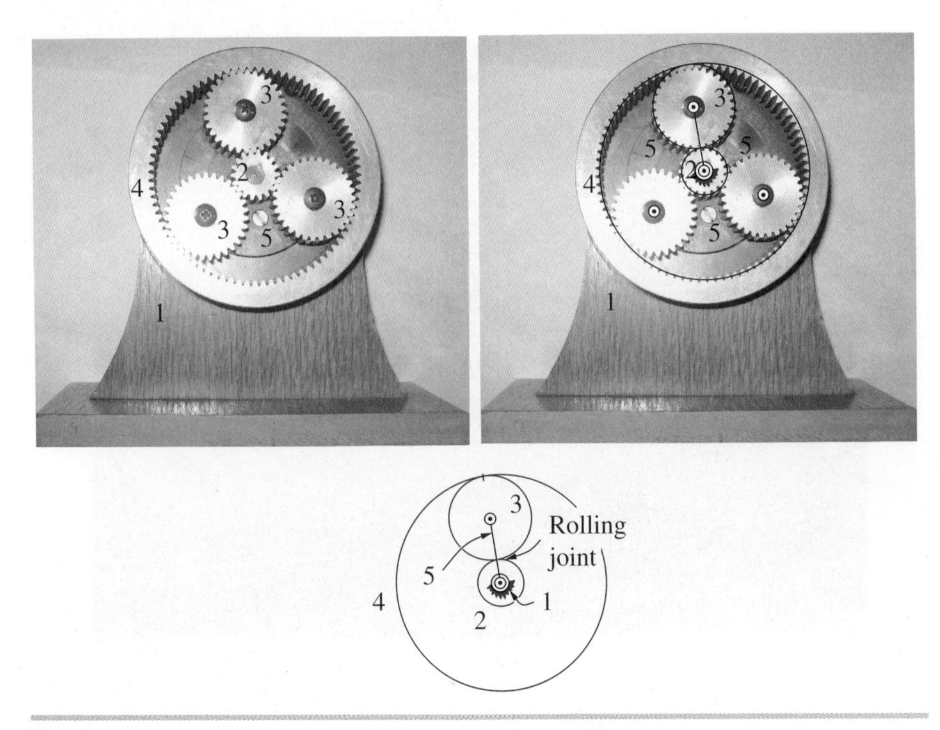

FIGURE 1.24 Planetary gear train

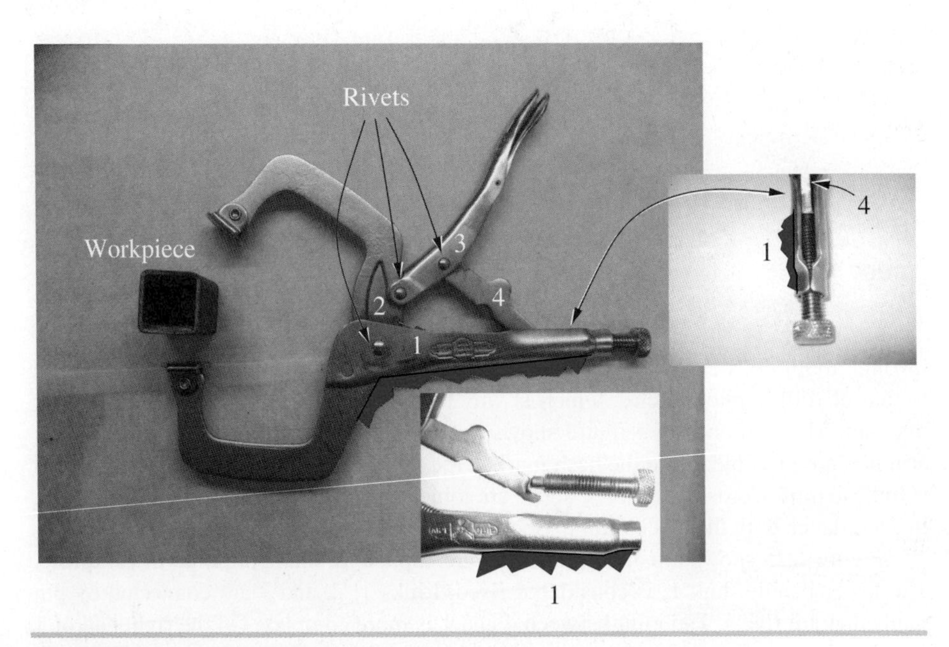

FIGURE 1.25 Clamping pliers

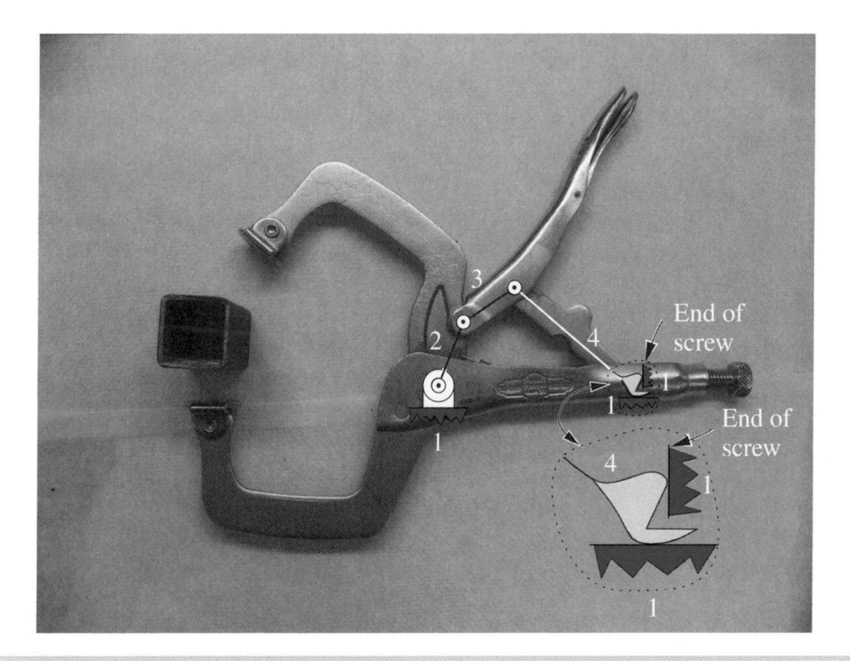

FIGURE 1.26 Skeleton diagram of pliers

end of 4 is curved and has a notch in it. The tip of the notch makes a point contact with the end of the screw, and the lower curved portion of 4 makes a point contact with the inside of 1. Figure 1.26 shows the tool with its skeleton diagram overlayed. The skeleton diagram shows the two point contacts between 4 and 1. The contacts between 4 and 1 are shown in a zoom-in. Both of these point contacts are slipping joints since there is rubbing at each of them. The top of Figure 1.27 shows the skeleton diagram alone. Below that on the left, the joints between 4 and 1 are enlarged so that the two slipping joints are clearly visible. Each slipping joint eliminates one degree of freedom, so the two together eliminate two degrees of freedom. From the point of view of degrees of freedom, the two slipping joints between 4 and 1 are equivalent to the pin joint connection on the right that eliminates two degrees of freedom. The difference in the two situations is that the center of rotation for the motion of 4 relative to 1 on the left moves, while for the case on the right it is stationary. For small motions of 4 relative to 1, the movement of the center of rotation is very small, and in this case the two situations are very similar. If we approximate the actual case with a pin joint, then we see at the bottom that we have a four bar mechanism.

Figure 1.28 shows the tool with its jaws (links 1 and 2) clamped onto a workpiece. When pliers are clamped, it is no longer necessary to squeeze the handles together to maintain a grip on the workpiece. We will study this clamping more when we discuss mechanical advantage in Section 7.2. We will see that when links 3 and 4 become inline, a force or torque applied to 3 has an infinite mechanical advantage over a force or torque applied to 2. The screw is an adjustment that allows this condition to occur for workpieces of different sizes. When 3 and 4 are inline, due to the infinite mechanical advantage, zero force or torque needs to be applied to link 3 to overcome any finite force or torque acting on 2. The problem is that with no force or torque applied to link 3, this

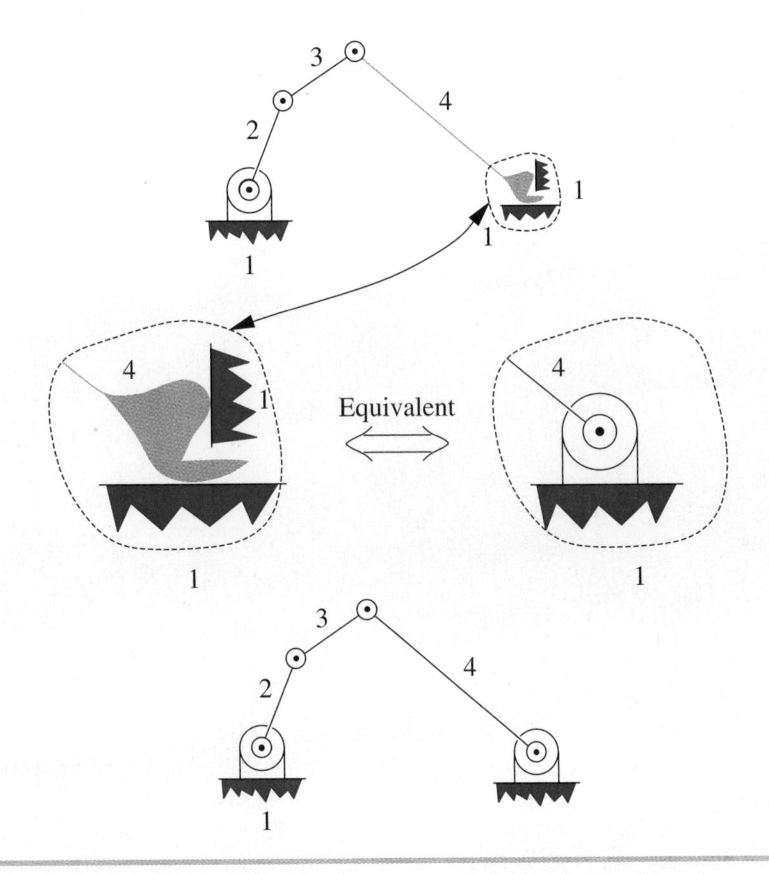

FIGURE 1.27 Pin joint equivalency

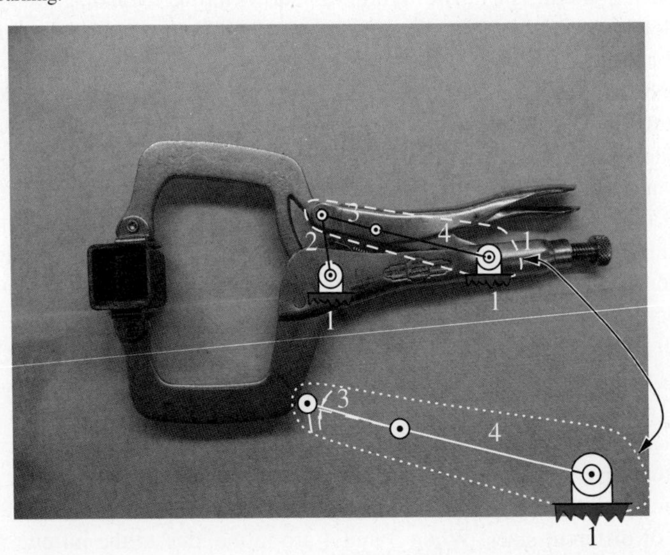

FIGURE 1.28 Pliers when clamped

inline configuration is unstable. Link 3 wants either to pop back open or to pop farther closed. To keep the workpiece clamped in a stable fashion, link 3 is pushed past this inline condition, and it then rests against the raised portion of link 4, which is seen in Figures 1.25 and 1.26. The little toggle link at the end of 3 is used to push link 3 back through the inline condition, which then releases the clamping force on the workpiece, and the jaws open.

Figure 1.29 shows the drive mechanism for the rear windshield wiper of an automobile. It consists of a four bar mechanism, links 1, 2, 3, and 4. The motor rotates

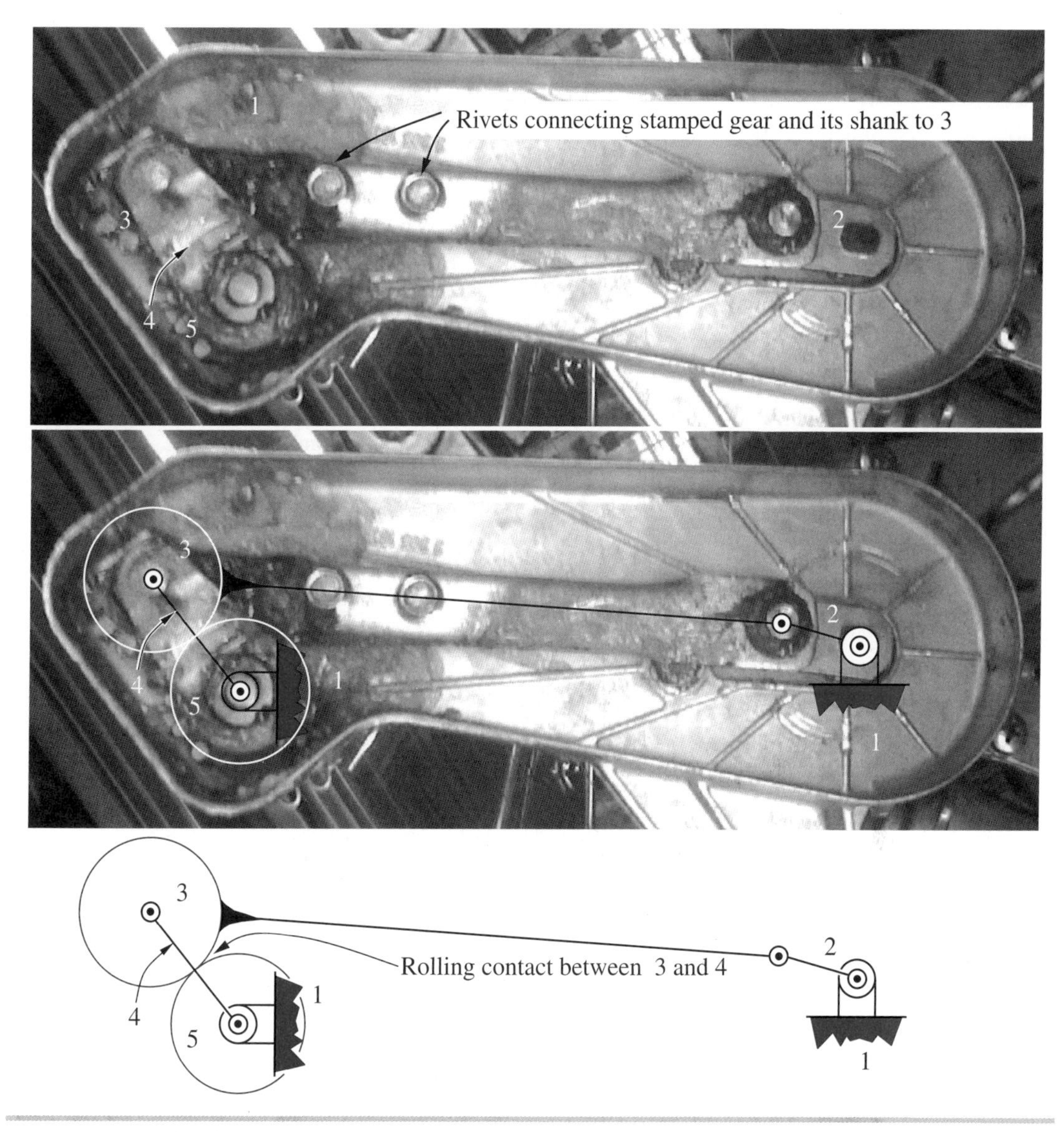

FIGURE 1.29 Windshield wiper drive mechanism

link 2 continuously (a crank), which causes 4 to oscillate (a rocker). Links 2 and 4 are stamped. Link 3 consists of two parts. The right side of 3 is a square tube, which is flattened at the far right end where it connects to 2 with a pin joint. The left side of 3 is a stamped gear with a square shank that slides into the tubular part of 3, and the two parts of 3 are then connected by two rivets, which are indicated in the figure. The gear portion of 3 drives a gear 5 whose pin joint to 1 is coincident with the pin joint between 4 and 1. The windshield wiper rotates with gear 5 and has a large oscillation of nearly 180°.

1.3 MECHANISMS AND MACHINES

We have been using the words *mechanism* and *machine* pretty interchangeably up to this point. In this section we will frame some further definitions and ultimately make a distinction between mechanisms and machines.

DEFINITIONS:

Planar Kinematic Chain *A system of links undergoing planar motion that are interconnected by joints.*

Figure 1.30(a) shows a closed kinematic chain. The connected links form a closed loop. Figure 1.30(b) shows an open kinematic chain. The connected links form an open loop. Figure 1.30(c) shows a hybrid open/closed kinematic chain that contains both a closed loop portion and an open loop portion. All the joints in Figure 1.30 happen to be pin joints, but they could have been the other types of joints as well. As in Figure 1.4, the use of the small black block in Figure 1.30(c) indicates that link 2 is solid across the pin joint that connects link 5 to link 2.

Planar Mechanism *A planar kinematic chain that has one fixed link.*

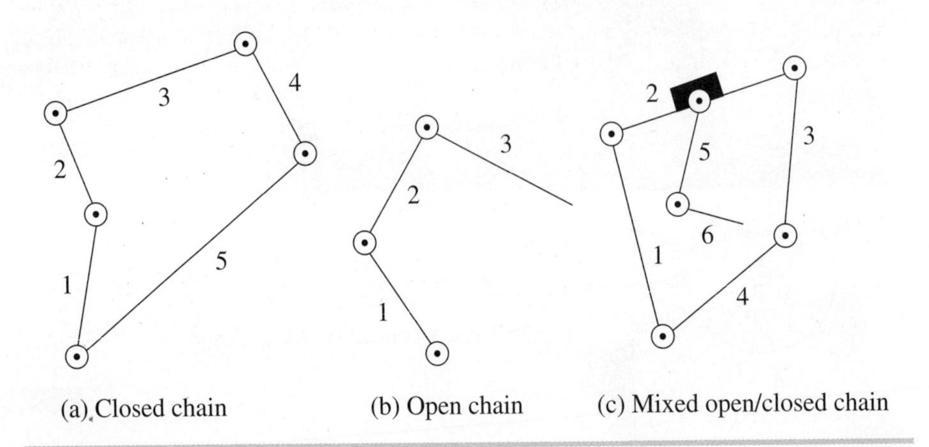

FIGURE 1.30 Examples of kinematic chains

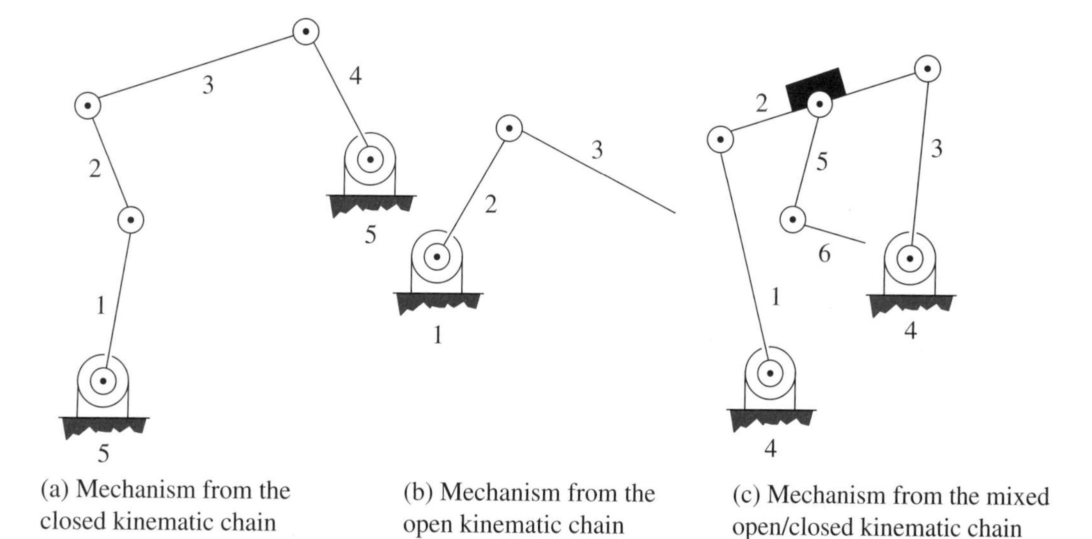

FIGURE 1.31 Examples of mechanisms
© Cengage Learning.

If a kinematic chain did not have a fixed link, then any force or torque applied to one of its links would slide and rotate the kinematic chain in the plane of motion, and no definite relative motion of the links would occur. One link must be held fixed. The three planar kinematic chains in Figure 1.30 become the planar mechanisms shown in Figure 1.31 by fixing links 5, 1, and 4, respectively.

Kinematic Inversions *Mechanisms that are derived from the same kinematic chain but have a different link fixed to ground.*

The *relative* motions of the links are the same in kinematic inversions (i.e., the motions at the joints are the same), but the *absolute* motions of the links are different, since they are being referenced to different links. Figure 1.32 shows two of the four possible

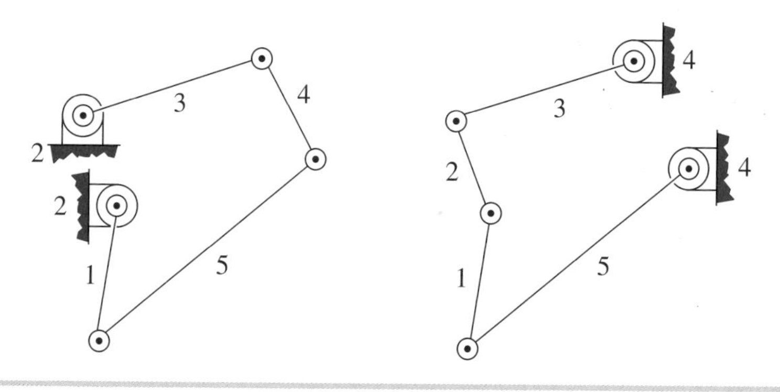

FIGURE 1.32 Two of the four possible kinematic inversions of the mechanism in Figure 1.31(a)
© Cengage Learning.

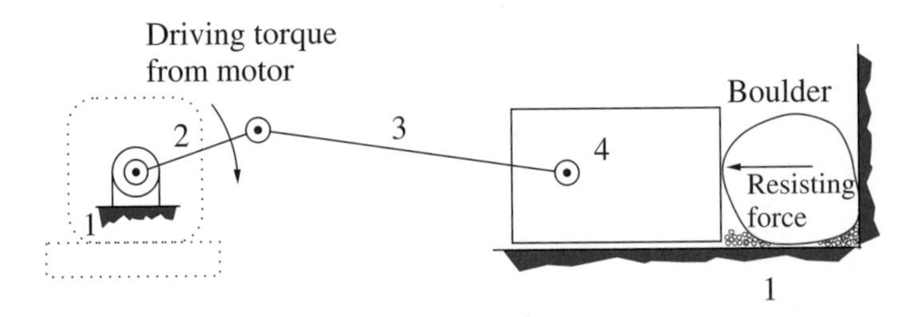

FIGURE 1.33 A crushing machine

kinematic inversions of the mechanism in Figure 1.31(a). They have links 2 and 4 as the fixed links instead of link 5.

Machine *A mechanism used to transmit power from a source (or sources) to a load (or loads).*

Figure 1.33 shows a machine being driven by a torque applied to a shaft connected to the crank, link 2. It is overcoming a resisting force acting against the slider, link 4. The mechanism within this machine is a crank-slider mechanism. The mechanism is a machine because it transfers power from the motor driving the crank to the boulder being crushed. After enough energy is absorbed by the boulder, it shatters. The front loader in Figure 1.16 is a machine. Power from the hydraulic cylinders is being used to lift loads with the bucket. The exercise machines in Figures 1.20 and 1.22 are machines that transfer power from the human operator into some type of energy dissipater.

In some cases mechanisms are used to create a desired relative motion of the links, and power transmission is not a major concern. Examples of this include the hood mechanisms in Figures 1.13 and 1.14 and the mechanism in the door of the racquetball court, Figure 1.19. However, even in these examples there is some very modest level of power transmission. A person would be doing work when lifting the hood or when opening the door. When the levels of power transfer through a mechanism become significantly high, we refer to the mechanism as a machine. But what is "significantly high"? That has never been quantified. For this reason the words are used somewhat interchangeably, but formally, a machine is a mechanism that transmits power.

1.4 GRUEBLER'S CRITERION AND DEGREES OF FREEDOM

The goal of this section is to develop *Gruebler's Criterion*. Gruebler's Criterion calculates the theoretical number of degrees of freedom within a mechanism, which is also known as the mechanism's F number. First we make another definition.

DEFINITION:

Degrees of Freedom *The minimum number of independent joint variables in a mechanism that must be specified in order to completely define the mechanism's configuration.*

In our discussion of a link moving in a plane in Figure 1.1, we concluded that an unconstrained link had three degrees of freedom. Thus a system of N *disconnected* links has $3N$ dof. For the system to become a mechanism, the N links must first be interconnected by P_1 and P_2 joints to form a kinematic chain. As seen in Sections 1.1.1 and 1.1.2, each P_1 joint eliminates two degrees of freedom, and each P_2 joint eliminates one degree of freedom from the system. In Section 1.3 we saw that in order for the kinematic chain to become a mechanism, it must have one link fixed to the ground (made immobile). This eliminates three dof from the system.

The above discussion leads to Gruebler's Criterion, which computes the theoretical number of dof in a mechanism, the F number. According to Gruebler's Criterion, the theoretical number of degrees of freedom in a mechanism is given by

$$F = 3(N - 1) - 2P_1 - P_2 \tag{1.2}$$

where: N is the number of links
P_1 is the number of P_1 joints
P_2 is the number of P_2 joints

▶ EXAMPLE 1.1

Compute the theoretical number of degrees of freedom in the truss shown in Figure 1.34.

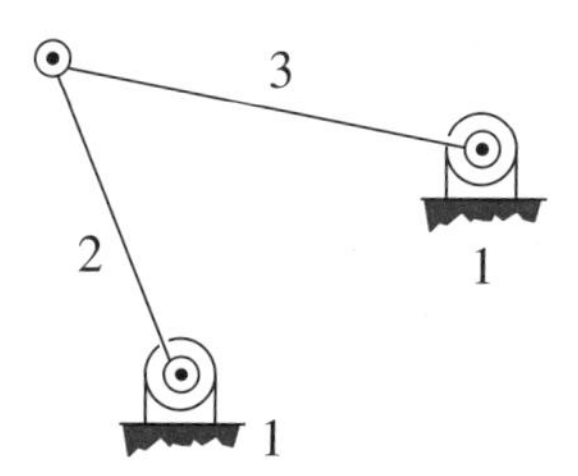

FIGURE 1.34 A truss

Solution

This system has three links, which have been numbered, and it has three pin joints. There are no point contacts, so there is no possibility of slipping joints. Thus $N = 3$, $P_1 = 3$, and $P_2 = 0$, and according to Equation (1.2),

$$F = 3(N - 1) - 2P_1 - P_2 = 3(3 - 1) - 2(3) - 0 = 0.$$

This result makes sense. There are three independent joint variables: the rotations at the three pin joints. None of these needs to be specified to define the configuration of the system since the system is immobile.

▶ EXAMPLE 1.2

Compute the theoretical number of degrees of freedom in the planar robot arm shown in Figure 1.35.

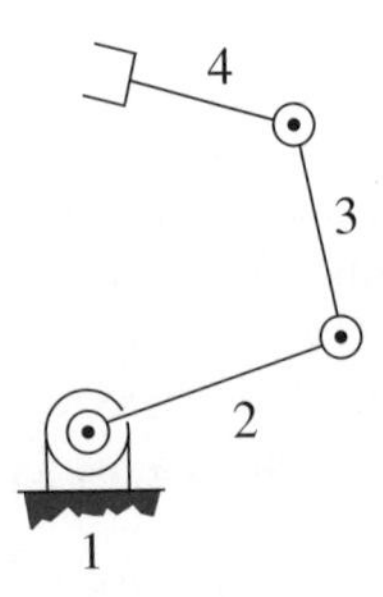

FIGURE 1.35 A planar robot

Solution

This mechanism consists of four links, $N = 4$, and three pin joints, $P_1 = 3$. There are no point contacts, so there is no possibility of slipping joints and $P_2 = 0$. According to Equation (1.2),

$$F = 3(N - 1) - 2P_1 - P_2 = 3(4 - 1) - 2(3) - 0 = 3.$$

The result makes sense. The rotations at each of the pin joints must be specified to define the configuration of the robot arm.

▶ EXAMPLE 1.3

Compute the theoretical number of degrees of freedom in the statically indeterminate structure shown in Figure 1.36.

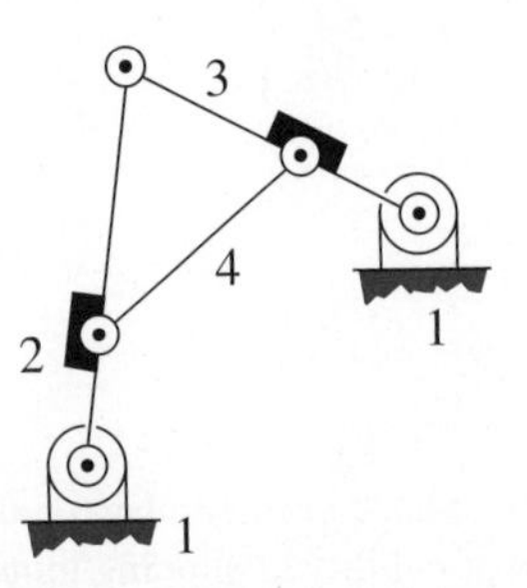

FIGURE 1.36 A statically indeterminate structure

Solution

This system has 4 links, $N = 4$, and 5 pin joints, $P_1 = 5$. There are no point contacts, so there is no possibility of slipping joints and $P_2 = 0$. From Equation (1.2),

$$F = 3(N - 1) - 2P_1 - P_2 = 3(4 - 1) - 2(5) - 0 = -1.$$

When $F = 0$ we have a determinate structure. When $F < 0$, however, the structure is statically indeterminate, and the degree of indeterminacy is equal to the magnitude of F.

▶ EXAMPLE 1.4

Compute the theoretical number of degrees of freedom in the mechanism shown in Figure 1.13.

Solution

This mechanism has six links, $N = 6$, and seven pin joints, $P_1 = 7$. There are no point contacts, so there is no possibility of slipping joints and $P_2 = 0$. From Equation (1.2),

$$F = 3(N - 1) - 2P_1 - P_2 = 3(6 - 1) - 2(7) - 0 = 1.$$

Lifting the hood sets the angle between links 3 and 4, and this angle is all that is necessary to define the configuration of the mechanism.

▶ EXAMPLE 1.5

Compute the theoretical number of degrees of freedom in the mechanism shown in Figures 1.14 and 1.15. These are both the well-known four bar mechanism.

Solution

In these four bar mechanisms there are four links, $N = 4$, and there are four pin joints, $P_1 = 4$, so according to Equation (1.2),

$$F = 3(N - 1) - 2P_1 - P_2 = 3(4 - 1) - 2(4) - 0 = 1.$$

If, for example, the joint variable (the angle at the pin joint between links 1 and 2) is specified, then the configuration of the mechanism is defined. That is, the positions of links 3 and 4 are known.

▶ EXAMPLE 1.6

Compute the theoretical number of degrees of freedom in the front loader shown in Figure 1.16, whose skeleton diagram is shown in Figure 1.17.

Solution

Referring to Figure 1.17 we see there are nine links, so $N = 9$. There are nine pin joints (the pin joint connecting links 6, 7, and 8 counts as two pin joints) and two sliding joints, so $P_1 = 11$. There are no point contacts, so there is no possibility of a P_2 joint. According to Equation (1.2),

$$F = 3(N - 1) - 2P_1 - P_2 = 3(9 - 1) - 2(11) - 0 = 2.$$

In this machine the joint variables in the sliding joints between 2 and 3, and between 5 and 6, are actuated by hydraulic cylinders. Specifying these two joint variables defines the configuration of the machine and, more important, the elevation and orientation of the shovel (link 9).

▶ EXAMPLE 1.7

Compute the theoretical number of degrees of freedom in the mechanism shown in Figure 1.18.

Solution

In order for this to be a mechanism, we must consider one of the links to be fixed. Let this be link 1. Figure 1.18 shows a skeleton diagram. We see that $N = 3$ and there are two pin joints, so $P_1 = 2$. There is also a point contact between the circle at the end of 2 and the slot in 3. Since there is not a hash mark there, this is a slipping joint and $P_2 = 1$. According to Equation (1.2),

$$F = 3(N-1) - 2P_1 - P_2 = 3(3-1) - 2(2) - 1 = 1.$$

▶ EXAMPLE 1.8

Compute the theoretical number of degrees of freedom in the mechanism shown in Figure 1.19.

Solution

Figure 1.19 shows the skeleton diagram. This mechanism is called an inverted crank-slider mechanism. We see that $N = 4$ and there are three pin joints and one sliding joint so $P_1 = 4$. There are no point contacts, so there is no possibility of a slipping joint and $P_2 = 0$. According to Equation (1.2),

$$F = 3(N-1) - 2P_1 - P_2 = 3(4-1) - 2(4) - 0 = 1.$$

▶ EXAMPLE 1.9

Compute the theoretical number of degrees of freedom in the exercise mechanism shown in Figure 1.21.

Solution

Figure 1.21 shows the skeleton diagram of this exercise machine. We see that $N = 6$. There are six pin joints (do not forget the pin joint between 3 and 4). There is one point contact and it is between 4 and 1. There is a hash mark there to indicate that this is a rolling joint and not a slipping joint, so $P_1 = 7$. According to Equation (1.2),

$$F = 3(N-1) - 2P_1 - P_2 = 3(6-1) - 2(7) - 0 = 1.$$

Since this machine has one degree of freedom, the angle between the arm, 6, and the ground, 1, defines the configuration of the machine and, more important, the position of the human's foot that sits in the pad on link 3.

▶ EXAMPLE 1.10

Compute the theoretical number of degrees of freedom in the exercise machine shown in Figure 1.23.

Solution

This machine has $N = 7$. There are eight pin joints (the pin joint connecting 1, 4, and 7 counts as two pin joints), so $P_1 = 8$. According to Equation (1.2),

$$F = 3(N-1) - 2P_1 - P_2 = 3(7-1) - 2(8) - 0 = 2.$$

This machine has two degrees of freedom, so two independent joint variables define the configuration of the machine or, more important, the position of the human's foot. These variables are

1. the angular position of the handle, 7, relative to the ground, 1
2. the angular position of link 4 relative to the ground

The machine is very interesting. It operates as a stair stepping machine when the handle does not rotate relative to the ground. In this case the human's foot, which sits in the pad on link 6, moves primarily in the vertical direction. When the handle is moved relative to the ground, the human's foot acquires a lateral movement along with the vertical movement, and the human appears to be striding or running. The stride of the run is increased as the movement of the handle is increased.

In the exercise machine of Example 1.9, where there was one degree of freedom, there was only one possible motion of the machine. Since this machine has two degrees of freedom, an infinite number of possible motions exist that vary between a stair stepping motion and a long-stride running motion.

▶ EXAMPLE 1.11

Compute the theoretical number of degrees of freedom in the clamping tool shown in Figure 1.26, which has the skeleton diagram at the top of Figure 1.27.

Solution

In the mechanism shown at the top of Figure 1.27 we have $N = 4$. There are three pin joints. There are also two point contacts between 4 and 1, shown more clearly in the zoom-in at the middle left side of the figure. These point contacts do not have a hash mark, which means they are slipping joints, so $P_2 = 2$. According to Equation (1.2),

$$F = 3(N - 1) - 2P_1 - P_2 = 3(4 - 1) - 2(3) - 2 = 1.$$

1.5 MOBILITY

It is important to realize that the F number is a *theoretical* result. Gruebler's Criterion can be fooled. Figure 1.37 shows such a special case that occurs commonly. It is a pair of circular links pinned to ground at their respective centers and rolling upon one another. This is an idealization of a pair of gears. The rolling circles are known as the "pitch

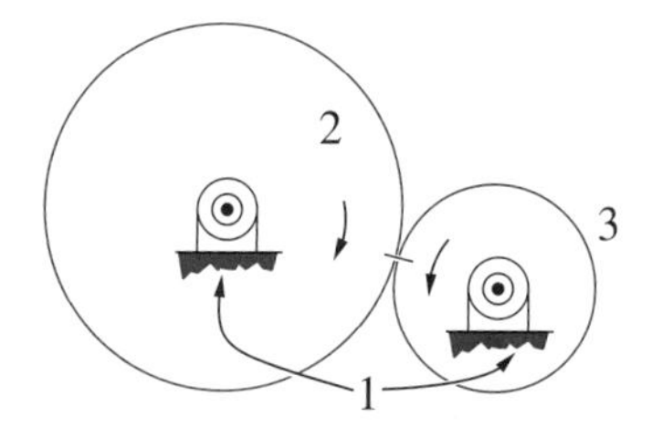

FIGURE 1.37 A pair of externally rolling gears

circles" of the gears. For this system $N = 3$, $P_1 = 3$ (2 pin joints and 1 rolling joint), and $P_2 = 0$. From Equation (1.2),

$$F = 3(N-1) - 2P_1 - P_2 = 3(3-1) - 2(3) - 1(0) = 0. \tag{1.3}$$

It is visually obvious that the gears counter-rotate and the system has one dof.

The problem is that Gruebler's Criterion is unaware of two special geometric conditions:

1. Links 2 and 3 are circular.
2. Links 2 and 3 are pinned to ground at their respective centers.

Gruebler's Criterion thinks the system is the pair of random shapes pinned to ground with a rolling contact between them, as shown in Figure 1.38. This mechanism is immovable and has zero dof. If the joint between links 2 and 3 in Figure 1.38 were a slipping joint instead of a rolling joint (which would have been communicated in the drawing by a *lack* of the hash mark shown where 2 contacts 3), then we would have had $P_1 = 2$ (2 pin joints), $P_2 = 1$ (1 slipping joint), and Gruebler's Criterion would have given $F = 1$. Figure 1.39 shows two examples of over-constrained mechanisms that have $F = 0$ and are yet movable. Again, Gruebler's Criterion is unaware of the special geometry.

The *actual* number of dof in a mechanism is known as the mobility, or M number, and it can be found only by inspection. M is always greater than or equal to F:

$$M \geq F.$$

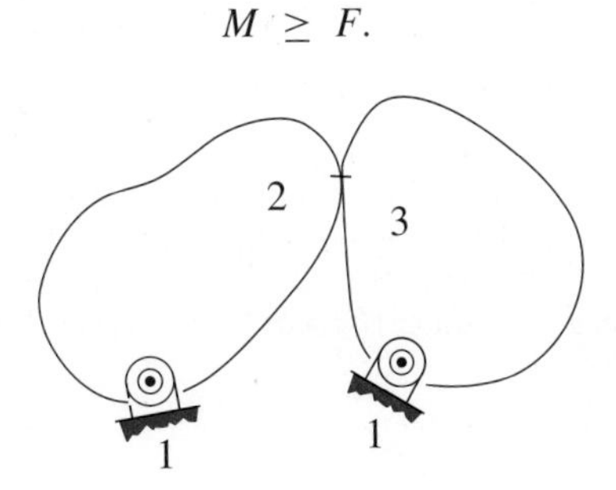

FIGURE 1.38 A pair of externally rolling eccentric gears

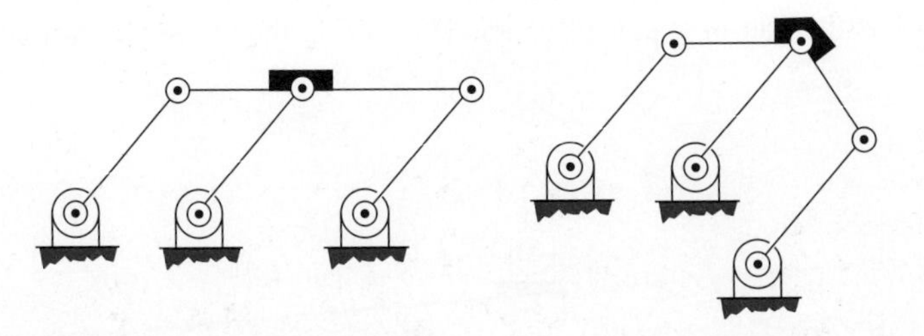

FIGURE 1.39 A pair of over-constrained mechanisms

DEFINITION:

Mobility *The actual number of dof in a mechanism.*

WORKING DEFINITION:

Mobility *The minimum number of independent joint variables that must be removed from a mechanism in order to reduce it to a determinate structure.*

The easiest way to eliminate independent joint variables is to "freeze" or "lock out" P_1 joints. In this way the working definition of mobility changes as follows.

ALTERNATIVE WORKING DEFINITION:

Mobility *The minimum number of P_1 joints that must be removed from a mechanism in order to reduce it to a determinate structure.*

▶ EXAMPLE 1.12

Determine the mobility of the four bar mechanism shown in Figure 1.14.

Solution

If we lock out (remove) the pin joint between links 1 and 2, making them a single link, the four bar mechanism becomes a three link truss, which is a structure. Hence $M = 1$, which agrees with the F number calculation.

▶ EXAMPLE 1.13

Determine the mobility of the indeterminate structure shown in Figure 1.36.

Solution

The system is already immovable, so no P_1 joints need to be frozen to make the system a structure. Hence $M = 0$, which disagrees with $F = -1$. The justification for this is that the structure is statically indeterminate and Gruebler's Criterion is unaware of this.

▶ EXAMPLE 1.14

Determine the mobility of the mechanism shown in Figure 1.37.

Solution

In order to make the mechanism a structure, we need to freeze any one of the three P_1 joints. Hence $M = 1$, which disagrees with $F = 0$. The justification for this is that Gruebler's Criterion is unaware of the two geometric conditions cited after Equation (1.3).

▶ EXAMPLE 1.15

Determine the mobility of the mechanism shown in Figure 1.23.

Solution

If we lock out the pin joint between 1 and 2, then 2 becomes fixed. If 2 is fixed, then links 3 and 4 form a truss that is immovable. Links 5, 6, and 7 can still move, however, as a four bar mechanism. If we now lock out the pin joint between 5 and 4, then 5 no longer moves, 6 and

7 form a truss, and the entire mechanism is immovable. We had to lock out two P_1 joints to make the mechanism immovable, so $M = 2$. This result agrees with Gruebler's Criterion in Example 1.10.

In most situations, the values of F and M are the same, but it is always worthwhile to determine M.

1.6 GRASHOF'S CRITERION

The four bar mechanism in Figure 1.40 is the most common and useful mechanism. You have seen examples of it in Figures 1.14, 1.15, 1.27, and 1.29. In this section we will consider the gross motion capabilities of four bar mechanisms. We answer the basic question of whether particular links within a given four bar mechanism are capable of continuous rotation, i.e., are fully rotatable.

1.6.1 Cranks and Rockers

The two ground pivoted links in a four bar mechanism, bodies 2 and 4 in Figure 1.40, are generically referred to as levers. Typically, one of the levers is an input rotation, the other the output rotation. The two levers are connected to each other through the coupler link, body 3, and the ground, body 1. A lever can be either a "crank" or a "rocker." (There are instances when the coupler is the input. Figure 1.25 is an example. The handle of the plier, which is the coupler of the four bar mechanism, is the input rotation.)

DEFINITIONS:

Crank *A lever that is capable of continuous rotations.*

Rocker *A lever that can only oscillate between limit positions and is not capable of continuous rotations.*

Cranks are convenient as inputs since most drivers (engines, motors, etc.) provide a continuous rotation about a fixed axis.

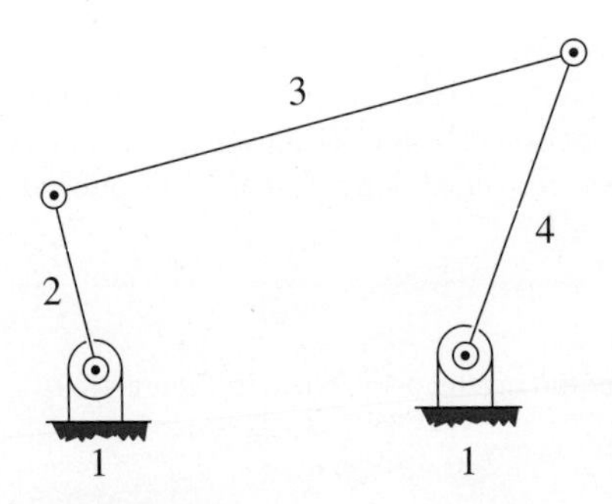

FIGURE 1.40 A four bar mechanism

When discussing four bar mechanisms it is customary to refer to the mechanism as (input lever type)-(output lever type). So, for example, a crank-rocker is a four bar mechanism that has a crank input and a rocker output, and a rocker-crank is a four bar mechanism that has a rocker input and a crank output.

The dimension (length) of each link will affect the *relative* motion of all the links. Based upon the link lengths, *Grashof's Criterion* will enable us to determine whether the levers of the four bar mechanism are cranks or rockers.

1.6.2 Grashof Four Bar Mechanisms

In 1883 F. Grashof published a criterion that enables one to determine whether the input and output levers of a four bar mechanism are cranks or rockers. His derivation proceeds as follows.

Consider a crank-rocker (crank input, rocker output) four bar mechanism such as the one shown in Figure 1.41(a). Link 2 is the input (the crank) and link 4 is the output (the rocker). Let L_i represent the distance between the two pin joints on any link i ($i = 1, 2, 3, 4$). This is the length (dimension) of link i, regardless of what the actual shape of link i is.

Since link 4 is a rocker, it has two limit positions between which it will oscillate. When link 4 reaches one of its limit positions, it momentarily dwells and then returns in the opposite direction until it reaches its other limit position, where again it momentarily dwells and returns in the opposite direction. In one limit position links 2 and 3 are in line and fully extended, and in the other limit position the links are in line but overlapped, as shown in Figures 1.41(b) and (c). In each of these limit positions a triangle is formed

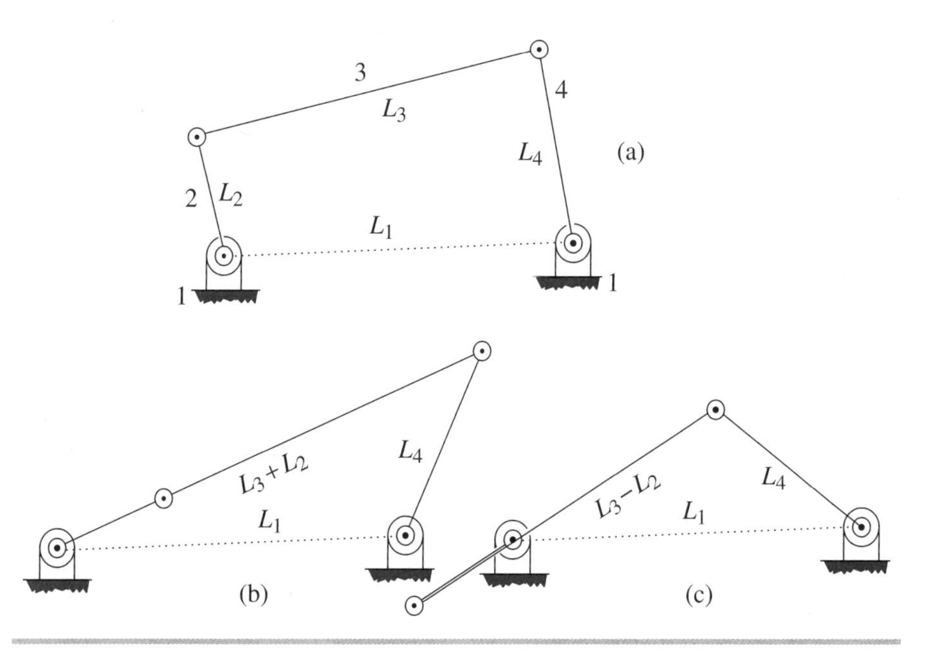

FIGURE 1.41 A crank-rocker four bar mechanism

by the links. In a triangle, any one side must be shorter in length than the sum of the remaining two sides. So for the triangle in Figure 1.41(c),

$$L_1 < (L_3 - L_2) + L_4 \tag{1.4}$$

$$L_4 < (L_3 - L_2) + L_1 \tag{1.5}$$

and for the triangle in Figure 1.41(b),

$$(L_3 + L_2) < L_1 + L_4. \tag{1.6}$$

Adding the left hand sides (lhs) and the right hand sides (rhs) of Equations (1.4) and (1.5) gives

$$L_2 < L_3. \tag{1.7}$$

Adding the lhs and rhs of Equations (1.5) and (1.6) gives

$$L_2 < L_1. \tag{1.8}$$

Adding the lhs and rhs of Equations (1.4) and (1.6) gives

$$L_2 < L_4. \tag{1.9}$$

These three inequalities show that in a crank-rocker four bar mechanism, the crank must be the shortest link.

Let L_{max} and L_{min} represent the longest and shortest links, respectively, in this crank-rocker mechanism. As we have shown, for link 2 to be a crank, L_2 must be L_{min}. Any of the other three links (L_1, L_3, or L_4) can be L_{max}. Let L_a and L_b represent the lengths of the intermediate-length links, and consider all three possibilities. If L_1 is L_{max}, and L_3 and L_4 are L_a and L_b, then from Equation (1.4),

$$\boxed{L_{max} + L_{min} < L_a + L_b} \tag{1.10}$$

If L_4 is L_{max}, and L_1 and L_3 are L_a and L_b, then Equation (1.5) also reduces to Equation (1.10). Finally, if L_3 is L_{max}, then Equation (1.6) also reduces to Equation (1.10). All three possibilities lead to Equation (1.10), which is known as Grashof's Criterion. The crank-rocker four bar mechanism in Figure 1.41 must satisfy Equation (1.10). Such a mechanism is known as a Grashof four bar mechanism, and the kinematic chain from which the mechanism is derived is known as a Grashof four bar kinematic chain. Figure 1.42 shows an animation of the crank-rocker four bar mechanism. Notice from the animation that the short link (the crank) is making a complete revolution relative to each of the remaining three links, whereas the remaining three links only oscillate relative to one another. This leads us to a further refinement of Grashof's Criterion. In a Grashof four bar kinematic chain, *the shortest link is capable of making a complete revolution*

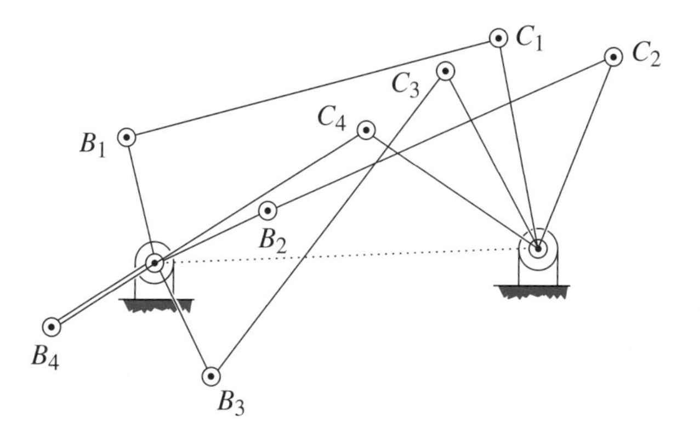

FIGURE 1.42 Animation of a crank-rocker four bar mechanism

relative to each of the three longer links. The three longer links can only oscillate relative to each other.

A Grashof four bar kinematic chain can result in three different types of mechanisms, depending on the location of the fully rotatable (shortest) link, as illustrated in Figures 1.43(b) through (e). Figure 1.43(a) shows a Grashof four bar kinematic chain. Figure 1.43(b) shows the mechanism that results from that chain when the shortest link is ground. Since this link makes a complete revolution *relative to* the remaining three links and this link is stationary, the three moving links must be fully rotatable, and the mechanism is a crank-crank. Notice that during one cycle of motion the coupler link also makes a complete revolution. Figures 1.43(c) and (d) are either a crank-rocker or a rocker-crank, depending on which of L_{min} and L_b is the input and which is the output. Figure 1.43(e) has the shortest link as the coupler. It always results in a rocker-rocker. The coupler is capable of making a complete revolution during one cycle of motion.

1.6.3 Non-Grashof Four Bar Mechanisms

When the sum of the lengths of the longest and shortest links is *greater than* the sum of the lengths of the intermediate pair of links, the condition in Equation (1.10) is not satisfied, and the four bar kinematic chain is referred to as non-Grashof. In non-Grashof four bar kinematic chains, no link is capable of making a complete revolution relative to any of the other three remaining links. All four links can only oscillate relative to each other, so non-Grashof four bar kinematic chains *always* result in rocker-rocker mechanisms.

Finally, there is a "borderline" case, where the sum of the lengths of the longest and shortest links equals the sum of the lengths of the intermediate links. The mechanisms that come from this anomaly are not sure what they want to be. The chain can be flattened, or folded onto itself, so that all the links are overlapping. Such mechanisms are

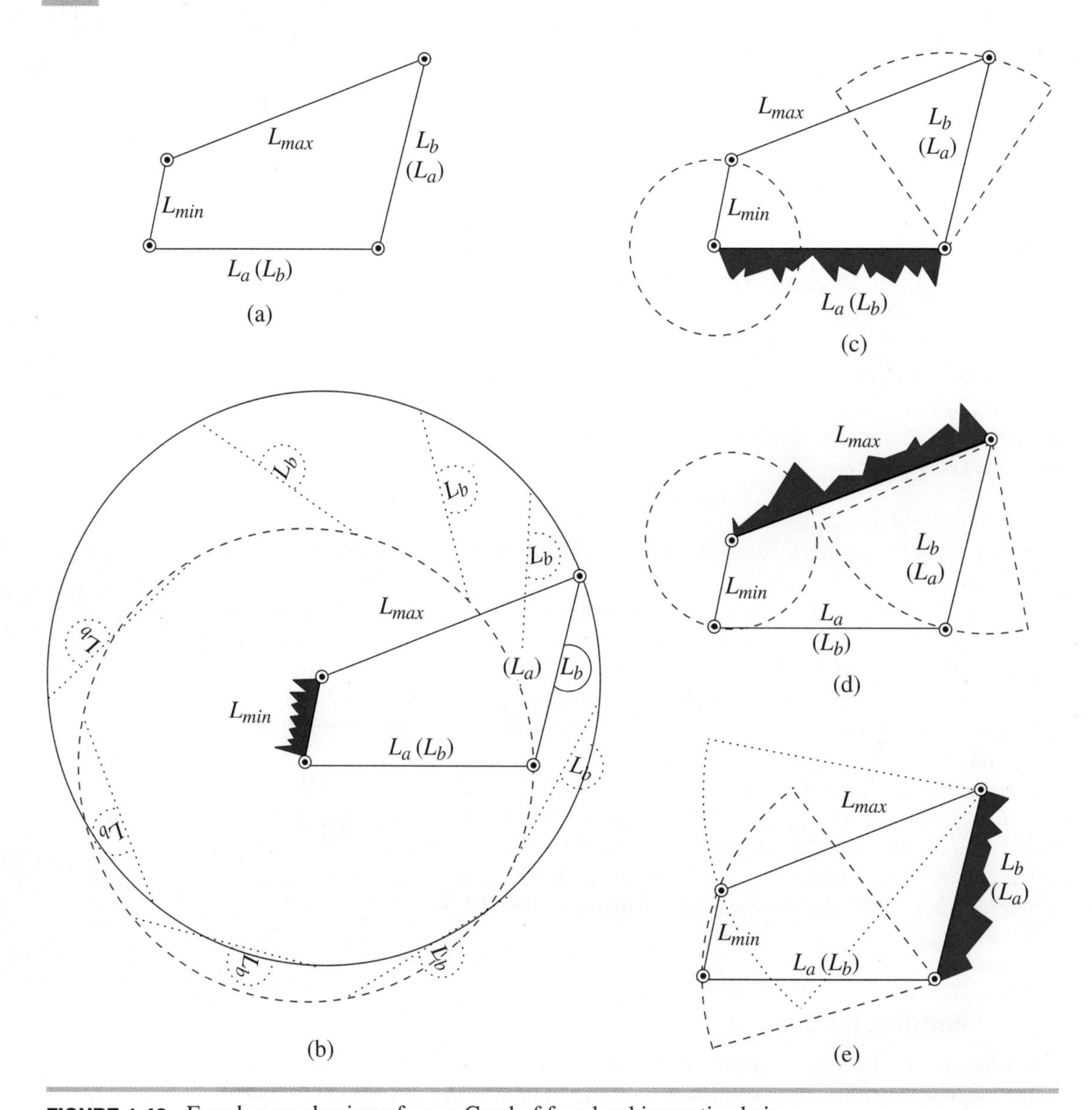

FIGURE 1.43 Four bar mechanisms from a Grashof four bar kinematic chain

typically avoided; however, they do appear in situations as parallelogram four bar mechanisms or isosceles four bar mechanisms, as shown in Figures 1.44 and 1.45. As with most things, they have their advantages and disadvantages. For example, to its advantage, the mechanism in Figure 1.44 has a constant 1:1 ratio of input to output speed. To its disadvantage, when it is completely folded, it instantaneously has two degrees of freedom! Be careful with these borderline cases.

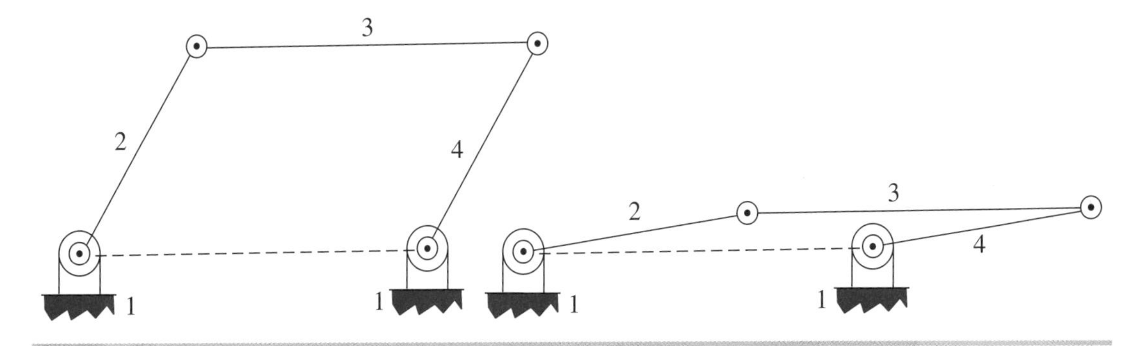

FIGURE 1.44 Example of a parallelogram four bar mechanism (a.k.a. borderline case)

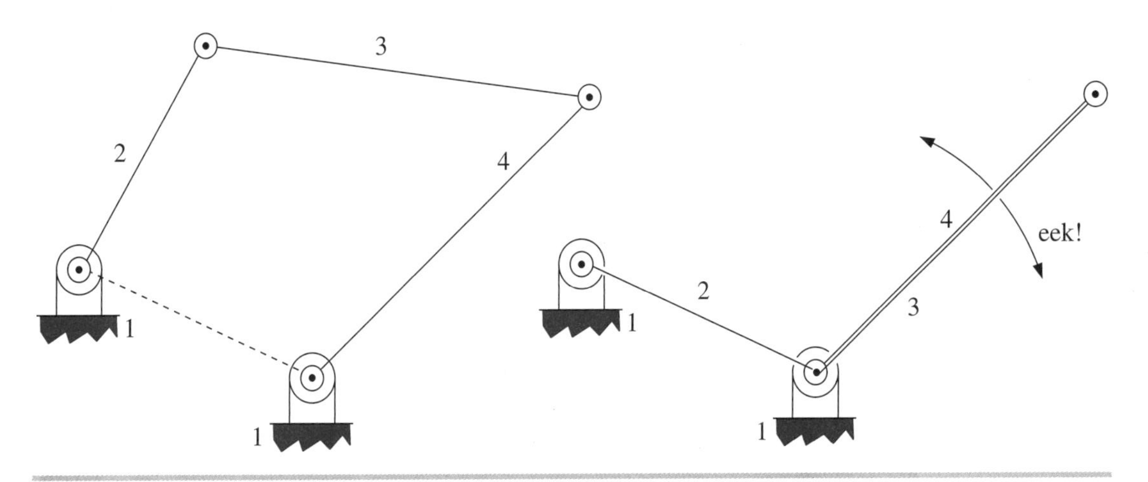

FIGURE 1.45 Example of an isosceles four bar mechanism (a.k.a. borderline case)

1.6.4 Grashof's Criterion Applied to a Four Bar Kinematic Chain with a Sliding Contact

The left hand side of Figure 1.46 shows a body x and a body y connected by an offset sliding joint. The motion of x relative to y is a translation in the direction of the indicated straight line. Relative to body y, all points on x travel on straight lines parallel to this direction, and vice versa.

A straight line is an arc on a circle of infinite radius. Thus the sliding joint between x and y can be replaced by a pin joint connection at infinity along a direction that is perpendicular to the relative translation.

The left side of Figure 1.47 shows a four link kinematic chain that contains a sliding joint. The chain has three link lengths: L_2, L_3, and the offset, e. The right side of the figure shows an equivalent kinematic chain composed of all revolute joints. The lengths of links 2 and 3 are the dimensions L_2 and L_3. The length of link 1 is $L_1 = L + e$, and the length of link 4 is $L_4 = L$, where L is infinite. This is the equivalent four bar mechanism to which we can apply Grashof's Criterion, Equation (1.10). In the equivalent four

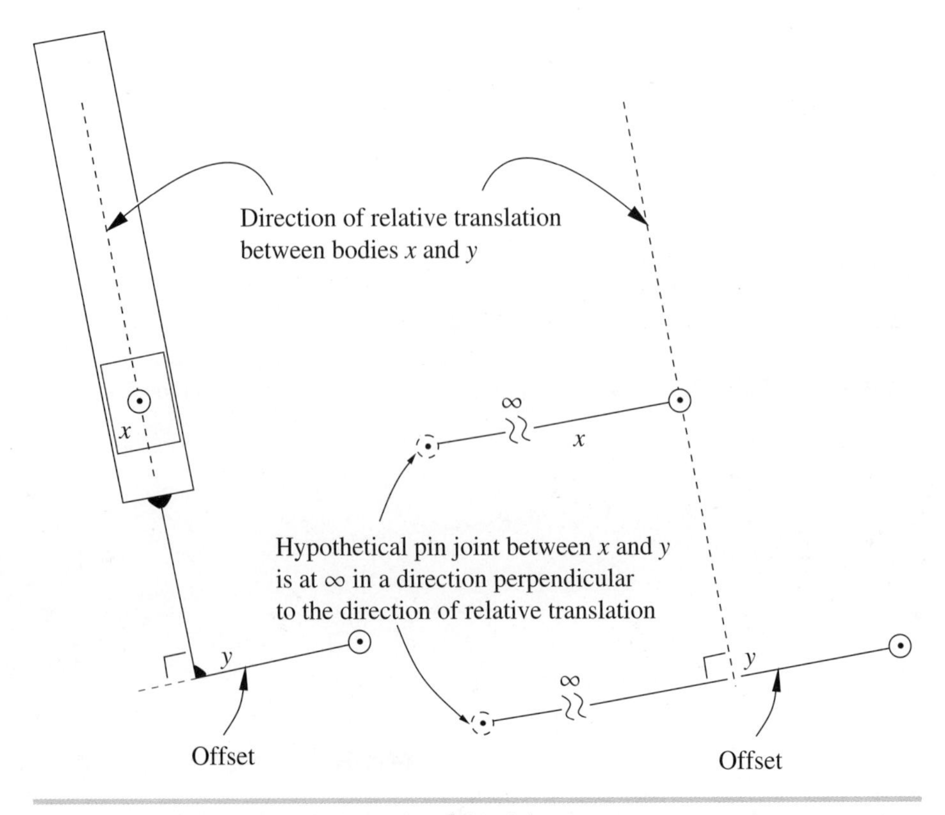

FIGURE 1.46 Pin joint equivalent of a sliding joint

bar mechanism the longest link is link 1, so $L_{max} = L_1 = L + e$ ($L = \infty$). The shortest link will be either link 2 or link 3, so L_{min} is either L_2 or L_3, whichever is shorter, and the other link is then an intermediate link L_a. Link 4 will definitely be an intermediate-length link, because it is just shorter than link 1 but definitely longer than link 2 or 3. In this case Equation (1.10) gives the condition for a Grashof four bar kinematic chain to be

$$\underbrace{(\not{L} + e)}_{L_{max}} + \underbrace{L_{min}}_{\text{smaller of } L_2 \text{ and } L_3} < \underbrace{\not{L}}_{L_a} + \underbrace{L_b}_{\text{larger of } L_2 \text{ and } L_3} \tag{1.11}$$

$$(e) + (\text{the smaller of } L_2 \text{ and } L_3) < (\text{the larger of } L_2 \text{ and } L_3). \tag{1.12}$$

When the condition in Equation (1.12) is met, the four bar kinematic chain being discussed here with a sliding joint is a Grashof kinematic chain, and the shortest link (i.e., the shorter of L_2 and L_3) makes a complete revolution relative to the remaining three links. The three remaining links only oscillate relative to one another. In the case of the sliding joint, an oscillation is a reciprocating sliding action.

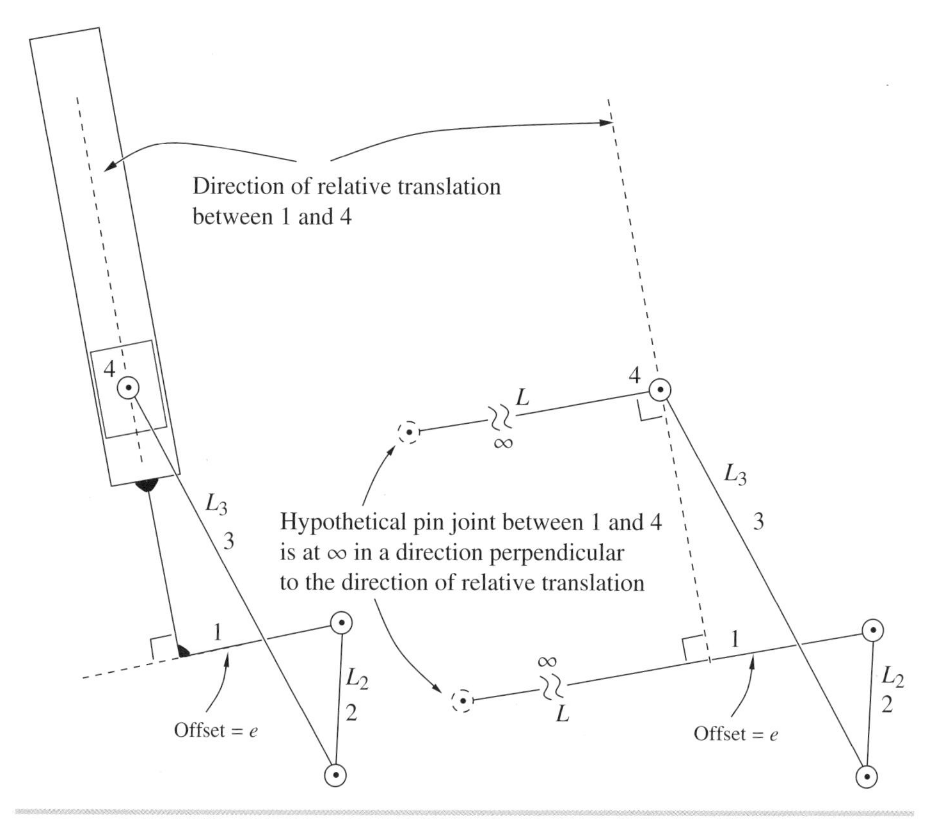

FIGURE 1.47 A four link kinematic chain with a sliding joint

Figure 1.48 shows the four kinematic inversions of a Grashof four bar kinematic chain with a sliding joint. L_2 is smaller than L_3 in this particular Grashof four bar kinematic chain. The figure indicates the type of motions of the input and output levers for each mechanism.

Mechanisms from non-Grashof four bar kinematic chains with a sliding joint will always have input and output levers that are rockers.

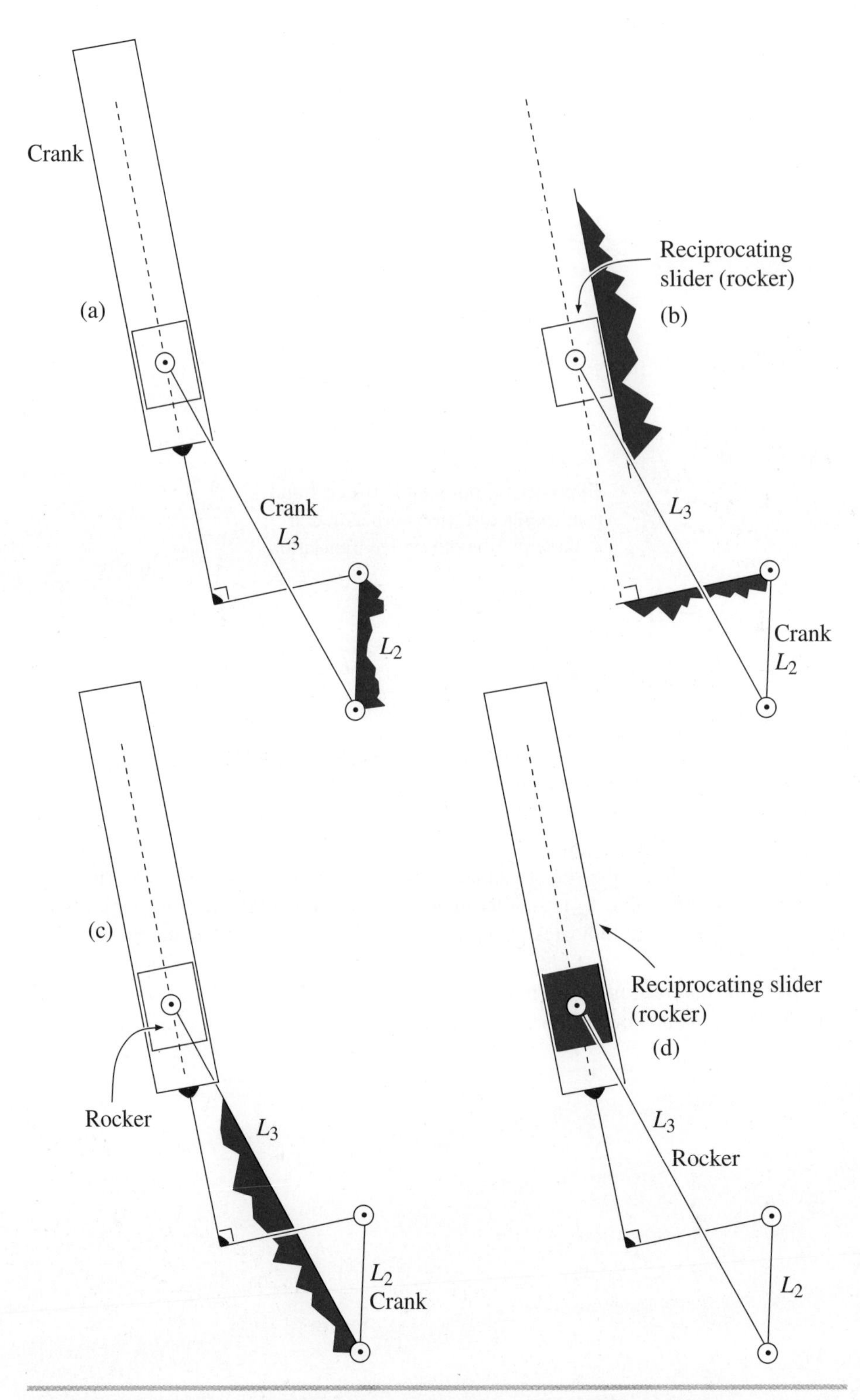

FIGURE 1.48 The four kinematic inversions of the crank-slider mechanism

1.7 PROBLEMS

For the mechanisms shown in Problems 1.1–1.20, find the theoretical number of dof (the F number) using Gruebler's Criterion and the actual number of dof (the M number) by inspection. Justify any difference between F and M.

Problem 1.1

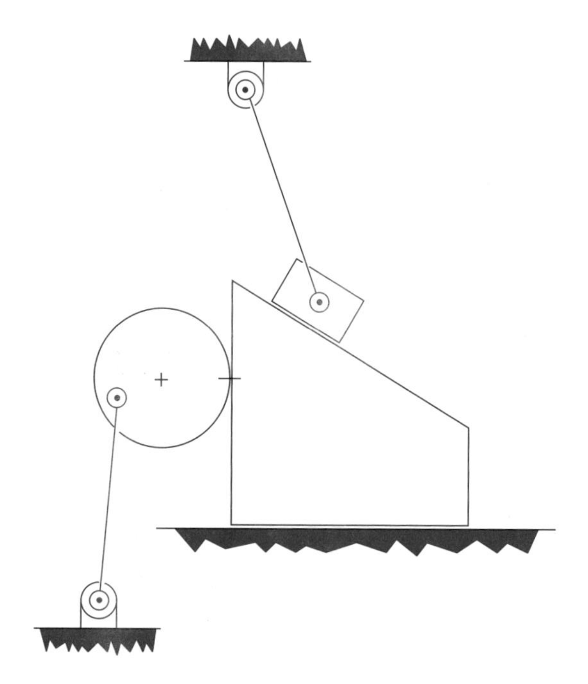

Problem 1.2

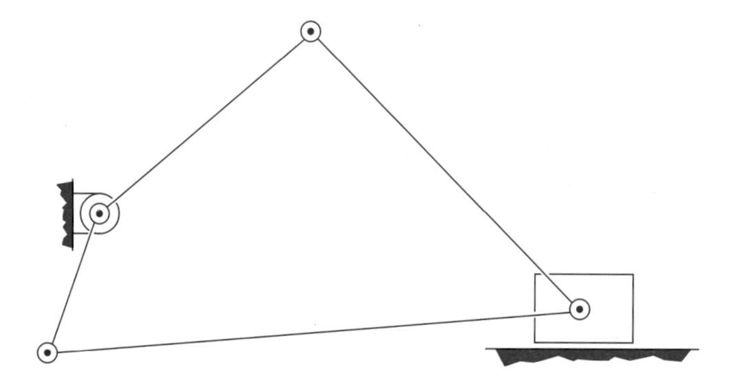

Problem 1.3

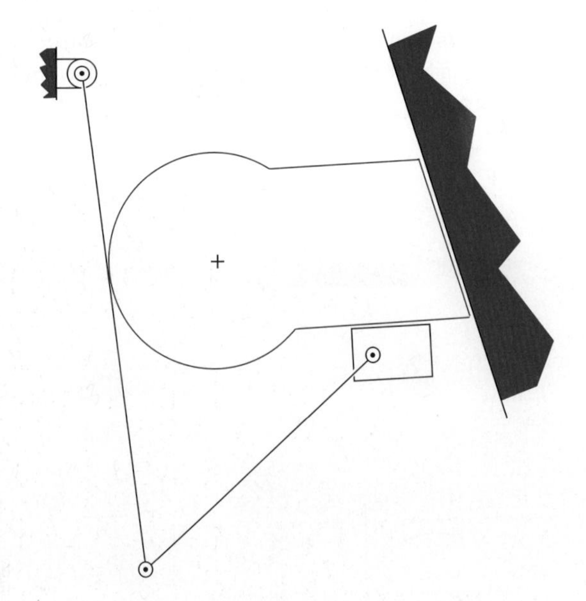

Problem 1.4

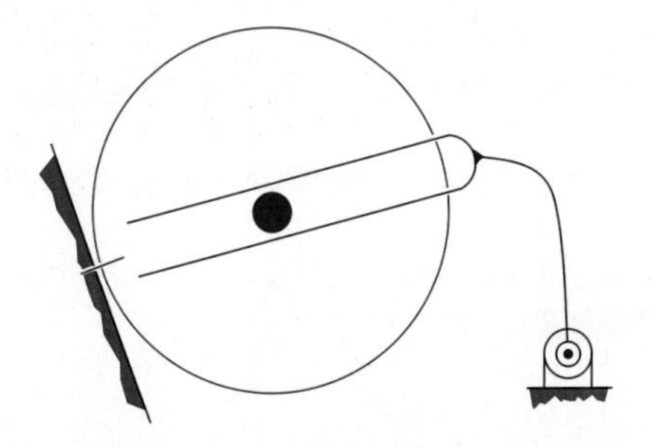

Problem 1.5

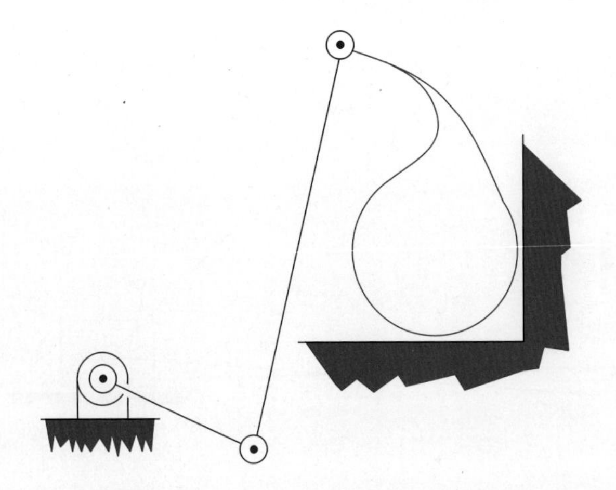

Problem 1.6

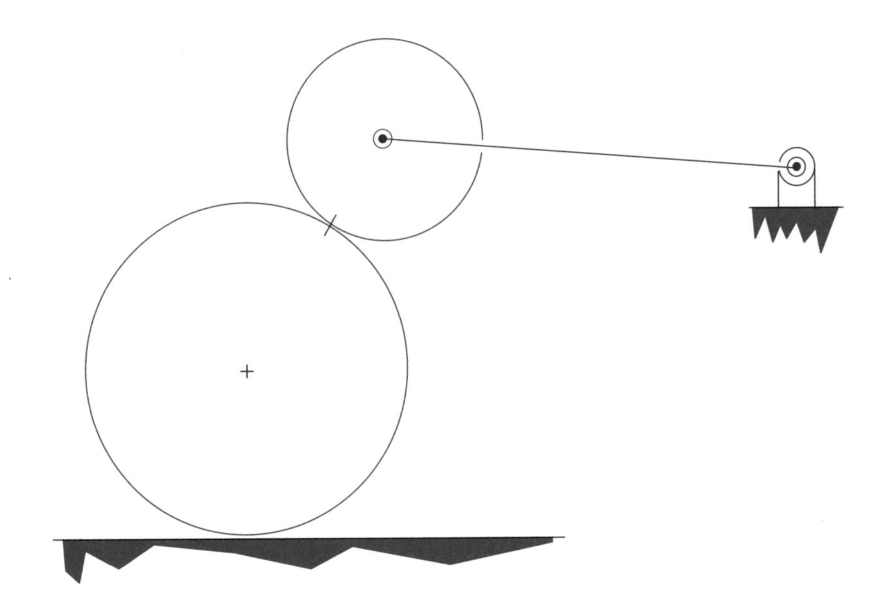

Problem 1.7

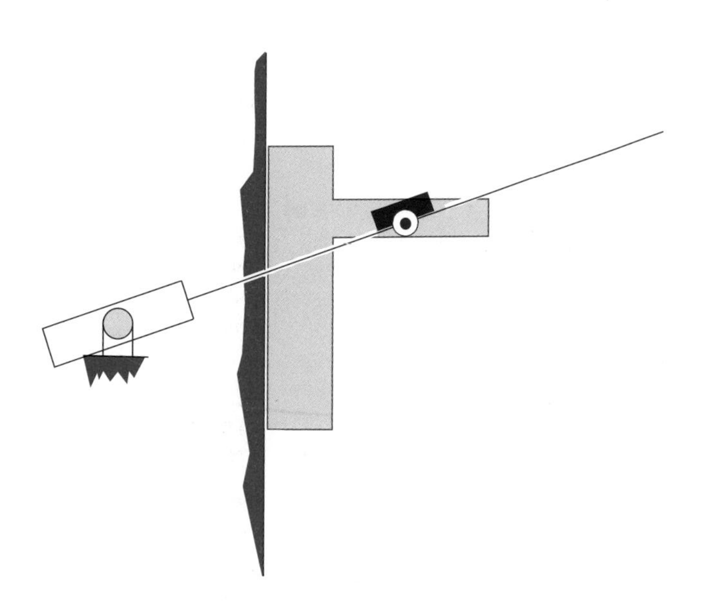

Problem 1.8

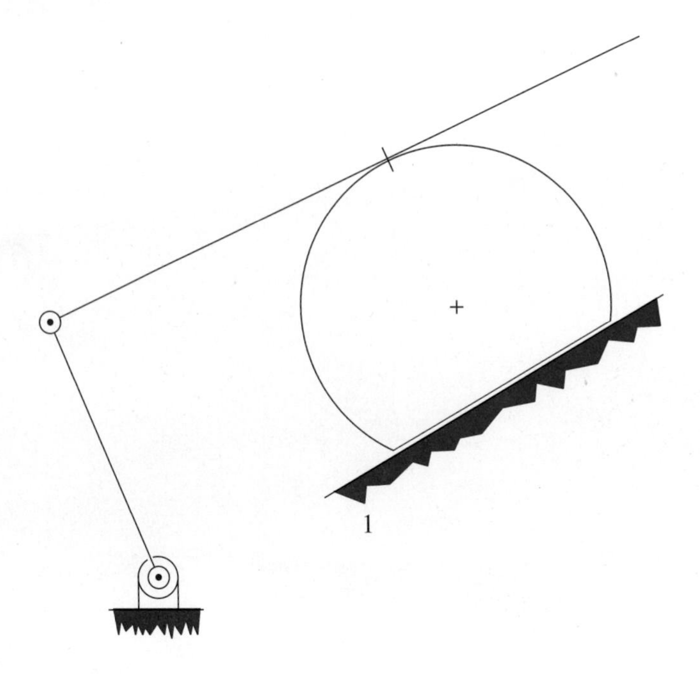

Problem 1.9

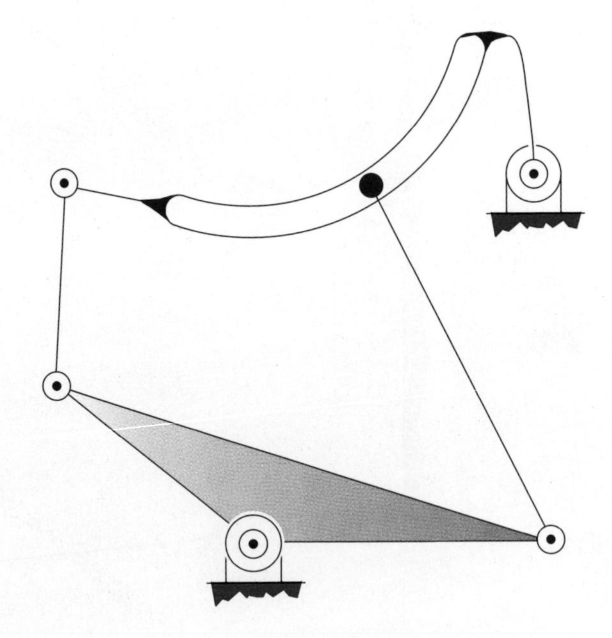

Problem 1.10

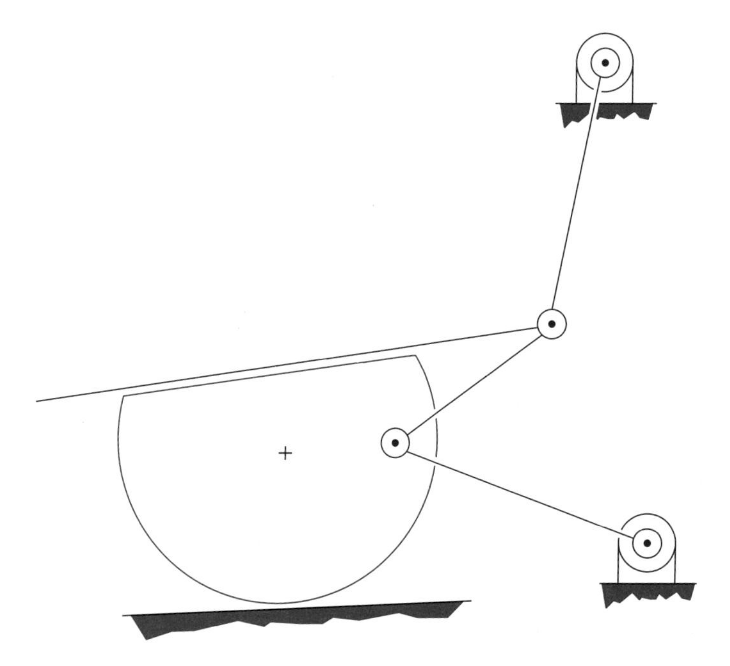

Problem 1.11

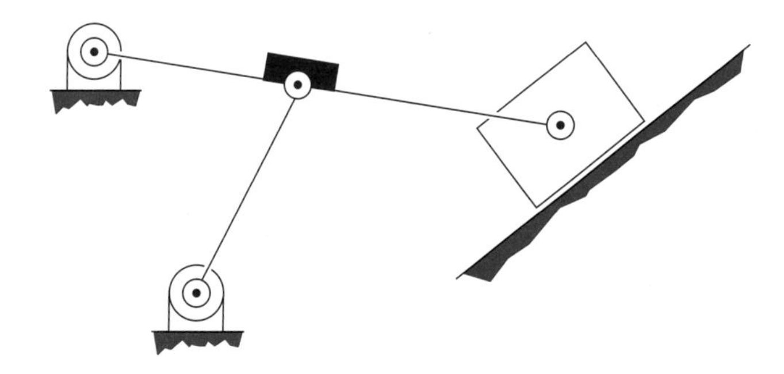

Problem 1.12

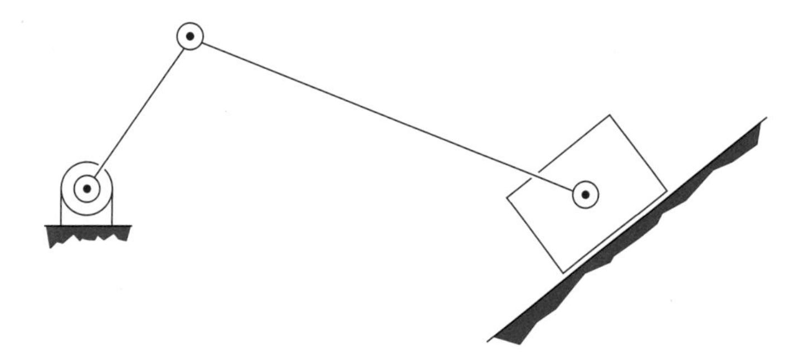

Problem 1.13

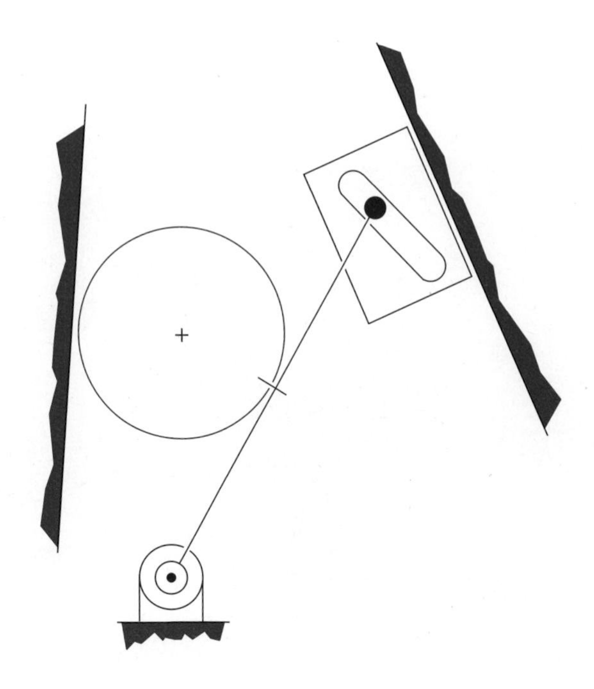

Problem 1.14

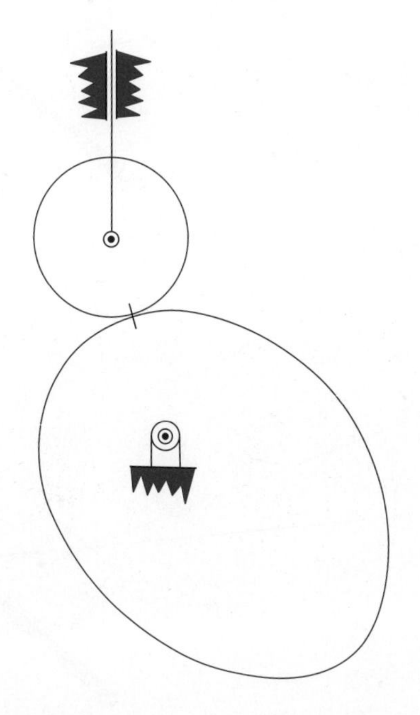

Problem 1.15

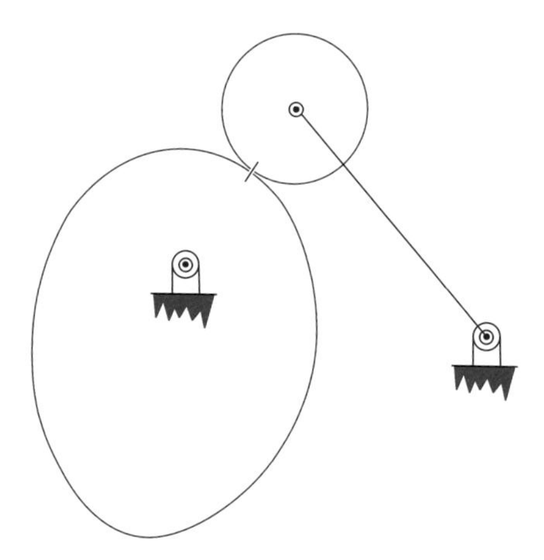

Problem 1.16

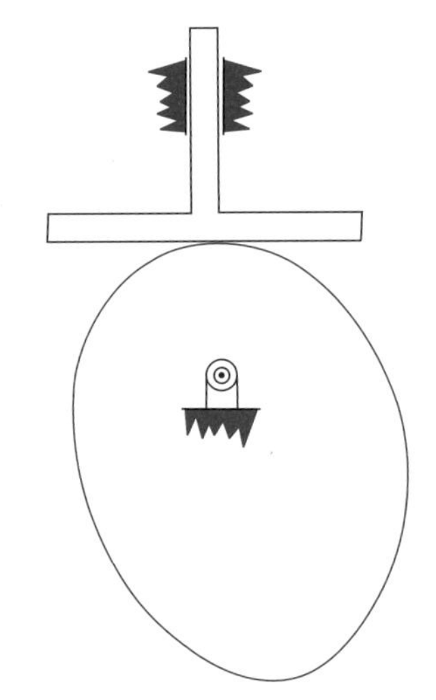

Problem 1.17

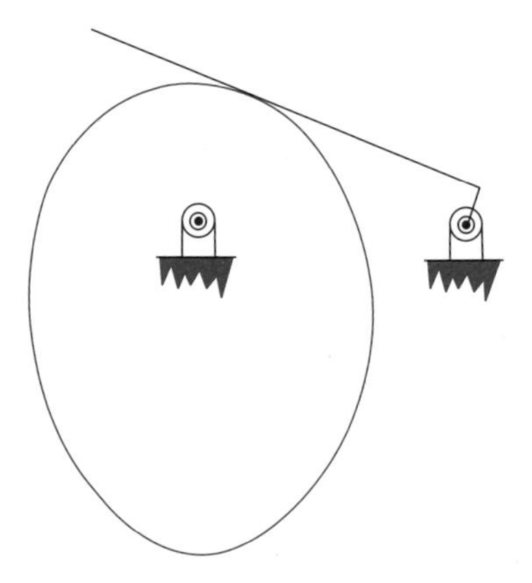

Problem 1.18

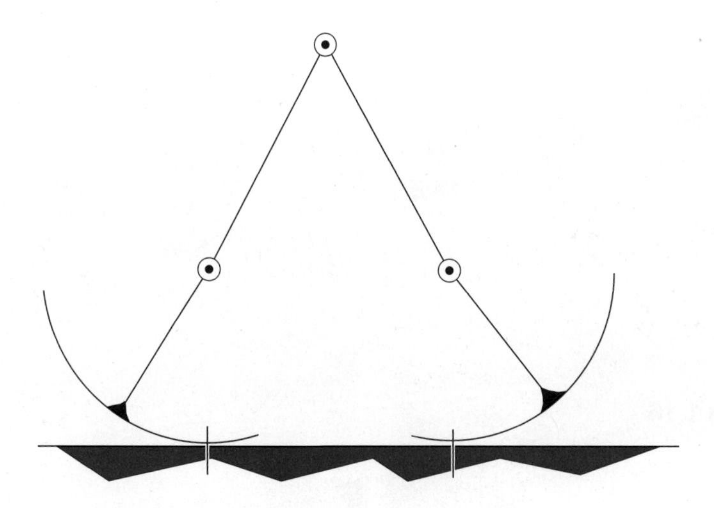

Problem 1.19

Problem 1.20

What type of joint must exist between links 2 and 3 for the mechanism to have one degree of freedom?

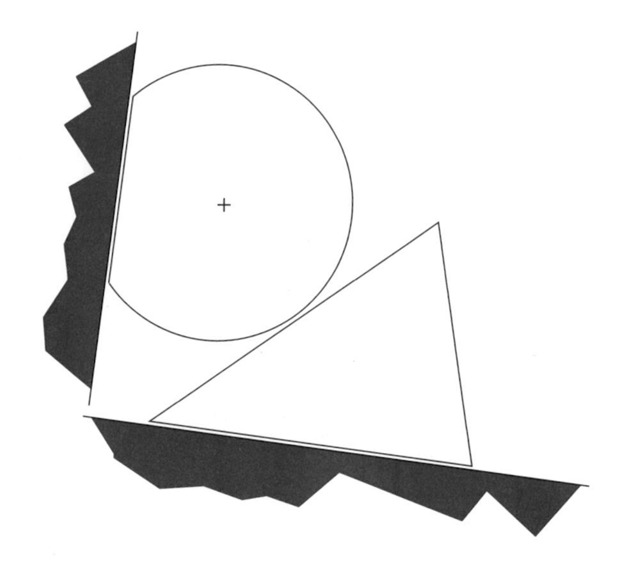

Problem 1.21

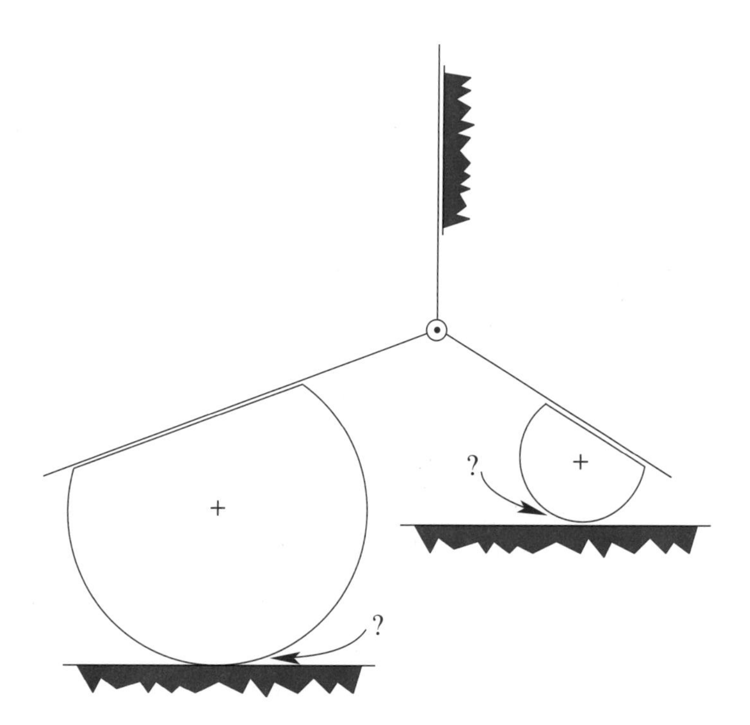

Problem 1.22

Give all the kinematic inversions of the mechanism shown in Figure 1.31(a).

Problem 1.23

Give the mechanism that results from Figure 1.30(c) when link 5 is grounded. Determine the theoretical dof (F) and mobility (M) of that mechanism.

Problem 1.24

The figure below shows a pair of gears in mesh. The smaller gear is referred to as the pinion, and the larger gear is referred to as the gear. Typically, the pinion is the driver and the gear is driven. Since the pinion is smaller than the gear, the gear rotates more slowly than the pinion, but the torque is increased. Generally speaking, that is the purpose of a gear pair: to reduce speed and to increase torque. This is because most prime movers (motors, engines, etc.) run at high speeds but produce relatively low torques. You will see all this later when we study gears.

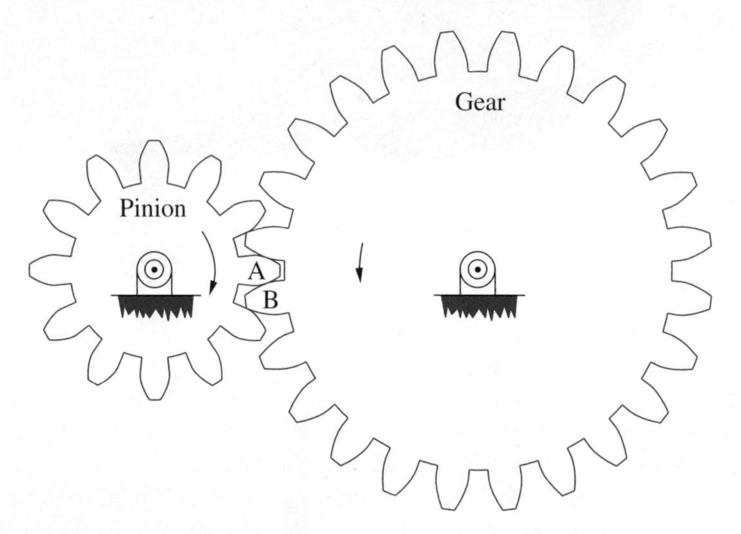

Now focus on the meshing action between gear tooth A belonging to the pinion as it drives gear tooth B belonging to the gear.

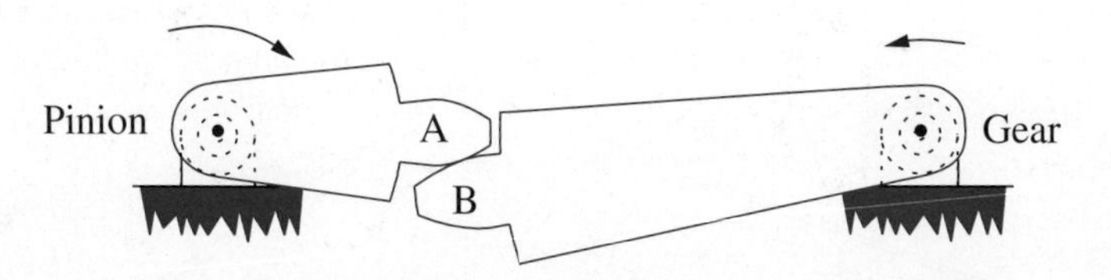

What type of joint must exist between these two gear teeth so that the system is movable, with one dof?

Problem 1.25

Look around campus or your neighborhood and find a machine. Draw a skeleton diagram of that machine, and determine both its theoretical and its actual number of degrees of freedom. Prepare a concise 2–3-page technical report of your findings. Be certain that all drawings are neat and professional looking.

CHAPTER 2

Kinematic Analysis Part I: Vector Loop Method

Kinematic analysis of mechanisms assumes that a mechanism and its dimensions are given. For one-degree-of-freedom mechanisms, kinematic analysis solves the following problem.

Given: the state of motion (position, velocity, acceleration) of one link (the "input" link).
Determine: the state of motion (position, velocity, acceleration) of all other links and particular points of interest, such as mass centers.

Several methods are helpful in finding the solution to this problem. Your first mechanics course treated a part of the problem (the velocity and acceleration analysis) as a preliminary to dynamics. The position problem, a complex nonlinear problem, was not dealt with there. The vector loop method presented here treats all three aspects of this problem and is well suited to computerization. This chapter introduces the vector loop method.

The vector loop method relies on being able to assign correct vector loops to a mechanism. Learning to make that correct assignment of vector loops is the primary goal of this chapter. The vector loops lead to a set of position equations that describe the changing configuration of a mechanism. These equations are nonlinear and generally have no closed-form (algebraic) solution. We develop Newton's method for solving these nonlinear position equations. In later chapters we apply the vector loop method to calculating velocities and accelerations, describing limit positions and dead positions, determining mechanical advantage, analyzing forces, and (ultimately) simulating dynamics.

2.1 KINEMATIC ANALYSIS AND THE VECTOR LOOP METHOD

Kinematic analysis by the vector loop method consists of the following steps.

Step 1: Attach a vector to each link of the mechanism, thus creating a series of vectors forming a closed loop (or closed loops). Each vector should be defined such that the vector's length and direction either correspond to a link dimension or relate to a joint variable within the mechanism. For example, if a chosen vector has a fixed length and a variable direction, the fixed length should indicate a dimension of the link, and the variable direction should relate to a rotational joint variable within the mechanism.

Step 2: Write the vector loop equation(s). This is a statement that the sum of the vectors in the loop is zero.

Step 3: Choose a fixed *X-Y* coordinate system whose origin is at a fixed point in the vector loop. This should be a right handed coordinate system where *X* crossed into *Y* equals *Z* and *Z* points out of the page.

Step 4: Decompose the vector loop equation(s) into scalar *X* and *Y* components; these are known as the position equations.

Step 5: List the scalar knowns and the scalar unknowns that appear in the position equations. Compare the number of unknown scalars to the number of position equations.

Note:

a. If the number of scalar unknowns exceeds the number of equations, the mechanism is movable, and the amount of the excess is equal to the number of degrees of freedom in the mechanism. The number of degrees of freedom indicates how many of the unknowns need to be given (i.e., are needed as inputs) in order that the remaining unknowns can be computed from the position equations.
b. If the number of scalar unknowns equals the number of equations, the system is not a mechanism. The system is a statically determinate structure.
c. If the number of equations exceeds the number of scalar unknowns, the system is not a mechanism. The system is a statically indeterminate structure, and the amount of the excess is equal to the degree of indeterminacy.

Step 6: Solve the position equations for the *position unknowns*. This is the solution to the position problem.

Step 7: Differentiate the position equations with respect to time, generating the velocity equations. Solve the velocity equations for the *velocity unknowns*. This is the solution to the velocity problem.

Step 8: Differentiate the position equations a second time with respect to time, generating the acceleration equations. Solve the acceleration equations for the *acceleration unknowns*. This is the solution to the acceleration problem.

A number of details involved in carrying out the above steps are best revealed through examples. Before engaging the examples, consider Figure 2.1, which shows a vector $\bar{r}$. To reference the components of this vector in the shown coordinate system, it will always be the convention to denote the direction of the vector with an angle θ measured counterclockwise positive from the positive direction of the *X* axis to the positive direction of the vector. The vector's components are then given by

$$\bar{r} = r\begin{bmatrix} \cos\theta \\ \sin\theta \end{bmatrix}, \tag{2.1}$$

where r is the magnitude of the vector $\bar{r}$, and θ is its direction as defined in Figure 2.1. This applies only in a right-handed coordinate system that has its *Z* axis coming out

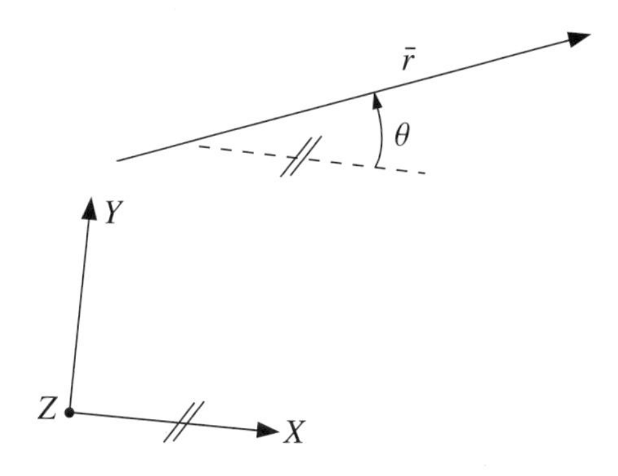

FIGURE 2.1 Convention for defining vector directions

of the plane of motion as in Figure 2.1. In the following discussions this convention is followed for defining the directions of all vectors.

▶ EXAMPLE 2.1

Figure 2.2 shows a four bar mechanism such as that in Figures 1.14 and 1.15. Follow steps 1–5 in the procedure outlined in Section 2.1 to derive the mechanism's position equations and deduce the number of degrees of freedom.

Solution

Step 1: Assign to each link a vector that goes from one pin joint to the next, as illustrated in Figure 2.3. Regardless of the actual physical shape of a link in this four bar mechanism (the link may be curved, for example, like link 3 in Figure 1.13), the vector attached to that link will span directly across the two pin joints joined by the link. Try to name each vector after the link to which it is attached. This is possible in the case of a four bar mechanism, but it may not always be possible because in some cases there will be multiple vectors associated with a single link. Each vector in Figure 2.3 can be taken in the direction shown or in the opposite direction.

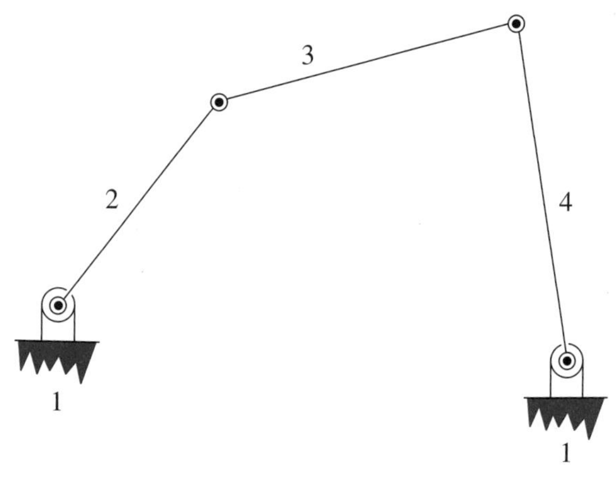

FIGURE 2.2 The four bar mechanism

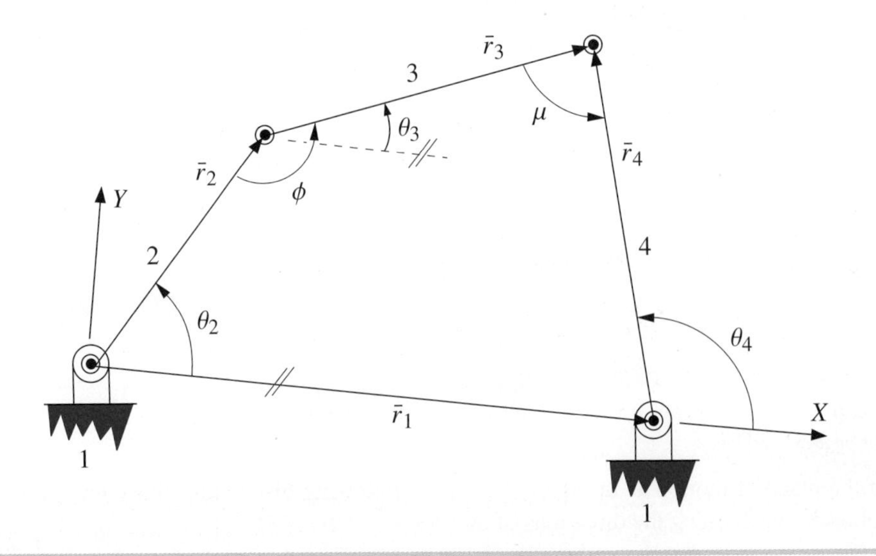

FIGURE 2.3 The four bar mechanism and its vector loop

Before proceeding to step 2, we should ask ourselves if the assigned vectors are appropriate. For the assigned vectors to be appropriate, they must define either a dimension of the mechanism or a joint variable within the mechanism. In this vector loop the magnitude of each vector is known, and it represents the length (dimension) of the link to which it is attached. The direction θ_1 is a known and is a somewhat meaningless dimension that defines the orientation of the entire mechanism. Changing θ_1 rotates the whole mechanism but has no effect on the relative motions of the links. However, as we will see when we analyze forces and dynamics, θ_1 is important in that the direction of gravity and weight forces are defined relative to it. Directions θ_2, θ_3, and θ_4 are unknowns that define the angular orientation of links 2, 3, and 4, respectively. They also relate to the joint variables associated with the pin joints. θ_2 is the joint variable at the pin joint connecting 2 to 1, and θ_4 is the joint variable at the pin joint connecting 4 to 1. θ_3 is not a joint variable directly, but in combination with θ_2 and θ_4 it is a joint variable. In Figure 2.3 the angles ϕ and μ are the joint variables associated with the pin joints connecting 2 to 3 and 3 to 4, respectively. In terms of the directions of the assigned vectors, these two joint variables are given by

$$\phi = \pi - (\theta_2 - \theta_3) \quad \text{and} \quad \mu = \theta_4 - \theta_3. \tag{2.2}$$

θ_2, θ_3, and θ_4 relate to the joint variables of the pin joints in the mechanism, so we conclude that the assigned vectors are appropriate.

Step 2: The vector loop equation (VLE) is

$$\bar{r}_2 + \bar{r}_3 - \bar{r}_4 - \bar{r}_1 = \bar{0}. \tag{2.3}$$

Step 3: Assign a fixed coordinate system whose origin is at the pin joint between 2 and 1, which is a fixed point in the vector loop. (The pin joint between 4 and 1 is also a fixed point in the vector loop and is also a suitable location for the origin.) Although it is not necessary, if we align one of the coordinate axes with the fixed vector $\bar{r}_1$, then the position equations will be slightly simplified because $\bar{r}_1$ will have a component only in that direction. Align the *X* axis with the direction of $\bar{r}_1$.

Step 4: The VLE has scalar components

$$r_2\cos\theta_2 + r_3\cos\theta_3 - r_4\cos\theta_4 - r_1 = 0 \quad (2.4)$$

$$r_2\sin\theta_2 + r_3\sin\theta_3 - r_4\sin\theta_4 = 0. \quad (2.5)$$

$\theta_1 = 0°$ has been incorporated into these equations. Equations (2.4) and (2.5) are the position equations.

Step 5: Within the position equations there are

scalar knowns: $r_2, r_3, r_4, r_1, \theta_1 = 0$

and

scalar unknowns: θ_2, θ_3, and θ_4.

Since there are three unknowns in these two equations, one must be given so that the remaining two can be calculated. This means the mechanism has one degree of freedom, which agrees with Gruebler's Criterion.

▶ EXAMPLE 2.2

Figure 2.4 shows an offset crank-slider mechanism. The term *offset* is used because the line of action of the pin joint between the connecting rod, 3, and the slider, 4, does not pass through the the pin joint connecting the crank, 2, and the ground, 1. Following steps 1–5 in Section 2.1, derive the mechanism's position equations and determine the number of degrees of freedom.

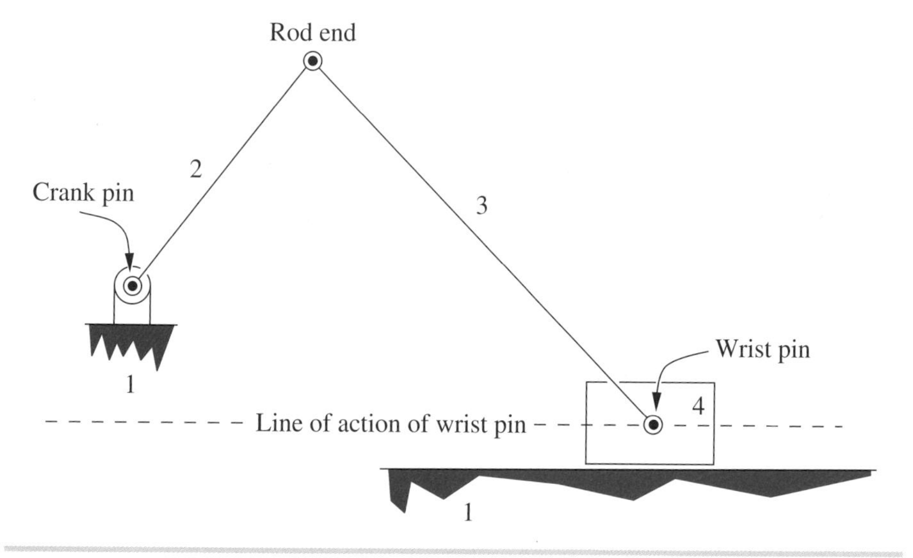

FIGURE 2.4 The offset crank-slider mechanism

Solution

Step 1: Figure 2.5 shows a vector loop. Vectors were assigned to links 2 and 3 as in the four bar mechanism. The question is how the vectors associated with bodies 1 and 4 and the sliding joint between them should be assigned. That is, what vectors should be used to get from the crank pin through the sliding joint between 1 and 4 and over to the wrist pin? The answer comes from understanding that the unique relationship between a point and a line is the perpendicular distance from the point to the line. This is the shortest distance from the point to the line, as shown in Figure 2.6. In light of this, to close the loop, use the perpendicular vectors $\bar{r}_1$ and $\bar{r}_4$ shown in Figure 2.5.

r_1 and θ_1 are dimensions. Once again, θ_1 is a meaningless dimension that defines the orientation of the mechanism in the plane. It has no bearing on the relative motion of the links. r_4 is the joint variable associated with the sliding joint between 1 and 4 and θ_4 is another known and meaningless dimension like θ_1. r_2 and r_3 are known dimensions of links 2 and 3, respectively. θ_2 and θ_3 define the orientation of links 2 and 3. θ_2 is the joint variable associated with the pin joint between 1 and 2. θ_3 is the joint variable associated with the pin joint between

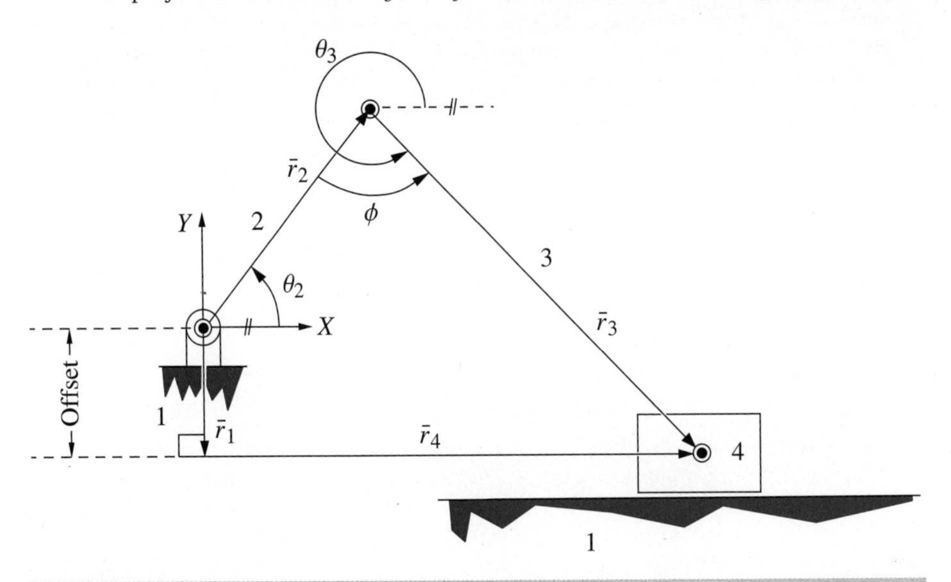

FIGURE 2.5 The offset crank-slider mechanism and its vector loop

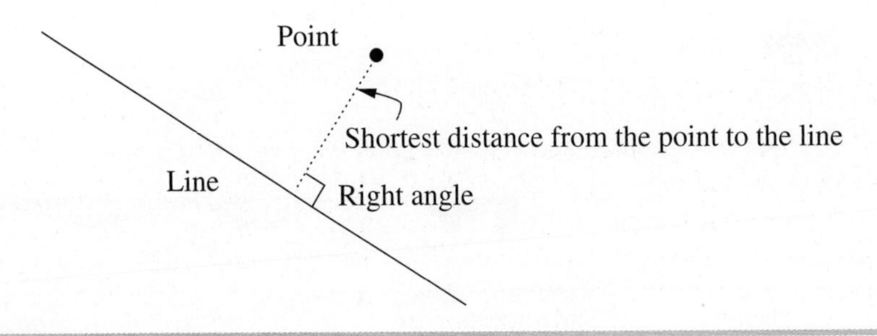

FIGURE 2.6 The shortest distance from a point to a line

3 and 4. In Figure 2.5, ϕ is defined as the joint variable associated with the pin joint between 2 and 3, and it is given by

$$\phi = \theta_3 - \theta_2 - \pi. \tag{2.6}$$

The magnitudes and directions of the assigned vectors are meaningful, so the vector assignments are correct.

Step 2: From Figure 2.5, the VLE is

$$\bar{r}_2 + \bar{r}_3 - \bar{r}_4 - \bar{r}_1 = \bar{0}.$$

Step 3: Take the origin of the fixed coordinate system to be at the pin joint between 1 and 2, and to simplify the position equations, align the axes with the vectors $\bar{r}_1$ and $\bar{r}_4$. The origin of the coordinate system could also have been placed where $\bar{r}_1$ and $\bar{r}_4$ connect; this is also a fixed point in the vector loop.

Step 4: The VLE has scalar components

$$\begin{aligned} r_2\cos\theta_2 + r_3\cos\theta_3 - r_4\cos\theta_4 - r_1\cos\theta_1 &= 0 \\ r_2\sin\theta_2 + r_3\sin\theta_3 - r_4\sin\theta_4 - r_1\sin\theta_1 &= 0. \end{aligned}$$

In these equations $\theta_1 = -\frac{\pi}{2}$ and $\theta_4 = 0$, which simplifies the position equations to

$$r_2\cos\theta_2 + r_3\cos\theta_3 - r_4 = 0 \tag{2.7}$$

$$r_2\sin\theta_2 + r_3\sin\theta_3 + r_1 = 0. \tag{2.8}$$

Step 5: Equations (2.7) and (2.8) are the position equations. They contain

scalar knowns: $r_2, r_3, r_1, \theta_1 = -\frac{\pi}{2}, \theta_4 = 0$

and

scalar unknowns: θ_2, θ_3, and r_4.

The two position equations contain three unknowns, which means that one of the scalar unknowns must be given so that the remaining two can be calculated from the position equations. This means the system has one degree of freedom, which agrees with Gruebler's Criterion.

Before continuing with more examples, let's consider the different types of joints that we saw in Sections 1.1.1 and 1.1.2 and discuss how one would correctly assign vectors to the links that are connected by these joints.

2.2 HINTS FOR CHOOSING VECTORS

There are correct and incorrect ways of assigning vectors in a mechanism. One can determine the correctness of a vector assignment by considering the meaning of the vector's direction and magnitude, as we did in the previous two examples. A known vector magnitude or direction should correspond to a dimension of the mechanism. An unknown vector magnitude or direction should correspond to a joint variable within the mechanism. As we saw in the previous example, many times an unknown vector direction describes the orientation of a link, and by combination with other vector directions they define a rotational joint variable within the mechanism. See, for example,

Equations (2.2) and (2.6). These statements are very important. We repeat: If a vector is correctly assigned in a vector loop, then

1. A known magnitude or direction of the vector corresponds to a dimension of the mechanism.
2. An unknown magnitude or direction of the vector corresponds to a joint variable of the mechanism or defines the orientation of a link within the mechanism.

As you assign a vector to a loop, ask yourself if the magnitude and the direction of that vector satisfy the criteria above. If the answer is no, then that vector assignment is incorrect because it contains no useful information. In the following subsections the vectors associated with the different types of joints are presented. Incorrect vector assignments are shown as dashed vectors.

2.2.1 The Straight Sliding Joint

Figure 2.7 shows the straight sliding joint where 3 slides on 2. What series of vectors should be used to work from 1 through 2 and 3, which have the sliding joint between them, and over to 4? We should use a set of vectors that describe the dimension of link 2 (the offset) and capture the sliding motion of 3 relative to 2.

Look at the correct vector assignment first. The vector $\bar{r}_3$ is assigned to be along the straight line of motion of the pin joint between 2 and 3. Vector $\bar{r}_2$ goes from the pin joint between 1 and 2 to the vector $\bar{r}_3$, meeting it at a right angle. Vectors $\bar{r}_2$ and $\bar{r}_3$ rotate together with links 2 and 3. Their unknown directions describe the orientations of bodies 2 and 3, which rotate together. Because they rotate together, there is a geometric constraint relating their directions. In this case $\theta_2 = \theta_3 + \frac{\pi}{2}$. Magnitude r_3 is an unknown

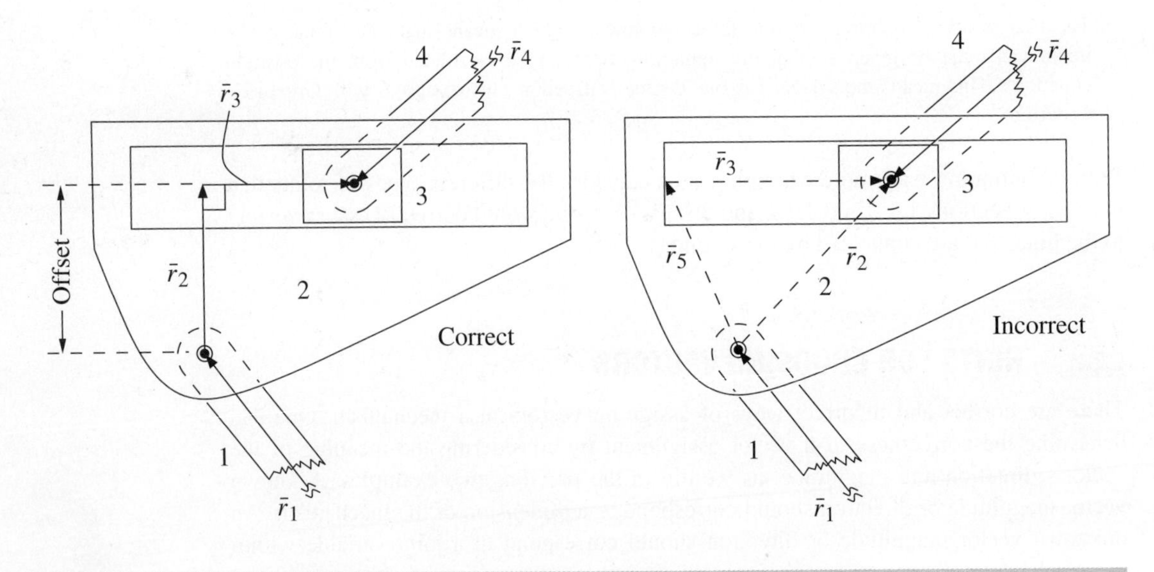

FIGURE 2.7 Correct and incorrect vectors associated with the straight sliding joint

that is the joint variable of the sliding joint between 3 and 2. Magnitude r_2 is a known that is the offset of link 2, a dimension.

Now look at the incorrect vector assignments. In one of the incorrect cases, a vector $\bar{r}_2$ is assigned going directly from the pin between 1 and 2 over to the pin between 3 and 4. The vector $\bar{r}_2$ has unknown direction θ_2 and magnitude r_2. θ_2 does not describe the orientation of link 2 or 3, and r_2 is not a joint variable. This vector $\bar{r}_2$ is meaningless and incorrect.

In the second incorrect case, vectors $\bar{r}_5$ and $\bar{r}_3$ are used. Magnitude r_3 is the joint variable for the sliding joint between 3 and 2, and θ_5 (or θ_3) defines the orientation of 2 and 3. Magnitude r_5, however, is not the dimension that defines the line of action of the slider relative to the pin between 1 and 2. That dimension is the perpendicular distance shown in the left image. Magnitude r_5 is meaningless, which makes $\bar{r}_5$ meaningless and incorrect.

2.2.2 The Circular Sliding Joint

Figure 2.8 shows a circular slider 3 moving in a circular track on body 2 whose center is indicated by the "+" sign. Consider the correct assignment of vectors first. To work from 1 over to 4 use vector $\bar{r}_2$, which goes from the pin between 1 and 2 to the center on body 2. Then vector $\bar{r}_3$ goes from this center to the pin on the circular slider between 3 and 4. r_2 and r_3 are knowns and are dimensions of 2. θ_2 defines the orientation of 2, θ_3 defines the orientation of 3, and $(\theta_3 - \theta_2)$ is the the joint variable measuring the motion of 3 relative to 2 at the circular sliding joint.

In one incorrect assignment of vectors, a vector $\bar{r}_2$ is used to go directly from the pin between 1 and 2 to the pin between 3 and 4. This vector's magnitude and direction

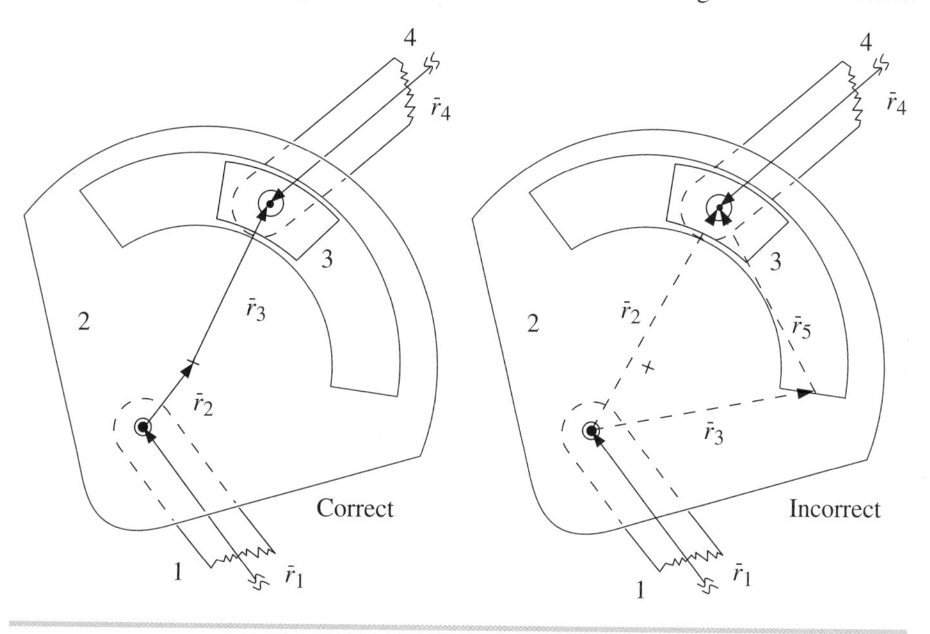

FIGURE 2.8 Correct and incorrect vectors associated with a circular sliding joint

are both unknown, and they do not correspond to any joint variable associated with the joint. The vector also does not rotate with any of the bodies. The vector is meaningless and incorrect.

The second incorrect assignment uses vectors $\bar{r}_3$ and $\bar{r}_5$. Vector $\bar{r}_3$ is attached to link 2, so θ_3 is an unknown that defines the orientation of 2. Magnitude r_3 is known but it is not a dimension of link 2. Magnitude r_5 and angle θ_5 change and are not joint variables associated with the joint.

2.2.3 The Straight Pin in a Slot Joint

Figure 2.9 shows a straight pin in a slot joint. The pin is a cylindrical protrusion of link 3, and this pin moves in a straight slot in link 2. The discussion here is the same as for the straight sliding joint. Consider the correct assignment of vectors first. The vector $\bar{r}_4$ is along the center of the straight slot. It locates where the center of the pin is in the slot. r_4 is an unknown that corresponds to a joint variable associated with this slipping joint. θ_4 is an unknown which defines the orientation of link 2. The vector $\bar{r}_2$ goes from the pin joint between 1 and 2 to vector $\bar{r}_4$, meeting at a right angle. r_2 is a known that corresponds to a dimension of link 2 (the offset). θ_2 is an unknown that also defines the orientation of 2. Both $\bar{r}_2$ and $\bar{r}_4$ rotate with 2, so the angle between them is constant and there is a geometric constraint relating their directions, $\theta_2 = \theta_4 + \frac{\pi}{2}$.

In one of the incorrect vector assignments, a vector $\bar{r}_2$ goes directly from the pin joint between 1 and 2 to the pin on 3. The vector does not rotate with any link. The magnitude and direction of this vector are unknown, and neither represents a joint variable associated with the slipping joint.

The other incorrect assignment uses vectors $\bar{r}_3$ and $\bar{r}_5$. r_5 is an unknown that locates the pin in the slot and is one of the joint variables associated with the slipping joint. θ_3 (or θ_5) is an unknown that defines the orientation of 2. The problem is that r_3, which is a known, is not a dimension of the mechanism. r_3 is the distance from the pin joint between 1 and 2 to a random point in the slot.

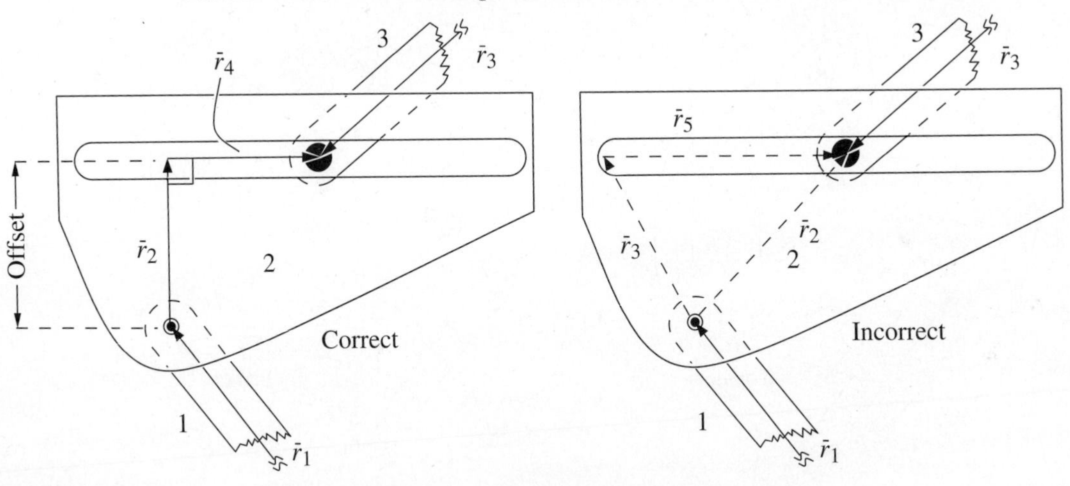

FIGURE 2.9 Correct and incorrect vectors associated with the straight pin in a slot joint

2.2.4 The Pin in a Circular Slot Joint

Figure 2.10 shows a circular pin in a slot joint between bodies 2 and 3. The "+" marks the center of the circular slot on 2. In the correct assignment of vectors, $\bar{r}_2$ goes from the pin joint between 1 and 2 to the center of the circular slot on 2, and vector $\bar{r}_4$ goes from the center of the slot to the center of the pin on 3. θ_2 defines the orientation of 2. r_2 is a known dimension of 2. r_4 is a known dimension of link 2 (the radius of the slot). θ_4 is an unknown that locates the pin in the slot and is a joint variable of the pin in a slot joint. The magnitudes and directions of these two vectors are meaningful.

In one of the incorrect vector assignments, the vector $\bar{r}_2$ goes directly from the pin joint on 2 to the center of the pin on 3. Both r_2 and θ_2 are unknown, and neither is a joint variable of a pin in a slot joint.

The other incorrect assignment uses vectors $\bar{r}_3$ and $\bar{r}_5$. Unknown θ_3 defines the orientation of 2, but r_3, r_5, and θ_5 are unknowns and are not joint variables associated with the joint.

2.2.5 Externally Contacting Circular Bodies

Figure 2.11 shows externally contacting circular bodies 2 and 3. The contact can be identified as an external contact because the point of contact lies between the centers of the circular shapes of 2 and 3. A rule you should always remember is that *when circular bodies are in contact, there will always be a vector connecting their centers.*

In the correct vector assignment, vector $\bar{r}_5$ connects the centers of 2 and 3. Vector $\bar{r}_2$ goes from the pin joint between 1 and 2 to the center of 2. Vector $\bar{r}_3$ goes from

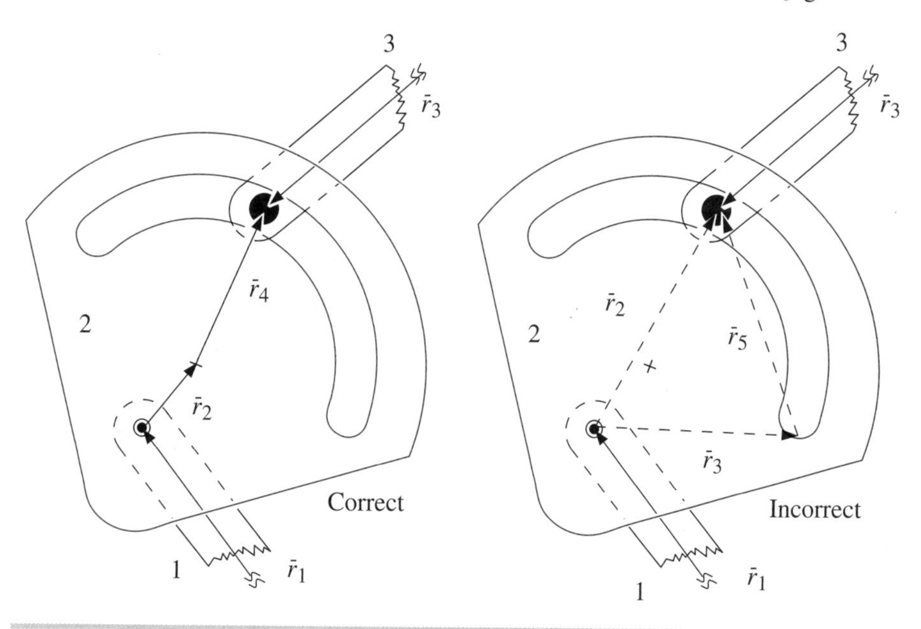

FIGURE 2.10 Correct and incorrect vectors associated with the circular pin in a slot joint

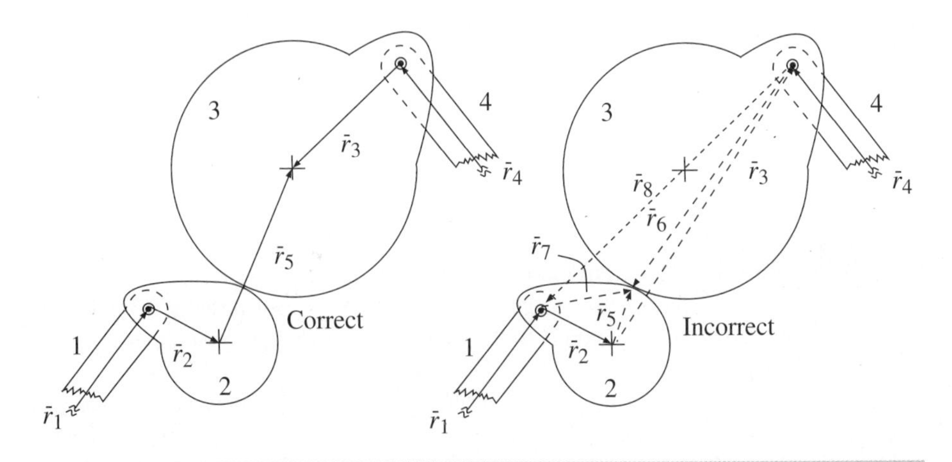

FIGURE 2.11 Correct and incorrect vectors associated with externally contacting circular bodies

the pin joint between 3 and 4 to the center of 3. Vectors $\bar{r}_2$ and $\bar{r}_3$ are attached to 2 and 3, respectively. r_2 and r_3 are known dimensions, and θ_2 and θ_3 are unknowns defining the orientation of 2 and 3, respectively. r_5 is a known dimension, and θ_5 is an unknown joint variable that indicates the motion of the center of 3 relative to the center of 2. In a machine it is quite likely that there would be an "arm" maintaining the distance between these centers, and $\bar{r}_5$ can be thought of as being attached to this arm.

In the incorrect assignment of vectors, several common errors are shown. The first uses a vector $\bar{r}_3$ that goes from the center of 2 directly to the pin joint between 3 and 4. When this is done there is no longer a vector that rotates with body 3. Additionally, r_3 and angle θ_3 are unknowns, yet they are not joint variables. They are meaningless.

Another incorrect choice uses either $\bar{r}_6$ and $\bar{r}_5$ or $\bar{r}_6$ and $\bar{r}_7$ and works through the point of contact. With either pair of vectors, the unknown magnitudes and directions of the vectors are meaningless, there also is no vector that rotates with 3.

The last case uses a vector $\bar{r}_8$ that goes directly from pin to pin. Both r_8 and θ_8 are unknowns and are not joint variables. Furthermore, there is no vector rotating with 2 or 3.

2.2.6 Internally Contacting Circular Bodies

Figure 2.12 shows internally contacting circular bodies 2 and 3. The contact can be identified as internal contact because the point of contact lies outside the centers of the circular shapes of 2 and 3. The vectors are labeled so that the discussion in Section 2.2.5 directly applies here.

2.2.7 Circular Bodies Pinned at Their Centers

If either circular body has a pin connection that should fall on its center, as shown for link 3 in Figure 2.13, a vector $\bar{r}_3$ attached to 3 must be introduced to capture its rotation.

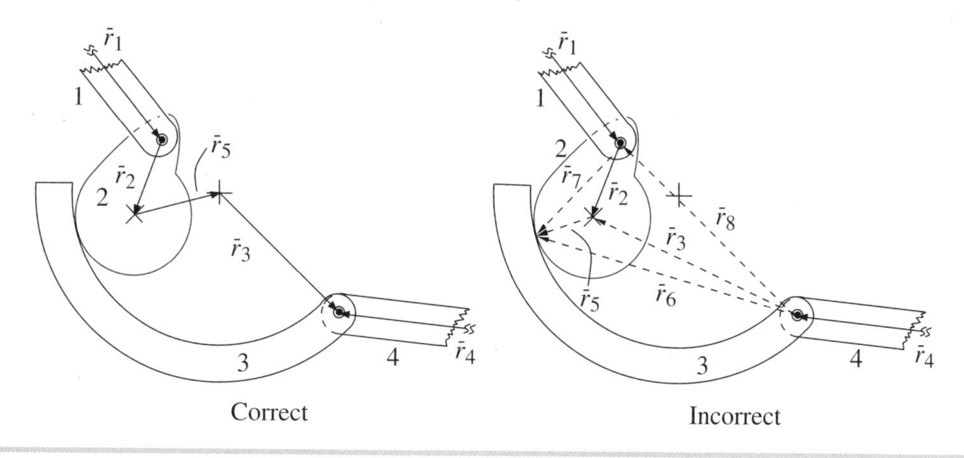

FIGURE 2.12 Correct and incorrect vectors associated with internally contacting circular bodies

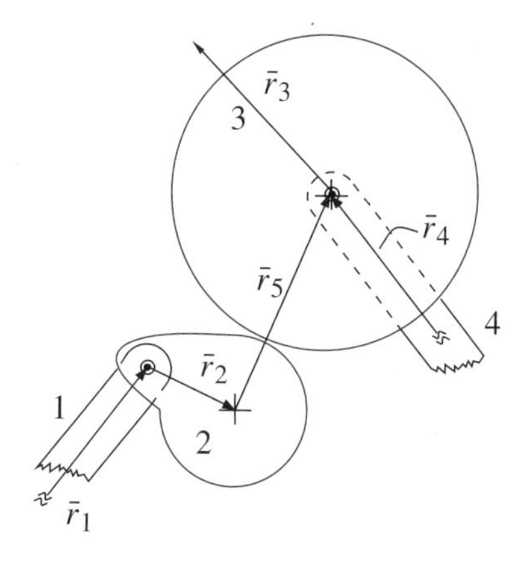

FIGURE 2.13 Additional vector associated with a circular body with a pin joint at its center

This vector will not be a part of the vector loop equation. It might be logical to let its magnitude be the radius of the circular shape, but in fact its magnitude is arbitrary.

2.2.8 Evaluating Vector Loops

When assigning vectors and vector loops to a mechanism, whether there are enough vector loops and whether good vectors have been selected can be determined by answering two questions.

1. Do the numbers of unknown magnitudes and directions exceed the number of scalar position equations by the number of degrees of freedom in the system given by its mobility?

2. Do the known and unknown magnitudes and directions of your vectors satisfy one of the criteria for good vector assignments given in Section 2.2?

If the answer to both questions is yes, then you have a sufficient number of loops, and you have chosen appropriate vectors.

▶ EXAMPLE 2.3

Figure 2.14 shows a four link mechanism that, according to Gruebler's Criterion, has two degrees of freedom. Derive the mechanism's position equations, and from them determine the degrees of freedom by following steps 1–5 in Section 2.1.

Solution

In the remaining examples we will follow the five steps outlined in Section 2.1 without identifying them individually. Figure 2.15 shows the appropriate vector loop. Assigning vectors to links 1 and 4 is straightforward, pin to pin. The question is: What vectors should be assigned to links 2 and 3 to get from the pin joint between 1 and 2 around to the pin joint between 4 and 3? The pin joints on bodies 2 and 3 are clear reference points for vectors. In addition to them, there are the centers of the circular shapes on bodies 2 and 3. If a vector is introduced between these centers, the vector loop in Figure 2.15 is developed. This completes step 1. The VLE is

$$\bar{r}_2 + \bar{r}_5 + \bar{r}_3 - \bar{r}_4 - \bar{r}_1 = \bar{0}.$$

This completes step 2. Choose the origin of the X-Y system to be at the fixed pin between 1 and 2, and align the X axis with $\bar{r}_1$. This completes step 3. The VLE has the scalar components

$$r_2\cos\theta_2 + r_5\cos\theta_5 + r_3\cos\theta_3 - r_4\cos\theta_4 - r_1 = 0 \tag{2.9}$$

$$r_2\sin\theta_2 + r_5\sin\theta_5 + r_3\sin\theta_3 - r_4\sin\theta_4 = 0, \tag{2.10}$$

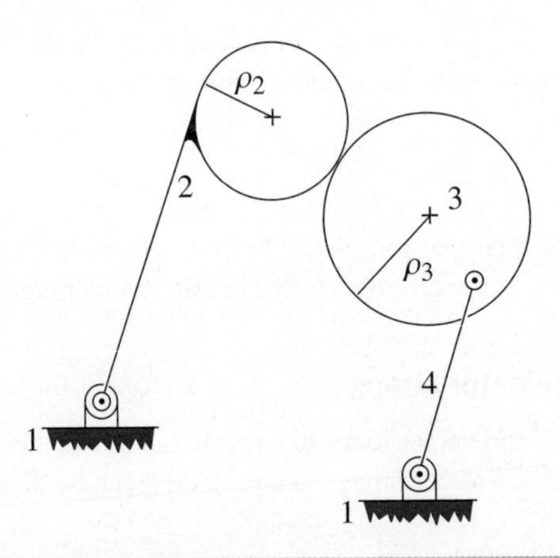

FIGURE 2.14 A three link mechanism with two degrees of freedom

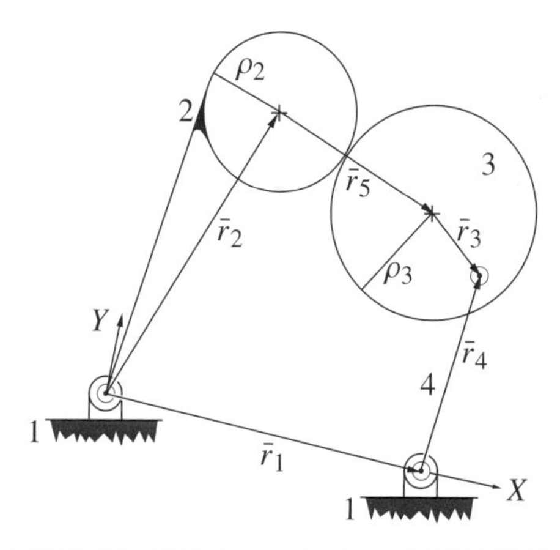

FIGURE 2.15 A three link mechanism and its vector loop

which completes step 4. Equations (2.9) and (2.10) are the position equations. They contain

scalar knowns: $r_2, r_5, r_3, r_4, r_1, \theta_1 = 0$

and

scalar unknowns: $\theta_2, \theta_5, \theta_3, \theta_4$.

This is a system of two equations in four unknowns, so the mechanism has two degrees of freedom. Two of the four scalar unknowns must be given in order that the remaining two can be calculated through the position equations. This completes step 5. The number of degrees of freedom agrees with Gruebler's Criterion.

▶ EXAMPLE 2.4

Consider the mechanism shown in Figure 2.16. Derive the mechanism's position equations, and from them deduce the number of degrees of freedom.

Solution

This is a two loop mechanism, 1-2-3-4 and 1-4-3-5-6. As a result of these two loops, a third loop 1-2-3-5-6 exists. However, it is a consequence of the first two loops and is not an independent loop. If one of the first two loops is removed, this loop also disappears. Mechanisms consisting of more than one independent loop require an equal number of independent vector loops for their kinematic analysis.

Figure 2.17 shows appropriate vector loops. There are two independent vector loop equations since there are two independent loops in the mechanism:

$$\bar{r}_2 + \bar{r}_3 - \bar{r}_4 - \bar{r}_1 = \bar{0} \tag{2.11}$$

$$\bar{r}_8 + \bar{r}_6 - \bar{r}_5 - \bar{r}_7 - \bar{r}_4 - \bar{r}_1 = \bar{0}. \tag{2.12}$$

Place a fixed coordinate systems at the pin joint between bodies 1 and 2, and align the X axis with the vector $\bar{r}_1$.

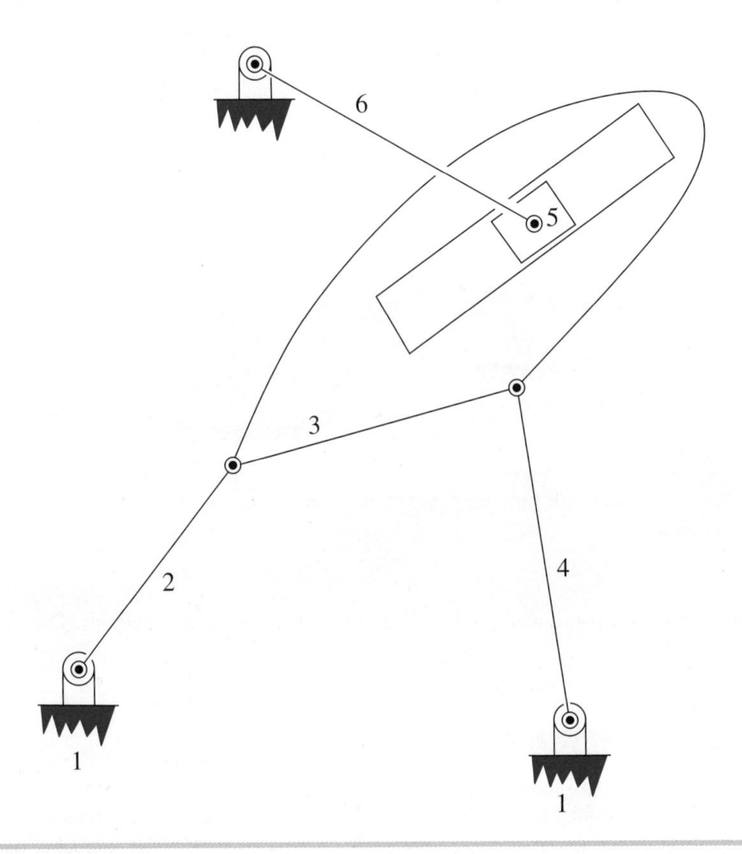

FIGURE 2.16 A two loop mechanism

A third vector loop exists,

$$\bar{r}_8 + \bar{r}_6 - \bar{r}_5 - \bar{r}_7 - \bar{r}_3 - \bar{r}_2 = \bar{0},$$

which is not independent. It comes from subtracting Equation (2.11) from Equation (2.12). Any two of these three VLEs can be used.

Use Equations (2.11) and (2.12). The scalar components of Equation (2.11) are

$$r_2\cos\theta_2 + r_3\cos\theta_3 - r_4\cos\theta_4 - r_1 = 0 \tag{2.13}$$

$$r_2\sin\theta_2 + r_3\sin\theta_3 - r_4\sin\theta_4 = 0, \tag{2.14}$$

and the scalar components of Equation (2.12) are

$$r_8\cos\theta_8 + r_6\cos\theta_6 - r_5\cos\theta_5 - r_7\cos\theta_7 - r_4\cos\theta_4 - r_1 = 0 \tag{2.15}$$

$$r_8\sin\theta_8 + r_6\sin\theta_6 - r_5\sin\theta_5 - r_7\sin\theta_7 - r_4\sin\theta_4 = 0. \tag{2.16}$$

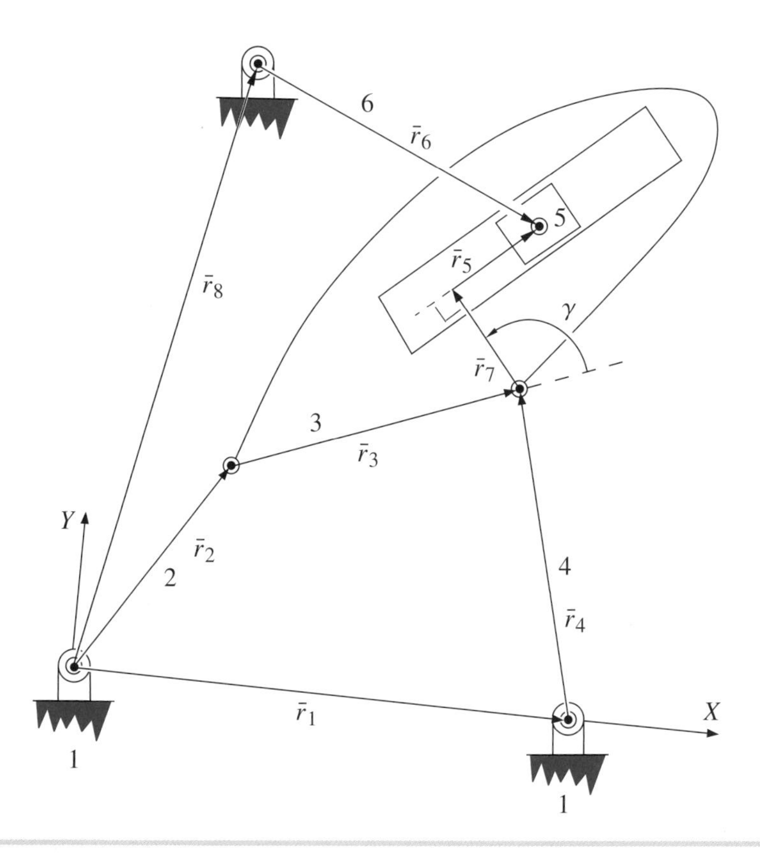

FIGURE 2.17 A two loop mechanism and its vector loops

In these four equations we have

scalar knowns: $r_2, r_3, r_4, r_1, \theta_1 = 0°, r_8, r_6, r_7, \theta_8$

and

scalar unknowns: $\theta_2, \theta_3, \theta_4, \theta_6, r_5, \theta_5, \theta_7$.

There are seven unknowns, which means three degrees of freedom. But Gruebler's Criterion predicts the system has one degree of freedom. There is no special geometry in this mechanism, so one would expect the result from Gruebler's Criterion to be correct. What's the problem? What has not been accounted for in the position equations?

Observe that the vectors $\bar{r}_3$, $\bar{r}_7$, and $\bar{r}_5$ each rotate with body 3, which means they maintain fixed directions relative to each other. In other words the unknown directions of $\bar{r}_3$, $\bar{r}_7$, and $\bar{r}_5$ are related. Specifically,

$$\theta_3 + \gamma = \theta_7 \Longrightarrow \theta_3 + \gamma - \theta_7 = 0 \tag{2.17}$$

$$\theta_5 + \frac{\pi}{2} = \theta_7 \Longrightarrow \theta_5 + \frac{\pi}{2} - \theta_7 = 0. \tag{2.18}$$

The left hand side of Equation (2.17) is developed by observing that the direction of vector $\bar{r}_3$, rotated counterclockwise by angle γ, gives the direction of vector $\bar{r}_7$. The right hand side rewrites this condition in a homogeneous form. Likewise the left hand side of Equation (2.18) is developed by observing that the direction of vector $\bar{r}_5$, rotated counterclockwise by angle $\frac{\pi}{2}$, gives the direction of vector $\bar{r}_7$ and again the right hand side rewrites this condition in a homogeneous form. We refer to Equations (2.17) and (2.18) as geometric constraints. Combining Equations (2.17) and (2.18) gives the relationship between θ_3 and θ_5. But this is not an independent geometric constraint. The two additional scalar equations now give a system of six position equations, Equations (2.13)–(2.18), which contain seven unknowns. Hence we have a one-degree-of-freedom mechanism, as predicted by Gruebler's Criterion. We must update our list of knowns to include the angle γ:

scalar knowns, updated: $r_2, r_3, r_4, r_1, \theta_1 = 0°, r_8, r_6, r_7, \theta_8, \gamma$

Another characteristic of a multiloop mechanism is that the loops are not always unique. In Figure 2.17 the path of the slider is referred to the pin between 3 and 4. It could just as well have been referred to the pin between 2 and 3, as shown in Figure 2.18. This leads to the alternative loop formulations shown there. These loops are as acceptable as those in Figure 2.17. There would be two independent VLEs and the two geometric constraints

$$\theta_3 + \gamma - \theta_5 = 0 \quad \text{and} \quad \theta_5 + \frac{\pi}{2} - \theta_7 = 0.$$

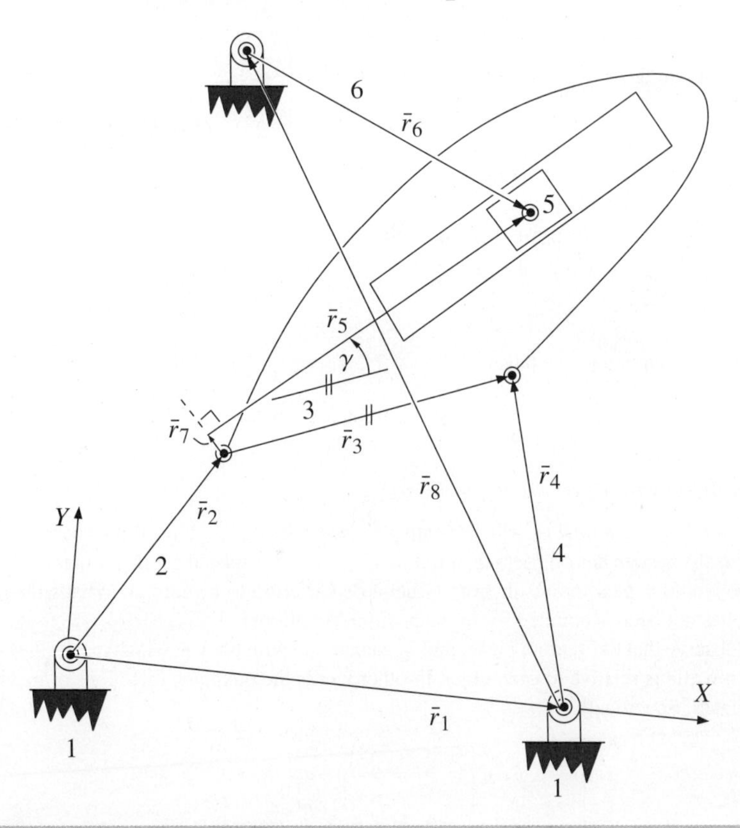

FIGURE 2.18 Alternative vector loops for the two loop mechanism of Example 2.4

▶ EXAMPLE 2.5

Consider the mechanism shown in Figure 2.19. Derive the mechanism's position equations, and from these deduce the number of degrees of freedom.

Solution

Observe that 1-2-3-4 form a closed loop and 1-4-5 form a second closed loop. A third loop also exists, 1-2-3-4-5. This third loop is not an independent loop. It exists as a consequence of the first two loops. Thus this is a two loop mechanism. So this mechanism must have two independent vector loop equations. Figure 2.20 shows a correct assignment of vectors.

The vector loop equations are

$$\bar{r}_5 + \bar{r}_6 + \bar{r}_7 - \bar{r}_9 = \bar{0}$$
$$\bar{r}_9 - \bar{r}_8 - \bar{r}_4 - \bar{r}_3 - \bar{r}_{10} - \bar{r}_2 - \bar{r}_1 = \bar{0}. \tag{2.19}$$

Take a coordinate system with origin at the the pin joint between 1 and 5, and align the X axis with $\bar{r}_1$. The simplified scalar components are

$$r_5\cos\theta_5 + r_6\cos\theta_6 + r_7\cos\theta_7 - r_9\cos\theta_9 = 0 \tag{2.20}$$
$$r_5\sin\theta_5 + r_6\sin\theta_6 + r_7\sin\theta_7 - r_9\sin\theta_9 = 0 \tag{2.21}$$
$$r_9\cos\theta_9 - r_8\cos\theta_8 - r_4\cos\theta_4 - r_3\cos\theta_3 - r_{10}\cos\theta_{10} - r_1 = 0 \tag{2.22}$$
$$r_9\sin\theta_9 - r_8\sin\theta_8 - r_4\sin\theta_4 - r_3\sin\theta_3 - r_{10}\sin\theta_{10} + r_2 = 0. \tag{2.23}$$

Note the sign of r_2 in Equation (2.23) is positive, whereas $\bar{r}_2$ in Equation (2.19) was negative. This is because $\theta_2 = -(\pi/2)$ so $-r_2\sin\theta_2 = r_2$. There are also four geometric constraints, one of which involves vector lengths:

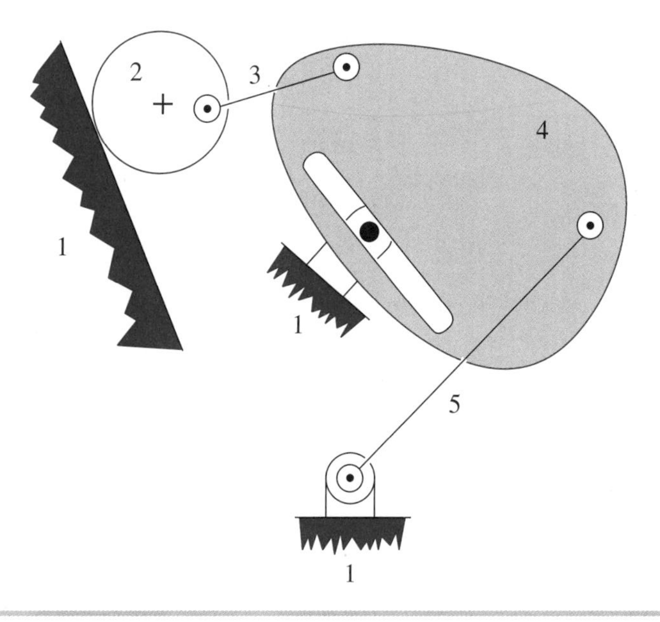

FIGURE 2.19 A two loop mechanism

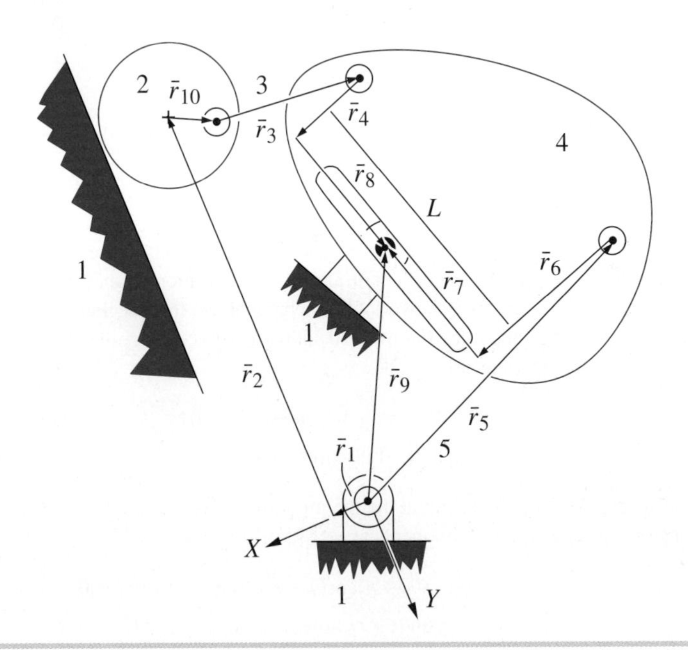

FIGURE 2.20 Appropriate vector loops

$$\theta_8 - \theta_7 + \pi = 0 \tag{2.24}$$

$$\theta_4 - \theta_8 + \frac{\pi}{2} = 0 \tag{2.25}$$

$$\theta_7 - \theta_6 + \frac{\pi}{2} = 0 \tag{2.26}$$

$$r_8 + r_7 = L. \tag{2.27}$$

Equations (2.20) through (2.27) are the position equations. They contain

scalar knowns: $\theta_2 = -\pi/2$, r_{10}, r_3, r_4, r_6, r_5, r_1, $\theta_1 = 0$, r_9, θ_9 (and do not forget) L

and

scalar unknowns: r_2, θ_{10}, θ_3, θ_4, θ_8, θ_7, r_8, r_7, θ_6, θ_5.

This is a system of eight equations in ten unknowns, which means two degrees of freedom.

The vector loops in Figure 2.20 were chosen to illustrate a case where there is a geometric constraint between vector lengths. It is a fairly rare case but it does occasionally occur. As with many multiloop mechanisms, the vector loops are not unique. Figure 2.21 shows what is probably a better pair of vector loops for that mechanism. There are two VLEs,

$$\bar{r}_1 + \bar{r}_9 + \bar{r}_2 + \bar{r}_3 - \bar{r}_4 - \bar{r}_5 = \bar{0}$$
$$\bar{r}_8 - \bar{r}_7 - \bar{r}_6 - \bar{r}_5 = \bar{0},$$

and two geometric constraints,

$$\theta_7 + \frac{\pi}{2} - \theta_6 = 0$$
$$\theta_4 + \gamma - \theta_6 = 0.$$

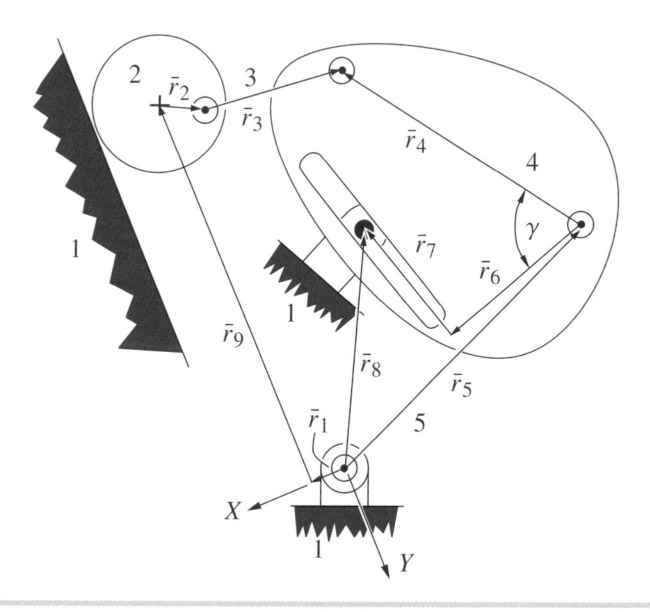

FIGURE 2.21 Another set of appropriate vector loops for Example 2.5

This is a system of six scalar equations with the eight unknowns $r_9, \theta_2, \theta_3, \theta_4, \theta_5, \theta_6, \theta_7$, and r_7, indicating that this is a two-degree-of-freedom mechanism. Given two of the eight unknowns, the remaining six could be calculated through the six position equations.

▶ EXAMPLE 2.6

Figure 2.22 is the skeleton diagram of the exercise machine we saw in Figure 1.23. Derive the mechanism's position equations, and from these deduce the number of degrees of freedom.

Solution

This is a two loop mechanism. One loop consists of 1-2-3-4. The second loop consists of 4-5-6-7. There is a third loop, 1-2-3-4-5-6-7, but this loop is a consequence of the first two loops and is not independent. Thus there should be two vector loop equations. The right side of the figure shows appropriate vectors. This case is simple since we are dealing with only pin joints. The vector loop equations are

$$\bar{r}_1 + \bar{r}_2 + \bar{r}_3 + \bar{r}_8 = \bar{0}$$
$$\bar{r}_4 + \bar{r}_5 - \bar{r}_6 - \bar{r}_7 = \bar{0}.$$

Take a coordinate system with origin at the pin joint between 1 and 5, and align the X axis with $\bar{r}_1$. The simplified scalar components are

$$r_1 + r_2\cos\theta_2 + r_3\cos\theta_3 + r_8\cos\theta_8 = 0 \tag{2.28}$$
$$r_2\sin\theta_2 + r_3\sin\theta_3 + r_8\sin\theta_8 = 0 \tag{2.29}$$
$$r_4\cos\theta_4 + r_5\cos\theta_5 - r_6\cos\theta_6 - r_7\cos\theta_7 = 0 \tag{2.30}$$
$$r_4\sin\theta_4 + r_5\sin\theta_5 - r_6\sin\theta_6 - r_7\sin\theta_7 = 0. \tag{2.31}$$

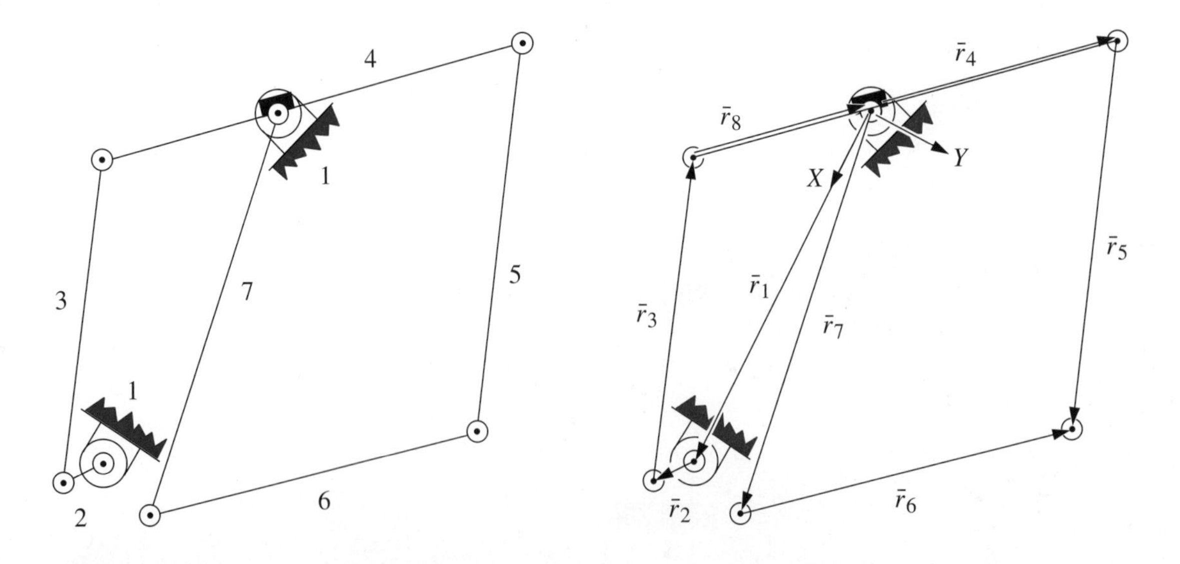

FIGURE 2.22 Vector loops for the exercise machine in Figure 1.23

$\bar{r}_4$ and $\bar{r}_8$ are both attached to link 4 so they rotate together. Consequently, there is a geometric constraint between θ_4 and θ_8:

$$\theta_8 = \theta_4 \longrightarrow \theta_8 - \theta_4 = 0. \tag{2.32}$$

Equations (2.28) through (2.32) are the position equations. They have

scalar knowns: $\theta_1 = 0, r_1, r_2, r_3, r_4, r_5, r_6, r_7, r_8,$

and

scalar unknowns: $\theta_2, \theta_3, \theta_4, \theta_5, \theta_6, \theta_7, \theta_8.$

This is a system of five equations in seven unknowns, which means two degrees of freedom. This agrees with Gruebler's Criterion in Example 1.10. Having two degrees of freedom means that two of the seven unknowns must be given so that the remaining five unknowns can be calculated through the five position equations. After making these calculations, one knows the position of each link in the mechanism (i.e., one could draw the mechanism). This means that the position problem has been solved.

▶ EXAMPLE 2.7

Figure 2.23 is the skeleton diagram of the front loader we saw in Figure 1.16. Derive the mechanism's position equations, and from these deduce the number of degrees of freedom.

Solution

This is a three loop mechanism. One loop consists of 1-2-3-4, the second loop consists of 4-5-6-8, and the third loop consists of 4-8-7-9. A fourth loop consisting of 4-5-6-7-9 exists as a consequence of the second and third loops, so this fourth loop is not independent. Thus there should be three vector loop equations. The bottom of the figure shows appropriate vectors.

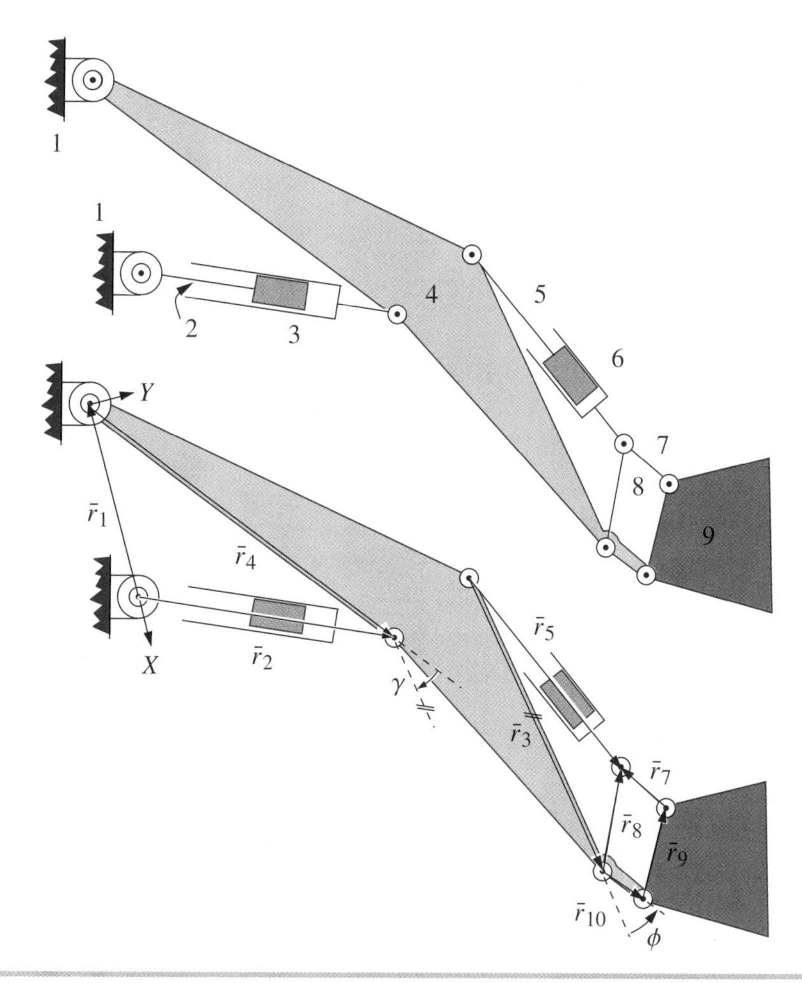

FIGURE 2.23 Vector loops for the front loader in Figure 1.17

$$\bar{r}_1 + \bar{r}_4 - \bar{r}_2 = \bar{0} \tag{2.33}$$
$$\bar{r}_5 - \bar{r}_8 - \bar{r}_3 = \bar{0}$$
$$\bar{r}_8 - \bar{r}_7 - \bar{r}_9 - \bar{r}_{10} = \bar{0}$$

Take a coordinate system with origin at the pin joint between 1 and 5, and align the X axis with $\bar{r}_1$. The simplified scalar components are

$$-r_1 + r_4\cos\theta_4 - r_2\cos\theta_2 = 0 \tag{2.34}$$
$$r_4\sin\theta_4 - r_2\sin\theta_2 = 0 \tag{2.35}$$
$$r_5\cos\theta_5 - r_8\cos\theta_8 - r_3\cos\theta_3 = 0 \tag{2.36}$$
$$r_5\sin\theta_5 - r_8\sin\theta_8 - r_3\sin\theta_3 = 0 \tag{2.37}$$
$$r_8\cos\theta_8 - r_7\cos\theta_7 - r_9\cos\theta_9 - r_{10}\cos\theta_{10} = 0 \tag{2.38}$$
$$r_8\sin\theta_8 - r_7\sin\theta_7 - r_9\sin\theta_9 - r_{10}\sin\theta_{10} = 0. \tag{2.39}$$

Notice that in Equation (2.34) r_1 is negative, whereas in Equation (2.33) $\bar{r}_1$ was positive. This is because with the chosen direction of the X axis, $\theta_1 = \pi$ and $r_1\cos\theta_1 = -r_1$.

$\bar{r}_4$ and $\bar{r}_3$ are both attached to link 4 so they rotate together. The angle between them is a constant, and this introduces a geometric constraint between θ_4 and θ_3:

$$\theta_4 - \gamma = \theta_3 \longrightarrow \theta_4 - \gamma - \theta_3 = 0. \tag{2.40}$$

Likewise $\bar{r}_3$ and $\bar{r}_{10}$ are both attached to link 4 and they rotate together, introducing the geometric constraint between θ_3 and θ_{10}:

$$\theta_3 + \phi = \theta_{10} \longrightarrow \theta_3 + \phi - \theta_1 = 0. \tag{2.41}$$

Equations (2.34) through (2.41) are the position equations. They contain

scalar knowns: $\theta_1 = \pi, r_1, r_3, r_4, r_7, r_8, r_9, r_{10}, \gamma, \phi$

and

scalar unknowns: $r_2, \theta_2, \theta_3, \theta_4, r_5, \theta_5, \theta_7, \theta_8, \theta_9, \theta_{10}$,

which is a system of eight equations in ten unknowns, which means two degrees of freedom. This result agrees with the calculation from Gruebler's Criterion in Example 1.6.

2.3 CLOSED-FORM SOLUTIONS TO THE POSITION EQUATIONS

A closed-form solution to a system of equations is one where the unknowns can be calculated directly without the use of any iterative or numerical techniques. This is generally limited to systems of linear equations, although in some cases there can be closed-form solutions to a system of nonlinear equations as well. The position equations of mechanisms are typically nonlinear since they involve sines and cosines of the unknowns. The position equations of simple mechanisms such as the four bar or crank-slider mechanism are examples of nonlinear equations with closed-form solutions. The following example develops a closed-form solution to the position equations of a four bar mechanism.

▶ EXAMPLE 2.8

Consider the four bar mechanism of Example 2.1. The position equations are given by Equations (2.4) and (2.5). Of the unknowns θ_2, θ_3, and θ_4, take θ_2 as the input, which leaves a system of two equations in the two unknowns θ_3 and θ_4. The equations are nonlinear and are not amenable to ordinary linear algebra. However, a closed-form solution is possible.

Solution

Equations (2.4) and (2.5) can be reduced to a single equation with only θ_4 as the unknown. To do so, isolate the unknown θ_3 on the left hand side,

$$r_3\cos\theta_3 = -r_2\cos\theta_2 + r_4\cos\theta_4 + r_1 \tag{2.42}$$

$$r_3\sin\theta_3 = -r_2\sin\theta_2 + r_4\sin\theta_4, \tag{2.43}$$

and then square both sides of each equation and add the results together:

$$r_3^2 = r_2^2 + r_4^2 + r_1^2 - 2r_2r_4\cos\theta_2\cos\theta_4 + 2r_1r_4\cos\theta_4 - 2r_2r_1\cos\theta_2 - 2r_2r_4\sin\theta_2\sin\theta_4. \tag{2.44}$$

Equation (2.44) contains only one unknown, θ_4. Unknown θ_3 has been eliminated. Equation (2.44) is known as the input-output displacement relationship of a four bar mechanism. Given the link lengths and the value of θ_2, the goal is to obtain a closed-form solution for θ_4 from this equation.

Begin by rewriting Equation (2.44) as

$$A\cos\theta_4 + B\sin\theta_4 = C, \tag{2.45}$$

where A, B, and C are knowns computed from the dimensions (r_1, r_2, r_3, and r_4) and the value of the input θ_2:

$$A = 2r_4(r_1 - r_2\cos\theta_2) \tag{2.46}$$

$$B = -2r_2r_4\sin\theta_2 \tag{2.47}$$

$$C = r_3^2 - r_2^2 - r_4^2 - r_1^2 + 2r_2r_1\cos\theta_2. \tag{2.48}$$

Equation (2.45) can be solved in closed form for θ_4 by using the "tangent of the half angle formulas." In case you are not familiar with these, they are derived for you in the appendix to this chapter. The tangent of the half angle formulas works like this:

Define u_4 as the tangent of the half angle of θ_4 — that is,

$$u_4 = \tan\left(\frac{\theta_4}{2}\right). \tag{2.49}$$

Then, as shown in the appendix,

$$\cos\theta_4 = \frac{1 - u_4^2}{1 + u_4^2} \quad \text{and} \quad \sin\theta_4 = \frac{2u_4}{1 + u_4^2}. \tag{2.50}$$

Substituting Equation (2.50) into Equation (2.45) yields a quadratic for u_4,

$$(C + A)u_4^2 - 2Bu_4 + (C - A) = 0,$$

which can be solved with the quadratic equation, giving two values of u_4:

$$u_4 = \frac{B \pm [A^2 + B^2 - C^2]^{\frac{1}{2}}}{C + A}, \tag{2.51}$$

where A, B, and C are computed from Equations (2.46) through (2.48). Each value of u_4 from Equation (2.51) is converted into a value of θ_4 using Equation (2.49),

$$\theta_4 = 2\tan^{-1}u_4. \tag{2.52}$$

In Equation (2.52), for one of the values of u_4, the $\tan^{-1}u_4$ will yield two angles that differ by π. These two angles are then multiplied by a factor of 2 to give two values of θ_4 that differ by 2π. Two angles that differ by 2π are the same angle. The conclusion is that there is one value of θ_4 for each value of u_4. There are two values of u_4 from Equation (2.51), so the problem has two answers for θ_4. *These two solutions are referred to as the geometric inversions of the mechanism that correspond to a value of* θ_2.

For each value of θ_4, Equations (2.42) and (2.43) yield values of $\cos\theta_3$ and $\sin\theta_3$:

$$\cos\theta_3 = \frac{-r_2\cos\theta_2 + r_4\cos\theta_4 + r_1}{r_3}$$

$$\sin\theta_3 = \frac{-r_2\sin\theta_2 + r_4\sin\theta_4}{r_3},$$

which give one value of θ_3 for each value of θ_4. In a computer program this would be done using the `atan2` function,

$$\texttt{t3} = \texttt{atan2(st3,ct3)} \tag{2.53}$$

where Equation (2.53) is a line from a computer program which has a variable `t3` that is the angle θ_3 in radians and variables `st3` and `ct3` that are the values of $\sin\theta_3$ and $\cos\theta_3$ respectively.

2.4 NUMERICAL SOLUTIONS TO POSITION EQUATIONS VIA NEWTON'S METHOD

For most mechanisms the nonlinear position equations do not have a closed-form solution. Newton's Method is well suited to solving systems of nonlinear equations. Its application extends far beyond mechanism analysis. We will develop Newton's Method.

Consider a system of n homogeneous nonlinear equations represented by the functions f_i $(i = 1, 2, \ldots, n)$, containing n unknowns x_j $(j = 1, 2, \ldots, n)$ — that is,

$$\begin{aligned} f_1(x_1, \ldots, x_j, \ldots, x_n) &= 0 \\ &\vdots \\ f_i(x_1, \ldots, x_j, \ldots, x_n) &= 0. \\ &\vdots \\ f_n(x_1, \ldots, x_j, \ldots, x_n) &= 0 \end{aligned} \tag{2.54}$$

The values of x_j that satisfy $f_i = 0$ in Equation (2.54) are called the "roots" of Equation (2.54). Newton's Method begins with a guess for the roots x_j. Call these guesses x_j^0 $(j = 1, 2, \ldots, n)$. Now write a first-order Taylor's series approximation of Equation (2.54), expanded about the point x_j^0. To keep the series compact, we define constants a_{ij} as

$$a_{ij} = \left.\frac{\partial f_i}{\partial x_j}\right|_{\bar{x}=\bar{x}^0} \qquad (i, j = 1, 2, \ldots, n),$$

where $\bar{x}$ is a vector of the unknown roots of Equation (2.54) and $\bar{x}^0$ is a vector of the guesses for unknown roots of Equation (2.54),

$$\bar{x} = \begin{bmatrix} x_1 \\ \vdots \\ x_j \\ \vdots \\ x_n \end{bmatrix} \quad \text{and} \quad \bar{x}^0 = \begin{bmatrix} x_1^0 \\ \vdots \\ x_j^0 \\ \vdots \\ x_n^0 \end{bmatrix}. \tag{2.55}$$

The first-order Taylor series approximation of Equation (2.54) is given by

$$f_1(\bar{x}) \approx f_1\big|_{\bar{x}=\bar{x}^0} + a_{11}\,(x_1 - x_1^0) + \cdots + a_{1j}\,(x_j - x_j^0) + \cdots + a_{1n}\,(x_n - x_n^0) = 0$$

$$\vdots$$

$$f_i(\bar{x}) \approx f_i\big|_{\bar{x}=\bar{x}^0} + a_{i1}\,(x_1 - x_1^0) + \cdots + a_{ij}\,(x_j - x_j^0) + \cdots + a_{in}\,(x_n - x_n^0) = 0 \qquad (2.56)$$

$$\vdots$$

$$f_n(\bar{x}) \approx f_n\big|_{\bar{x}=\bar{x}^0} + a_{n1}\,(x_1 - x_1^0) + \cdots + a_{nj}\,(x_j - x_j^0) + \cdots + a_{nn}\,(x_n - x_n^0) = 0.$$

By using a first-order Taylor series approximation, we convert the original system of nonlinear equations, Equation (2.54), into a system of *approximate* linear equations Equation (2.56), which can be solved for the unknowns x_j by standard methods of linear algebra.

To keep the equations compact, it helps to write the system of Equation (2.56) in a matrix form. Define an n-dimensional column vector of the functions, $\bar{f}(\bar{x})$:

$$\bar{f}(\bar{x}) = \begin{bmatrix} f_1(\bar{x}) \\ \vdots \\ f_i(\bar{x}) \\ \vdots \\ f_n(\bar{x}) \end{bmatrix}. \qquad (2.57)$$

With this, Equations (2.54) become the vector equation

$$\bar{f}(\bar{x}) = \bar{0}, \qquad (2.58)$$

to which we are seeking the solution, or roots, $\bar{x}$. Define an $n \times n$ matrix that contains partial derivatives of $\bar{f}(\bar{x})$ with respect to each element of $\bar{x}$. The standard name for this matrix is the Jacobian matrix $J(\bar{x})$:

$$J(\bar{x}) = \left[\, (\partial\bar{f}/\partial x_1) \,\ldots\, (\partial\bar{f}/\partial x_j) \,\ldots\, (\partial\bar{f}/\partial x_n) \,\right].$$

Finally, define a matrix of constants A that is the Jacobian evaluated at the guess $\bar{x} = \bar{x}^0$.

$$A = J\big|_{\bar{x}=\bar{x}^0} \qquad (2.59)$$

Then Equation (2.56) can be written as the linear system of equations

$$\bar{f}(\bar{x}) \approx \bar{f}\big|_{\bar{x}=\bar{x}^0} + A\left(\bar{x} - \bar{x}^0\right) = 0, \qquad (2.60)$$

which can be solved for the roots $\bar{x}$:

$$\bar{x} = -A^{-1}\,\bar{f}\big|_{\bar{x}=\bar{x}^0} + \bar{x}^0. \qquad (2.61)$$

Since this solution for $\bar{x}$ is computed from a first-order Taylor series approximation of $\bar{f}(\bar{x})$, it is not exact. But it will be a better solution than the original guess, so this solution is then used as a new guess, and the process is repeated until the changes in the solution

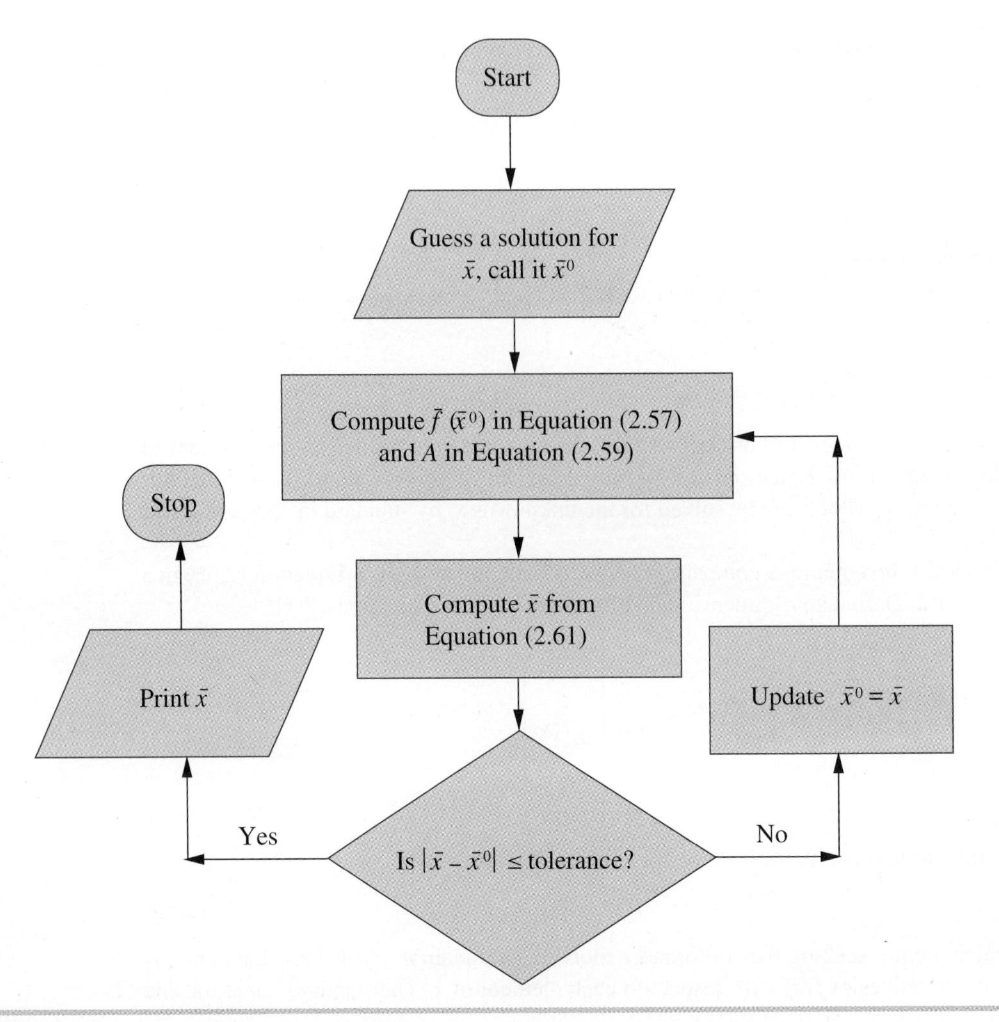

FIGURE 2.24 Flowchart of Newton's Method

are small — i.e., until there is convergence. Figure 2.24 shows a flowchart of the logic behind Newton's Method as it would be used in a computer program.

2.4.1 Graphically Checking the Solution to the Position Problem

The question of whether your closed-form solution or Newton's Method algorithm is working properly can be answered with a graphical check. For a particular value of the input variable, make a graphical construction of the mechanism and measure the unknown variables from the drawing. Do this for three to four values of the input variable, and if your graphical results are close to the computed result, then you can be confident that your closed-form solution or Newton's Method algorithm is working properly.

If your results do not agree, the most common source of errors is in units. Any components of $\bar{x}$ in Equation (2.61) that are vector magnitudes are in the same units used to describe all known vector magnitudes (dimensions), and any components of $\bar{x}$ in Equation (2.61) that are vector directions are in units of radians. Errors in the latter occur more frequently. The following example illustrates the use of Newton's Method on a system of polynomial (nonlinear) equations.

▶ EXAMPLE 2.9

Find a solution to the following system of three equations in three unknowns.

$$\begin{aligned} x_1^2 + x_2 + x_3 &= 6 \\ x_1 + x_2^2 + x_3 &= 8 \\ x_1 + x_2 + x_3^2 &= 12 \end{aligned} \tag{2.62}$$

Solution

In this problem we have a vector of unknowns,

$$\bar{x} = \begin{bmatrix} x_1 \\ x_2 \\ x_3 \end{bmatrix},$$

and the system of homogeneous equations to which we are seeking the roots is given by

$$\bar{f}(\bar{x}) = \begin{bmatrix} f_1(x_1, x_2, x_3) \\ f_2(x_1, x_2, x_3) \\ f_3(x_1, x_2, x_3) \end{bmatrix} = \begin{bmatrix} x_1^2 + x_2 + x_3 - 6 \\ x_1 + x_2^2 + x_3 - 8 \\ x_1 + x_2 + x_3^2 - 12 \end{bmatrix} = \bar{0} \tag{2.63}$$

and

$$J = \begin{bmatrix} (\partial\bar{f}/\partial x_1) & (\partial\bar{f}/\partial x_2) & (\partial\bar{f}/\partial x_3) \end{bmatrix} = \begin{bmatrix} 2x_1 & 1 & 1 \\ 1 & 2x_2 & 1 \\ 1 & 1 & 2x_3 \end{bmatrix}. \tag{2.64}$$

The solutions to Equation (2.62) are found by implementing Equations (2.63) and (2.64) in a computer program following the logic outlined in the flowchart in Figure 2.24. The process begins with the guess at a solution,

$$\bar{x}^0 = \begin{bmatrix} 0 \\ 0 \\ 0 \end{bmatrix}.$$

Substitute $\bar{x}^0$ into Equations (2.63) and (2.64), compute $\bar{f}\big|_{\bar{x}=\bar{x}^0}$ and $A = J\big|_{\bar{x}=\bar{x}^0}$, and substitute these into Equation (2.61) to compute $\bar{x}$:

$$\underbrace{\bar{x} = \begin{bmatrix} 7 \\ 5 \\ 1 \end{bmatrix}.}_{\text{1st iteration}}$$

Now use this as a new guess $\bar{x}^0$, and repeat this process four more times to obtain the following sequence of values for $\bar{x}$.

$$\underbrace{\bar{x} = \begin{bmatrix} 3.4961 \\ 2.6055 \\ 3.4492 \end{bmatrix}}_{\text{2nd iteration}}, \quad \underbrace{\bar{x} = \begin{bmatrix} 1.9167 \\ 1.9119 \\ 2.8091 \end{bmatrix}}_{\text{3rd iteration}}, \quad \underbrace{\bar{x} = \begin{bmatrix} 1.2411 \\ 1.9471 \\ 2.9691 \end{bmatrix}}_{\text{4th iteration}}, \text{ and } \underbrace{\bar{x} = \begin{bmatrix} 1.0270 \\ 1.9947 \\ 2.9965 \end{bmatrix}}_{\text{5th iteration}}$$

The actual solution to this system of equations is known to be

$$\bar{x} = \begin{bmatrix} 1.0000 \\ 2.0000 \\ 3.0000 \end{bmatrix}.$$

The following example illustrates the use of Newton's Method on a system of transcendental (nonlinear) equations that arise from vector loops. In this example there were five iterations. The actual number of iterations required is determined by the value of the tolerance in Figure 2.24

▶ EXAMPLE 2.10

This example continues with the two loop mechanism of Example 2.4. The purpose is to outline Newton's Method for solving the mechanism's position equations, given a value of $\theta_2 = 58° = 1.01$ radians. The mechanism's dimensions are given as

$$\gamma = 110° = 1.92 \text{ rad.}, \quad r_2 = 3.0, \quad r_3 = 3.6, \quad r_4 = 3.9, \quad r_1 = 6.0,$$
$$r_6 = 3.9, \quad r_7 = 1.4, \quad r_8 = 7.35, \text{ and } \theta_8 = 80° = 1.40 \text{ rad},$$

with units of inches for all vector lengths.

Solution

In Example 2.4 we derived the mechanism's six position Equations, (2.13) through (2.18). We put these into a vector of functions,

$$\bar{f} = \begin{bmatrix} f_1 \\ f_2 \\ f_3 \\ f_4 \\ f_5 \\ f_6 \end{bmatrix} = \begin{bmatrix} r_2\cos\theta_2 + r_3\cos\theta_3 - r_4\cos\theta_4 - r_1 \\ r_2\sin\theta_2 + r_3\sin\theta_3 - r_4\sin\theta_4 \\ r_8\cos\theta_8 + r_6\cos\theta_6 - r_5\cos\theta_5 - r_7\cos\theta_7 - r_4\cos\theta_4 - r_1 \\ r_8\sin\theta_8 + r_6\sin\theta_6 - r_5\sin\theta_5 - r_7\sin\theta_7 - r_4\sin\theta_4 \\ \theta_3 - \theta_7 + \gamma \\ \theta_5 - \theta_7 + \frac{\pi}{2} \end{bmatrix}. \quad (2.65)$$

Given the value of θ_2, we arrange the remaining six scalar unknowns into a vector of unknowns, $\bar{x}$:

$$\bar{x} = \begin{bmatrix} \theta_3 \\ \theta_4 \\ r_5 \\ \theta_5 \\ \theta_6 \\ \theta_7 \end{bmatrix}.$$

We must derive the mechanism's Jacobian matrix, J:

$$J(\bar{x}) = \begin{bmatrix} \partial\bar{f}/\partial\theta_3 & \partial\bar{f}/\partial\theta_4 & \partial\bar{f}/\partial r_5 & \partial\bar{f}/\partial\theta_5 & \partial\bar{f}/\partial\theta_6 & \partial\bar{f}/\partial\theta_7 \end{bmatrix}$$

$$= \begin{bmatrix} -r_3\sin\theta_3 & r_4\sin\theta_4 & 0 & 0 & 0 & 0 \\ r_3\cos\theta_3 & -r_4\cos\theta_4 & 0 & 0 & 0 & 0 \\ 0 & r_4\sin\theta_4 & -\cos\theta_5 & r_5\sin\theta_5 & -r_6\sin\theta_6 & r_7\sin\theta_7 \\ 0 & -r_4\cos\theta_4 & -\sin\theta_5 & -r_5\cos\theta_5 & r_6\cos\theta_6 & -r_7\cos\theta_7 \\ 1 & 0 & 0 & 0 & 0 & -1 \\ 0 & 0 & 0 & 1 & 0 & -1 \end{bmatrix}. \tag{2.66}$$

Equations (2.65) and (2.66) are implemented with Equation (2.61) in a Matlab® code that uses the logic of the flowchart in Figure 2.24. The Matlab code is shown in an appendix to this chapter. For an initial guess we refer to Figure 2.4. We could eyeball fairly precise values, but to demonstrate the robustness of Newton's Method, let's use the following very rough guess:

$$\bar{x}^0 = \begin{bmatrix} 0^\circ \\ 90^\circ \\ 1.5\text{ inches} \\ 0^\circ \\ 0^\circ \\ 90^\circ \end{bmatrix} = \begin{bmatrix} 0.0\text{ radians} \\ 1.571\text{ radians} \\ 1.5\text{ inches} \\ 0.0\text{ radians} \\ 0\text{ radians} \\ 1.571\text{ radians} \end{bmatrix}.$$

The Matlab codes generates the following sequence of solutions for $\bar{x}$.

$$\bar{x} = \underbrace{\begin{bmatrix} 0.377\text{ radian} \\ 1.78\text{ radians} \\ 1.00\text{ inch} \\ 0.726\text{ radian} \\ -0.218\text{ radian} \\ 2.30\text{ radians} \end{bmatrix}}_{\text{1st iteration}}, \underbrace{\begin{bmatrix} 0.347\text{ radian} \\ 1.84\text{ radians} \\ 1.06\text{ inches} \\ 0.696\text{ radian} \\ -0.446\text{ radian} \\ 2.27\text{ radians} \end{bmatrix}}_{\text{2nd iteration}}, \underbrace{\begin{bmatrix} 0.346\text{ radian} \\ 1.84\text{ radians} \\ 0.802\text{ inch} \\ 0.695\text{ radian} \\ -0.504\text{ radian} \\ 2.27\text{ radians} \end{bmatrix}}_{\text{3rd iteration}},$$

$$\underbrace{\begin{bmatrix} 0.346\text{ radian} \\ 1.84\text{ radians} \\ 0.785\text{ inch} \\ 0.695\text{ radian} \\ -0.508\text{ radian} \\ 2.27\text{ radians} \end{bmatrix}}_{\text{4th iteration}}, \underbrace{\begin{bmatrix} 0.346\text{ radians} \\ 1.84\text{ radians} \\ 0.785\text{ inches} \\ 0.695\text{ radians} \\ -0.508\text{ radians} \\ 2.27\text{ radians} \end{bmatrix}}_{\text{5th iteration}}$$

The solution has converged to three significant figures in the 4th iteration. In this example there were five iterations. The actual number of iterations required is determined by the value of the tolerance in Figure 2.24

2.5 THE MOTION OF POINTS OF INTEREST

Many times in mechanism analysis the position, velocity, or acceleration of a particular point on one of the links needs to be calculated. Examples of such "points of interest" are mass centers, whose accelerations would need to be determined in order to estimate driving torques or bearing forces. Another example of a point of interest is a point of application of a driving force or load force. This would be done in order to find mechanical advantage, which is discussed in Chapter 7. The procedure for describing the motion of a point of interest is best illustrated by an example.

▶ EXAMPLE 2.11

Consider the offset crank-slider mechanism from Example 2.2. The mechanism has one degree of freedom. Given a value of θ_2, we have two position equations, Equations (2.7) and (2.8), through which we can compute θ_3 and r_4. Suppose the position problem is solved, and we have computed the value of θ_3 and r_4 that corresponds to a value of θ_2.

Figure 2.25 shows that same crank-slider mechanism, now with a point on its coupler connected to ground with a spring. If we want to describe the force in the spring as a function of θ_2, it is necessary to describe the position of Q as a function of θ_2. Q is an example of a "point of interest." Derive the system of equations that would be solved for the position of point Q.

Solution

The coordinates of Q are given by the vector loop

$$\bar{r}_q = \bar{r}_2 + \bar{r}_5,$$

where $\bar{r}_5$ is a known vector that locates Q on the coupler. This VLE has X and Y components,

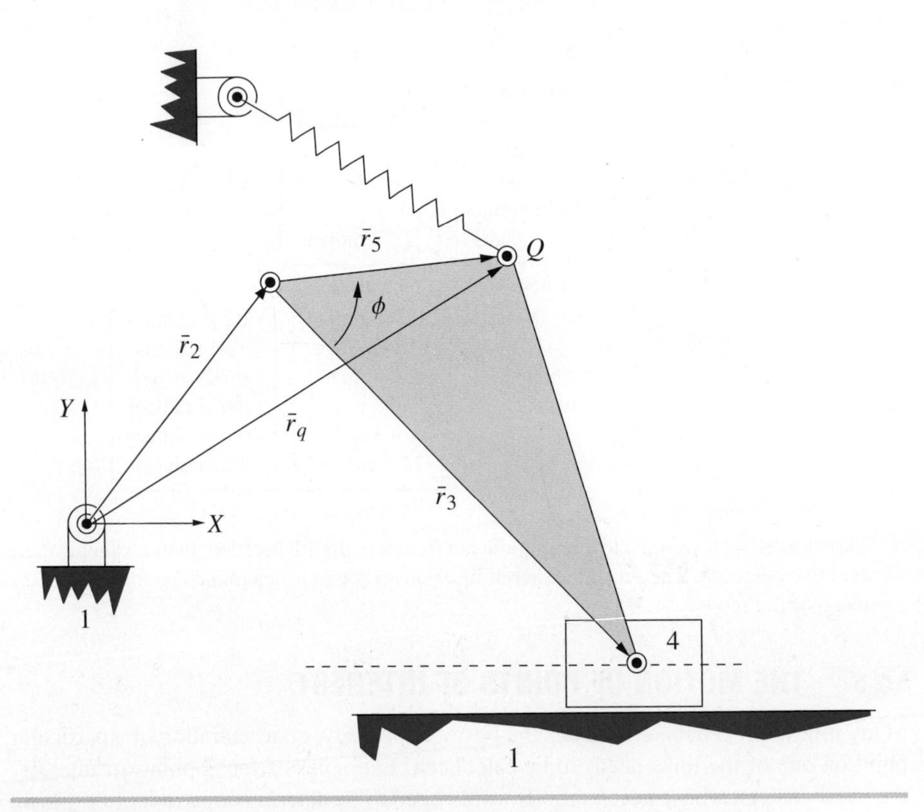

FIGURE 2.25 A vector loop locating Q, a point of interest on the coupler

$$r_{q_x} = r_2 \cos\theta_2 + r_5 \cos(\theta_3 + \phi) \quad (2.67)$$

$$r_{q_y} = r_2 \sin\theta_2 + r_5 \sin(\theta_3 + \phi), \quad (2.68)$$

where r_5 and ϕ are known dimensions that locate Q on the coupler relative to $\bar{r}_3$, as shown. After the position problem has been solved and the values of θ_2 and θ_3 are known, Equations (2.67) and (2.68) can be used to compute the coordinates of Q.

2.6 PROBLEMS

For the mechanisms shown in Problems 2.1–2.16:

1. Draw an appropriate vector loop.
2. Write out the vector loop equation(s), or VLE(s).
3. Write the X and Y components of the VLE(s) in their simplest form.
4. Write down all geometric constraints.
5. Summarize the scalar knowns and the scalar unknowns.
6. From all the above, deduce the number of degrees of freedom in the system.
7. Check your result in part 6 against Gruebler's Criterion.

Problem 2.1

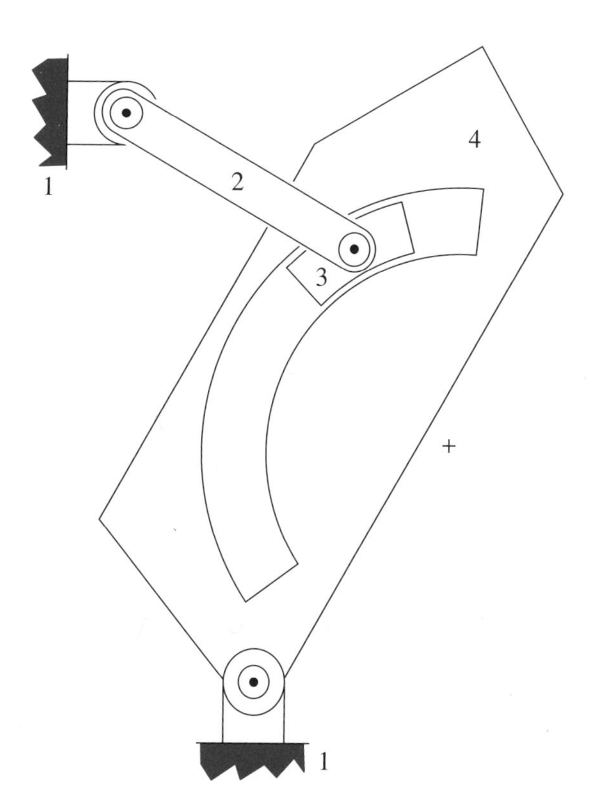

Problem 2.2

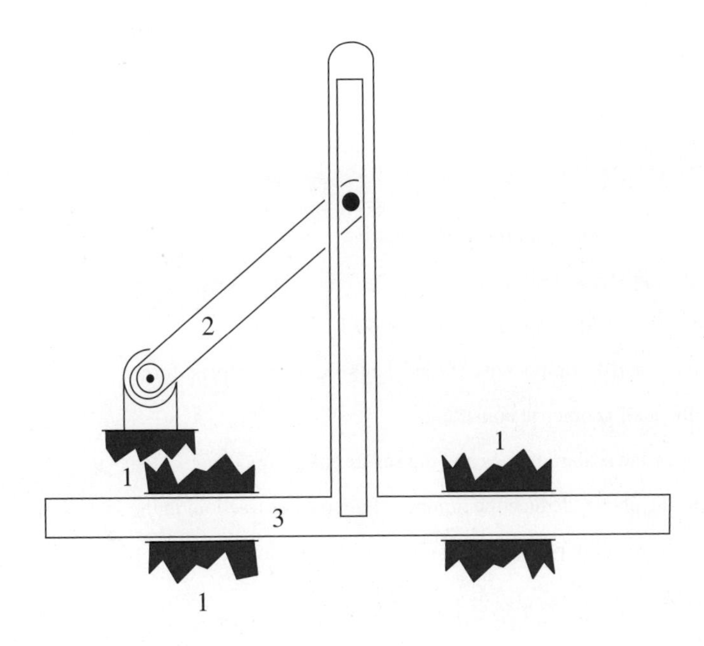

Problem 2.3

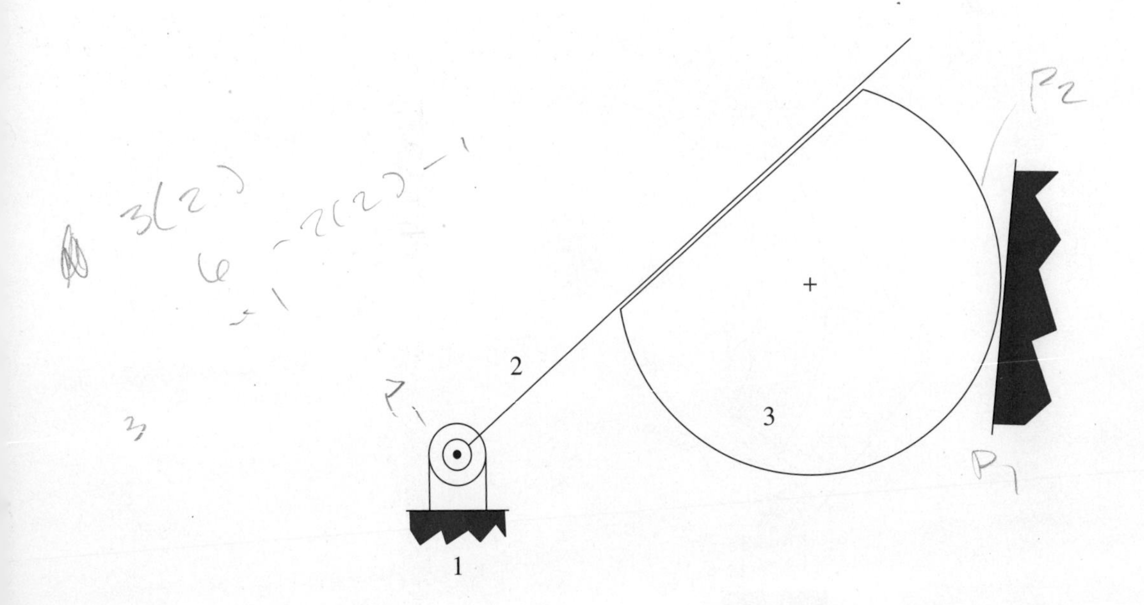

Problem 2.4

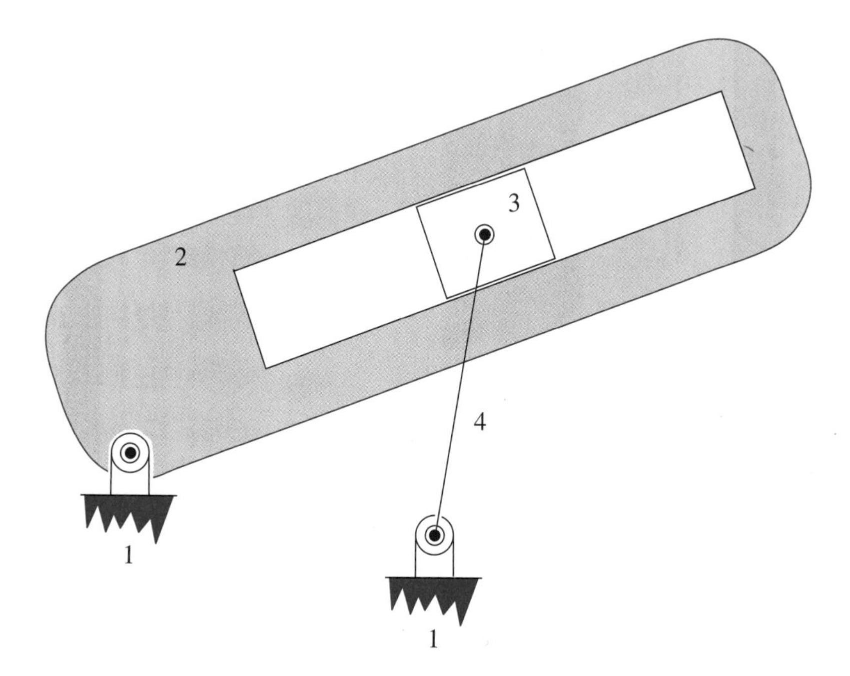

Problem 2.5

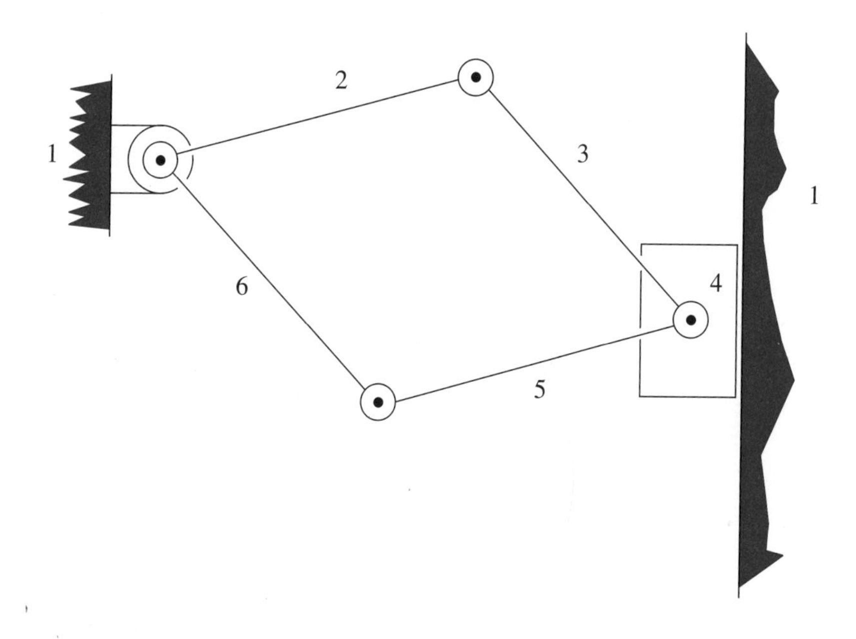

Problem 2.6

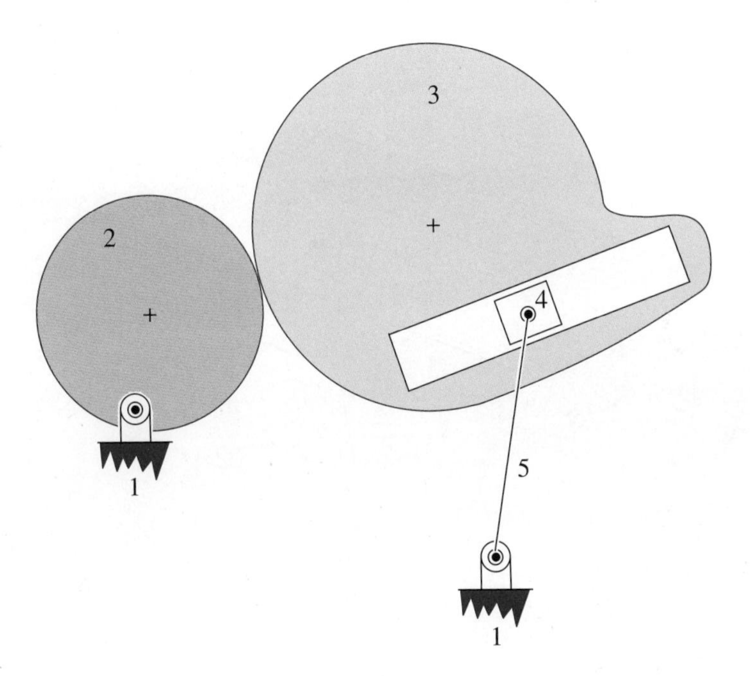

Problem 2.7

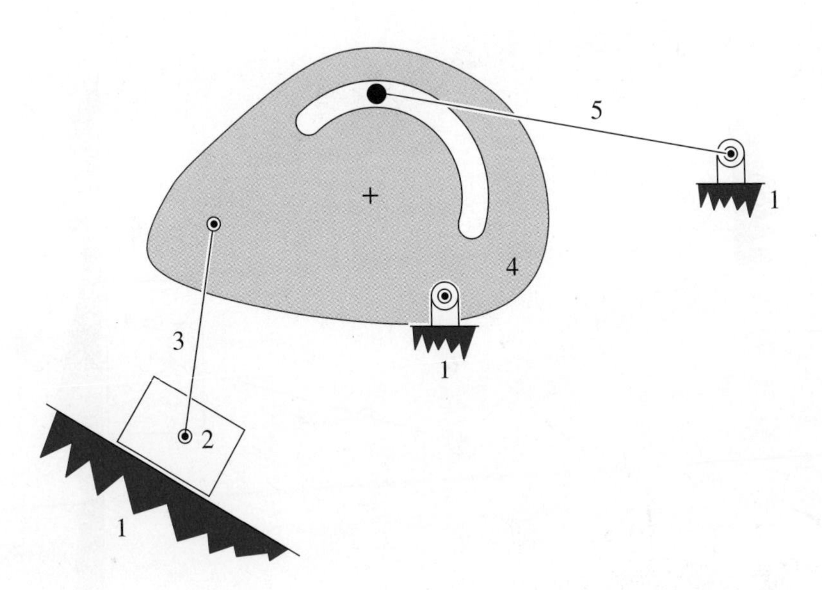

Problem 2.8

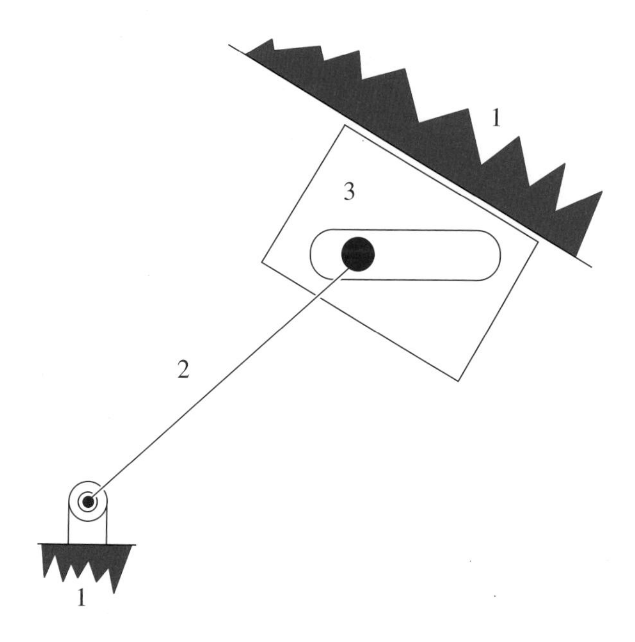

Problem 2.9

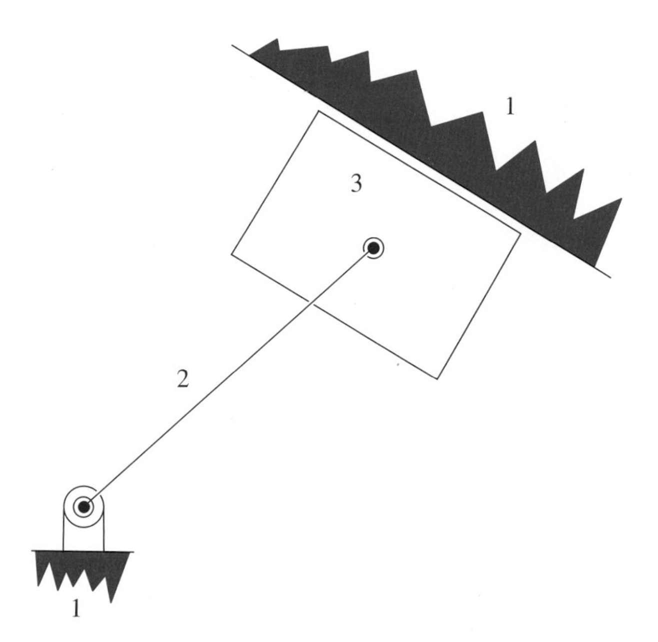

Problem 2.10

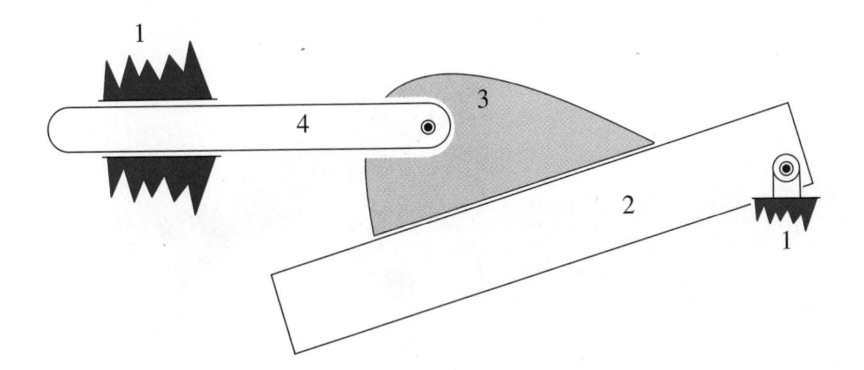

Problem 2.11

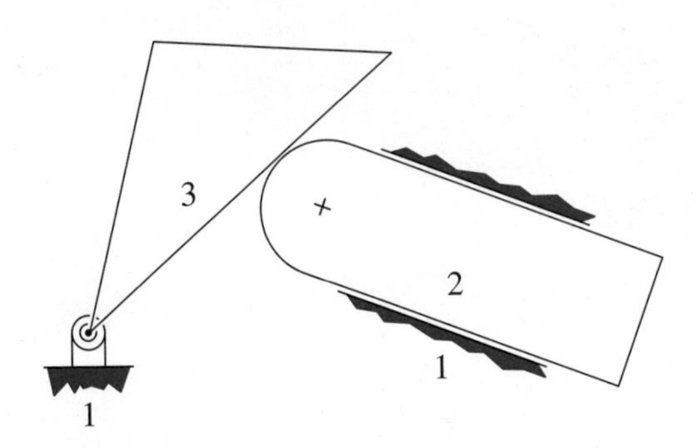

Problem 2.12

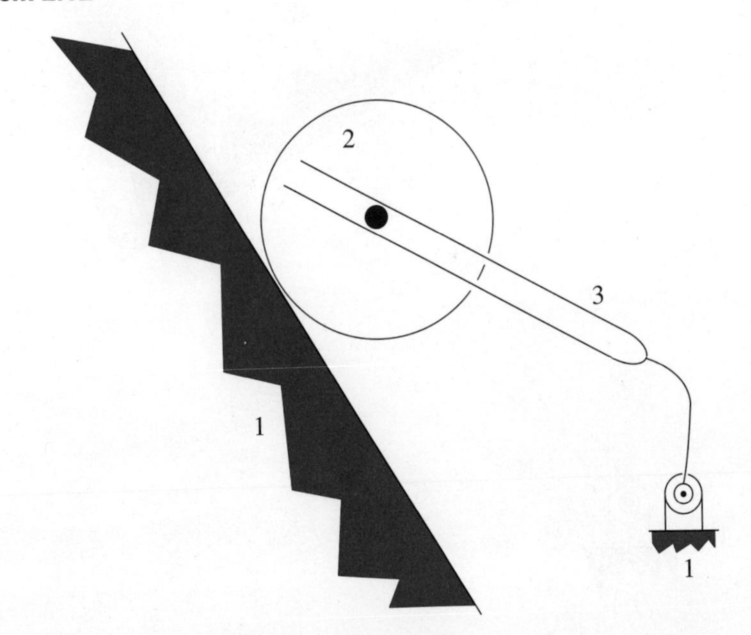

Problem 2.13

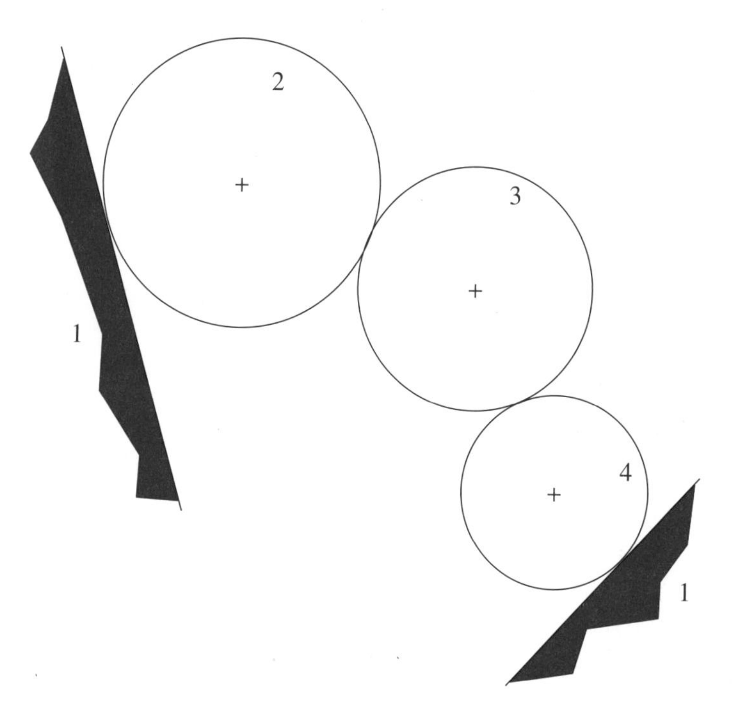

Problem 2.14

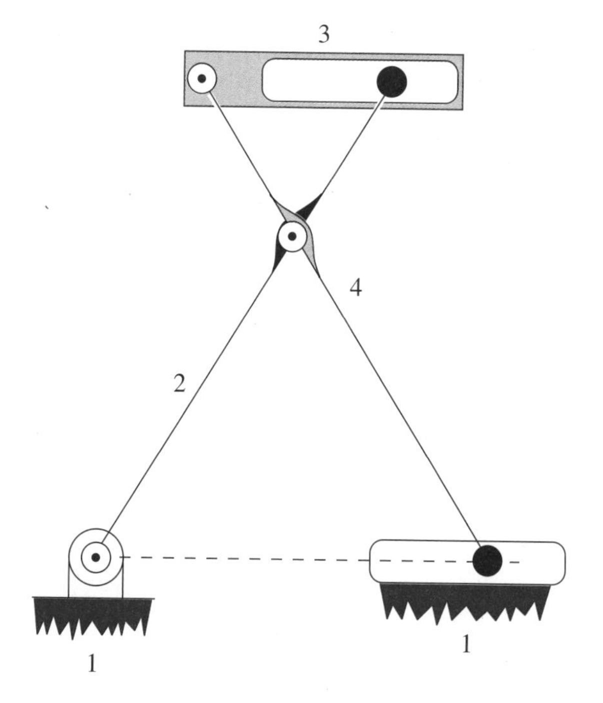

Problem 2.15

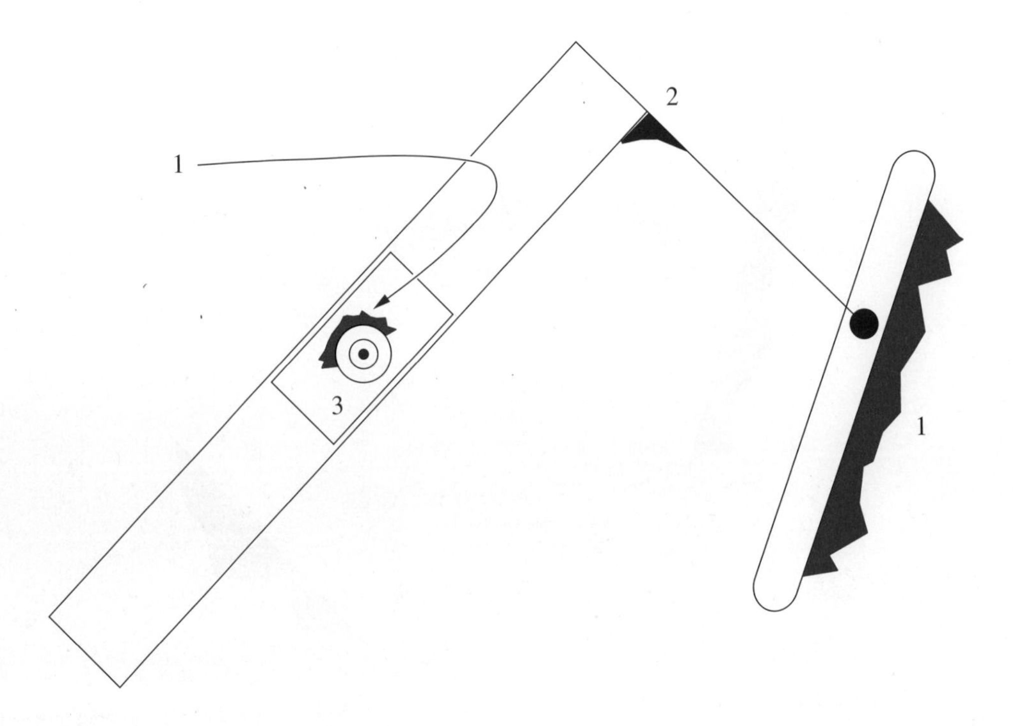

Problem 2.16

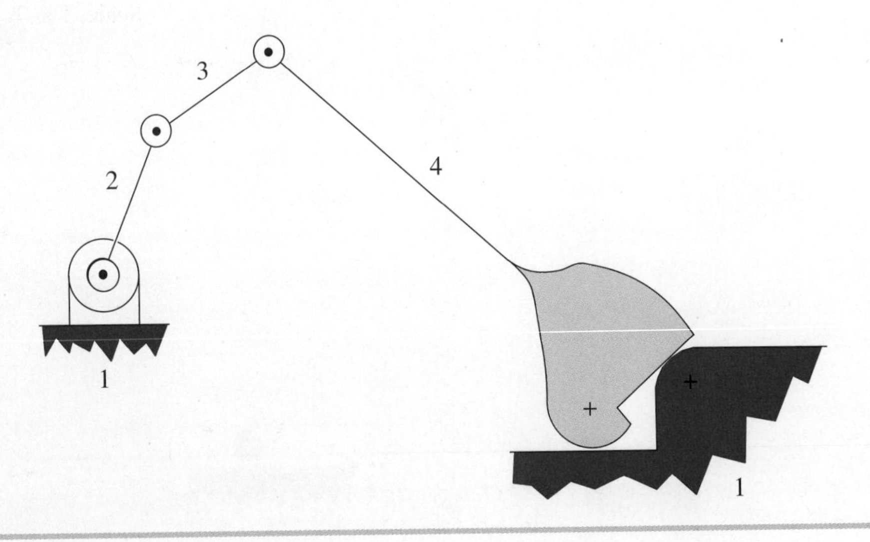

Skeleton diagram of pliers in Figure 1.27

Problem 2.17

Consider the mechanism below and its two vector loops.

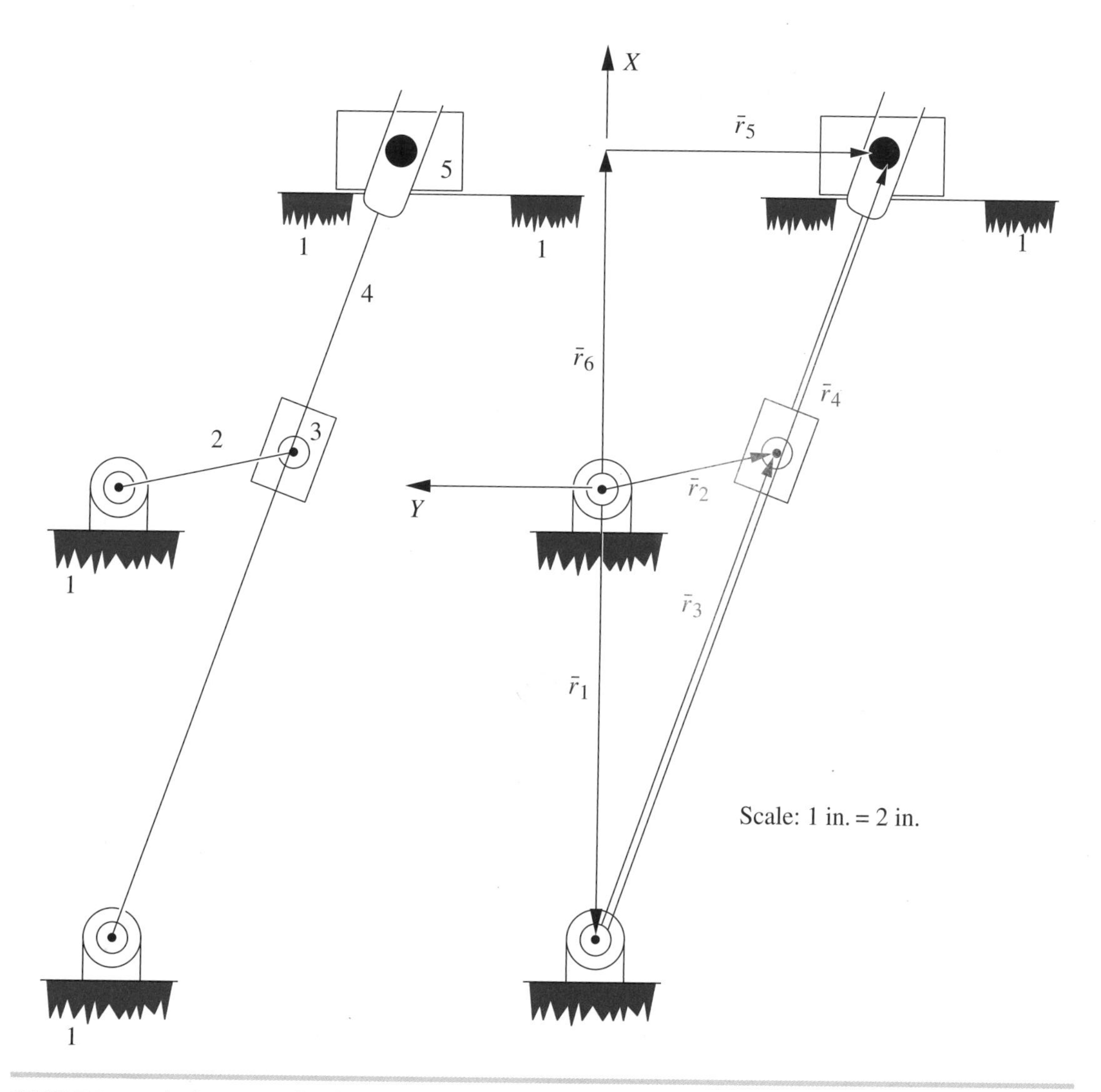

FIGURE 2.26 A shaper mechanism

The vector loop equations and geometric constraint are

$$
\begin{aligned}
\bar{r}_2 - \bar{r}_3 - \bar{r}_1 &= \bar{0} \\
\bar{r}_6 + \bar{r}_5 - \bar{r}_4 - \bar{r}_1 &= \bar{0} \\
\theta_4 - \theta_3 &= 0,
\end{aligned}
\tag{2.69}
$$

which give a system of five scalar equations in six unknowns: $\theta_2, \theta_3, r_3, \theta_4, r_4$, and r_5.

$$\begin{aligned} r_2\cos\theta_2 - r_3\cos\theta_3 + r_1 &= 0 \\ r_2\sin\theta_2 - r_3\sin\theta_3 &= 0 \\ r_6 - r_4\cos\theta_4 + r_1 &= 0 \\ -r_5 - r_4\sin\theta_4 &= 0 \\ \theta_4 - \theta_3 &= 0 \end{aligned} \tag{2.70}$$

scalar knowns: $r_1, \theta_1 = -\pi, r_2, r_6, \theta_6 = 0, \theta_5 = -\frac{\pi}{2}$

scalar unknowns: $\theta_2, r_3, \theta_3, r_5, r_4, \theta_4$

1. Take θ_2 as the known input. Outline all equations needed to implement Newton's Method to solve for the remaining five unknowns, as was done in Section 2.4, using

$$\bar{x} = \begin{bmatrix} r_3 \\ r_4 \\ r_5 \\ \theta_3 \\ \theta_4 \end{bmatrix}.$$

2. Create a flowchart to show how these equations would be used in a computer program to solve for the output variables.

2.7 PROGRAMMING PROBLEMS

For the following problems, you may use any programming language or script of your choice.

Programming Problem 1

This problem continues with Problem 2.17. The mechanism drawing in Problem 2.17 has a scale of 1″ = 2″. Use the dimensions

$$r_1 = 4.8'', \quad r_2 = 2.0'', \quad \text{and} \quad r_6 = 3.65''.$$

The input has a value of $\theta_2 = 283° = 4.94$ radians. Use an initial guess at the solution of

$$\bar{x}^0 = \begin{bmatrix} 3'' \\ 11'' \\ 4'' \\ 300° \\ 300° \end{bmatrix} = \begin{Bmatrix} 3'' \\ 11'' \\ 4'' \\ 5.2360 \text{ rad} \\ 5.2360 \text{ rad} \end{Bmatrix}.$$

Print out the initial value of $\bar{x}^0$ given above, along with the new value of $\bar{x}$ that comes after each of five iterations through Newton's Method, as in Example 2.9.

The program you write here will be used in the first programming problem 5 in Section 7.7, so save your code.

Programming Problem 2

The two four bar mechanisms in Figures 2.27 and 2.28 are geometric inversions. Plot θ_3 vs. θ_2 and θ_4 vs. θ_2 for each mechanism.

- Be sure that the angles are plotted in units of degrees (not radians).
- Be sure that the plot has a title, the axes are labeled, and the units are given on the axes.

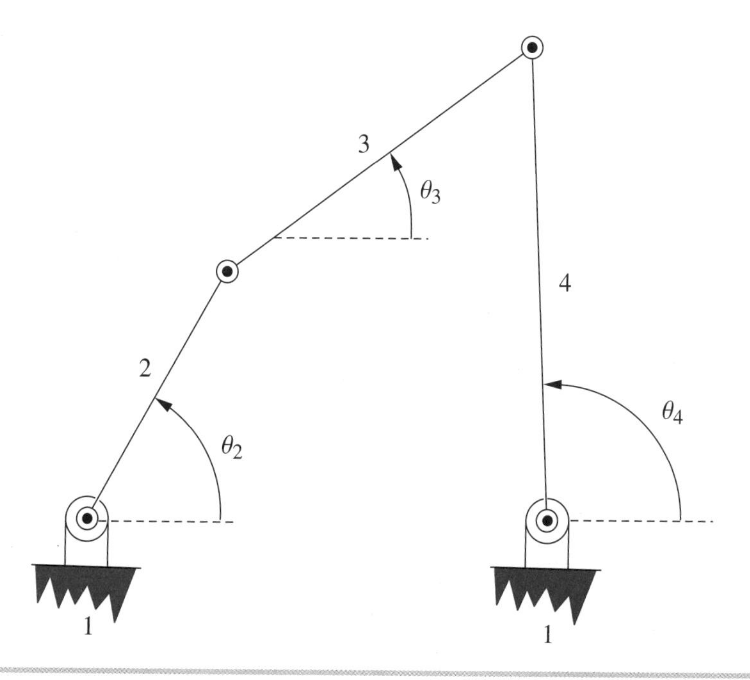

FIGURE 2.27 Open four bar mechanism

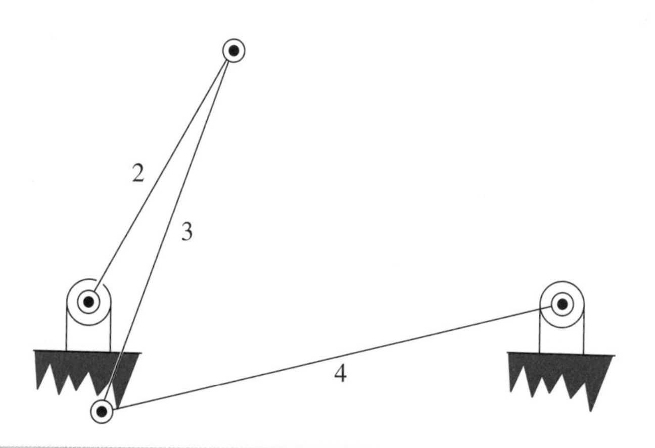

FIGURE 2.28 Crossed four bar mechanism

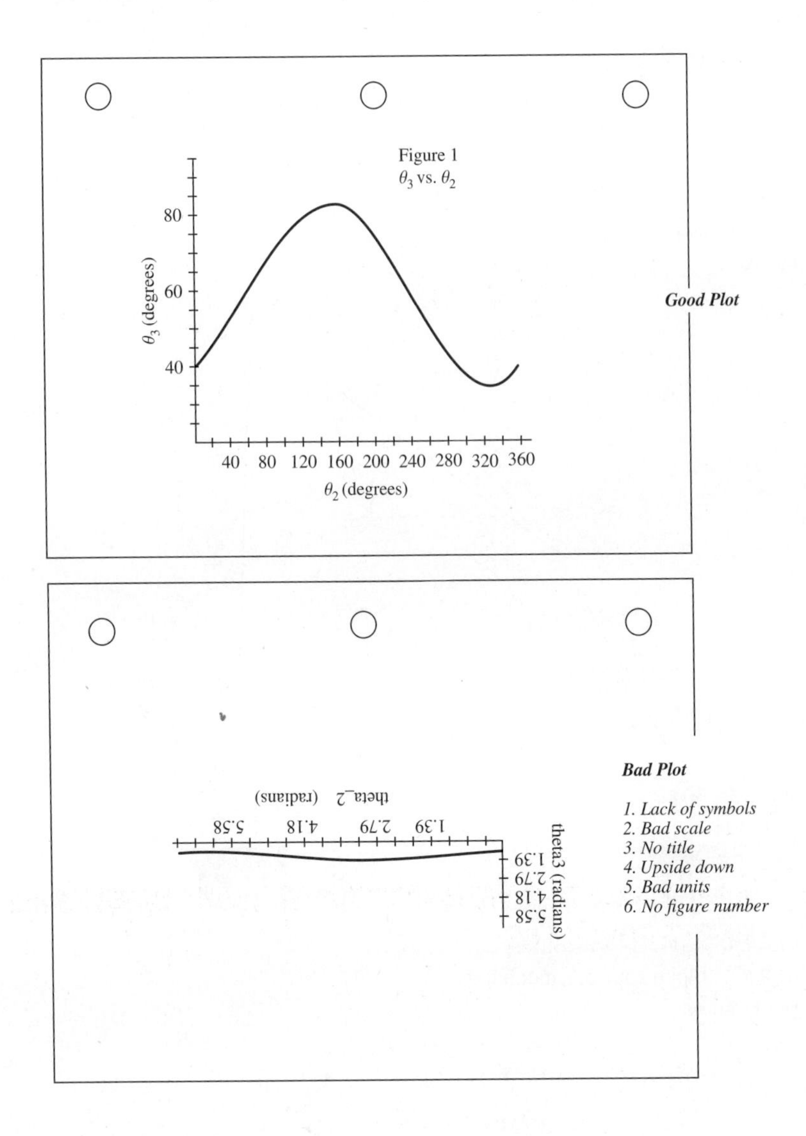

- Be sure the plot is not upside down.
- Link dimensions are, link 1: 10 inches, link 2: 6 inches, link 3: 8 inches, and link 4: 10 inches.

Check your results graphically for at least one of the cases to be certain your program is working correctly. You will need this program to complete programming problem 3 in Section 2.7 and programming problem 4 in Section 4.8.

Programming Problem 3

For the open four bar mechanism in programming problem 2 of Section 2.7, consider the point Q on the coupler shown on the following page. Plot the trajectory of Q, and on this plot show where Q is at the instant shown in the figure.

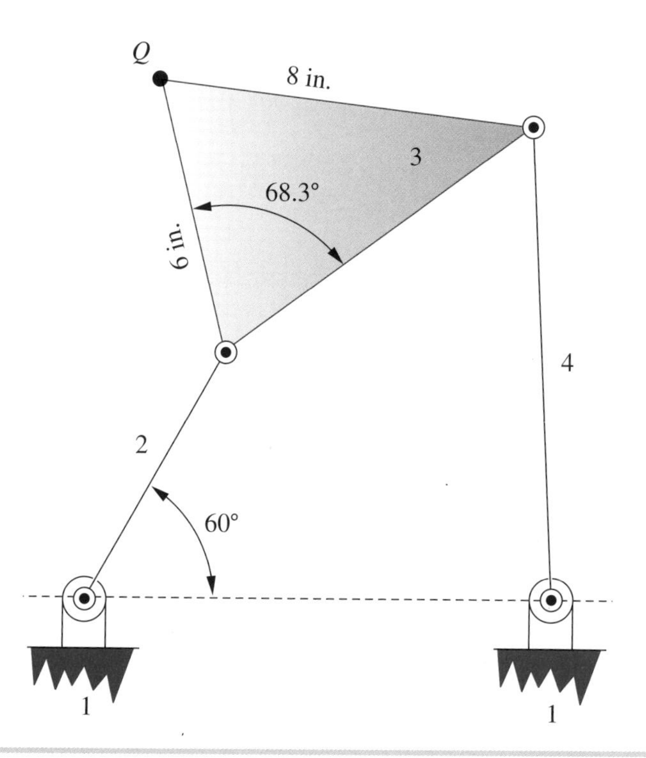

FIGURE 2.29 Coupler point Q

2.8 APPENDIX I: DERIVATION OF THE TANGENT OF THE HALF ANGLE FORMULAS

Here it will be proved that

$$\cos\theta = \frac{1-u^2}{1+u^2} \quad \text{and} \tag{2.71}$$

$$\sin\theta = \frac{2u}{1+u^2}, \tag{2.72}$$

where u is defined as the tangent of the half angle—that is,

$$u = \tan\left(\frac{\theta}{2}\right). \tag{2.73}$$

The proof relies on the trigonometric double angle formulas:

$$\cos(a \pm b) = \cos a \cos b \mp \sin a \sin b \tag{2.74}$$

$$\sin(a \pm b) = \sin a \cos b \pm \cos a \sin b. \tag{2.75}$$

Begin with

$$\cos\theta = \cos\left(\frac{\theta}{2} + \frac{\theta}{2}\right). \tag{2.76}$$

In light of Equation (2.74), the right hand side of Equation (2.76) can be written as

$$\cos\theta = \cos^2\left(\frac{\theta}{2}\right) - \sin^2\left(\frac{\theta}{2}\right) \tag{2.77}$$

$$= \frac{\cos^2\left(\frac{\theta}{2}\right) - \sin^2\left(\frac{\theta}{2}\right)}{\cos^2\left(\frac{\theta}{2}\right)}\cos^2\left(\frac{\theta}{2}\right) \tag{2.78}$$

$$= \left[1 - \tan^2\left(\frac{\theta}{2}\right)\right]\cos^2\left(\frac{\theta}{2}\right). \tag{2.79}$$

The term $\cos^2\left(\frac{\theta}{2}\right)$ on the right hand side of Equation (2.79) can be written as

$$\cos^2\left(\frac{\theta}{2}\right) = \frac{\cos^2\left(\frac{\theta}{2}\right)}{\cos^2\left(\frac{\theta}{2}\right) + \sin^2\left(\frac{\theta}{2}\right)} = \frac{1}{1 + \tan^2\left(\frac{\theta}{2}\right)}. \tag{2.80}$$

Substituting (2.80) into the right hand side of Equation (2.79) gives

$$\cos\theta = \frac{1 - \tan^2\left(\frac{\theta}{2}\right)}{1 + \tan^2\left(\frac{\theta}{2}\right)} = \frac{1 - u^2}{1 + u^2},$$

and Equation (2.71) has been derived.

To get (2.72), begin with

$$\sin\theta = \sin\left(\frac{\theta}{2} + \frac{\theta}{2}\right). \tag{2.81}$$

In light of Equation (2.75), the right hand side of Equation (2.81) can be written as

$$\sin\theta = 2\sin\left(\frac{\theta}{2}\right)\cos\left(\frac{\theta}{2}\right) = 2\tan\left(\frac{\theta}{2}\right)\cos^2\left(\frac{\theta}{2}\right). \tag{2.82}$$

Substituting Equation (2.80) into the right hand side of (2.82) gives

$$\sin\theta = \frac{2\tan\left(\frac{\theta}{2}\right)}{1 + \tan^2\left(\frac{\theta}{2}\right)} = \frac{2u}{1 + u^2},$$

and Equation (2.72) has been derived.

2.9 APPENDIX II: MATLAB® CODE USED IN EXAMPLE 2.10 DEMONSTRATING NEWTON'S METHOD

```
pi=4.0*atan(1.0);
gamma=110*pi/180;
% values of scalar knowns (dimensions)
r2=3.0;
r3=3.6;
r4=3.9;
r1=6.0;
r6=3.9;
```

```
r7=1.4;
r8=7.35;
t8=80*pi/180;
% set value of input angle t2 (given joint variable)
t2=58*pi/180;
% guess values of scalar unknowns the (remaining joint
   variables)
t3=0*pi/180;
t4=90*pi/180;
r5=1.5;
t5=0*pi/180;
t6=0*pi/180;
t7=90*pi/180;
% define vector of initial guesses
x=[t3;t4;r5;t5;t6;t7];
for i=1:5
% compute the necessary sines and cosines of angles t2-t8
ct2=cos(t2);
st2=sin(t2);
ct3=cos(t3);
st3=sin(t3);
ct4=cos(t4);
st4=sin(t4);
ct5=cos(t5);
st5=sin(t5);
ct6=cos(t6);
st6=sin(t6);
ct7=cos(t7);
st7=sin(t7);
ct8=cos(t8);
st8=sin(t8);
%compute the functions at the guessed value
f1=r2*ct2+r3*ct3-r4*ct4-r1;
f2=r2*st2+r3*st3-r4*st4;
f3=r8*ct8+r6*ct6-r5*ct5-r7*ct7-r4*ct4-r1;
f4=r8*st8+r6*st6-r5*st5-r7*st7-r4*st4;
f5=t3-t7+gamma;
f6=t5-t7+(pi/2);
% define vector of functions computed at the guessed
  solution
f=[f1;f2;f3;f4;f5;f6];
% calculate the partials of f w.r.t. each element of x
dfdt3=[-r3*st3;r3*ct3;0;0;1;0];
dfdt4=[r4*st4;-r4*ct4;r4*st4;-r4*ct4;0;0];
dfdr5=[0;0;-ct5;-st5;0;0];
dfdt5=[0;0;r5*st5;-r5*ct5;0;1];
dfdt6=[0;0;-r6*st6;r6*ct6;0;0];
dfdt7=[0;0;r7*st7;-r7*ct7;-1;-1];
% define the A matrix
A = [dfdt3 dfdt4 dfdr5 dfdt5 dfdt6 dfdt7];
```

```
% use equation (2.67) to compute the solution x
x = x-inv(A)*f;
t3=x(1,1)
t4=x(2,1)
r5=x(3,1)
t5=x(4,1)
t6=x(5,1)
t7=x(6,1)
end
```

In this example there were five iterations. The actual number of iterations required is determined by the value of the tolerance in Figure 2.24

CHAPTER 3

Kinematic Analysis Part II: Rolling Contacts

This chapter examines the kinematics of mechanisms that contain rolling contacts (a.k.a. rolling joints). Rolling contacts with widely spaced centers of rotation are realized with chains and sprockets (positive drive) or with belts and pulleys (friction drive). When the centers of rotation are closer together, the rolling is realized with gears that have involute tooth profiles (positive drive) or with rolling wheels (friction drive).

The vector loop method is unaware of rolling joints. It sees rolling bodies as contacting circular shapes. The vector loop method is unaware of the fact that as a result of the rolling contacts (i.e., the no-slip conditions), the rotations of these bodies are related. The vector loop method needs to be supplemented with a *rolling contact equation* for each rolling joint, so that the relation between the rotations of the rolling bodies can be included in the kinematic analysis. This chapter develops the rolling contact equation and shows how it can be incorporated into the vector loop method. In addition to mechanisms with rolling contacts, this chapter also investigates mechanisms that consist of nothing but rolling contacts. These are referred to as geartrains or transmissions. Several examples of automotive applications are presented, and the relationship among power, torque, and angular velocity is studied.

In mechanisms, rolling contacts are most often realized through the use of gears whose teeth have involute tooth profiles. (See the appendix to this chapter for a development of the involute tooth profile.) The rolling circles in a skeleton diagram are the pitch circles of gears that are in mesh. When gears transfer power between them, it is done through a gear force that acts between the gear teeth that are in mesh. The gear force has a tangential component that transmits the torque and power, but since the gear tooth profile is inclined, there is also a radial component to the gear force that transmits no power and does nothing more than push the gears apart in a radial direction. This is called a separating force as it tends to drive the gears apart. To equilibrate the separating force and prevent separation of the gears, there is always a link connecting the centers of the gears that hold the gears in place. This link is called the arm. Kinematically the arm is not necessary, but it is required in practice. Thus, in many skeleton diagrams that involve rolling circles, an arm is shown connecting the centers of circles as in Figure 3.1.

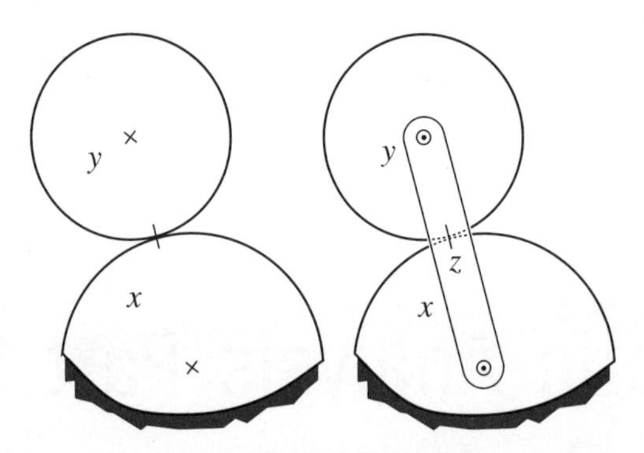

FIGURE 3.1 Pitch circle of gear y rolling around the fixed-pitch circle of gear x

On the left side we see the pitch circle of gear y rolling around the outside of the pitch circle of gear x, which is fixed. It is easy to visualize the motion. According to Gruebler's Criterion, Equation (1.2), $F = 1$; that is, there is one degree of freedom ($N = 2$, $P_1 = 1$). On the right side we see that same situation, but here an arm connects the centers of the pitch circles. The motion of y relative to x is unchanged, and now there is the arm z rotating as well. We see the system still has one degree of freedom, but according to Gruebler's Criterion (now $N = 3$ and $P_1 = 3$), $F = 0$ and the system is a structure. We discussed this paradox in Section 1.5. As shown in Figure 1.38, Gruebler's Criterion considers links x and y to be arbitrary shapes, and in that case the mechanism is a determinate structure, yet we know it is still movable. Do not rely on Gruebler's Criterion to give you the correct number of degrees of freedom for systems that have rolling contacts held together by arms.

3.1 EXTERNALLY CONTACTING ROLLING BODIES

The left hand side (l.h.s.) of Figure 3.2 shows two bodies numbered x and y that have an *external* rolling contact. This contact can be identified as external because the point of contact is on the line segment between the centers of x and y. Per Section 2.2.5, there are a pair of vectors $\bar{r}_x$ and $\bar{r}_y$ attached to bodies x and y, and a vector $\bar{r}_z$ is attached between their centers. ρ_x and ρ_y represent the radii of bodies x and y, respectively. The bodies are in these initial positions before any rolling occurs.

As rolling occurs, consider the *changes* in the directions of $\bar{r}_x$ and $\bar{r}_y$ *relative* to $\bar{r}_z$, denoted as $\Delta\theta_{x/z}$ and $\Delta\theta_{y/z}$ respectively. Motions relative to $\bar{r}_z$ are visualized by considering $\bar{r}_z$ as fixed. As x and y roll on each other they unwrap equal circumferences, indicated by the thickened arc length on the right hand side (r.h.s.) of Figure 3.2. This is the no-slip condition associated with a rolling contact. With $\bar{r}_z$ fixed the r.h.s shows the rolling action. The dashed vectors $\bar{r}_{x_i}$ and $\bar{r}_{y_i}$ were the initial positions of $\bar{r}_x$ and $\bar{r}_y$ from the l.h.s. Due to the no-slip condition, the vectors counter-rotate by the angles $\Delta\theta_{x/z}$ and $\Delta\theta_{y/z}$, respectively, and equal arc lengths are unrolled from the circumferences of x and y. This results in

$$\rho_x \Delta\theta_{x/z} = -\rho_y \Delta\theta_{y/z}. \tag{3.1}$$

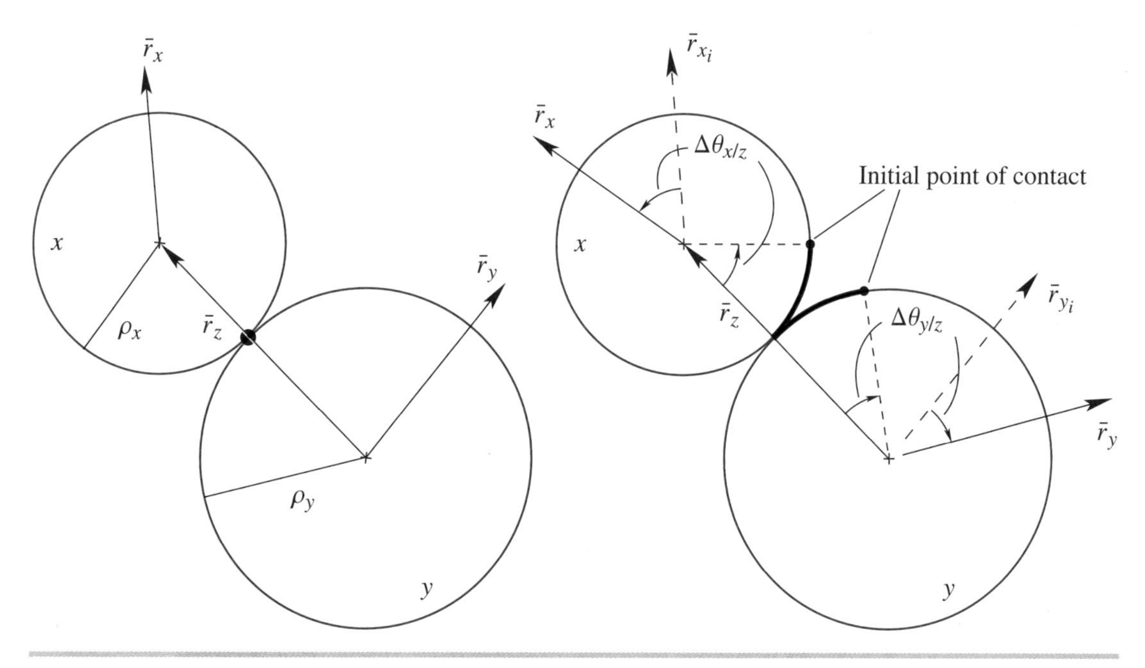

FIGURE 3.2 Externally contacting rolling bodies
© Cengage Learning.

This is the rolling contact equation for external rolling. The negative sign in Equation (3.1) indicates that $\bar{r}_x$ and $\bar{r}_y$ counter-rotate relative to $\bar{r}_z$. In order to relate the absolute angular positions of $\bar{r}_x$, $\bar{r}_y$, and $\bar{r}_z$, it is necessary to expand Equation (3.1). First we expand the relative changes in angular position in terms of the absolute changes,

$$\Delta\theta_{x/z} = \Delta\theta_x - \Delta\theta_z \quad \text{and} \quad \Delta\theta_{y/z} = \Delta\theta_y - \Delta\theta_z. \tag{3.2}$$

Then we expand the absolute changes in terms of current positions and known initial positions,

$$\Delta\theta_x = (\theta_x - \theta_{x_i}), \quad \Delta\theta_y = (\theta_y - \theta_{y_i}), \quad \Delta\theta_z = (\theta_z - \theta_{z_i}), \tag{3.3}$$

and substitute Equations (3.2) and (3.3) into Equation (3.1) to obtain the fully expanded form of the rolling contact equation,

$$\rho_x[(\theta_x - \theta_{x_i}) - (\theta_z - \theta_{z_i})] = -\rho_y[(\theta_y - \theta_{y_i}) - (\theta_z - \theta_{z_i})]. \tag{3.4}$$

3.2 INTERNALLY CONTACTING ROLLING BODIES

The l.h.s. of Figure 3.3 shows two bodies designated x and y that have an *internal* rolling contact. This contact is identified as internal because the point of contact is outside of the line segment between the centers of x and y. Per Section 2.2.6, a pair of vectors $\bar{r}_x$ and $\bar{r}_y$ is attached to x and y, and a vector $\bar{r}_z$ is attached between their centers. The same discussion applies here as in the previous section, except that now $\bar{r}_x$ and $\bar{r}_y$ rotate in the same direction relative to $\bar{r}_z$, and Equation (3.1) becomes

$$\rho_x\Delta\theta_{x/z} = \rho_y\Delta\theta_{y/z}. \tag{3.5}$$

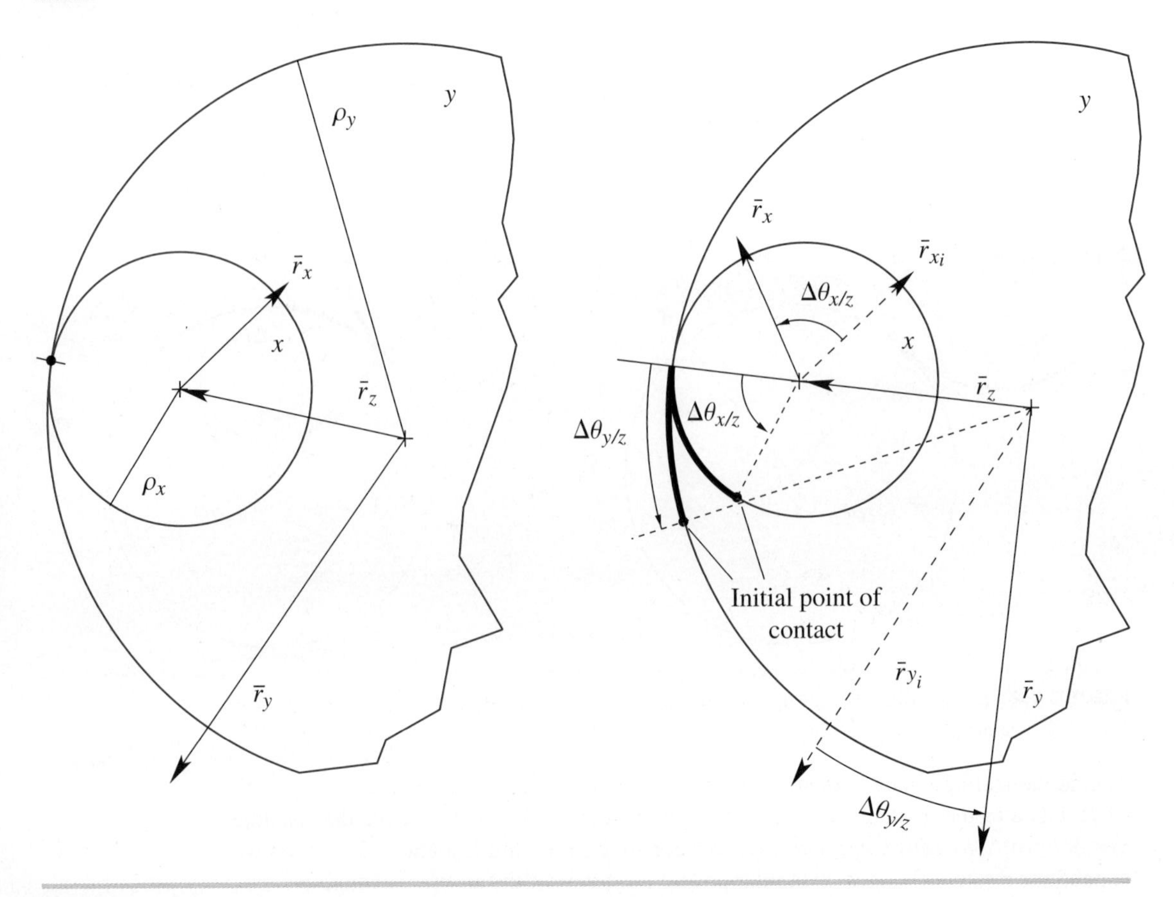

FIGURE 3.3 Internally contacting rolling bodies

Equation (3.5) is the rolling contact equation for the case of internal rolling. To relate the angular positions of $\bar{r}_x$, $\bar{r}_y$, and $\bar{r}_z$, expand Equation (3.5) just as Equation (3.1) was expanded:

$$\rho_x(\Delta\theta_x - \Delta\theta_z) = \rho_y(\Delta\theta_y - \Delta\theta_z)$$

$$\rho_x[(\theta_x - \theta_{x_i}) - (\theta_z - \theta_{z_i})] = \rho_y[(\theta_y - \theta_{y_i}) - (\theta_z - \theta_{z_i})]. \tag{3.6}$$

3.3 ONE BODY WITH A FLAT SURFACE

The case when one of the rolling elements, body y, is a flat surface is shown in Figure 3.4. In this case the rolling elements are referred to as the rack (body y) and the pinion (body x).

The entities in the figure are defined in a slightly different manner.

ρ_x represents the radius of x.
$\bar{r}_x$ is a vector attached to body x.
$\bar{r}_y$ is a vector attached to body y, whose length locates the point of contact.

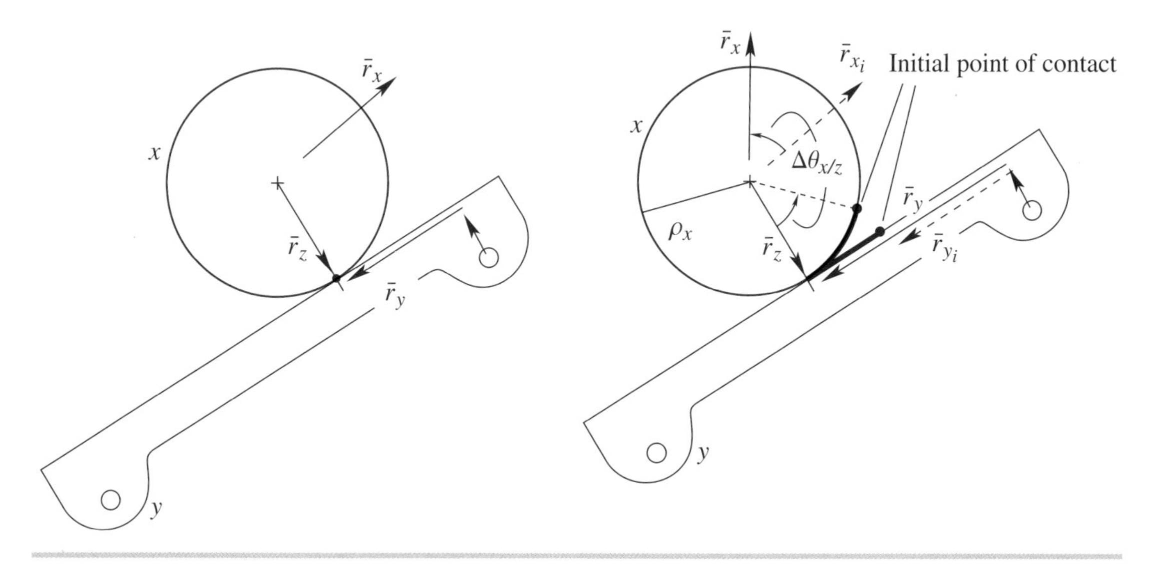

FIGURE 3.4 Rolling contact with a flat surface

$\bar{r}_z$ is a vector connecting the center of x to the point of contact, a hypothetical arm that keeps x in contact with y.

The l.h.s. of Figure 3.4 shows the rack and pinion in a known initial configuration. The r.h.s. shows the system after $\bar{r}_x$ has undergone some rotation relative to $\bar{r}_z$, $\Delta\theta_{x/z}$. This relative rotation causes a change in the length of vector $\bar{r}_y$. Due to the condition of rolling (no slip), the change in length of $\bar{r}_y$ is equal to the unrolled circumference of body x:

$$\Delta r_y = \rho_x \Delta\theta_{x/z}, \tag{3.7}$$

where

$$\Delta r_y = r_y - r_{y_i}.$$

To relate r_y to θ_x and θ_z, Equation (3.7) must be expanded:

$$r_y - r_{y_i} = \rho_x(\Delta\theta_x - \Delta\theta_z)$$

$$r_y - r_{y_i} = \rho_x[(\theta_x - \theta_{x_i}) - (\theta_z - \theta_{z_i})]. \tag{3.8}$$

Equation (3.7) is one of two possible forms of the rolling contact equation for rolling on a flat surface. An alternative form of the rolling contact equation is obtained if the vector $\bar{r}'_y$ is used in place of $\bar{r}_y$ (shown in Figure 3.5). Just like $\bar{r}_y$, vector $\bar{r}'_y$ is attached to body Y, but $\bar{r}'_y$ locates the point of rolling contact from the opposite side. Comparing the two figures, you can see that the sum of the magnitudes r_y and r'_y is a constant, so their changes in length are opposite of each other. As one lengthens, the other shortens. If $\bar{r}'_y$ had been used in place of $\bar{r}_y$, Equation (3.7) would have become

$$\Delta r'_y = -\rho_x \Delta\theta_{x/z}. \tag{3.9}$$

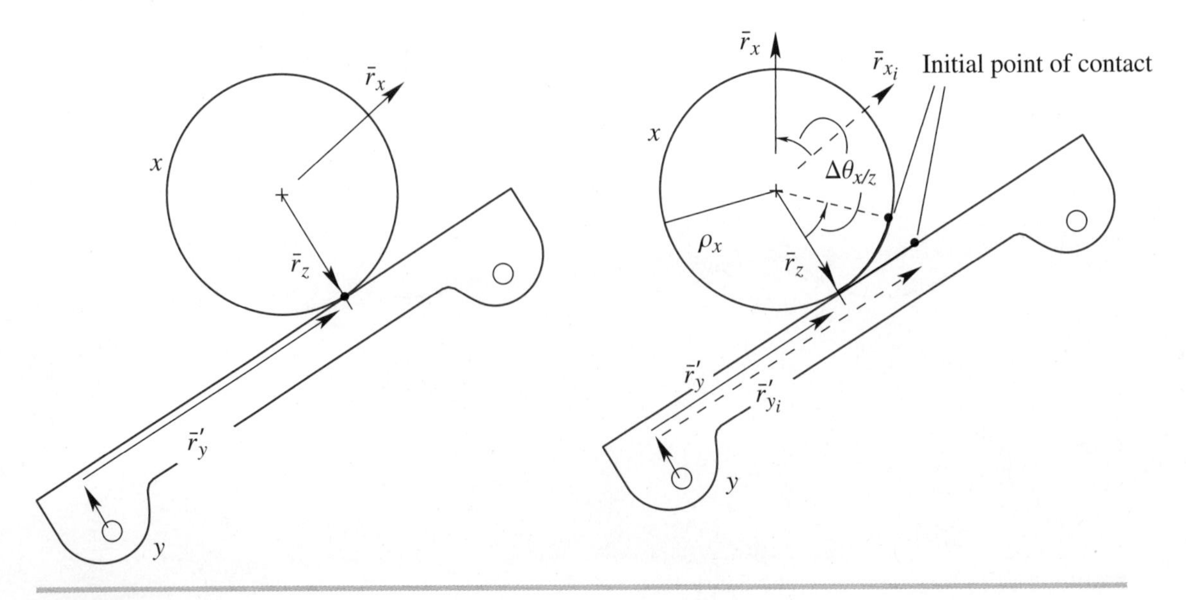

FIGURE 3.5 Alternative vector for rolling contact with a flat surface

Equation (3.9) is the other form of the rolling contact equation for rolling on a flat surface, and it can be expanded to

$$r'_y - r'_{y_i} = -\rho_x(\Delta\theta_x - \Delta\theta_z)$$
$$r'_y - r'_{y_i} = -\rho_x[(\theta_x - \theta_{x_i}) - (\theta_z - \theta_{z_i})]. \tag{3.10}$$

3.4 ASSEMBLY CONFIGURATION

The rolling contact Equations (3.1), (3.5), (3.7), and (3.9) are relationships between *changes* in vector directions and magnitudes. When they are fully expanded into Equations (3.4), (3.6), (3.8), and (3.10), the initial values of the vector variables become involved. These initial values are known as the *assembly configuration*—that is, the configuration in which the system was initially put together. Figure 3.6 illustrates the effect of assembly configuration for a simple case. Body 1 is the arm and it is fixed. Bodies 2 and 3 roll externally. Both systems were assembled with the same initial position of $\bar{r}_2$ but different initial positions of the vector $\bar{r}_3$. In both systems $\bar{r}_2$ was rotated by the same amount, causing a rotation of $\bar{r}_3$ that is the same in both cases. However, since the initial position of $\bar{r}_3$ is different in each system, the final position of $\bar{r}_3$ is also different. This is because the systems have different assembly configurations.

In order to perform the position analysis of mechanisms with rolling contacts, the assembly configuration must be known. To further illustrate this, consider the mechanism in Figure 3.7. On the left side is the system assembled with $\bar{r}_2$ and $\bar{r}_3$ pointing downward. On the right side is that mechanism after θ_2 was rotated $\frac{\pi}{2}$. Note the positions of bodies 4 and 5. The left side of Figure 3.8 shows the same system except that when it is assembled, $\bar{r}_3$ is horizontal, an assembly configuration different from that in

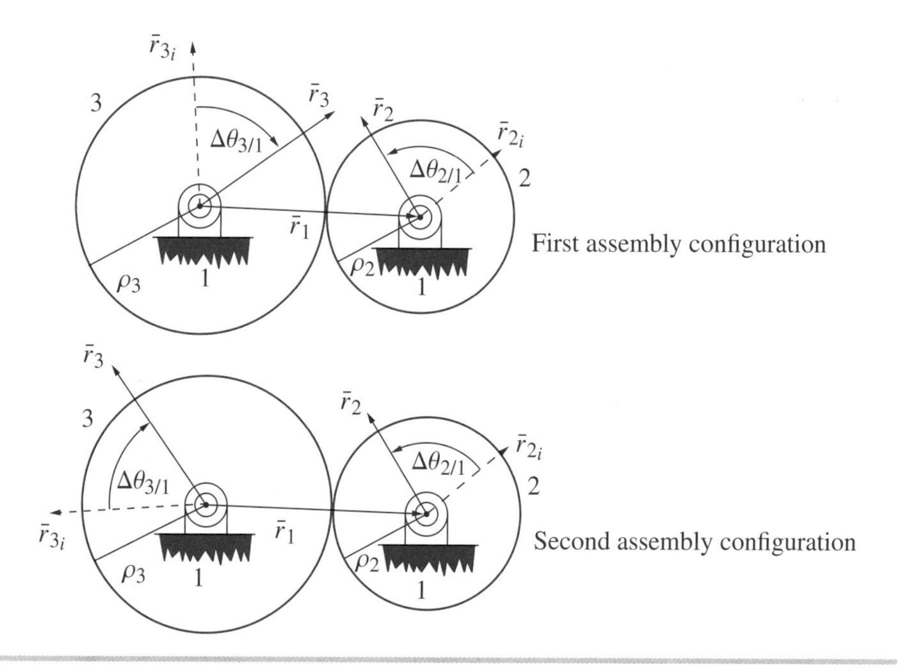

FIGURE 3.6 Simple example of assembly configuration effects

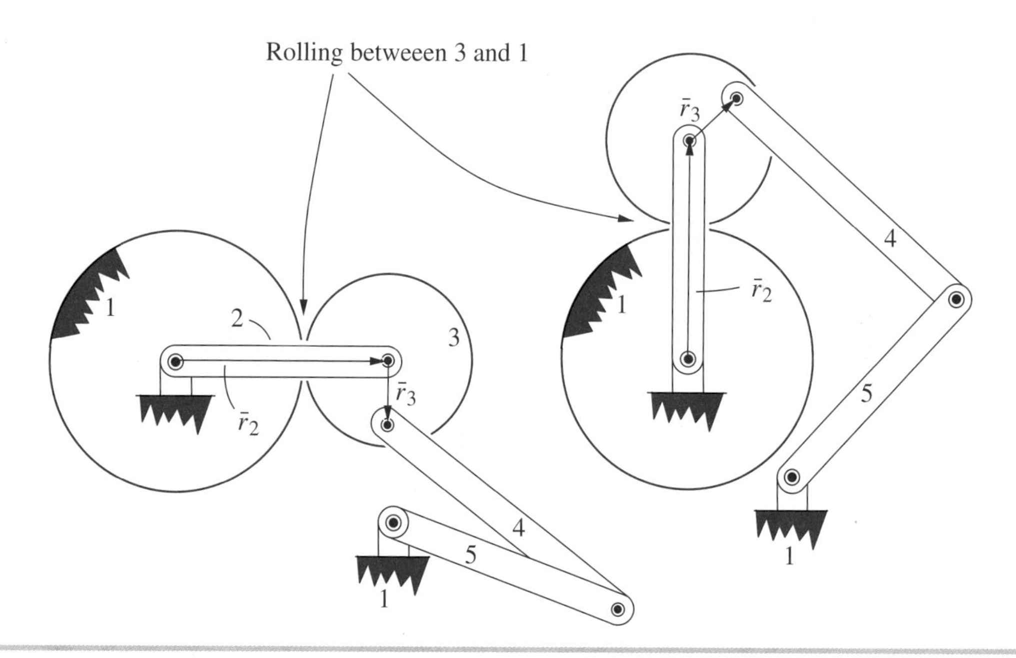

FIGURE 3.7 First assembly configuration

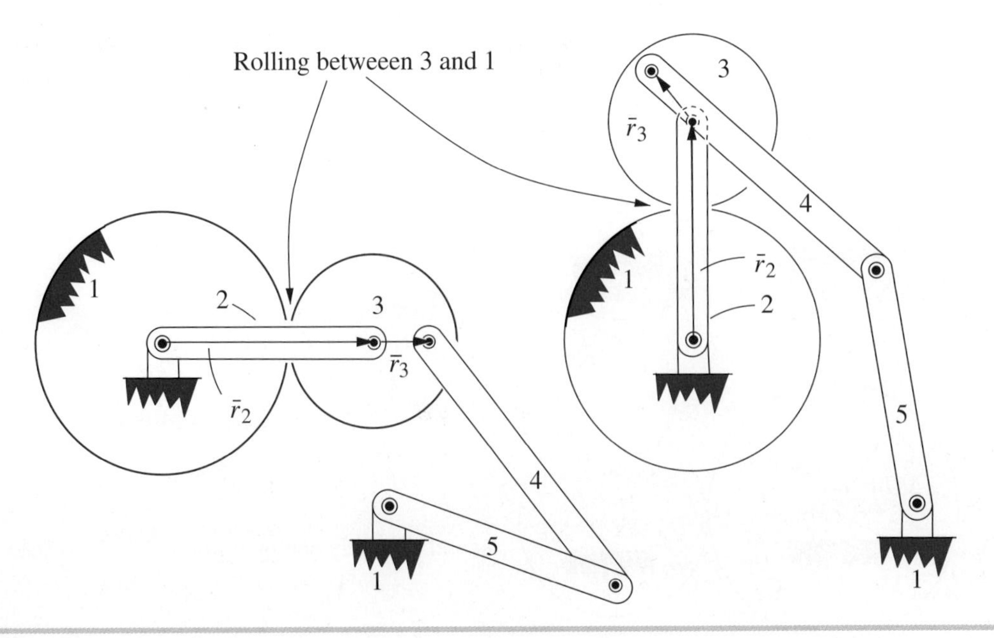

FIGURE 3.8 Second assembly configuration

Figure 3.7. On the right is that mechanism after θ_2 is again rotated $\frac{\pi}{2}$. Note the final positions of bodies 4 and 5. They are different than in Figure 3.7. As a result of different assembly configurations the two mechanisms move differently. In fact, as a result of the different assembly configurations they are two different mechanisms. The assembly configuration is effectively a dimension of the mechanism, and having different assembly configurations is the same as having different dimensions.

▶ EXAMPLE 3.1

Figure 3.9 shows a five bar mechanism where one of the coupler links, body 4, has a rolling contact with the ground. As discussed earlier, link 5, the "arm," theoretically is not necessary. Practically, it is needed in order to maintain contact between 4 and 1. Due to the presence of link 5, Gruebler's Criterion gives $F = 0$. With 5 removed, Gruebler's Criterion gives the correct value of $F = 1$.

Derive the mechanism's position equations, and from them deduce the number of degrees of freedom in the system.

Solution

Figure 3.10 shows a suitable vector loop. The vector loop equation is

$$\bar{r}_2 + \bar{r}_3 - \bar{r}_4 - \bar{r}_5 + \bar{r}_1 = \bar{0},$$

and noting that $\theta_1 = 0$, the simplified X and Y components are

$$r_2\cos\theta_2 + r_3\cos\theta_2 - r_4\cos\theta_4 - r_5\cos\theta_5 + r_1 = 0 \quad (3.11)$$

$$r_2\sin\theta_2 + r_3\sin\theta_2 - r_4\sin\theta_4 - r_5\sin\theta_5 = 0. \quad (3.12)$$

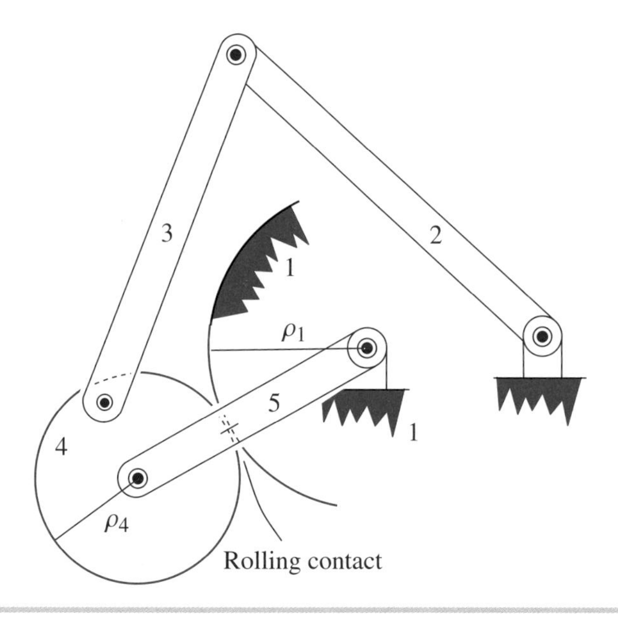

FIGURE 3.9 First example of a mechanism with a rolling contact

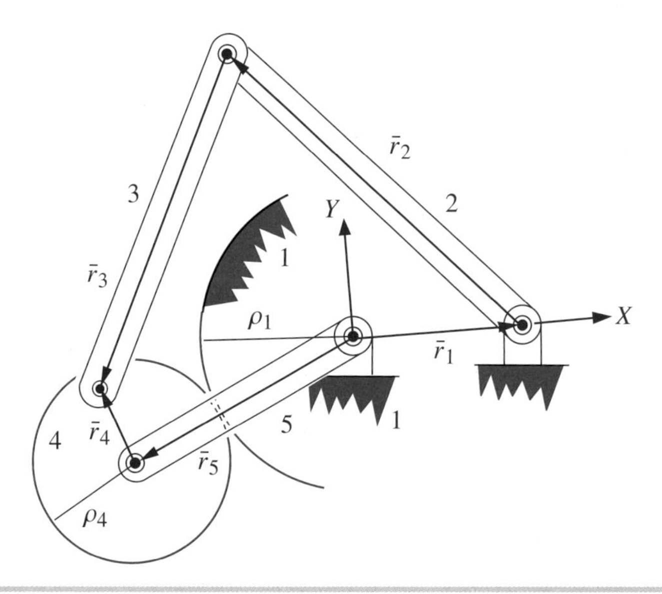

FIGURE 3.10 Vector loop for first example of a mechanism with a rolling contact

Body 4 rolls externally on body 1 and the arm is body 5. The vectors attached to those bodies are $\bar{r}_4$, $\bar{r}_1$, and $\bar{r}_5$, respectively. Referring to Figure 3.2, x is body 4, y is body 1, and z is body 5. The rolling contact equation for external rolling, Equation (3.1), relates the changes in the directions of vectors $\bar{r}_4$, $\bar{r}_1$ $\bar{r}_5$:

$$\rho_4 \Delta\theta_{4/5} = -\rho_1 \Delta\theta_{1/5};$$

that is,

$$\rho_4(\Delta\theta_4 - \Delta\theta_5) = -\rho_1(\Delta\theta_1 - \Delta\theta_5).$$

Since body 1 is fixed, $\Delta\theta_1 = 0$ and this equation simplifies to

$$\rho_4(\Delta\theta_4 - \Delta\theta_5) = \rho_1 \Delta\theta_5,$$

and incorporating the assembly configuration gives

$$\rho_4 \left[(\theta_4 - \theta_{4_i}) - (\theta_5 - \theta_{5_i})\right] = \rho_1(\theta_5 - \theta_{5_i}),$$

which in homogeneous form is

$$\rho_4(\theta_4 - \theta_{4_i}) - (\rho_1 + \rho_4)(\theta_5 - \theta_{5_i}) = 0. \tag{3.13}$$

Equations (3.11), (3.12), and (3.13) are a system of three position equations with

scalar knowns: $r_1,\ \theta_1 = 0,\ r_2,\ r_3,\ r_4,\ r_5,\ \rho_1,\ \rho_4,\ \theta_{4_i},\ \theta_{5_i}$

and

scalar unknowns: $\theta_2,\ \theta_3,\ \theta_4,\ \theta_5$.

Note that the radii of the rolling elements—ρ_1 and ρ_4—and the assembly configuration θ_{4_i} and θ_{5_i} are included in the scalar knowns, along with the other dimensions of the mechanism. This mechanism has a system of three position equations in four unknowns, making it a one-degree-of-freedom mechanism.

▶ EXAMPLE 3.2

Figure 3.11 shows a somewhat strange four link mechanism. Body 3 slides in and out of body 4, and body 4 rolls on body 1. Derive the mechanism's position equations, and from them deduce the number of degrees of freedom in the mechanism.

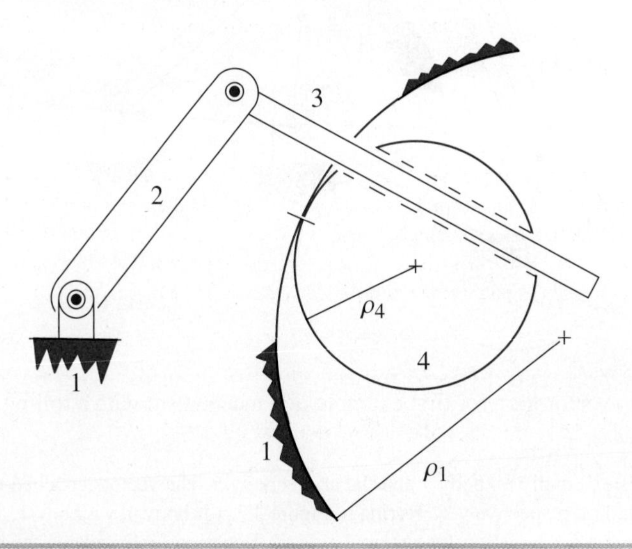

FIGURE 3.11 Second example of a mechanism with a rolling contact

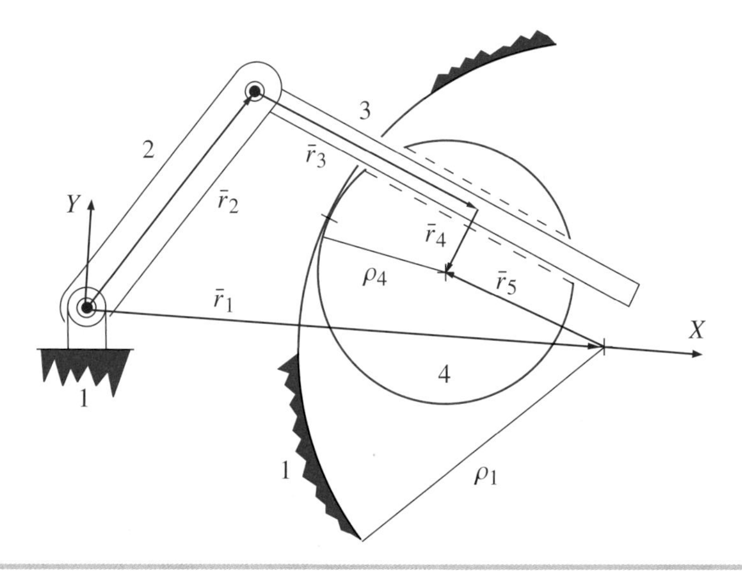

FIGURE 3.12 Vector loop for second example of a mechanism with a rolling contact

Solution

Figure 3.12 shows an appropriate vector loop. The vector loop equation is

$$\bar{r}_2 + \bar{r}_3 + \bar{r}_4 - \bar{r}_5 - \bar{r}_1 = \bar{0}.$$

Noting that $\theta_1 = 0$, the simplified scalar components are

$$r_2\cos\theta_2 + r_3\cos\theta_3 + r_4\cos\theta_4 - r_5\cos\theta_5 - r_1 = 0 \tag{3.14}$$

$$r_2\sin\theta_2 + r_3\sin\theta_3 + r_4\sin\theta_4 - r_5\sin\theta_5 = 0. \tag{3.15}$$

There is also the geometric constraint,

$$\theta_4 - \theta_3 + \frac{\pi}{2} = 0. \tag{3.16}$$

Body 4 rolls internally on body 1 and the vector $\bar{r}_5$ is the "arm." If this mechanism were transmitting power as a machine, there would be a body 5 connecting the centers of 4 and 1 in order to maintain their contact. There are vectors $\bar{r}_4$ and $\bar{r}_1$ attached to bodies 4 and 1, respectively, and the vector $\bar{r}_5$ between their centers. The rolling contact equation for internal rolling, Equation (3.5), relates the changes in the directions of $\bar{r}_1$, $\bar{r}_4$, and $\bar{r}_5$:

$$\rho_4 \Delta\theta_{4/5} = \rho_1 \Delta\theta_{1/5};$$

that is,

$$\rho_4(\Delta\theta_4 - \Delta\theta_5) = \rho_1(\Delta\theta_1 - \Delta\theta_5).$$

Since body 1 is fixed, $\Delta\theta_1 = 0$, so

$$\rho_4(\Delta\theta_4 - \Delta\theta_5) = -\rho_1 \Delta\theta_5.$$

Incorporating the assembly configuration gives

$$\rho_4\left[(\theta_4 - \theta_{4_i}) - (\theta_5 - \theta_{5_i})\right] = -\rho_1(\theta_5 - \theta_{5_i}),$$

which in homogeneous form is

$$\rho_4(\theta_4 - \theta_{4_i}) + (\rho_1 - \rho_4)(\theta_5 - \theta_{5_i}) = 0. \tag{3.17}$$

Equations (3.14), (3.15), (3.16), and (3.17) are a system of four equations with

scalar knowns: $r_1,\ \theta_1 = 0,\ r_2,\ r_4,\ r_5,\ \rho_1,\ \rho_4,\ \theta_{4_i},\ \theta_{5_i}$

and

scalar unknowns: $\theta_2,\ r_3,\ \theta_3,\ \theta_4,\ \theta_5,$

making this a one-degree-of-freedom mechanism.

3.5 GEARTRAINS

Geartrains are mechanisms that consist entirely of rolling contacts. In geartrains there are no links other than the gears and the arms that connect the centers of the gears. Because their position equations consist of only rolling contact equations, which are linear, geartrains have constant speed and torque ratios. The only reason geartrains are used is for their constant speed and torque ratio property. Conventional mechanisms (such as the four bar mechanism) can transmit far more torque and provide far greater mechanical advantage than a geartrain. But a conventional mechanism cannot provide a constant speed ratio.

DEFINITIONS:

Geartrain *A mechanism that consists of only rolling elements and the associated arms connecting their centers. The kinematics of the system is described by only rolling contact equations.*

Simple Geartrain *A geartrain whose rolling elements all have fixed arms.*

Since the arms are fixed in simple geartrains, the centers of all the gears do not move.

Stage of Gear Reduction *Each rolling contact in a simple geartrain is a stage of gear reduction.*

A geartrain with single-stage has one rolling contact, two stages imply two rolling contacts, etc.

Planetary Geartrain *A geartrain whose rolling elements have arms that are not fixed.*

Since the arms are not fixed, in a planetary geartrain some of the gear centers will move on circular orbits.

Hybrid Geartrain *A geartrain that is a combination of a simple and a planetary geartrain.*

The following definitions apply to geartrains with one degree of freedom.

Gear Ratio, *R* *The number of rotations of the input that causes one rotation of the output.*

If we let $\Delta\theta_{in}$ represent the rotation of the input and $\Delta\theta_{out}$ represent the rotation of the output, then R can be computed as,

$$R = \frac{\Delta\theta_{in}}{\Delta\theta_{out}}. \tag{3.18}$$

The gear ratio is frequently stated as "R to one" and many times is written as R:1 and at times it is referred to as simply the ratio. *Since the position equations are linear, R is constant.*

Torque Ratio n_t *Represents the torque increase (or decrease) created by a transmission.*

Let T_{in} represent the input (driving) torque applied to the input gear and T_{out} represent the output (load) torque applied to the output gear, then according to this definition,

$$n_t = \frac{T_{out}}{T_{in}} \longrightarrow T_{out} = n_t T_{in} \tag{3.19}$$

If we neglect losses due to friction and changes in kinetic or potential energy, then conservation of energy requires the net work done by these torques, ΔW, is zero,

$$\Delta W = T_{in}\Delta\theta_{in} + T_{out}\Delta\theta_{out} = 0 \tag{3.20}$$

From Equations (3.18), (3.19) and (3.20) we see that,

$$n_t = -R \tag{3.21}$$

Speed Ratio, n_s *Represents the speed decrease (or increase) created by a transmission.*

Let ω_{in} represent the angular velocity of the input gear and ω_{out} represent the angular velocity of the output gear, then according to this definition,

$$n_s = \frac{\omega_{out}}{\omega_{in}} \longrightarrow \omega_{out} = n_s\omega_{in} \tag{3.22}$$

In Equation (3.18) if we substitute $\Delta\theta_{in} = (\theta_{in} - \theta_{in_i})$ and $\Delta\theta_{out} = (\theta_{out} - \theta_{out_i})$, where θ_{in_i} and θ_{out_i} are initial values of θ_{in} and θ_{out}, we obtain,

$$(\theta_{in} - \theta_{in_i}) = R(\theta_{out} - \theta_{out_i}) \longrightarrow \dot{\theta}_{in} = R\dot{\theta}_{out} \longrightarrow R = \frac{\dot{\theta}_{in}}{\dot{\theta}_{out}} = \frac{\omega_{in}}{\omega_{out}} = \frac{1}{n_s} \tag{3.23}$$

which shows that

$$n_s = \frac{1}{R}. \tag{3.24}$$

In summary, from Equations (3.21) and (3.24), gear ratio, torque and speed ratio are related as,

$$R = -n_t = \frac{1}{n_s},$$

and from this equation any one of R, n_t *or* n_s gives the other two.

3.5.1 Simple Geartrains

Simple geartrains consist of rolling elements whose centers are all fixed. Figure 3.13 shows a simple geartrain with a single-stage of speed reduction.

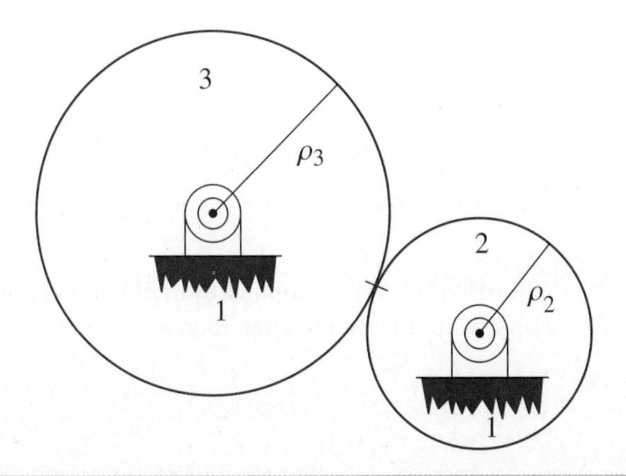

FIGURE 3.13 A simple geartrain with a single-stage

In a single-stage geartrain the smaller gear is usually the driver (the input), and the larger gear is driven (the load). This is because in most applications, speed is to be reduced while torque is increased. In these cases the smaller gear is referred to as "the pinion," and the larger gear is referred to as "the gear." This is not always the case. In rare instances the input is a high torque at a low speed. In these instances a large gear drives a smaller gear, increasing speed and reducing torque. Examples of this are the chain drive of a bicycle and the transmission of a large wind turbine.

► EXAMPLE 3.3

In the single-stage simple geartrain shown in Figure 3.13, 2 is the input and 3 is the output. Derive the position equations, deduce the number of degrees of freedom, and compute the gear ratio.

Solution

The single-stage simple geartrain shown in Figure 3.13 would have the vector loop shown in Figure 3.14. The vector loop is meaningless. It consists of one vector, $\bar{r}_1$, which goes between the fixed centers. The vector loop equation says $\bar{r}_1 - \bar{r}_1 = \bar{0}$. The only meaningful equations in geartrains are the rolling contact equations. We will no longer write vector loop equations for geartrains.

The rolling contact is external, and from Equation (3.1), the rolling contact equation is

$$\rho_3 \Delta\theta_{3/1} = -\rho_2 \Delta\theta_{2/1}$$

$$\rho_3(\Delta\theta_3 - \Delta\theta_1) = -\rho_2(\Delta\theta_2 - \Delta\theta_1).$$

Body 1 is fixed, meaning that $\Delta\theta_1 = 0$, and this equation becomes

$$\rho_3 \Delta\theta_3 = -\rho_2 \Delta\theta_2, \qquad (3.25)$$

$$\rho_3 \Delta\theta_3 + \rho_2 \Delta\theta_2 = 0.$$

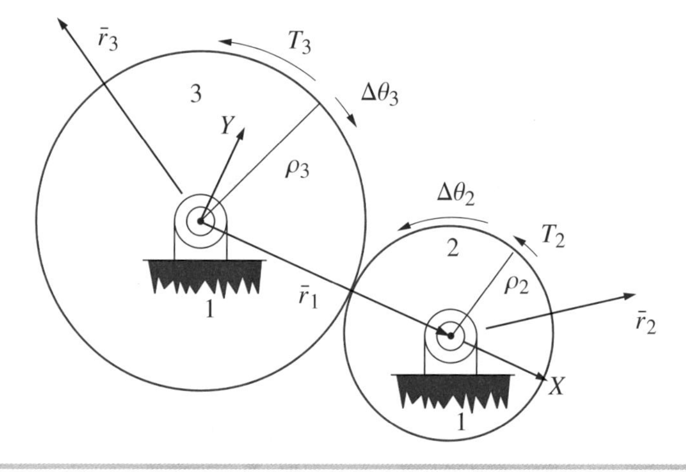

FIGURE 3.14 Vectors needed for kinematic analysis

Introducing the assembly configuration gives

$$\rho_3(\theta_3 - \theta_{3_i}) + \rho_2(\theta_2 - \theta_{2_i}) = 0.$$

This is the position equation for this geartrain. It has

scalar knowns: $\rho_3, \rho_2, \theta_{2i}, \theta_{3i}$

and

scalar unknowns: θ_2, θ_3,

which means there is one degree of freedom. Equation (3.25) gives R,

$$R = \frac{\Delta\theta_{in}}{\Delta\theta_{out}} = \frac{\Delta\theta_2}{\Delta\theta_3} = -\frac{\rho_3}{\rho_2}, \tag{3.26}$$

and the gear ratio of this geartrain is $R = (-\rho_3/\rho_2)$ and since $R < 0$, $\Delta\theta_2$ and $\Delta\theta_3$ are in opposite directions. $n_t = -R = (\rho_3/\rho_2)$ so $n_t > 0$ and T_2 (the driving torque) and T_3 (the load torque) are in the same direction. Figure 3.14 shows this.

In order for a pair of gears to mesh, their teeth must be able to engage, so they must have teeth of the same size. This means that the spacing of the teeth along the circumference of the pitch circles of the two gears must be the same. Thus, if N_x and N_y represent the numbers of teeth on gears x and y in order for x and y to have the same size teeth, then

$$\frac{2\pi\rho_x}{N_x} = \frac{2\pi\rho_y}{N_y} \quad \longrightarrow \quad \frac{\rho_x}{\rho_y} = \frac{N_x}{N_y}.$$

This is an important observation because using the rolling contact equation will give your results in terms of the ratios of the radii of the pitch circles, but in a machine the pitch circle radii are not commonly known. The number of teeth on each gear is known. If nothing else one can always count the number of teeth on a gear. After deriving results in terms of ratios of pitch circle radii, we replace these ratios with ratios of tooth numbers. Thus, in the previous example the gear ratio is $R = (-N_3/N_2)$. For a more detailed discussion of the geometry of gear teeth, refer to Appendix I of this chapter.

If an intermediate gear is introduced between 2 and 3, as shown in Figure 3.15, there are two external rolling contacts and two stages of gear reduction.

▶ EXAMPLE 3.4

A two-stage simple geartrain is shown in Figure 3.15. The input is 2 and the output is 3. Determine the geartrain's position equations, deduce the number of degrees of freedom, and determine the gear ratio.

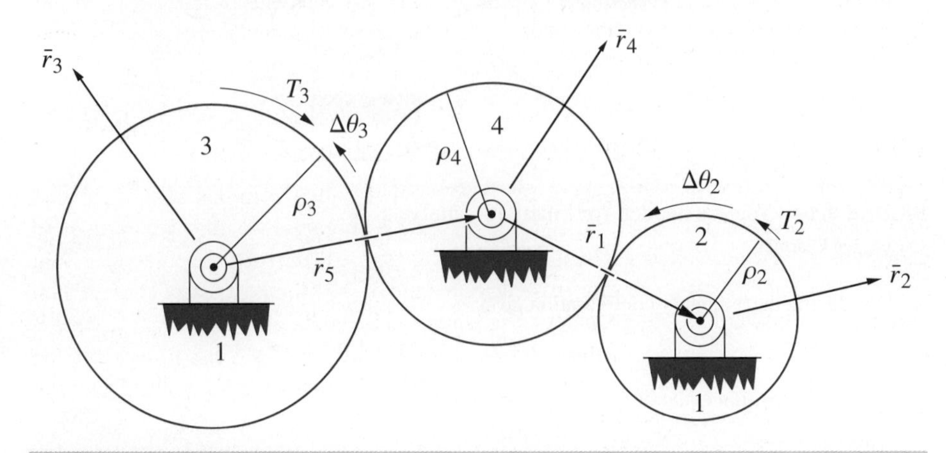

FIGURE 3.15 Simple geartrain with an idler gear

Solution

There are two rolling contact equations. From Equation (3.1),

$$\rho_3\Delta\theta_{3/5} = -\rho_4\Delta\theta_{4/5} \text{ and } \rho_4\Delta\theta_{4/1} = -\rho_2\Delta\theta_{2/1},$$

and since $\Delta\theta_5 = \Delta\theta_1 = 0$,

$$\rho_3\Delta\theta_3 = -\rho_4\Delta\theta_4 \text{ and } \rho_4\Delta\theta_4 = -\rho_2\Delta\theta_2. \tag{3.27}$$

Introducing the assembly configuration and writing Equations (3.27) in homogeneous form give the geartrain's position equations:

$$\rho_3(\theta_3 - \theta_{3_i}) + \rho_4(\theta_4 - \theta_{4_i}) = 0$$
$$\rho_4(\theta_4 - \theta_{4_i}) + \rho_2(\theta_2 - \theta_{2_i}) = 0.$$

In the position equations there are

scalar knowns: $\rho_3, \rho_2, \rho_4, \theta_{2_i}, \theta_{3_i}, \theta_{4_i}$

and

scalar unknowns: $\theta_2, \theta_3, \theta_4$.

Two position equations in three unknowns mean there is one degree of freedom. Eliminate $\Delta\theta_4$ between the two equations in Equation (3.27) to get

$$\rho_3\Delta\theta_3 = \rho_2\Delta\theta_2. \tag{3.28}$$

From Equation (3.28), the gear ratio for the system with the idler gear is given by

$$R = \frac{\Delta\theta_{in}}{\Delta\theta_{out}} = \frac{\Delta\theta_2}{\Delta\theta_3} = \frac{\rho_3}{\rho_2} = \frac{N_3}{N_2}. \tag{3.29}$$

Comparing Equations (3.26) and (3.29) shows that introducing gear 4 between gears 2 and 3 only changes the sign of the gear ratio, which is $R = (N_3/N_2)$. Since $R > 0$, $\Delta\theta_2$ and $\Delta\theta_3$ are in the same direction. According to Equation (3.19), in this geartrain $n_t < 0$, so T_2 and T_3 are in opposite directions. Gear 4 is known as an idler gear. The radius of its pitch circle and its number of teeth have no effect on the kinematics. The only consequence of 4 is the direction reversal in both the output rotation and torque. Figure 3.15 shows the directions of the input and output rotations and torques.

3.5.2 A Two-Stage Simple Geartrain with Compound Gears

A compound gear is several gears connected together with a common hub or spindle. The common hub causes the gears to rotate together. Figure 3.16 shows a compound gear consisting of two gears. Using a compound gear in place of the idler gear 4 in Figure 3.15 affects the gear ratio. In such geartrains it is difficult to draw the skeleton diagram by looking onto the plane of motion, as has been done up until now. Instead it is easier to take a side view of the geartrain as in Figure 3.17, where you now see the profiles of the gears. This is a view along the plane of motion—i.e., in this figure the plane of motion is perpendicular to the page.

Figure 3.17 shows two representations of the same two stage simple geartrain. On the left the gears are fixed to their shafts, and their shafts rotate relative to the ground. In this case, since the shafts are spinning they are known as spindles. On the right the gears rotate relative to their shafts while the shafts are fixed to ground. The figure shows the shafts passing through the hubs of the gears (the dashed lines), implying that relative rotation can occur. In this case, since the shafts do not spin they are known as axles.

The idler gear 4 in Figure 3.15 has been replaced by the compound gear 45 in Figure 3.17. 4 meshes with 2, and 5 meshes with 3, while 4 and 5 rotate together. It is easiest to think of 4 and 5 as a single gear 45, keeping in mind the meshings. In discussion of geartrains, you can assume that the number of teeth on each gear is known.

FIGURE 3.16 A compound gear

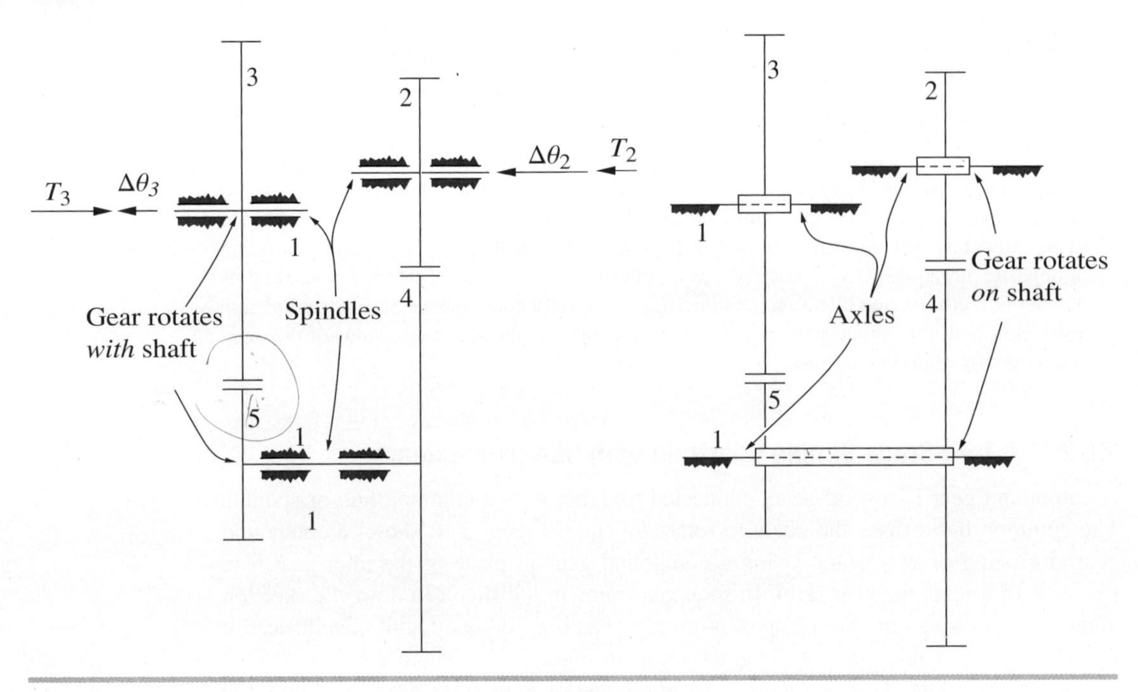

FIGURE 3.17 A two stage simple geartrain with a compound gear

▶ **EXAMPLE 3.5**

Derive the position equations, and deduce the number of degrees of freedom for the simple geartrain with two stages of reduction shown in Figure 3.17. Determine the gear ratio if 2 is the input and 3 is the output.

Solution

There are two rolling contact equations:

$$\rho_2 \Delta\theta_{2/1} = -\rho_4 \Delta\theta_{45/1} \quad \text{and} \quad \rho_3 \Delta\theta_{3/1} = -\rho_5 \Delta\theta_{45/1}.$$

Body 1 is fixed, so these two equations simplify to

$$\rho_2 \Delta\theta_2 = -\rho_4 \Delta\theta_{45} \quad \text{and} \quad \rho_3 \Delta\theta_3 = -\rho_5 \Delta\theta_{45}. \tag{3.30}$$

Incorporating the assembly configuration and writing these equations in homogeneous form give the two position equations,

$$\rho_2(\theta_2 - \theta_{2_i}) + \rho_4(\theta_{45} - \theta_{45_i}) = 0 \tag{3.31}$$

$$\rho_3(\theta_3 - \theta_{3_i}) + \rho_5(\theta_{45} - \theta_{45_i}) = 0. \tag{3.32}$$

Equations (3.31) and (3.32) are the position equations. They contain

scalar knowns: $\rho_3, \rho_2, \rho_4, \rho_5, \theta_{2_i}, \theta_{3_i}, \theta_{45_i}$

and

scalar unknowns: $\theta_2, \theta_3, \theta_{45}$.

Two position equations in three unknowns mean there is one degree of freedom. To find the gear ratio, eliminate $\Delta\theta_{45}$ from Equations (3.30):

$$R = \frac{\rho_3\rho_4}{\rho_5\rho_2} = \frac{N_3N_4}{N_5N_2}. \tag{3.33}$$

Since $R > 0$ $\Delta\theta_2$ and $\Delta\theta_3$ are in the same direction. According to Equation (3.19) $n_t < 0$ so T_2 and T_3 are in opposite directions. These are indicated in Figure 3.17 with arrows. The arrow represents the direction your thumb is pointing in a right hand rule operation, where your fingers are curling with the rotation or torque.

A Manual Automotive Transmission

A manual automotive transmission is similar to the two stage simple geartrain with a compound gear we just examined. In the manual transmission, the compound gear consists of more than two gears. In some cases there are up to six or seven gears on the compound gear of a manual transmission.

Before discussing the manual transmission of an automobile, we need to understand the process of shifting gears in a manual transmission. To do so, it is necessary to understand what a spline connection (shown in Figure 3.18) is and what it does. A spline connection consists of an externally splined shaft fitted with an internally splined bushing. In many cases the splines have involute profiles, like gear teeth. In a spline connection the bushing rotates at the same speed as the shaft, but the bushing has the ability to slide axially along the shaft.

Figure 3.19 shows a skeleton diagram of a manual four speed automotive transmission. The shaft on the right is the input rotation coming from the flywheel of the engine. A flywheel is a circular plate of rotating mass that serves to store energy and smooth the output torque of the engine, as well as giving up energy smoothly while the automobile's clutch is being engaged.

The automobile's clutch is engaged by lifting your foot and allowing the clutch pedal to rise. This causes the input shaft to the transmission and gear 4 (which is attached to the input shaft) to rotate with the flywheel of the engine. 4 in turn drives 5, which is part of a large compound gear 56789, known as the layshaft or countershaft. Gears 3, 2, and 1 spin on the output shaft at different speeds. 3 is driven by 6, 2 is driven by 7, and 1 and is driven by 9 (disregard 8 and 10 for the moment).

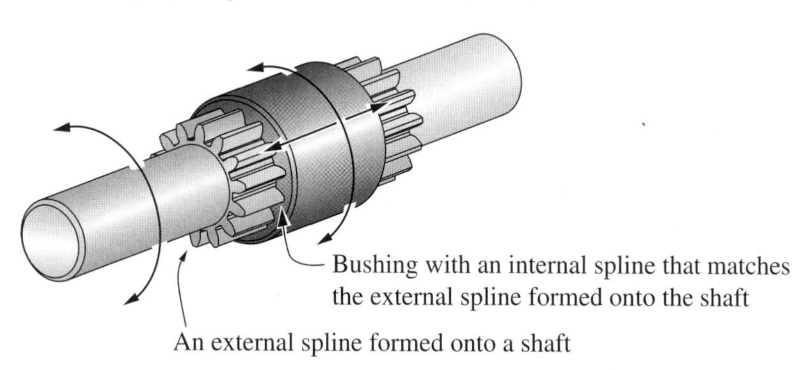

FIGURE 3.18 A splined shaft with fitting splined bushing

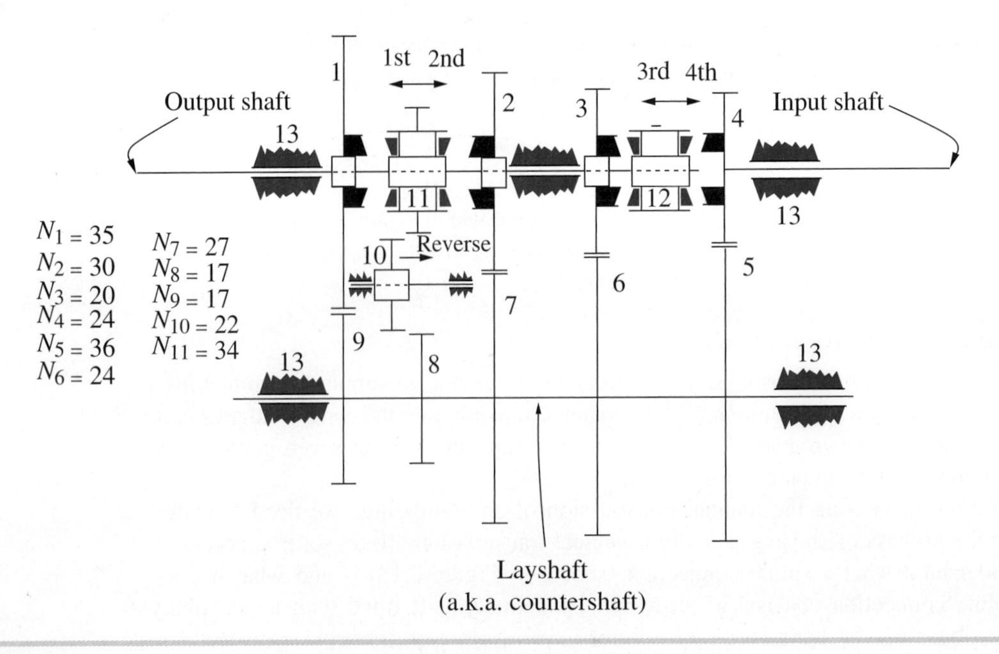

FIGURE 3.19 A four speed manual automotive transmission

One common misconception about manual transmissions is that when gears are shifted (such as when going from first gear to second gear), various gears within a manual transmission are being brought in and out of mesh. With the exception of reverse, this is not the case. The forward speed gears of a manual transmission (gears 1, 2, 3, and 4 in Figure 3.19) are all being driven by the input shaft when the clutch is engaged.

To shift gears, these manual transmissions use elements called shift selectors. These are elements 11 and 12 in Figure 3.19. Shift selectors are similar to splined bushings, except that they have one set of permanently meshed splines and another set of intermittently meshed splines.

The permanently meshed splines of the shift selectors are a spline connection between the shift selector and the output shaft. Consequently, the shift selectors rotate continuously with the output shaft but can move on it axially. These are indicated as the permanently meshed splines in Figure 3.20.

Figure 3.20 also indicates where the intermittently meshed splines of the shift selectors are brought in and out of mesh with external splines on the hubs of gears 1, 2, 3, and 4. Spline connections that are brought in and out of mesh are called "dog clutches." When the dog clutch meshes with the shift selector and gear hub, it causes the shift selector and output shaft to rotate at the same speed as the gear to which the dog clutch connection has been made. This determines the gear that the transmission is in. If more than one dog clutch is engaged, the output shaft will be forced to try to rotate at two different speeds, and the transmission will lock up. There are provisions in the design of the shifting mechanism connecting the stick shift to the axially sliding shift selectors that prevent this.

Before engaging one of the dog clutches, the clutch pedal is depressed, disengaging the automobile's clutch and releasing the connection between the input shaft and flywheel of the engine. Gears 4, 56789, 1, 2, and 3 are now all free spinning and are

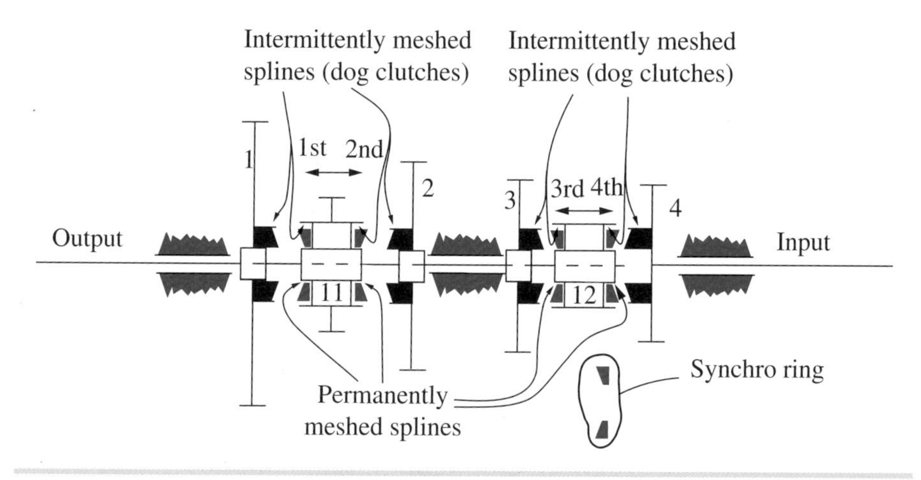

FIGURE 3.20 Close-up of the input shaft from the four speed manual automotive transmission

© Cengage Learning.

slowing down since they are in a viscous oil bath. If the automobile is standing still, the output shaft and shift selectors are not rotating. If the vehicle is rolling down the road, the output shaft and shift selectors are spinning, being back-driven by the drive wheels of the automobile. In either case, the rotational speeds of the dog clutch splines on the shift selector and the dog clutch splines on the hub of the gear to which the dog clutch connection is being made are asynchronous. Forcing them to mesh while they rotate at different speeds creates the grinding noise you hear when you shift improperly. The sound is not a "grinding of gears," as people like to say; rather, it is a grinding of the splines in the dog clutch.

For the splines in a dog clutch to connect or mesh without grinding, their rotational speeds must match (be synchronized). For this purpose, pairs of synchro rings are used, as shown conceptually in the close-up of the output shaft in Figure 3.20. This is referred to as synchromesh, since the synchro rings will *synchronize* (match) the rotational speeds of the shift selector and gear and allow the splines of the dog clutch to *mesh* (engage).

The synchro rings are the blue-colored elements on the shift selectors. They are typically made of a brass alloy. They are small conical clutches. They are splined on the inside and are permanently connected to a spline on the outside of the shift selector. This is shown as the meshed splines in Figure 3.21.

When the shift selector is moved into a gear's hub, first a conical region of the gear's hub contacts a corresponding conical region of the synchro ring and pushes the synchro ring back into the shift selector. When the synchro ring is pushed against the shift selector, it presses against the gear. The synchro ring acts as a conical clutch, and through friction, it very quickly causes the gear to spin up to the same speed as the shift selector (and the output shaft splined to it), synchronizing their rotational speeds. The dog clutch can now mesh without grinding, connecting the gear to the shift selector and output shaft. By engaging the automobile's clutch, the engine can now transfer power to the drive wheels. Synchro rings wear out over time, but in modern transmissions with good lubricants, their life is adequate.

In reverse, gear 10 is brought into mesh in between 8 and a gear that is a part of shift selector 11. 10 is an idler gear, and it reverses the direction of rotation of the output shaft.

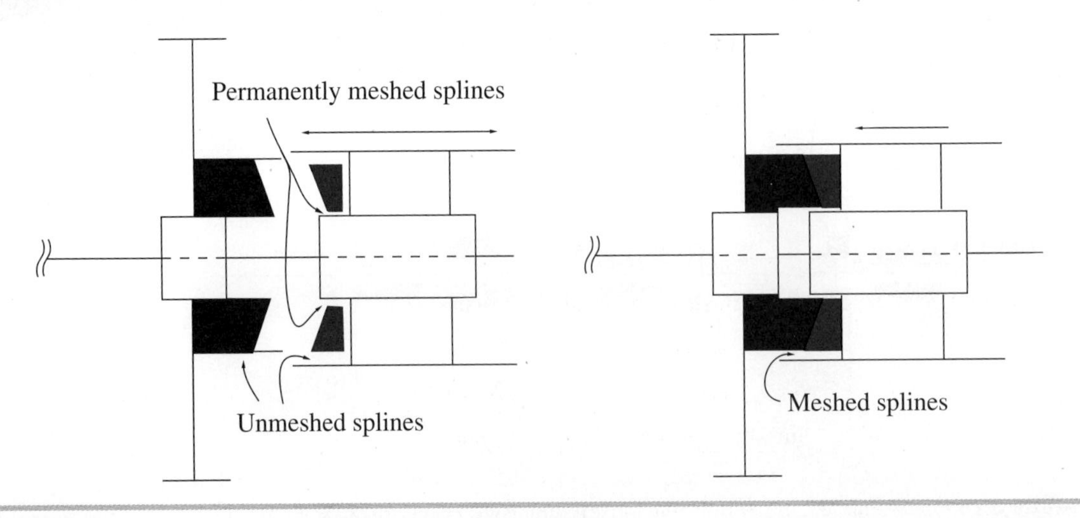

FIGURE 3.21 Action of the synchro rings in synchromesh

Reverse gear generally has no synchromesh, and if you shift into reverse too soon after depressing the clutch pedal (that is, if you have not allowed enough time for the layshaft to stop rotating), you will hear a grinding noise, and in this case it actually *is* a grinding of the gears.

▶ EXAMPLE 3.6

Calculate the gear ratio for 1st gear in the manual transmission shown in Figure 3.19. If the input to the transmission is power of 230 hp at a speed of 2000 rpm, calculate the transmission's output torque.

Solution

When in 1st gear, the shift selector 11 is pushed to the left and engaged to gear 1. This means the output shaft is rotating with gear 1, so $\Delta\theta_{out} = \Delta\theta_1$. The input shaft drives gear 4, so $\Delta\theta_{in} = \Delta\theta_4$. The transmission has two stages of gear reduction. In the first stage, 4 rolls externally on 5 (of compound gear 56789) and the arm is the ground, 13. From Equation (3.1),

$$\rho_4\Delta\theta_{4/13} = -\rho_5\Delta\theta_{56789/13}$$
$$\rho_4(\Delta\theta_4 - \Delta\theta_{13}) = -\rho_5(\Delta\theta_{56789} - \Delta\theta_{13}).$$

Since 13 is fixed, $\Delta\theta_{13} = 0$, so the above equation reduces to

$$\rho_4\Delta\theta_4 = -\rho_5\Delta\theta_{56789}. \tag{3.34}$$

In the second stage, 1 rolls externally on 9 (of compound gear 56789) and the arm is the ground, 13. From Equation (3.1),

$$\rho_1\Delta\theta_{1/13} = -\rho_9\Delta\theta_{56789/13}$$
$$\rho_1(\Delta\theta_1 - \Delta\theta_{13}) = -\rho_9(\Delta\theta_{56789} - \Delta\theta_{13}),$$

and, accounting for 13 being fixed, the above reduces to

$$\rho_1\Delta\theta_1 = -\rho_9\Delta\theta_{56789}. \tag{3.35}$$

Eliminating $\Delta\theta_{56789}$ from Equations (3.34) and (3.35) gives R from Equation (3.18):

$$R = \frac{\Delta\theta_{in}}{\Delta\theta_{out}} = \frac{\Delta\theta_4}{\Delta\theta_1} = \frac{\rho_5}{\rho_4}\frac{\rho_1}{\rho_9}.$$

Replacing the ratios of the pitch circles with ratios of the tooth numbers given in Figure 3.19 gives

$$R = \frac{N_5}{N_4}\frac{N_1}{N_9} = \frac{36}{24}\frac{35}{17} = 3.088. \tag{3.36}$$

To answer the second part of the question, use Equations (3.19) and (3.21):

$$T_{out} = n_t T_{in} = -RT_{in}. \tag{3.37}$$

Power is the product of torque and speed, so

$$T_{in} = \frac{P_{in}}{\omega_{in}} = \frac{230 \text{ hp}}{2,000 \frac{\text{rev}}{\text{min}}} = 0.115 \frac{\text{hp} \cdot \text{min}}{\text{rev}},$$

which is not a standard unit of torque. We must make unit conversions. Recall that

$$1 \text{ hp} = \frac{550 \text{ ft} \cdot \text{lb}_\text{f}}{\text{s}},$$

so

$$T_{in} = 0.115 \frac{\text{hp} \cdot \text{min}}{\text{rev}} \frac{550 \text{ ft} \cdot \text{lb}_\text{f}}{\text{hp} \cdot \text{s}} \frac{60 \text{ s}}{\text{min}} \frac{\text{rev}}{2\pi \text{ rad}} = 604 \text{ ft} \cdot \text{lb}_\text{f},$$

and from Equations (3.36) and (3.37),

$$T_{out} = -3.088\,(604 \text{ ft} \cdot \text{lb}_\text{f}) = -1,865 \text{ ft} \cdot \text{lb}_\text{f}.$$

The negative sign means the output torque is in the opposite direction from the input torque.

3.5.3 Planetary Geartrains

Planetary geartrains are geartrains in which the arm associated with the rolling contacts rotates, causing the centers of some of the gears in the geartrain to move on circular orbits. Nevertheless, like simple geartrains, their kinematic equations consist of only rolling contact equations and they have constant gear ratios.

Figure 3.22 shows a skeleton diagram of a planetary geartrain. On the right side of the figure is a view onto the plane of motion, and on the left side is a view along the plane of motion. The system has elements 1, 2, 3, 4, and 5. Elements 2, 3, 4, and 5 are referred to as the sun gear, planet gear, ring gear, and arm, respectively. Suppose that a vector $\bar{r}_i$ is attached to element i and that the number of teeth on each gear is known. The vector loop is trivial. There are two rolling contacts;

3 rolls externally on 2, and the arm is 5, so from Equations (3.1),

$$\rho_3 \Delta\theta_{3/5} = -\rho_2 \Delta\theta_{2/5}, \tag{3.38}$$

and 3 rolls internally on 4, and the arm is 5, so from Equation (3.5),

$$\rho_3 \Delta\theta_{3/5} = \rho_4 \Delta\theta_{4/5}. \tag{3.39}$$

In these equations,

$$\Delta\theta_{3/5} = \Delta\theta_3 - \Delta\theta_5, \quad \Delta\theta_{2/5} = \Delta\theta_2 - \Delta\theta_5, \quad \text{and} \quad \Delta\theta_{4/5} = \Delta\theta_4 - \Delta\theta_5. \tag{3.40}$$

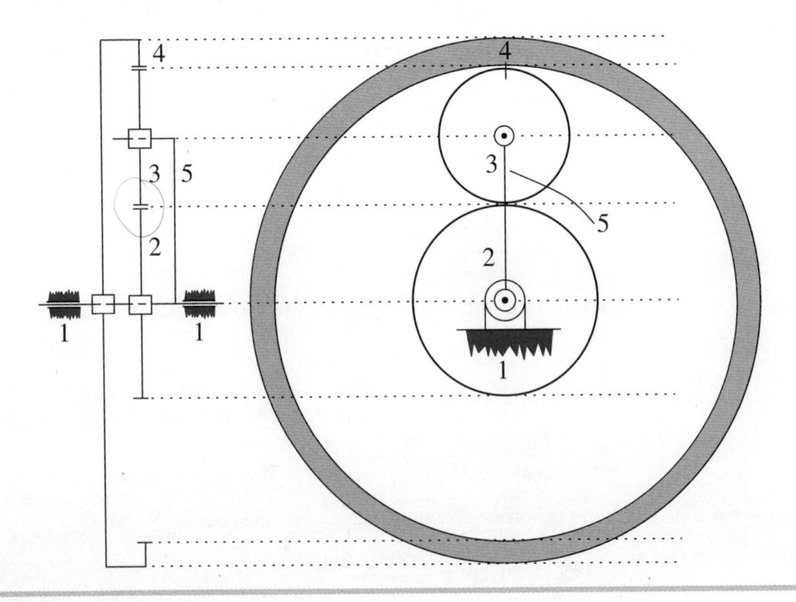

FIGURE 3.22 Planetary geartrain

Substituting Equation (3.40) into Equations (3.38) and (3.39) and expanding to include the assembly configuration give

$$\rho_3[(\theta_3 - \theta_{3_i}) - (\theta_5 - \theta_{5_i})] + \rho_2[(\theta_2 - \theta_{2_i}) - (\theta_5 - \theta_{5_i})] = 0 \tag{3.41}$$

$$\rho_3[(\theta_3 - \theta_{3_i}) - (\theta_5 - \theta_{5_i})] - \rho_4[(\theta_4 - \theta_{4_i}) - (\theta_5 - \theta_{5_i})] = 0. \tag{3.42}$$

Equations (3.41) and (3.42) are a system of two position equations with

scalar knowns: $\rho_2,\ \rho_3,\ \rho_4,\ \theta_{2_i},\ \theta_{3_i},\ \theta_{4_i},\ \theta_{5_i}$

and

scalar unknowns: $\theta_2,\ \theta_3,\ \theta_4,\ \theta_5$.

The planetary geartrain is a mechanism with two position equations in four unknowns, making it a two-degrees-of-freedom system. For a constrained motion, the system needs two inputs. In many cases one of the elements of a planetary geartrain is fixed. This adds a third constraint and reduces the mechanism to a one-degree-of-freedom system. Equations (3.41) and (3.42) show that the kinematic relations are in terms of *differences* in rotations, so planetary geartrains are often called differentials.

▶ EXAMPLE 3.7

Calculate the gear ratio for the planetary geartrain in Figure 3.22 if the ring gear 4 is fixed, the arm 5 is the output, and the sun gear 2 is the input.

Solution

Our goal is to solve Equations (3.38) and (3.39) for

$$R = \frac{\Delta\theta_{in}}{\Delta\theta_{out}} = \frac{\Delta\theta_2}{\Delta\theta_5}.$$

To do so, we eliminate $\Delta\theta_{3/5}$ from Equations (3.38) and (3.39) to get

$$-\rho_2\Delta\theta_{2/5} = \rho_4\Delta\theta_{4/5},$$

substitute Equation (3.40) into this, and set $\Delta\theta_4 = 0$ (the ring gear is fixed) to obtain

$$\rho_2\Delta\theta_2 = (\rho_2 + \rho_4)\Delta\theta_5,$$

from which we find R:

$$R = \frac{\Delta\theta_2}{\Delta\theta_5} = 1 + \frac{\rho_4}{\rho_2} = 1 + \frac{N_4}{N_2},$$

where N_2 and N_4 are the numbers of teeth on gears 2 and 4. The gear ratio here is always positive, so the input and output rotate in the same direction, and the output torque has the opposite direction from the input torque.

From the figure we see that $\rho_4 \sim 2.5\rho_2$, so $R \sim 3.5$ when 4 is fixed. If you compare this to the size of a simple geartrain that gives the same ratio, you will see that the planetary geartrain is about three times smaller than the simple geartrain. In general, planetary geartrains are more compact than simple geartrains, which is frequently the justification for using them.

Figure 3.23 shows a side view of a variation of the planetary geartrain that has a compound gear for the planet gear and uses two sun gears. As we will see in the next example, depending on the number of teeth on the rolling elements, the gear ratio can be either positive or negative. This interesting property is exploited in the design of automotive automatic transmissions.

▶ EXAMPLE 3.8

The elements of the geartrain shown in Figure 3.23 are the gears 2, 34 (the compound planet gear), and 5 and the arm 6. Element 5 is fixed. The input is 6 ($\Delta\theta_{in} = \Delta\theta_6$) and the output is 2 ($\Delta\theta_{out} = \Delta\theta_2$). Determine the gear ratio of the transmission.

Solution

The geartrain has two rolling contacts. In one rolling contact, 2 rolls externally with 3 (of compound gear 34) and the arm is 6. From Equation (3.1),

$$\rho_2\Delta\theta_{2/6} = -\rho_3\Delta\theta_{34/6}. \tag{3.43}$$

In the second rolling contact, 5 rolls externally with 4 (of compound gear 34) and the arm is 6. From Equation (3.1),

$$\rho_5\Delta\theta_{5/6} = -\rho_4\Delta\theta_{34/6}. \tag{3.44}$$

Eliminate $\Delta\theta_{34/6}$ from these equations to get

$$\frac{\rho_2}{\rho_3}\Delta\theta_{2/6} = \frac{\rho_5}{\rho_4}\Delta\theta_{5/6}.$$

Expand this equation as before ($\Delta\theta_{2/6} = \Delta\theta_2 - \Delta\theta_6$, etc.) and set $\Delta\theta_5 = 0$ (5 is fixed) to get

$$\frac{\rho_2}{\rho_3}(\Delta\theta_2 - \Delta\theta_6) = -\frac{\rho_5}{\rho_4}\Delta\theta_6,$$

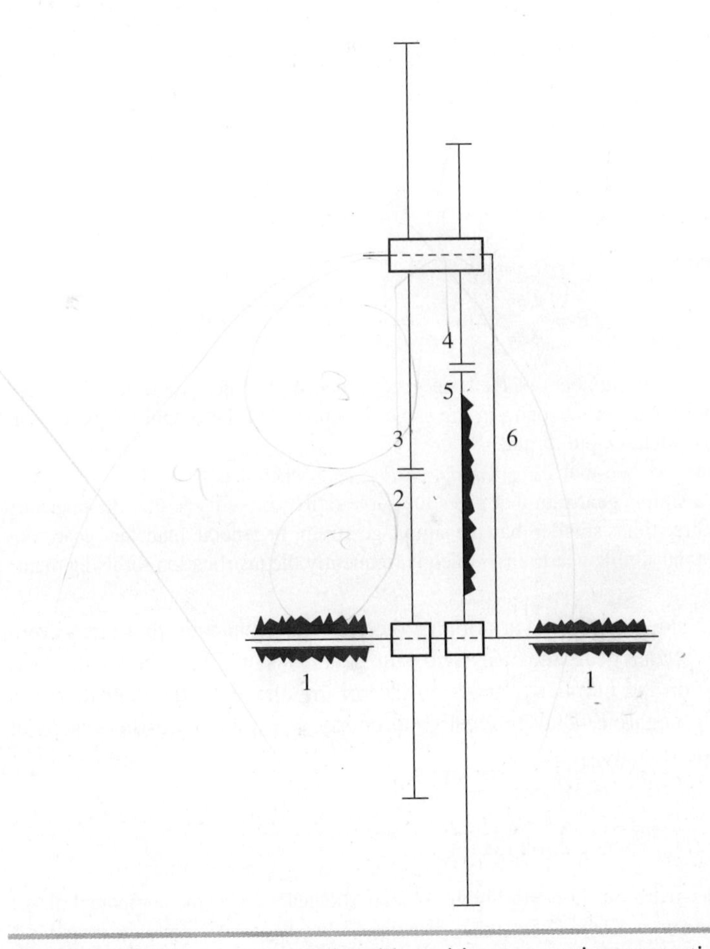

FIGURE 3.23 Planetary geartrain with positive or negative gear ratio

and solve this for R:

$$R = \frac{\Delta\theta_6}{\Delta\theta_2} = \frac{1}{1 - \frac{\rho_3}{\rho_2}\frac{\rho_5}{\rho_4}} = \frac{1}{1 - \frac{N_3}{N_2}\frac{N_5}{N_4}}, \tag{3.45}$$

where N_i represents the number of teeth on gear i. Equation (3.45) shows that this system can have an either positive or negative gear ratio.

$$\text{If} \quad \frac{N_3}{N_2}\frac{N_5}{N_4} > 1 \rightarrow \quad R < 1 \rightarrow \quad \text{negative gear ratio.}$$

$$\text{If} \quad \frac{N_3}{N_2}\frac{N_5}{N_4} < 1 \rightarrow \quad R > 1 \rightarrow \quad \text{positive gear ratio.}$$

This observation may have been the motivation for the first semiautomatic transmission that was developed for the Ford Model T. Visual inspection of the geartrain in Figure 3.23 shows that $\frac{\rho_3\rho_5}{\rho_2\rho_4} > 1$ so $R < 0$ in this case.

The Model T Semi-automatic Transmission

Figure 3.24 shows a skeleton diagram of the Ford Model T semi-automatic transmission. It was the forerunner of today's fully automatic transmission. The system has elements 345, 6, 7, 8, and 9. Element 345 is a compound planet gear. The shafts of gears 6, 7, and 8 are concentric, as indicated by the dashed lines. Element 345 rolls externally on 6, 7, and 8, and the arm is 9 in each case. From Equation (3.1), the rolling contact equation for each case is

$$\rho_3 \Delta\theta_{345/9} = -\rho_6 \Delta\theta_{6/9}$$
$$\rho_4 \Delta\theta_{345/9} = -\rho_7 \Delta\theta_{7/9}.$$
$$\rho_5 \Delta\theta_{345/9} = -\rho_8 \Delta\theta_{8/9}$$

Expanding these rolling contact equations by substituting

$$\Delta\theta_{345/9} = (\Delta\theta_{345} - \Delta\theta_9)$$
$$\Delta\theta_{6/9} = (\Delta\theta_6 - \Delta\theta_9)$$
$$\Delta\theta_{7/9} = (\Delta\theta_7 - \Delta\theta_9)$$
$$\Delta\theta_{8/9} = (\Delta\theta_8 - \Delta\theta_9)$$

yields

$$\rho_3(\Delta\theta_{345} - \Delta\theta_9) = -\rho_6(\Delta\theta_6 - \Delta\theta_9) \tag{3.46}$$
$$\rho_4(\Delta\theta_{345} - \Delta\theta_9) = -\rho_7(\Delta\theta_7 - \Delta\theta_9) \tag{3.47}$$
$$\rho_5(\Delta\theta_{345} - \Delta\theta_9) = -\rho_8(\Delta\theta_8 - \Delta\theta_9). \tag{3.48}$$

We realize that we can incorporate the assembly position (such as $\Delta\theta_6 = \theta_6 - \theta_{6_i}$, etc.) to end up with a system of three equations with

scalar knowns: $\rho_3,\ \rho_4,\ \rho_5,\ \rho_6,\ \rho_7,\ \rho_8,\ \theta_{345_i},\ \theta_{6_i},\ \theta_{7_i}\ \theta_{8_i},\ \theta_{9_i}$

and

scalar unknowns: $\theta_{345},\ \theta_6,\ \theta_7,\ \theta_8,\ \theta_9.$

With five unknowns, this is a two-degree of freedom geartrain.

Figure 3.24 shows that gears 7 and 8 each have an attached rotating drum. These drums are fitted with band brakes that can be used to stop the rotation of the drum and its associated gear. Figure 3.24 also shows gear 6 is attached to a plate clutch that when engaged would connect the input shaft directly to 6. The Model T transmission uses two foot pedals to engage the band brakes B_1 and B_2 and the plate clutch. Table 3.1 enumerates the state of the band brakes and plate clutch for each operating mode of the transmission. The condition for each operating mode adds a fourth equation to our system of three equations and reduces the geartrain to one degree of freedom. The input is the rotation of 9 and the output is the rotation of 6, so $\Delta\theta_{in} = \Delta\theta_9$ and $\Delta\theta_{out} = \Delta\theta_6$ in each operating mode.

▶ EXAMPLE 3.9

Calculate the gear ratio for the low gear operating mode of the Ford Model T semiautomatic transmission.

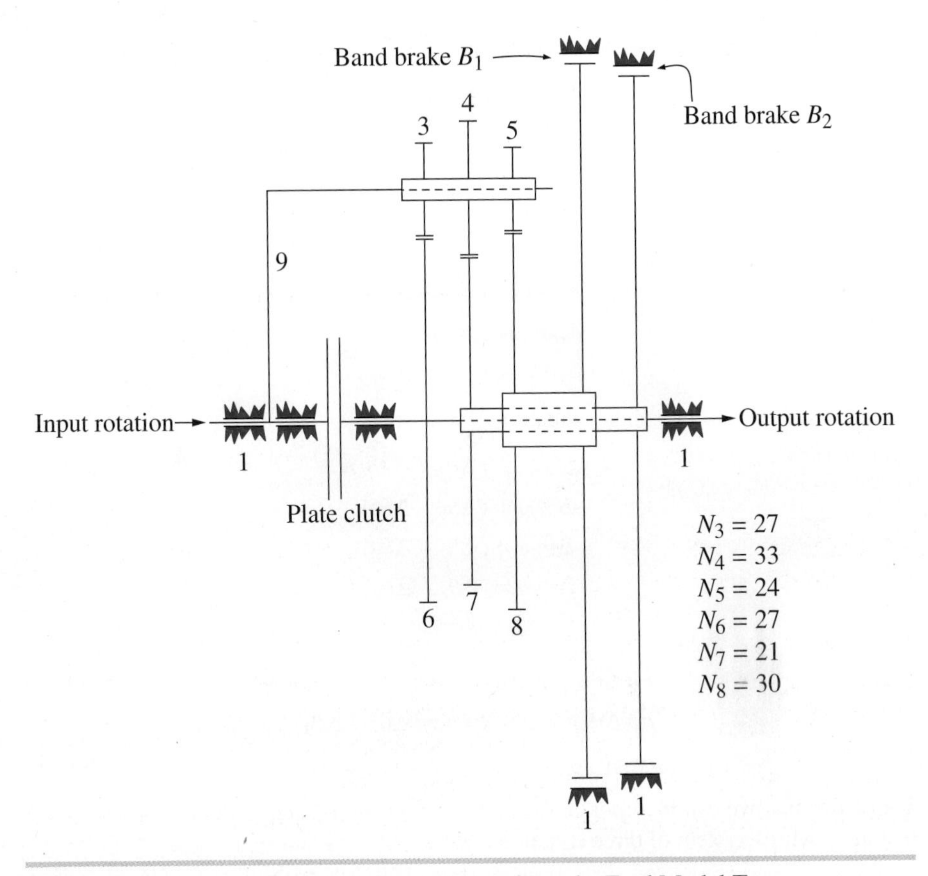

FIGURE 3.24 The semiautomatic transmission from the Ford Model T

Operating Mode	Plate Clutch	B_1	B_2
Low	Disengaged	Disengaged	Engaged
High	Engaged	Disengaged	Disengaged
Reverse	Disengaged	Engaged	Disengaged
Neutral	Disengaged	Disengaged	Disengaged

TABLE 3.1 Condition for Each Operating Mode of the Model

Solution

According to the table, band brake B_2 is engaged in the low gear condition. This causes the drum that rotates with 7 to be fixed, so

$$\Delta\theta_7 = 0. \tag{3.49}$$

This condition and the three conditions Equations (3.46), (3.47), and (3.48) enable us to solve for $R = \frac{\Delta\theta_9}{\Delta\theta_6}$. We substitute Equation (3.49) into Equation (3.47),

$$\rho_4(\Delta\theta_{345} - \Delta\theta_9) = \rho_7\Delta\theta_9, \tag{3.50}$$

and eliminate $(\Delta\theta_{345} - \Delta\theta_9)$ from Equations (3.46) and (3.50) to get

$$-\frac{\rho_6}{\rho_3}(\Delta\theta_6 - \Delta\theta_9) = \frac{\rho_7}{\rho_4}\Delta\theta_9,$$

which can be solved for R:

$$R = \frac{\Delta\theta_9}{\Delta\theta_6} = \frac{1}{1 - \frac{\rho_3}{\rho_6}\frac{\rho_7}{\rho_4}} = \frac{1}{1 - \frac{N_3}{N_6}\frac{N_7}{N_4}},$$

where N_i represents the number of teeth on gear i. Using the tooth numbers given in Figure 3.24 gives

$$R = \frac{1}{1 - \frac{27}{27}\frac{21}{33}} = 2.75,$$

so in the low gear operating mode, the output torque is 2.75 times greater than the input torque and, according to Equation (3.20), is in the opposite direction from the input torque.

Problem 3.21 asks you to compute the gear ratio in the high and the reverse speed conditions. The gear ratio in reverse should be negative.

A Two-Speed Automatic Automotive Transmission

Figure 3.25 shows a two-speed automatic transmission such as was found in automobiles in the 1960s. Today's automatic transmissions may have up to six speeds. The system has the elements 3, 4, 58, 6, 7, and 9. It is a planetary geartrain, and 9 is the arm for all rolling contacts. There are four rolling contacts.

4 rolls externally on 3 with 9 as the arm, so according to Equation (3.1),

$$\Delta\theta_{4/9} = -\frac{\rho_3}{\rho_4}\Delta\theta_{3/9}. \tag{3.51}$$

4 rolls internally on 58 with 9 as the arm, so according to Equation (3.5),

$$\Delta\theta_{4/9} = \frac{\rho_5}{\rho_4}\Delta\theta_{58/9}. \tag{3.52}$$

7 rolls externally on 58 with 9 as the arm, so according to Equation (3.1),

$$\Delta\theta_{7/9} = -\frac{\rho_8}{\rho_7}\Delta\theta_{58/9}. \tag{3.53}$$

7 rolls internally on 6 with 9 as the arm, so according to Equation (3.5),

$$\Delta\theta_{7/9} = \frac{\rho_6}{\rho_7}\Delta\theta_{6/9}. \tag{3.54}$$

We could expand the relative rotations (e.g., $\Delta\theta_{58/9} = \Delta\theta_{58} - \Delta\theta_9$) and introduce the assembly configuration (e.g., $\Delta\theta_{58} = (\theta_{58} - \theta_{58_i})$ to develop a system of four equations with

scalar knowns: $\rho_3,\ \rho_4,\ \rho_{58},\ \rho_6,\ \rho_7,\ \rho_8,\ \theta_{3_i},\ \theta_{4_i},\ \theta_{58_i}\ \theta_{6_i},\ \theta_{7_i},\ \theta_{9_i}$

and

scalar unknowns: $\theta_3,\ \theta_4,\ \theta_{58},\ \theta_6,\ \theta_7,\ \theta_9.$

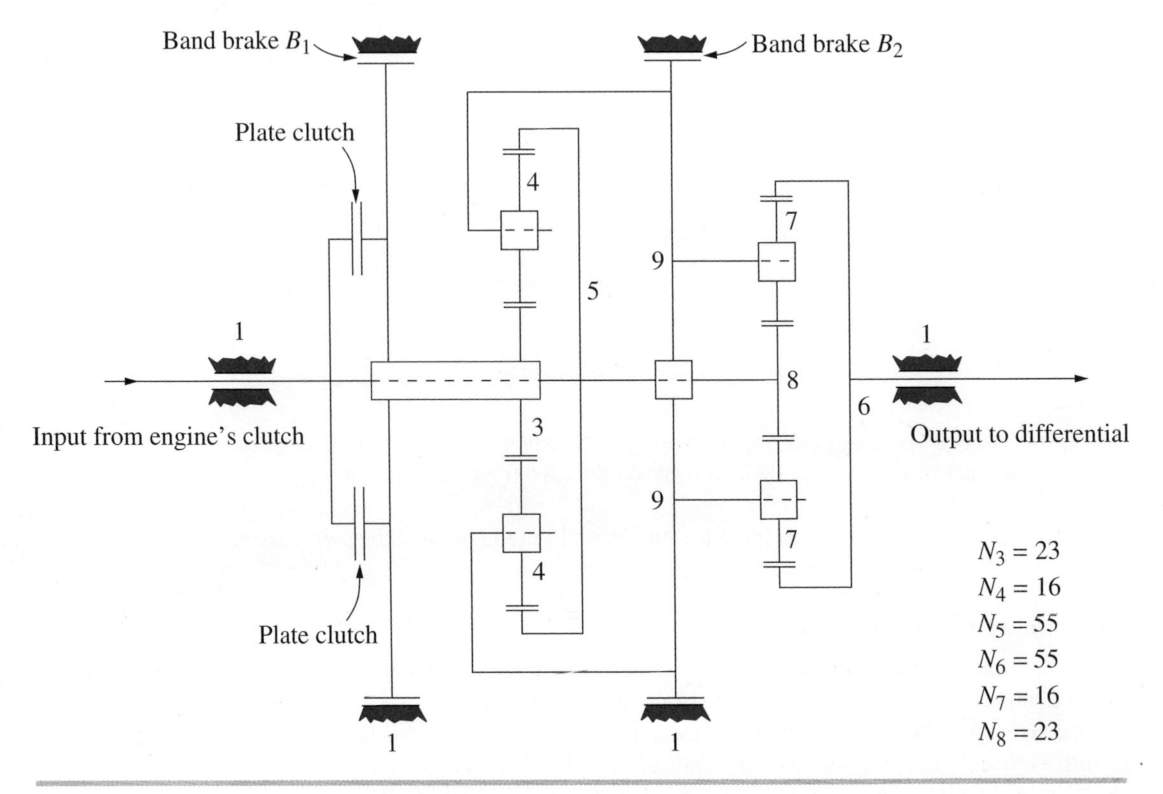

FIGURE 3.25 A two-speed automatic automotive transmission

Speed Condition	Plate Clutch	B_1	B_2
Low	Disengaged	Engaged	Disengaged
High	Engaged	Disengaged	Disengaged
Reverse	Disengaged	Disengaged	Engaged
Neutral	Disengaged	Disengaged	Disengaged

TABLE 3.2 Speed Conditions

With six unknowns in four position equations, this is a two-degrees-of-freedom geartrain. The input is the rotation of 58 ($\Delta\theta_{in} = \Delta\theta_{58}$) and the output is the rotation of 6 ($\Delta\theta_{out} = \Delta\theta_6$).

The gear ratio can be determined for each operating mode when the corresponding condition in Table 3.2 is enforced.

In the Model T transmission, the band brakes were actuated by foot pedals. In the automatic transmission, the band brakes are actuated by hydraulic cylinders that are activated by a control system that senses engine rpm and vehicle speed in order to decide when to transit between operating modes. The bands are typically flexible metal bands that need intermittent tightening to take up wear.

▶ EXAMPLE 3.10

Calculate the gear ratio for the low speed condition of the automatic transmission in Figure 3.25.

Solution

From Table 3.2, band brake B_1 is engaged in low gear. This causes

$$\Delta\theta_3 = 0. \tag{3.55}$$

Equation (3.55), along with Equations (3.51) through (3.54), is a system of five position equations in six unknowns, reducing this to a one-degree-of-freedom mechanism. To find the gear ratio in the low speed condition, eliminate $\Delta\theta_{4/9}$ from Equations (3.51) and (3.52):

$$\Delta\theta_{58/9} = -\frac{\rho_3}{\rho_5}\Delta\theta_{3/9}. \tag{3.56}$$

Expanding the relative rotations yields

$$(\Delta\theta_{58} - \Delta\theta_9) = -\frac{\rho_3}{\rho_5}(\Delta\theta_3 - \Delta\theta_9),$$

and setting $\Delta\theta_3$ to zero as in Equation (3.55) gives

$$(\Delta\theta_{58} - \Delta\theta_9) = \frac{\rho_3}{\rho_5}\Delta\theta_9. \tag{3.57}$$

Now eliminate $\Delta\theta_{7/9}$ from Equations (3.53) and (3.54),

$$\Delta\theta_{6/9} = -\frac{\rho_8}{\rho_6}\Delta\theta_{58/9}, \tag{3.58}$$

substitute Equation (3.56) into Equation (3.58),

$$\Delta\theta_{6/9} = \frac{\rho_8}{\rho_6}\frac{\rho_3}{\rho_5}\Delta\theta_{3/9},$$

and expand the relative rotations and again enforce Equation (3.55):

$$(\Delta\theta_6 - \Delta\theta_9) = -\frac{\rho_8}{\rho_6}\frac{\rho_3}{\rho_5}\Delta\theta_9. \tag{3.59}$$

Solve Equation (3.57) for $\Delta\theta_{58}$,

$$\Delta\theta_{58} = \left(1 + \frac{\rho_3}{\rho_5}\right)\Delta\theta_9, \tag{3.60}$$

and solve Equation (3.59) for $\Delta\theta_6$,

$$\Delta\theta_6 = \left(1 - \frac{\rho_8}{\rho_6}\frac{\rho_3}{\rho_5}\right)\Delta\theta_9. \tag{3.61}$$

From Equations (3.59) and (3.61), solve for R:

$$R = \frac{\Delta\theta_{58}}{\Delta\theta_6} = \frac{\cancel{\Delta\theta_9}\,(1 + \frac{\rho_3}{\rho_5})}{\cancel{\Delta\theta_9}\,(1 - \frac{\rho_8\rho_3}{\rho_6\rho_5})}.$$

Replacing the ratios of pitch circle radii with ratios of tooth numbers and substituting for the tooth numbers from the information given in Figure 3.25 give our result:

$$R = \frac{(1 + \frac{N_3}{N_5})}{(1 - \frac{N_8 N_3}{N_6 N_5})} = 1.72.$$

Problem 3.22 asks you to the compute the gear ratio in the high and the reverse speed conditions of this transmission.

An Automotive Differential

Figure 3.26 shows the differential found in a rear-wheel-drive automobile or a truck. In these vehicles, the engine is mounted so that its crankshaft is directed longitudinally—that is, in the front-to-back direction of the automobile. The view looking down from the top of the vehicle showing the internal gearing of the differential is shown in Figure 3.26. The input 2 is the driveshaft from the transmission. The differential has two outputs: the axle of the left wheel, 5, and the axle of the right wheel, 6. Its purpose is to allow the input $\Delta\theta_2$ to transmit power to both outputs $\Delta\theta_5$ and $\Delta\theta_6$ and to accommodate a relationship between $\Delta\theta_5$ and $\Delta\theta_6$, which is determined by the vehicle's wheel track w and the turning radius R_{turn}. This relationship can be derived referencing Figure 3.27 and enforcing a no-slip condition that requires the left wheel to roll on the circumference of a circle of radius $(R_{turn} - \frac{w}{2})$, while the right wheel rolls on the circumference of a circle

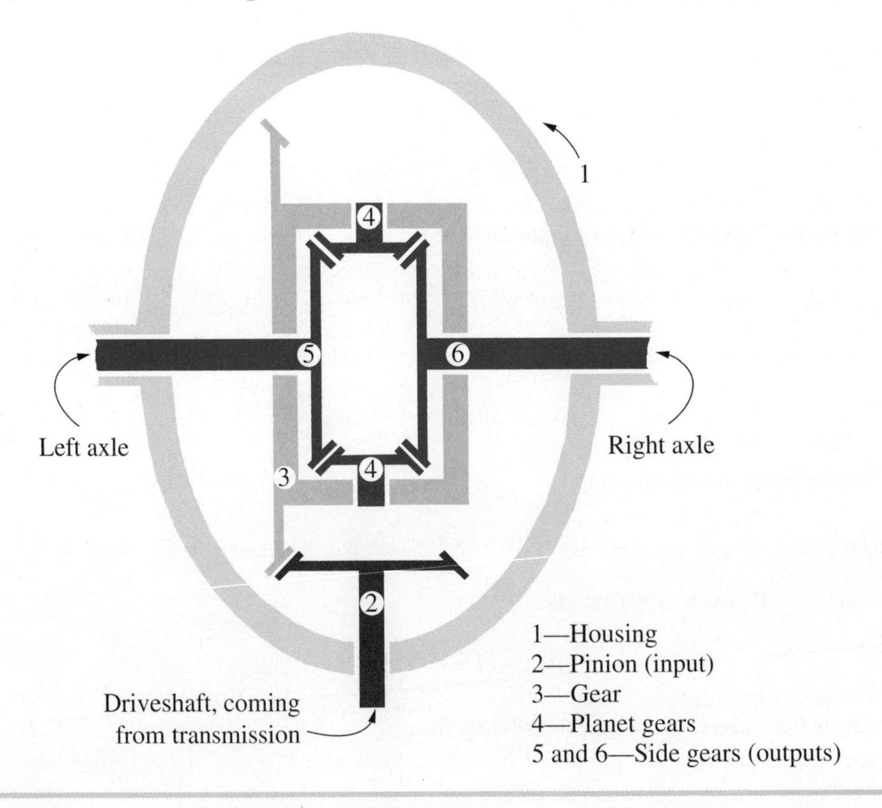

FIGURE 3.26 A typical automotive differential

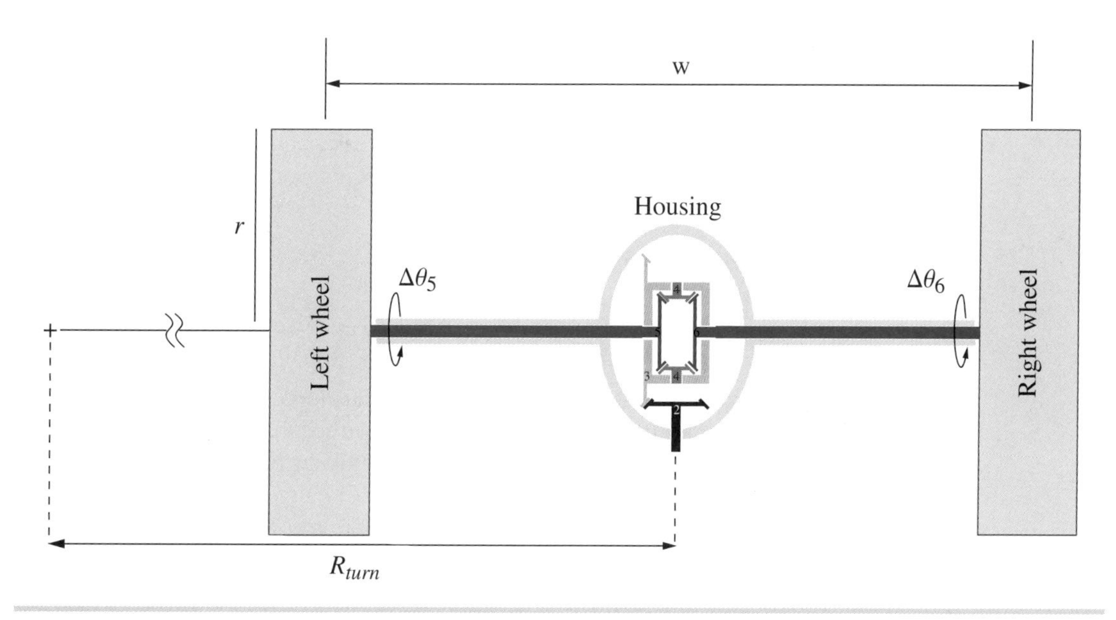

FIGURE 3.27 The typical differential accommodating a turn

of radius ($R_{turn} + \frac{w}{2}$). Assuming the tires have the same radius r,

$$\frac{\Delta\theta_5}{\Delta\theta_6} = \frac{R_{turn} - \frac{w}{2}}{R_{turn} + \frac{w}{2}}. \tag{3.62}$$

The rolling contact equation can be applied to the differential. The difficulty is that this is not a planar mechanism. Referring to Figure 3.26, the system's elements are 1, 2, 3, 4, 5, and 6. There are two geartrains here. Elements 1, 2, and 3 form a simple geartrain, where 2 rolls on 3 and the arm is 1, which is the fixed body. The rolling contact equation is

$$\rho_2 \Delta\theta_{2/1} = \rho_3 \Delta\theta_{3/1},$$

and since 1 is fixed, $\Delta\theta_1 = 0$ and the above reduces to

$$\rho_2 \Delta\theta_2 = \rho_3 \Delta\theta_3. \tag{3.63}$$

Elements 4, 5, and 6 roll on one another, and the arm is 3, forming a planetary geartrain:

$$\rho_4 \Delta\theta_{4/3} = (?) \rho_5 \Delta\theta_{5/3}, \tag{3.64}$$

$$\rho_4 \Delta\theta_{4/3} = (?) \rho_6 \Delta\theta_{6/3}. \tag{3.65}$$

The question mark is there because the sign is unknown due to the fact that the motion is not planar, and the sign conventions developed for planar rolling contacts do not apply here. But if $\rho_4 \Delta\theta_{4/3}$ is eliminated from Equations (3.64) and (3.65), the sign can be determined because we see that relative to 3, 5 and 6 counter rotate. Thus

$$\rho_5 \Delta\theta_{5/3} = -\rho_6 \Delta\theta_{6/3}.$$

In a differential, $\rho_5 = \rho_6 = \rho$ and this relation simplifies to

$$\Delta\theta_{5/3} = -\Delta\theta_{6/3}. \tag{3.66}$$

Equations (3.62), (3.63), and (3.66) are a system of three equations in the four unknowns, $\Delta\theta_2$, $\Delta\theta_3$, $\Delta\theta_5$, and $\Delta\theta_6$. Given $\Delta\theta_2$ (the output rotation of the transmission), R_{turn}, and w, we can compute $\Delta\theta_5$ and $\Delta\theta_6$.

The Gear Ratio of a Differential

When a vehicle is driving on a straight road, $R_{turn} = \infty$ and Equation (3.62) shows that

$$\Delta\theta_5 = \Delta\theta_6 \quad \longrightarrow \quad \Delta\theta_{5/3} = \Delta\theta_{6/3}. \tag{3.67}$$

Substituting the right side of Equation (3.67) into Equation (3.66) shows that $\Delta\theta_{5/3} = \Delta\theta_{6/3} = 0$. Then, from Equations (3.64) and (3.65), we see that $\Delta\theta_{4/3} = 0$. This means that when driving a straight line there is no rolling taking place between 4 and 5 or between 4 and 6, and $\Delta\theta_3 = \Delta\theta_4 = \Delta\theta_5 = \Delta\theta_6$, so elements 3, 4, 5, and 6 rotate together as a solid unit that looks like a solid axle that rotates an angle $\Delta\theta_3$ while being driven by 2.

In this case, the differential is said to have a gear ratio

$$R = \frac{\Delta\theta_2}{\Delta\theta_3},$$

and from Equation (3.63),

$$R = \frac{\Delta\theta_2}{\Delta\theta_3} = \frac{\rho_3}{\rho_2} = \frac{N_3}{N_2}.$$

In normal vehicles, $R \sim 2.7{:}1$. In drag racing vehicles, $R \sim 4.3{:}1$, which gives greater torque at the rear wheels and better acceleration.

Transaxles

In vehicles in which the driven wheels are on the same end of the vehicle as the engine (such as front-wheel-drive vehicles), the engines are mounted transversely, so that their crankshafts are directed in the side-to-side direction of the vehicle, and the transmission (manual or automatic) is built as one unit with the differential. There is no driveshaft. These are referred to as “transaxles.”

Comment

Mechanical engineering is predated only by civil engineering. Mechanical engineering predates the Industrial Revolution, but it was during the Industrial Revolution that mechanical engineering saw its most rapid development. As a result of its long history, mechanical engineering has its own culture and language.

The word *spindle* is used to describe a rotating shaft that has an object attached to it that rotates with it, such as the shafts on the left hand side of Figure 3.17. Generally, power is being transmitted through the shaft to the rotating object. The word *axle* is used to describe a fixed shaft upon which an object is rotating, such as the shafts on the right hand side of Figure 3.17. Generally, the rotating object is passive and not doing any work. (See the discussion of Figure 3.17.)

However, if you choose a career in the automotive industry, you will see just the opposite. In that industry, an axle rotates and a spindle is fixed. This is because the automotive industry is a legacy of the horse-drawn wagon and carriage industry. Automobiles were thought of as horseless carriages. In the wagon/carriage industry, the passive wheels were connected to axles, and since that industry evolved into the automotive industry, the word *axle* was also carried along, incorrectly. So do not be surprised when the automotive tech refers to the front axles of your front-wheel-drive automobile (rather than front spindles) as well when the same tech refers to the passive rear wheels as rotating on spindles (rather than axles).

3.5.4 Hybrid Geartrains

A hybrid geartrain is a combination of a simple and a planetary geartrain. In a hybrid geartrain some elements will have a fixed arm and are thus a part of a simple geartrain, while some elements will have a rotating arm and are thus a part of a planetary geartrain. Invariably, there will be one element that belongs to both and is the connection that forms the hybrid geartrain.

▶ EXAMPLE 3.11

In the geartrain in Figure 3.28, the input is the shaft on the left and the output is the shaft on the right. Assume the number of teeth on each gear is known. Derive an expression for the gear ratio in terms of tooth numbers.

Solution

The geartrain consists of the elements 1, 27, 34, 56, 89, and 10. The input rotation is $\Delta\theta_{27}$ and the output rotation is $\Delta\theta_{10}$, so

$$R = \frac{\Delta\theta_{27}}{\Delta\theta_{10}}.$$

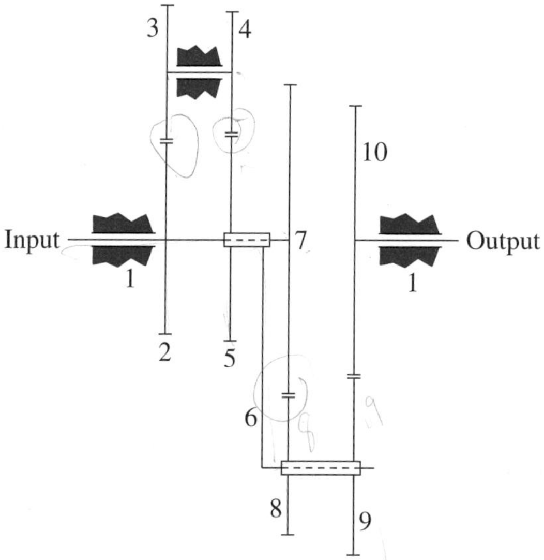

FIGURE 3.28 A hybrid geartrain

There are four rolling contacts, and all of them are external rolling contacts, so Equation (3.1) is applied to each case. The first two rolling contacts are a part of a simple geartrain because the arm for the rolling contacts is fixed.

34 rolls externally with 27, and 1 is the arm:

$$\rho_4 \Delta\theta_{34/1} = -\rho_2 \Delta\theta_{27/1}, \quad \Delta\theta_1 = 0 \longrightarrow \rho_4 \Delta\theta_{34} = -\rho_5 \Delta\theta_{27}. \tag{3.68}$$

34 rolls externally with 56, and 1 is the arm:

$$\rho_3 \Delta\theta_{34/1} = -\rho_5 \Delta\theta_{56/1}, \quad \Delta\theta_1 = 0 \longrightarrow \rho_3 \Delta\theta_{34} = -\rho_2 \Delta\theta_{56}. \tag{3.69}$$

The remaining two rolling contacts are part of a planetary geartrain because the arm is rotating.

27 rolls externally with 89, and 56 is the arm:

$$\rho_7 \Delta\theta_{27/56} = -\rho_8 \Delta\theta_{89/56}. \tag{3.70}$$

89 rolls externally with 10, and 56 is the arm:

$$\rho_9 \Delta\theta_{89/56} = -\rho_{10} \Delta\theta_{10/56}. \tag{3.71}$$

Notice that element 56 belongs to both the simple geartrain and the planetary geartrain that form this hybrid geartrain. If we expand the relative rotations and include the assembly configuration, we have a system of four equations with

scalar knowns: $\rho_2, \rho_3, \rho_4, \rho_5, \rho_7, \rho_8, \rho_9, \rho_{10}, \theta_{27_i}, \theta_{34_i}, \theta_{56_i}, \theta_{89_i}, \theta_{10_i}$

and

scalar unknowns: $\theta_{27}, \theta_{34}, \theta_{56}, \theta_{89}, \theta_{10}$.

Four position equations in five unknowns mean there is one degree of freedom. To solve for R, eliminate $\Delta\theta_{34}$ from Equations (3.68) and (3.69),

$$\Delta\theta_{56} = \frac{\rho_4}{\rho_5}\frac{\rho_2}{\rho_3}\Delta\theta_{27}, \tag{3.72}$$

eliminate $\Delta\theta_{89/56}$ from Equations (3.70) and (3.71),

$$\frac{\rho_{10}}{\rho_9}\Delta\theta_{10/56} = \frac{\rho_7}{\rho_8}\Delta\theta_{27/56},$$

and expand the relative rotations in this equation,

$$\frac{\rho_{10}}{\rho_9}\left(\Delta\theta_{10} - \Delta\theta_{56}\right) = \frac{\rho_7}{\rho_8}\left(\Delta\theta_{27} - \Delta\theta_{56}\right).$$

Substituting Equation (3.72) into the above equation gives

$$\frac{\rho_{10}}{\rho_9}\left(\Delta\theta_{10} - \frac{\rho_4}{\rho_5}\frac{\rho_2}{\rho_3}\Delta\theta_{27}\right) = \frac{\rho_7}{\rho_8}\left(\Delta\theta_{27} - \frac{\rho_4}{\rho_5}\frac{\rho_2}{\rho_3}\Delta\theta_{27}\right),$$

which can be solved for R,

$$R = \frac{\Delta\theta_{27}}{\Delta\theta_{10}} = \frac{\rho_{10}\rho_5\rho_3\rho_8}{\rho_7\rho_9\rho_5\rho_3 + \rho_{10}\rho_4\rho_2\rho_8 - \rho_7\rho_4\rho_2\rho_9},$$

or in terms of tooth numbers,

$$R = \frac{\Delta\theta_{27}}{\Delta\theta_{10}} = \frac{N_{10}N_5N_3N_8}{N_7N_9N_5N_3 + N_{10}N_4N_2N_8 - N_7N_4N_2N_9}.$$

3.6 PROBLEMS

In Problems 3.1–3.8:

1. Draw an appropriate vector loop.
2. Write out the vector loop equation(s), or VLE(s).
3. Write the X and Y components of the VLE(s) in their simplest form.
4. Write down all geometric constraints.
5. Write down all rolling contact equations.
6. Summarize the knowns and the unknowns.
7. From all the above, deduce the number of degrees of freedom in the system.

Problem 3.1

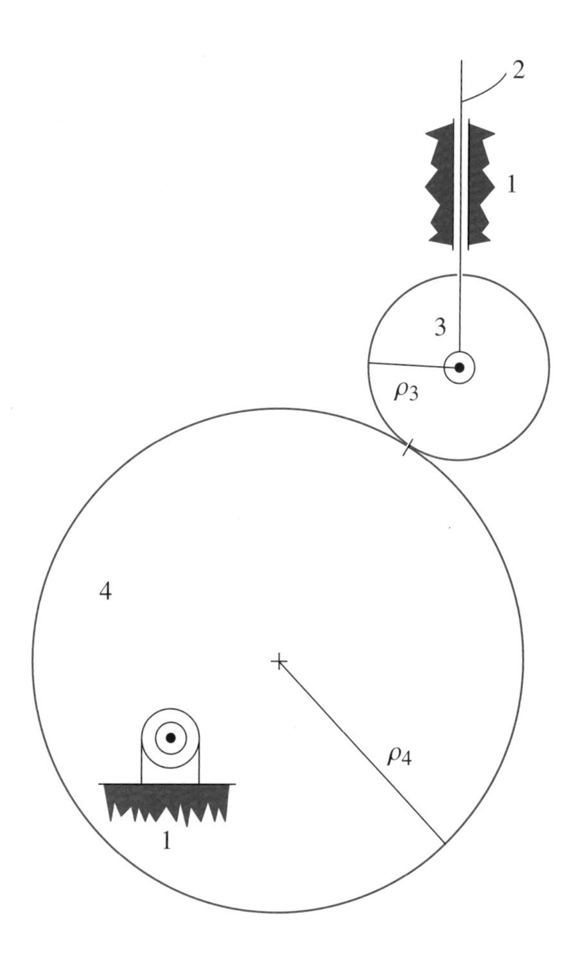

Problem 3.2

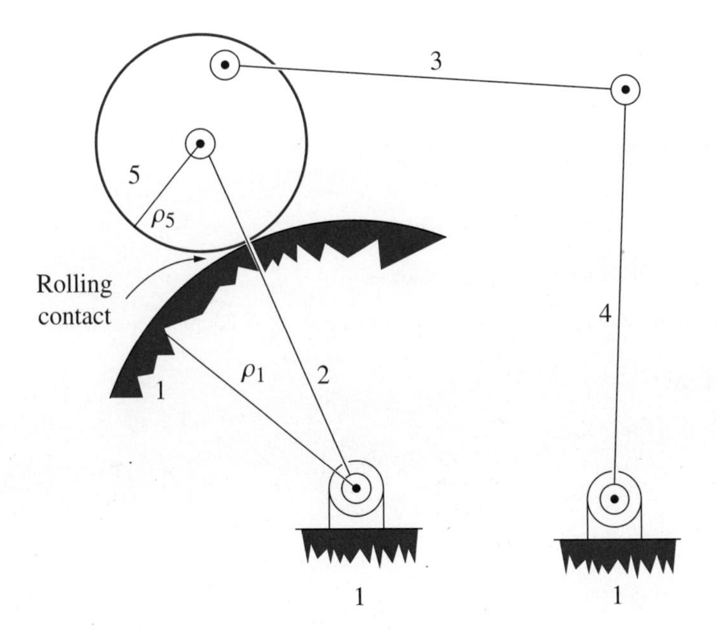

Problem 3.3

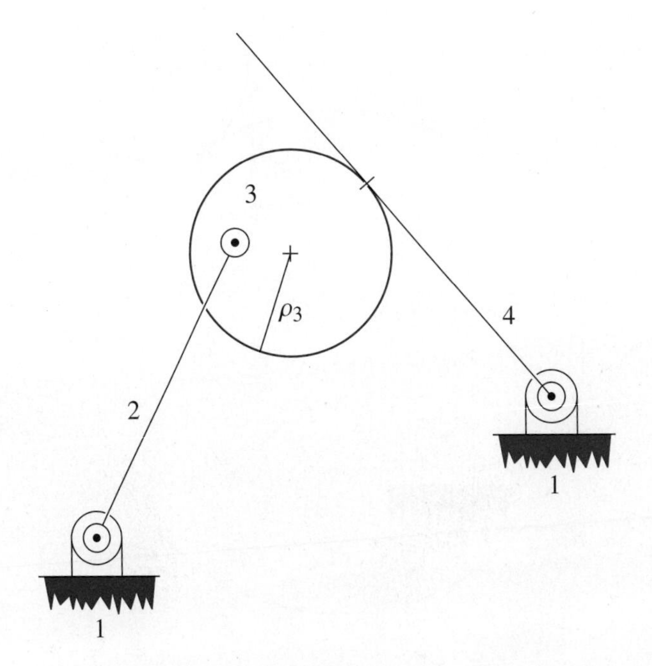

Problem 3.4

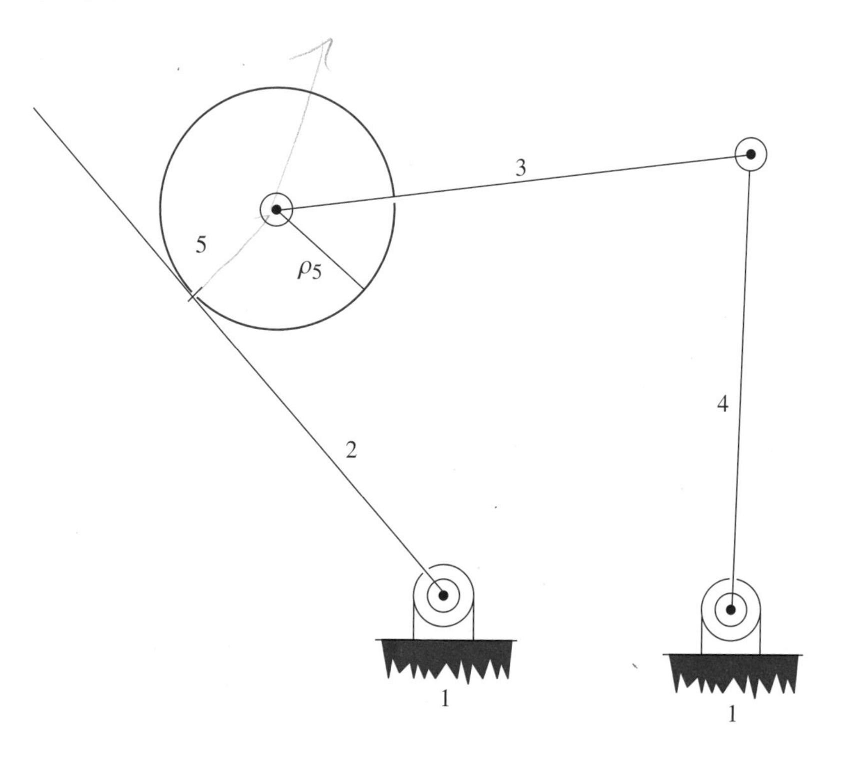

Problem 3.5

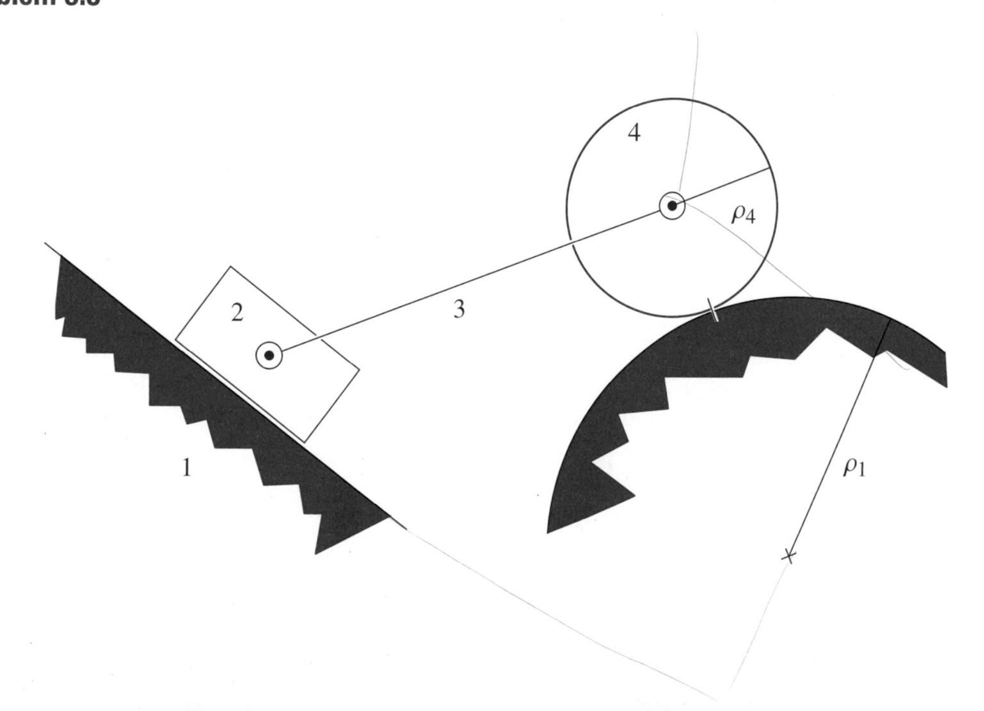

Problem 3.6

The mechanism below is a planar biped walking machine found in research labs developing humanoids. Its feet are circles whose centers are at the knees. The radii of the feet are the same, ρ. It has no ankles, and the hip is the pin joint at the top.

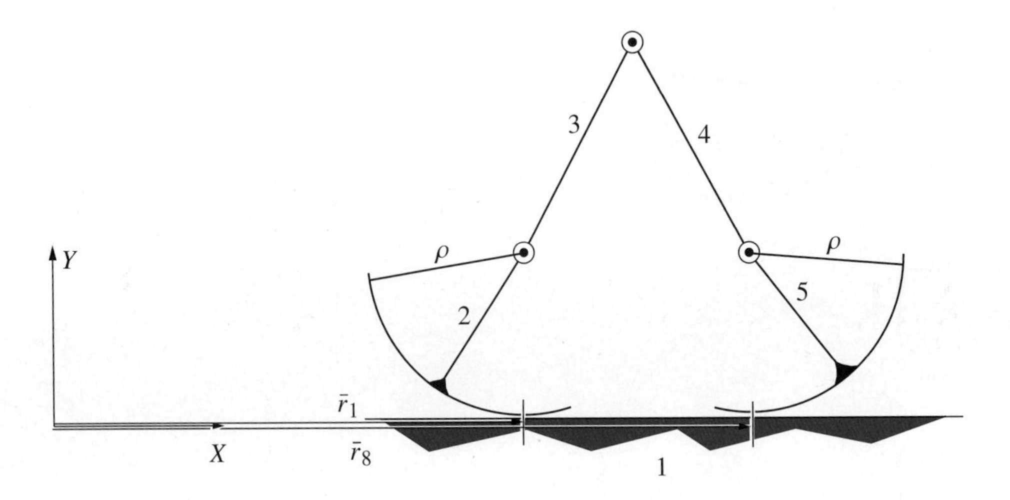

FIGURE 3.29

The mechanism is a bit unconventional because it has no fixed point to which we can attach a fixed X-Y coordinate system (see step 3 in Section 2.1). Take the coordinate system to be at the location shown, and use the vectors $\bar{r}_1$ and $\bar{r}_8$ to locate the point of contact between the ground and the left and right foot, respectively. Complete the vector loop, and write all equations necessary for a position analysis. Deduce the number of degrees of freedom in the system.

Problem 3.7

The mechanism below is the drive mechanism for the rear windshield wiper of an automobile, shown previously in Figure 1.29. A motor drives link 2, and the wiper blade oscillates with link 5.

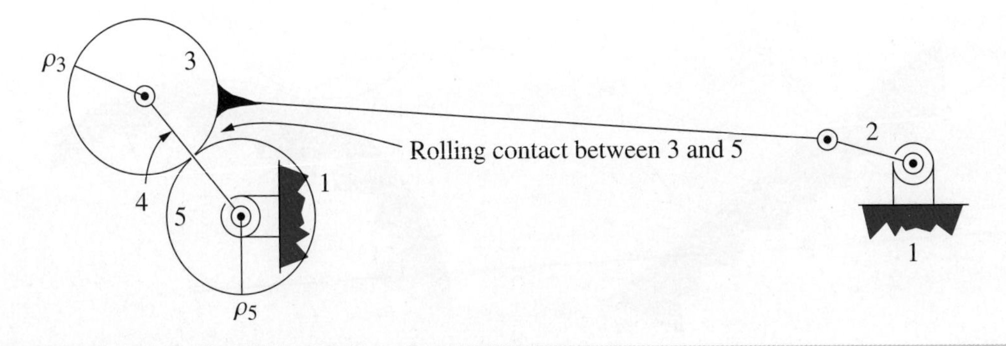

FIGURE 3.30 Drive mechanism for the rear windshield wiper of an automobile

Problem 3.8

The figure below shows a robot hand. Body 1 is the palm, and bodies 2 through 7 are the digits of the hand. Body 8 is a circular object being manipulated by being rolled among the digits 3, 5, and 7.

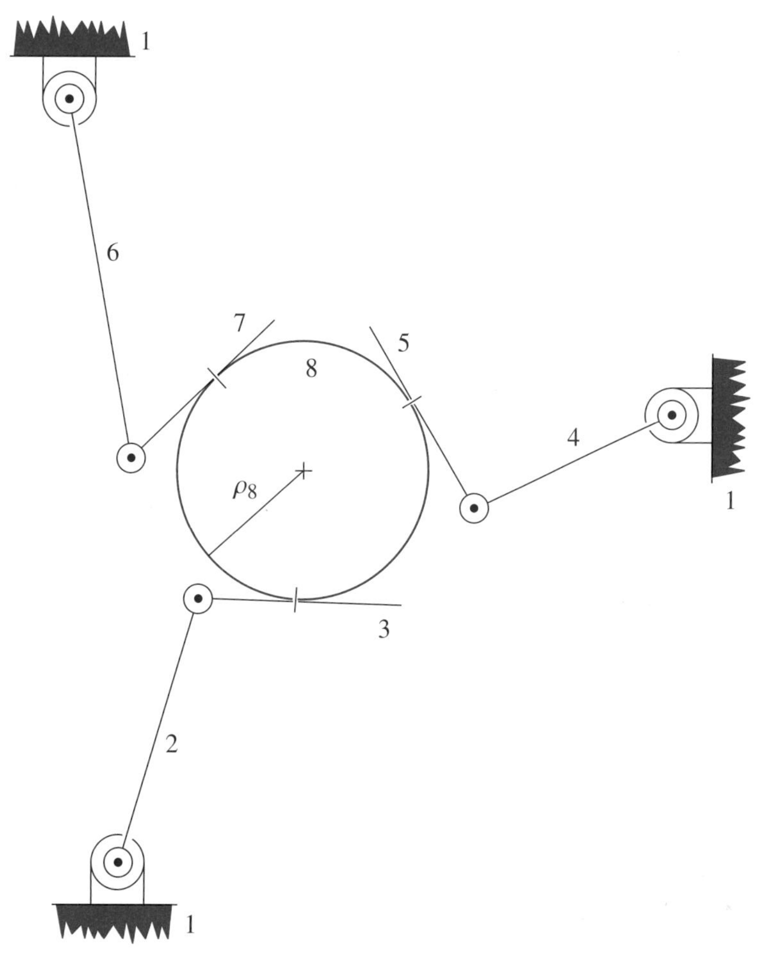

FIGURE 3.31 A robotic hand performing finger manipulation of an object

Problem 3.9

For the mechanism on the left, an appropriate vector loop is shown on the right (see the figures at the top of the following page).

The vector loop equation is

$$\bar{r}_2 + \bar{r}_5 - \bar{r}_4 - \bar{r}_1 = \bar{0}.$$

The X and Y components of this vector loop equation simplify to

$$r_2\cos\theta_2 + r_5\cos\theta_5 - r_4\cos\theta_4 - r_1 = 0 \tag{3.73}$$

$$r_2\sin\theta_2 + r_5\sin\theta_5 - r_4\sin\theta_4 = 0. \tag{3.74}$$

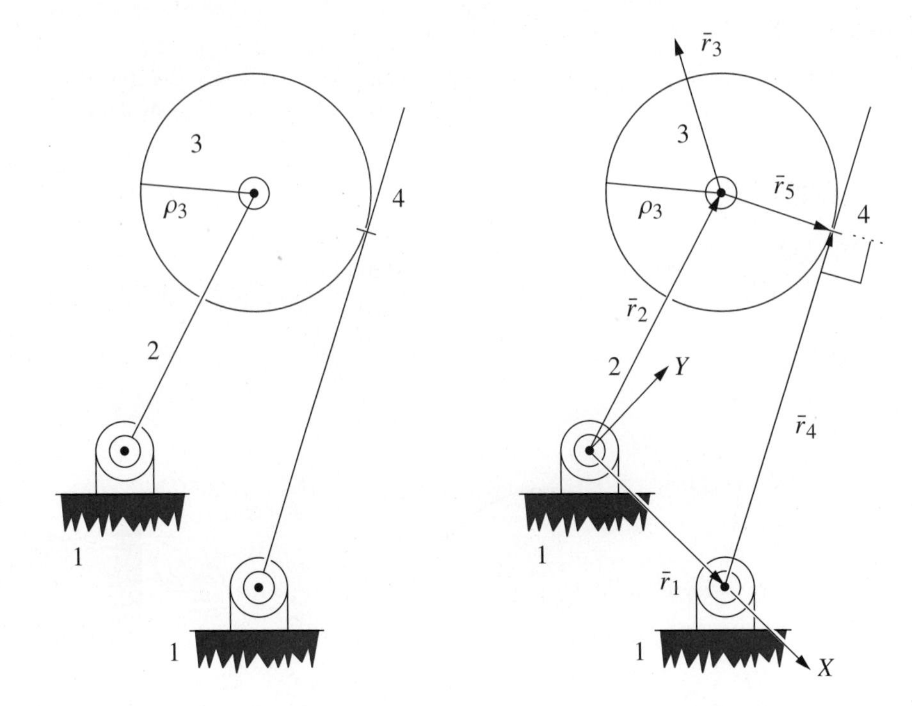

There is also the rolling contact equation,

$$\rho_3\left[(\theta_3-\theta_{3_i})-(\theta_5-\theta_{5_i})\right]+(r_4-r_{4_i})=0, \tag{3.75}$$

and the geometric constraint,

$$\theta_5+\frac{\pi}{2}-\theta_4=0, \tag{3.76}$$

Equations (3.73) through (3.76) contain

scalar knowns: $r_2, r_3, r_5, r_1, \theta_1 = 0°, \rho_3$

and

scalar unknowns: $\theta_2, \theta_3, \theta_4, r_4, \theta_5$.

Suppose that a value of θ_2 is given and the corresponding values of θ_3, θ_4, r_4, and θ_5 are to be computed. Define the vector of unknowns as

$$\bar{x}=\begin{bmatrix}\theta_3\\ \theta_4\\ r_4\\ \theta_5\end{bmatrix}.$$

1. Outline all equations necessary to implement Newton's Method.
2. Present a flowchart that shows the logic of Newton's Method, and in the flowchart, refer to the equations you have derived by their equation numbers.

Problem 3.10

Consider the mechanism shown below. An appropriate vector loop is shown on the right.

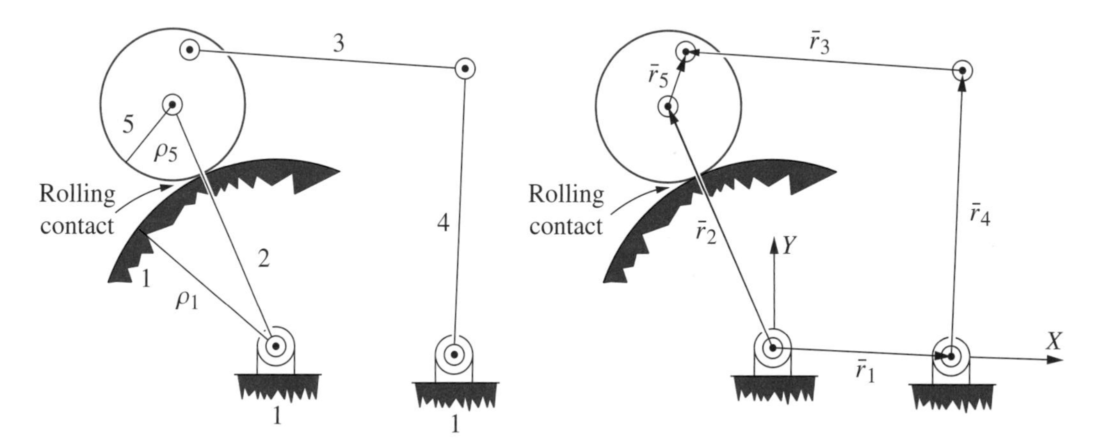

The scalar components of the vector loop equation $\bar{r}_1 + \bar{r}_4 + \bar{r}_3 - \bar{r}_5 - \bar{r}_2 = \bar{0}$ are

$$r_1 + r_4\cos\theta_4 + r_3\cos\theta_3 - r_5\cos\theta_5 - r_2\cos\theta_2 = 0$$
$$r_4\sin\theta_4 + r_3\sin\theta_3 - r_5\sin\theta_5 - r_2\sin\theta_2 = 0,$$

and the fully expanded rolling contact equation, $\rho_5(\Delta\theta_{5/2}) = -\rho_1(\Delta\theta_{1/2})$, is

$$\rho_5(\theta_5 - \theta_{5_i}) - (\rho_5 + \rho_1)(\theta_2 - \theta_{2_i}) = 0. \tag{3.77}$$

In this system of three position equations, there are

scalar knowns: $r_1, \theta_1 = 0, r_4, r_3, r_5, r_2, \rho_5, \rho_1$

and

scalar unknowns: $\theta_2, \theta_3, \theta_4, \theta_5$.

Suppose that a value of θ_2 is given and the corresponding values of θ_3, θ_4, and θ_5 are to be computed. Define the vector of unknowns, $\bar{x}$, as

$$\bar{x} = \begin{bmatrix} \theta_3 \\ \theta_4 \\ \theta_5 \end{bmatrix}.$$

1. Outline all equations necessary to implement Newton's Method.
2. Present a flowchart that shows the logic of Newton's Method, and in the flowchart, refer to the equations you have derived by their equation numbers.

Problem 3.11

For each mechanism shown on the next page, an appropriate vector loop is shown in dashed vectors. Each mechanism contains a rolling contact, as indicated. Next to each mechanism:

1. Write the rolling contact equation.
2. Completely expand the rolling contact equation and simplify it as much as possible.

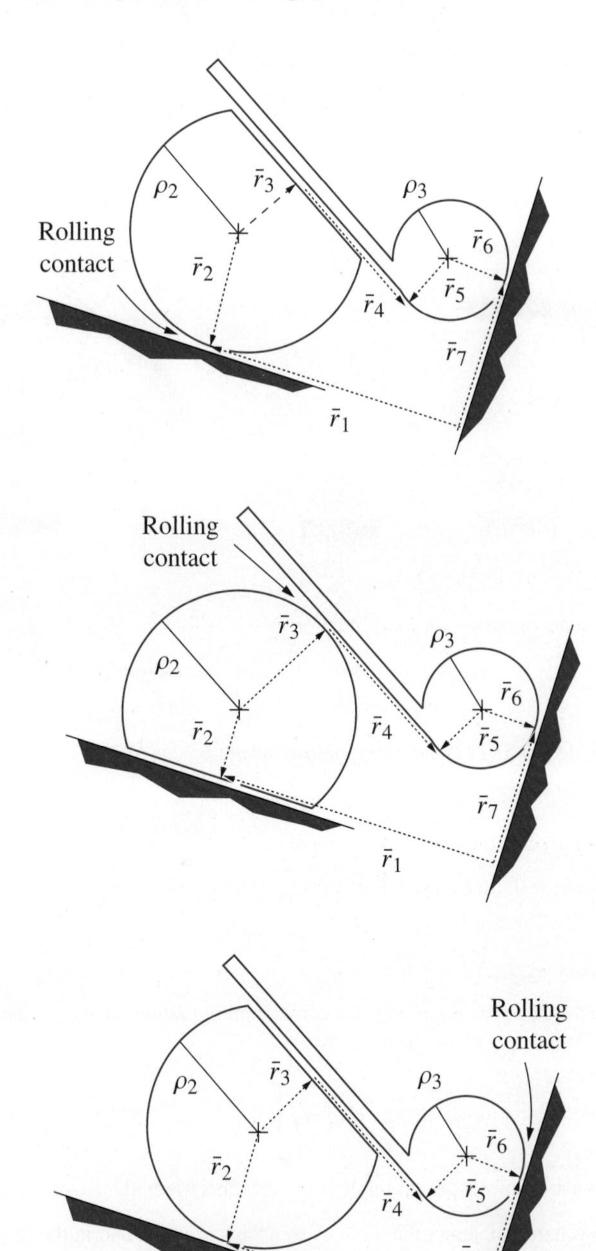
Rolling contact
ρ_2
$\bar{r}_3$
ρ_3
$\bar{r}_6$
$\bar{r}_2$
$\bar{r}_5$
$\bar{r}_4$
$\bar{r}_7$
$\bar{r}_1$
Rolling contact
$\bar{r}_3$
ρ_3
ρ_2
$\bar{r}_6$
$\bar{r}_4$
$\bar{r}_2$
$\bar{r}_5$
$\bar{r}_7$
$\bar{r}_1$
Rolling contact
$\bar{r}_3$
ρ_2
ρ_3
$\bar{r}_6$
$\bar{r}_2$
$\bar{r}_5$
$\bar{r}_4$
$\bar{r}_7$
$\bar{r}_1$

Problem 3.12

Below each diagram, write out the rolling contact equation for that case. Vectors $\bar{r}_4$ and $\bar{r}_3$ are attached to bodies 4 and 3, respectively.

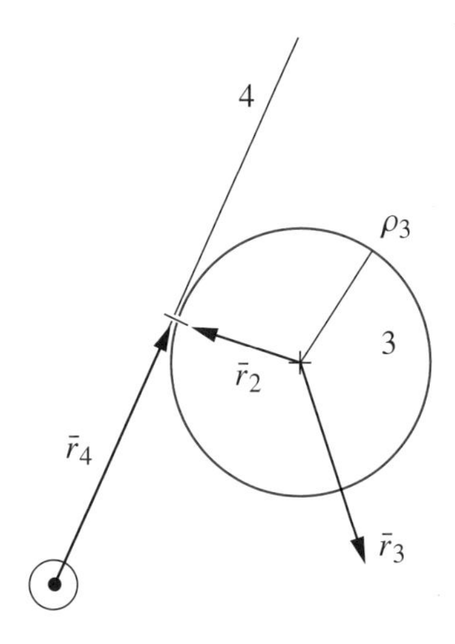

Case i

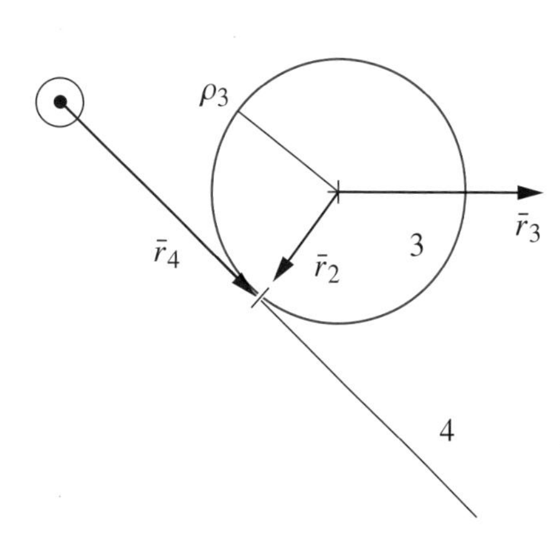

Case ii

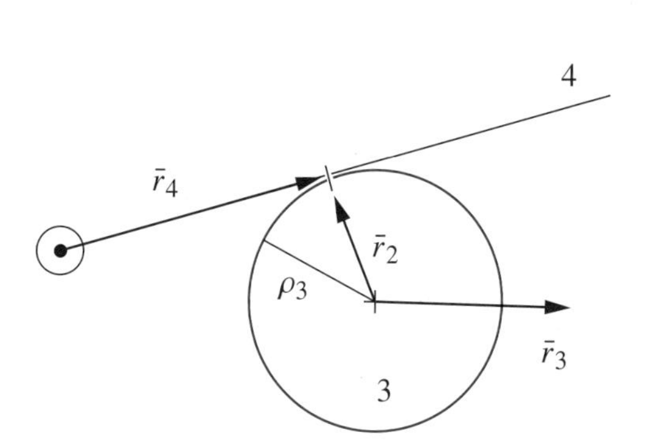

Case iii

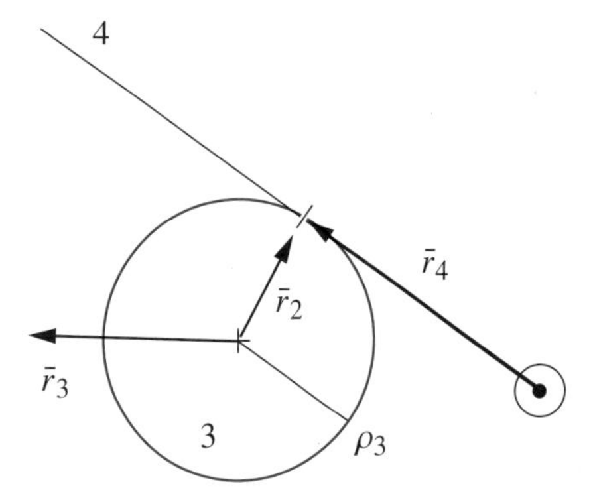

Case iv

Problem 3.13

For the following planetary geartrain, 1 is the input and 5 is the output.

1. Compute the gear ratio.
2. Indicate what similarity this transmission has to that studied in Example 3.8.
3. Compute the speed ratio.
4. Compute the figure ratio.

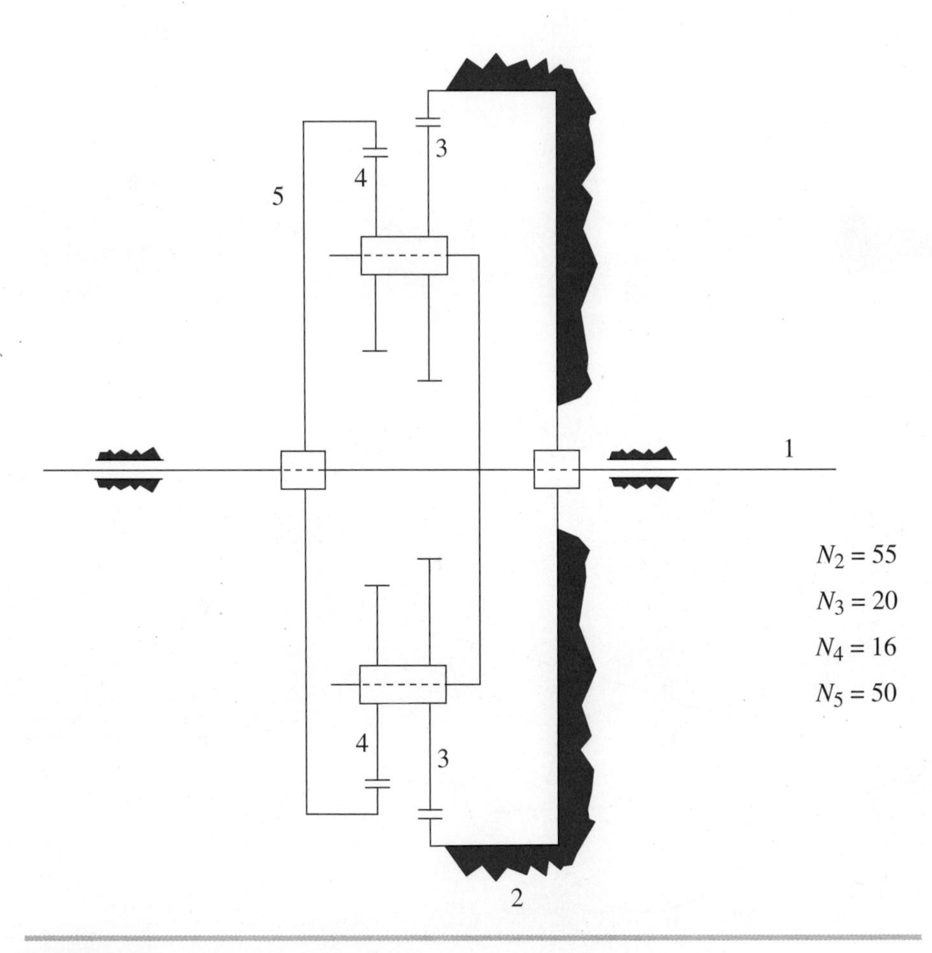

FIGURE 3.32 A planetary geartrain

Problem 3.14

In the following transmission, 2 is the input and 9 is the output.

1. Compute the gear ratio.
2. Compute the speed ratio.
3. Compute the torque ratio.

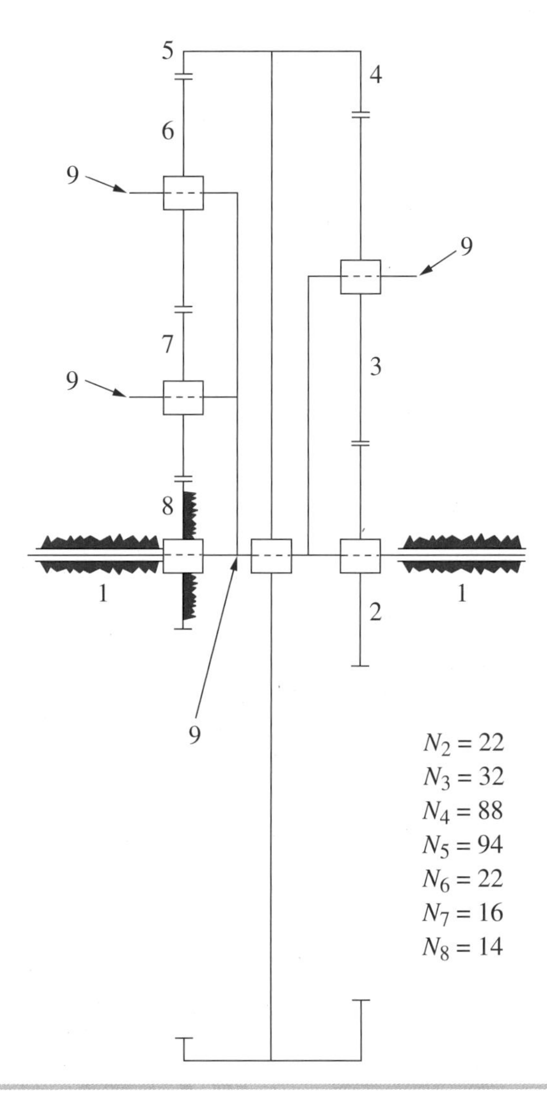

FIGURE 3.33 A planetary geartrain with an idler planet gear

Problem 3.15

In the following transmission, 6 is the input and 9 is the output.

1. Compute the gear ratio.
2. Compute the speed ratio.
3. Compute the torque ratio.

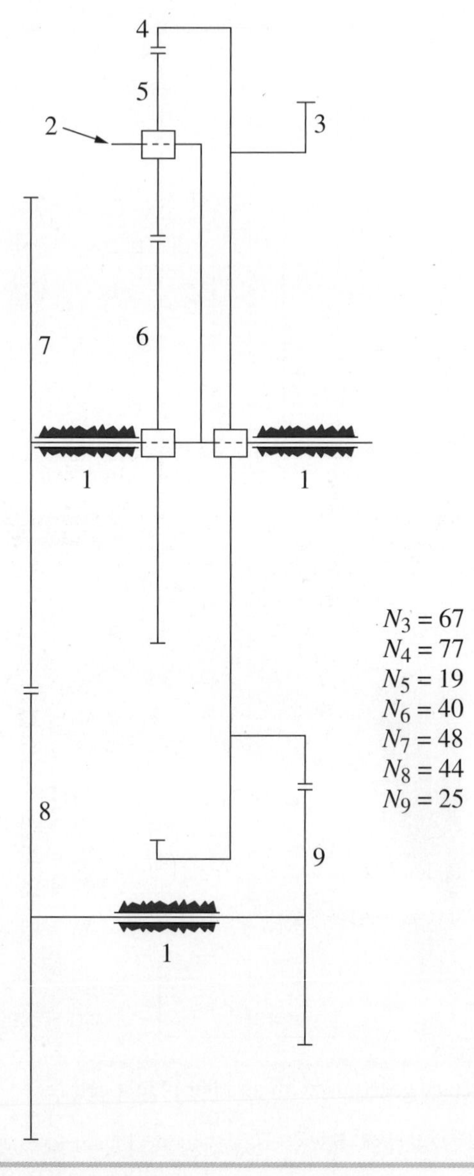

FIGURE 3.34 A hybrid (mixed) planetary and simple geartrain

Problem 3.16

In the transmission in Figure 3.35, the compound gear consisting of 9 and 3 is the output, and 2 is the input.

1. Compute the gear ratio.
2. Compute the speed ratio.
3. Compute the torque ratio.

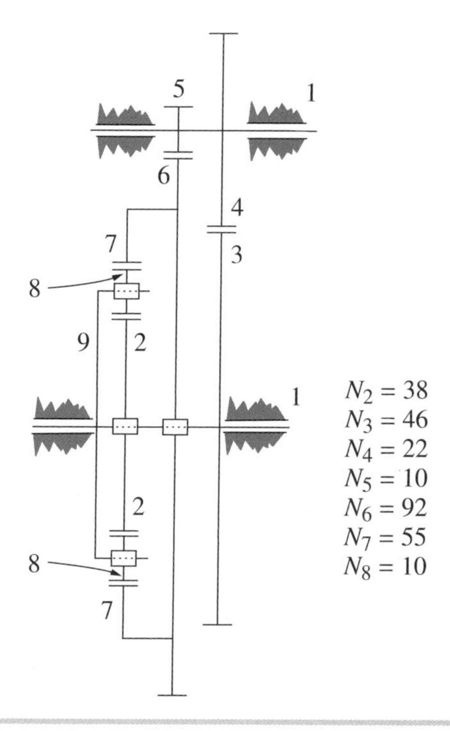

FIGURE 3.35 A hybrid (mixed) planetary and simple geartrain

Problem 3.17

Consider the geartrain in Figure 3.36. Let 2 be the input and 7 be the output.

The input is 8.5 hp at 3200 rpm. Answer the questions below.

1. What is the gear ratio?
2. What is the speed ratio?
3. What is the torque ratio?
4. What is the output torque?
5. Is the output torque in the same direction as the input torque or in the opposite direction?
6. What is the output speed?
7. Is the output speed in the same direction as the input speed or in the opposite direction?

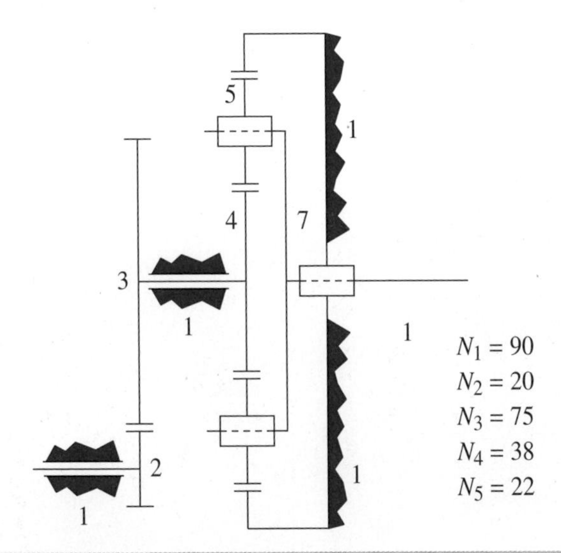

FIGURE 3.36 A hybrid planetary and simple geartrain

Problem 3.18

For the manual transmission in Section 3.5.2, calculate the gear ratio for 2nd, 3rd, 4th, and reverse gears.

Problem 3.19

Continuing with Problem 3.18, the input from the engine is 220 hp at 3800 rpm. Calculate the output torque of the transmission in first gear.

Problem 3.20

Continuing with Problem 3.19, suppose the automobile is driving in a straight line and the differential has the gear ratio R = 3.8:1. The diameter of the rear tires is 27 inches. What force is produced to accelerate the vehicle?

Problem 3.21

For the semiautomatic Model T transmission in Section 3.5.3, compute the gear ratio in high speed and reverse speed conditions.

Problem 3.22

For the two speed automatic transmission in Section 3.5.3, compute the gear ratio in high and reverse speed conditions.

Problem 3.23

In the automatic transmission in Section 3.5.3, the input from the engine is 200 hp at 3200 rpm. In Section 3.5.3 the gear ratio in the low speed condition was computed in Example 3.10 as 1.72:1. Calculate the output torque of the transmission in the low speed condition.

Problem 3.24

For the two speed automatic transmission found in Section 3.5.3, explain how neutral works.

Problem 3.25

A common type of continuously variable transmission (cvt) uses split pulleys for both the driving (input) pulley and the driven (output) pulley. The pulleys have a Vee belt connecting them. In response to (a) their rotational speed, (b) the torque they are transmitting, or (c) direct electronic actuation, the two halves of the driving pulley move apart and the two halves of the driven pulley move closer together, causing the belt to move outward on the driven pulley and inward on the driven pulley. This increases the speed ratio of the cvt $n_{S_{cvt}}$, where

$$n_{S_{cvt}} = \frac{\omega_{\text{driven pulley}}}{\omega_{\text{driving pulley}}}.$$

A cvt provides only a positive speed ratio. The transmission in Figure 3.37 incorporates a cvt into a planetary geartrain and results in an infinitely variable transmission (ivt) where the overall speed ratio of the ivt has a range that is both positive and negative.

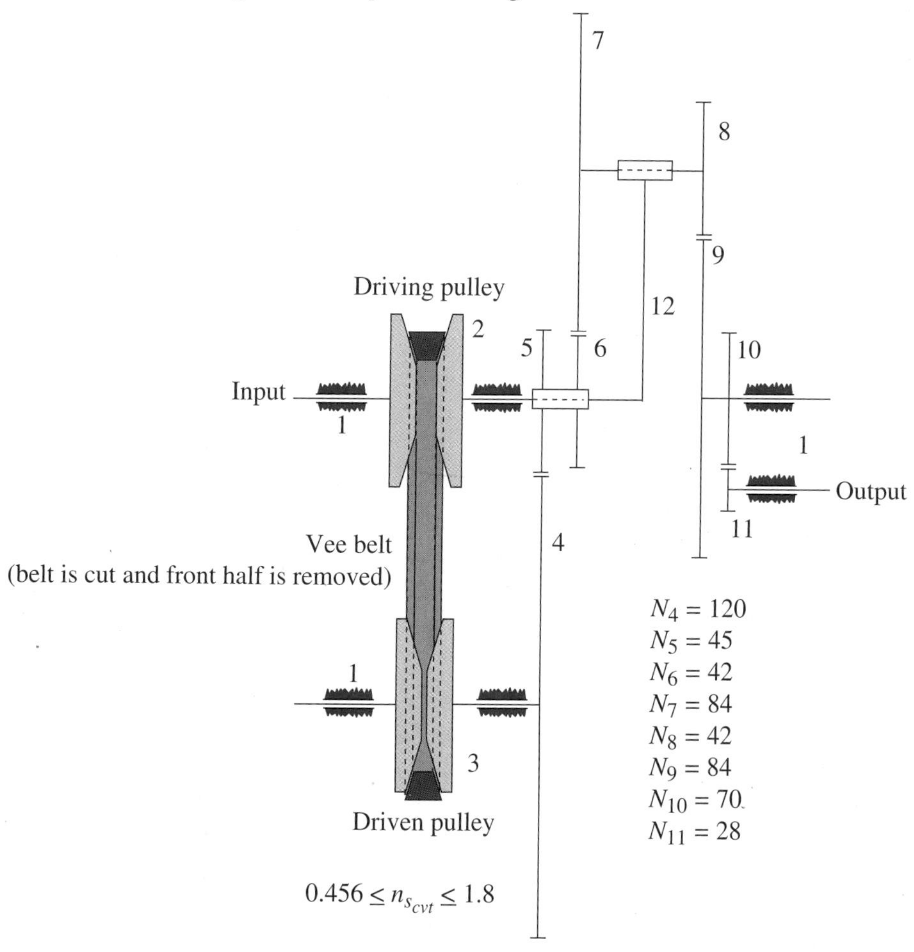

FIGURE 3.37 An infinitely variable transmission

Plot the transmission's speed ratio, $n_s = \frac{\omega_{out}}{\omega_{in}} = \frac{\omega_{11}}{\omega_2}$, as a function of $n_{s_{cvt}}$ for the range of speed ratios produced by this cvt, $0.456 \leq n_{s_{cvt}} \leq 1.8$. Then identify:

1. The transmission's maximum positive speed ratio.
2. The transmission's minimum negative speed ratio.
3. The value of $n_{s_{cvt}}$ at which the transmission's speed ratio is zero.

3.7 APPENDIX I: THE INVOLUTE TOOTH PROFILE

Mechanisms typically do not provide a constant speed ratio. For example, in a four bar mechanism, if the input is rotated at a constant speed, the output speed will vary, and in some cases (rocker outputs) the output will even stop and reverse its rotation. In many machines, such as an automotive transmission, a constant speed ratio is desirable, and for those applications something other than a mechanism must be used.

For constant speed ratio applications, mechanical engineers have developed gears. The basic question is: What type of tooth geometry (i.e., tooth profile/shape) will provide this constant speed ratio effect? The goal of the following discussion is to develop this tooth geometry.

3.7.1 Mechanics Review: Relative Velocity of Two Points on the Same Rigid Body

Consider two points A and B belonging to a moving rigid body (link), as illustrated in Figure 3.38. The XYZ coordinate system is a fixed coordinate system (indicated by the cross hatching). The xyz coordinate system is attached to the moving body. The moving body (and xyz) has an angular velocity $\bar{\omega}$. Let the unit vectors $\hat{i}$, $\hat{j}$, and $\hat{k}$ be aligned with the x, y, and z axes, respectively. The position of B relative to A is given by the vector $\bar{r}_{b/a}$. Since this is a rigid body, the length of $\bar{r}_{b/a}$ does not change; however, as the body moves and rotates, the direction of $\bar{r}_{b/a}$ changes. The velocity of B relative to A is

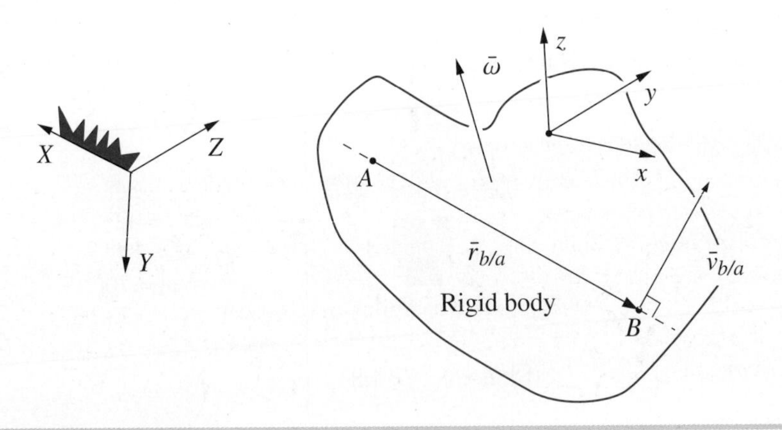

FIGURE 3.38 Relative velocity of two points on the same body

represented by the vector $\bar{v}_{b/a}$, and by definition,

$$\bar{v}_{b/a} = \frac{d}{dt}(\bar{r}_{b/a}). \tag{3.78}$$

Relative to the moving coordinate system, the vector $\bar{r}_{b/a}$ is given by

$$\bar{r}_{b/a} = (r_{b/a})_x\hat{i} + (r_{b/a})_y\hat{j} + (r_{b/a})_z\hat{k}, \tag{3.79}$$

and because the body is rigid, the scalar components $(r_{b/a})_x$, $(r_{b/a})_y$, and $(r_{b/a})_z$ are constants. Substituting Equation (3.79) into Equation (3.78) and applying the product rule for differentiation gives

$$\bar{v}_{b/a} = \underbrace{\frac{d(r_{b/a})_x}{dt}}_{\bar{0}}\hat{i} + (r_{b/a})_x\frac{d\hat{i}}{dt} + \underbrace{\frac{d(r_{b/a})_y}{dt}}_{\bar{0}}\hat{j} + (r_{b/a})_y\frac{d\hat{j}}{dt} + \underbrace{\frac{d(r_{b/a})_z}{dt}}_{\bar{0}}\hat{k} + (r_{b/a})_z\frac{d\hat{k}}{dt}. \tag{3.80}$$

The time derivatives of the unit vectors in Equation (3.80) are given by

$$\frac{d\hat{i}}{dt} = \omega \times \hat{i}, \quad \frac{d\hat{j}}{dt} = \omega \times \hat{j} \quad \text{and} \quad \frac{d\hat{k}}{dt} = \omega \times \hat{k}, \tag{3.81}$$

so

$$\bar{v}_{b/a} = (r_{b/a})_x\,(\bar{\omega} \times \hat{i}) + (r_{b/a})_y\,(\bar{\omega} \times \hat{j}) + (r_{b/a})_z\,(\bar{\omega} \times \hat{k}). \tag{3.82}$$

Factoring out $\bar{\omega}$ gives

$$v_{b/a} = \bar{\omega} \times \left\{(r_{b/a})_x\hat{i} + (r_{b/a})_y\hat{j} + (r_{b/a})_z\hat{k}\right\}, \tag{3.83}$$

and from Equation (3.79),

$$\bar{v}_{b/a} = \bar{\omega} \times \bar{r}_{b/a}. \tag{3.84}$$

Geometric Interpretation for Planar Motion

Planar motion is defined as "a motion where all points on all bodies move on parallel planes, known as the plane of motion, and all bodies rotate about an axis that is perpendicular to the plane of motion." It is conventional to take the *XY* plane as the plane of motion and the *Z* axis as the direction of all rotations.

Equation (3.84) is for the general case of three-dimensional (a.k.a. spatial) motion. Restricting ourselves to planar motion means that in Equation (3.84), $\bar{r}_{b/a}$ will have only *X* and *Y* components,

$$\bar{r}_{b/a} = \begin{Bmatrix} (r_{b/a})_x \\ (r_{b/a})_y \\ 0 \end{Bmatrix}, \tag{3.85}$$

and the angular velocity vector $\bar{\omega}$ will have only a Z component,

$$\bar{\omega} = \begin{Bmatrix} 0 \\ 0 \\ \omega \end{Bmatrix}. \tag{3.86}$$

Substituting Equations (3.85) and (3.86) into Equation (8.26) gives, for $\bar{v}_{BA}$,

$$\bar{v}_{b/a} = \begin{Bmatrix} \omega(r_{b/a})_y \\ -\omega(r_{b/a})_x \\ 0 \end{Bmatrix}. \tag{3.87}$$

From the right hand side of Equation (3.87), observe that the magnitude of $\bar{v}_{b/a}$, denoted as $v_{b/a}$, is given by

$$v_{b/a} = r_{b/a}\,\omega, \tag{3.88}$$

where $r_{b/a}$ is the magnitude of $\bar{r}_{b/a}$ and is the distance from A to B.

First Conclusion:
$v_{b/a}$ is proportional to both $r_{b/a}$ and ω.

Also observe from Equations (3.87) and (3.85) that

$$\bar{v}_{b/a} \cdot \bar{r}_{b/a} = 0. \tag{3.89}$$

And from Equation (3.89), observe that $\bar{v}_{b/a}$ is perpendicular to $\bar{r}_{b/a}$.

Second Conclusion:
$\bar{v}_{b/a}$ is perpendicular to the line connecting points A and B.

3.7.2 Meshing Theory

Figure 3.39 shows a pair of gear teeth that are in mesh. The gear teeth belong to body x and body y. The gear teeth happen to be in the process of disengaging with the pinion x driving the gear y. The point P is called the pitch point. The pitch point locates a pair of coincident points, P_x on x and P_y on y, whose relative velocity is zero, i.e. $\bar{v}_{p_x/p_y} = \bar{0}$. By definition of a rolling contact, this would be the hypothetical point where x and y have a rolling contact. Our goal here is to show that such a pitch point exists and that there is a very specific shape for the profile of the gear teeth that will cause the pitch point to remain stationary.

A stationary pitch point results in the relative motion of x and y appearing to be a rolling contact, where the pitch point is the point of rolling contact. Let us now consider the geometric conditions that define the location of P. It turns out that there are two of them.

Figure 3.40 shows abstractions of bodies (gears) x and y and their pin joint connections to fixed body 1. Suppose that instantaneously x and y have angular velocities ω_x and ω_y. The goal is to find what geometric conditions exist on the location of P as a consequence of x and y having ground pivots O_x and O_y, respectively.

Assume a candidate point for P is arbitrarily chosen, as shown in Figure 3.40. The candidate P defines coincident points P_x and P_y on bodies x and y, respectively, which by the definition of P should have the property that

$$\bar{v}_{p_x/p_y} = \bar{0}. \tag{3.90}$$

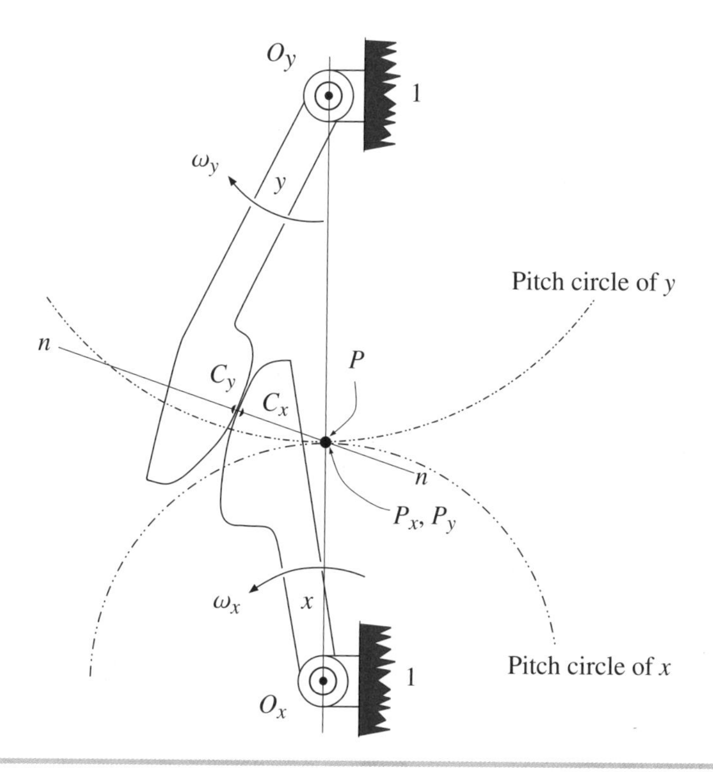

FIGURE 3.39 A pair of gear teeth in mesh

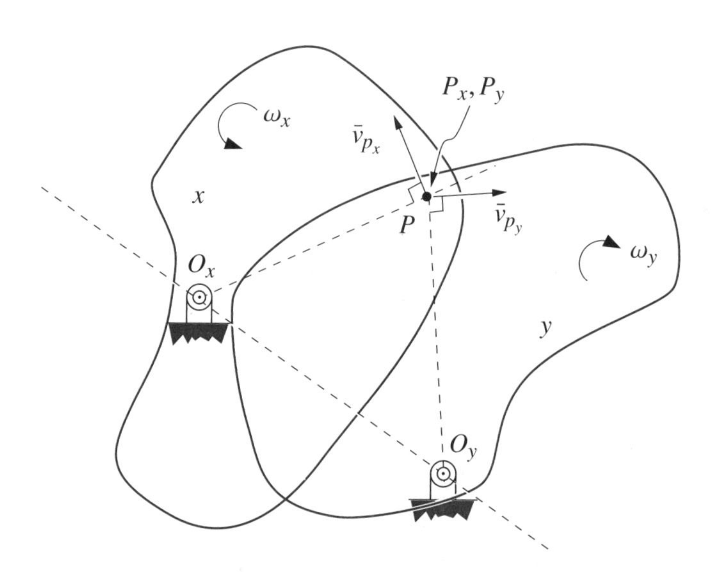

FIGURE 3.40 Links X and Y and their pin joint connections to ground

In order for this to be possible it must be that $\bar{v}_{p_x} = \bar{v}_{p_y}$, so $\bar{v}_{p_x}$ and $\bar{v}_{p_y}$ must be in the same direction. Now, according to the Second Conclusion in Section 3.7.1, in Figure 3.40 $\bar{v}_{p_x}$ must be perpendicular to the line from O_x to P_x, and $\bar{v}_{p_y}$ must be perpendicular to the line from O_y to P_y. In light of the Second Conclusion in Section 3.7.1, in order for $\bar{v}_{p_x}$ and $\bar{v}_{p_y}$ to be in the same direction,

> *P* must lie on the line containing O_x and O_y, which is known as the line of centers.

This is the first geometric condition on the location of *P*.

In order that the mechanism in Figure 3.39 be movable, the contact between *x* and *y* must be a slipping contact. In general, gear teeth rub (slip) against each other. Consider the consequences of the slipping contact between *x* and *y* for the location of *P*.

Figure 3.41 shows two bodies, *x* and *y*, with a slipping contact between them. The instantaneously coincident pair of points on bodies *x* and *y* at the point of contact are labeled E_x and E_y, respectively.

It is important to understand that by virtue of the slipping contact, ω_x and ω_y are independent of one another. Due to the slipping contact, the bodies rotate independently, while rubbing on each other at their point of contact. Due to the rubbing, in general the relative velocity $\bar{v}_{e_x/e_y} \neq \bar{0}$, so E_x and E_y do not define the location of *P*. What, then, can we say about *P*?

Bodies *x* and *y* are rigid, and they cannot penetrate each other or separate. This means that $\bar{v}_{e_x/e_y}$ must be directed along the common tangent to bodies *x* and *y* at their point of contact. Consider now an arbitrary candidate point *P*. Assign a coordinate system with its *X* and *Y* axes instantaneously aligned with the common tangent and common normal of bodies *x* and *y* and the origin at the point of contact. Relative to this system, P_x and P_y are located by the vector $\bar{r}_{p/e}$, which has a magnitude $r_{p/e}$ and a direction of θ. The goal here is to determine the conditions on $r_{p/e}$ and θ that must be satisfied if *P* is to be

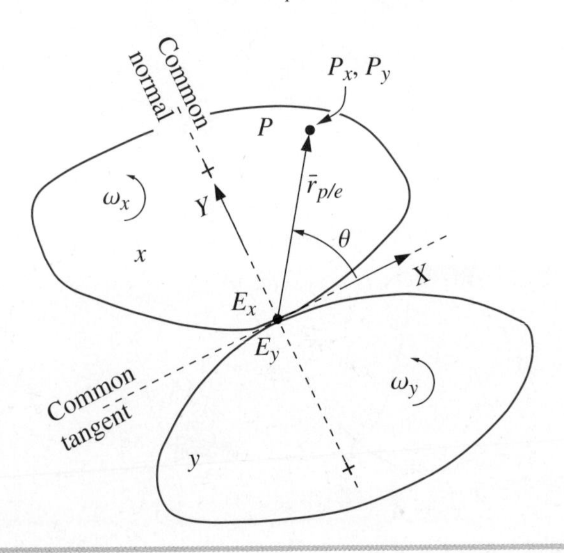

FIGURE 3.41 Condition on location of *P* resulting from slipping contact at gear teeth

the pitch point. These conditions are found from the definition of P; specifically, it is necessary that $\bar{v}_{p_x/p_y} = \bar{0}$ where P_x and P_y are the coincident points on x and y located at P.

Since E_x and P_x both belong to body x,

$$\bar{v}_{p_x/e_x} = \bar{\omega}_x \times \bar{r}_{p/e} = \begin{Bmatrix} 0 \\ 0 \\ \omega_x \end{Bmatrix} \times \begin{Bmatrix} r_{p/e}\cos\theta \\ r_{p/e}\sin\theta \\ 0 \end{Bmatrix} = \begin{Bmatrix} -r_{p/e}\omega_x\sin\theta \\ r_{p/e}\omega_x\cos\theta \\ 0 \end{Bmatrix}. \quad (3.91)$$

Since E_y and P_y both belong to body y,

$$\bar{v}_{p_y/e_y} = \bar{\omega}_y \times \bar{r}_{p/e} = \begin{Bmatrix} 0 \\ 0 \\ \omega_y \end{Bmatrix} \times \begin{Bmatrix} r_{p/e}\cos\theta \\ r_{p/e}\sin\theta \\ 0 \end{Bmatrix} = \begin{Bmatrix} -r_{p/e}\omega_y\sin\theta \\ r_{p/e}\omega_y\cos\theta \\ 0 \end{Bmatrix}. \quad (3.92)$$

The absolute velocities of P_x and P_y can be expressed in terms of the absolute velocities of E_x and E_y respectively:

$$\bar{v}_{p_x} = \bar{v}_{e_x} + \bar{v}_{p_x/e_x} \quad (3.93)$$

$$\bar{v}_{p_y} = \bar{v}_{e_y} + \bar{v}_{p_y/e_y}. \quad (3.94)$$

Thus,

$$\begin{aligned} \bar{v}_{p_x/p_y} &= \bar{v}_{p_x} - \bar{v}_{p_y} = \bar{0} \\ &= (\bar{v}_{e_x} + \bar{v}_{p_x/e_x}) - (\bar{v}_{e_y} + \bar{v}_{p_y/e_y}) = \bar{0} \\ &= (\bar{v}_{e_x} - \bar{v}_{e_y}) + \bar{v}_{p_x/e_x} - \bar{v}_{p_y/e_y} = \bar{0} \\ &= \bar{v}_{e_x/e_y} + \bar{v}_{p_x/e_x} - \bar{v}_{p_y/e_y} = \bar{0}. \end{aligned} \quad (3.95)$$

In Equation (3.95), $\bar{v}_{e_x/e_y}$ is in the direction of the common tangent in Figure 3.41; that is,

$$\bar{v}_{e_x/e_y} = \begin{Bmatrix} v_{e_x/e_y} \\ 0 \\ 0 \end{Bmatrix}. \quad (3.96)$$

Substituting the right hand sides of Equations (3.91), (3.92), and (3.96) into the right hand side of Equation (3.95) yields

$$\bar{v}_{p_x/p_y} = \begin{Bmatrix} v_{e_x/e_y} \\ 0 \\ 0 \end{Bmatrix} + \begin{Bmatrix} -r_{p/e}\omega_x\sin\theta \\ r_{p/e}\omega_x\cos\theta \\ 0 \end{Bmatrix} - \begin{Bmatrix} -r_{p/e}\omega_y\sin\theta \\ r_{p/e}\omega_y\cos\theta \\ 0 \end{Bmatrix} = \bar{0}. \quad (3.97)$$

From the Y component of Equation (3.97),

$$r_{p/e}(\omega_x - \omega_y)\cos\theta = 0. \quad (3.98)$$

In general, $r_{p/e}$ cannot be zero because E is the point of slipping contact. In general, the term $(\omega_x - \omega_y)$ is not zero because in a slipping joint, ω_x and ω_y are independent of each other. So it must be the case that in general $\cos\theta$ is zero, so $\theta = (\frac{\pi}{2}, \frac{3\pi}{2})$. This gives the result that in order for the points P_x and P_y to be at the pitch point P,

> P must lie on the common normal to x and y at their point of contact, which is the second geometric condition on the location of P.

The conclusion is that P lies on the intersection of the line of centers with the line of the common normal to x and y at their point of contact. In order for the relative motion of x and y to be equivalent to that of rolling pitch circles, the pitch point must remain fixed. This leads us to the *law of conjugate gear tooth action*:

As a pair of gears in mesh rotate, and the teeth undergo engagement then disengagement, the common normal to the tooth surfaces at the point of contact must always intersect the line of centers at a fixed pitch point P.

If a pair of meshed gears have teeth with a geometry that satisfies this law, then as the teeth on the gears come in and out of engagement, they all have common normals at their points of contact which pass through the same pitch point, and the gears then appear as two rolling pitch, circles, as shown in Figure 3.39.

The law of conjugate gear tooth action can be satisfied by various tooth profiles, but the only one important in machinery is the *involute tooth profile* or, more precisely, *the involute of a circle*. The involute tooth profile was proposed as the shape of a gear tooth by Leonhard Euler. A much less used shape is that of a cycloid, which is the shape generated by a point on a circle as it rolls around the circumference of another circle. Cycloidal teeth are used only in situations where the teeth are miniature, in which case cycloidal teeth have a strength advantage over involute teeth. Cycloidal teeth may be found in watches and clocks, for example.

The Involute Tooth Profile

Figure 3.42 shows the involute of a circle. It is generated by a point P on a cord being unwrapped from a circle (involute means "rolled inward"). The circle from which the cord is unwrapped is known as the base circle.

At any instant, as the cord wraps or unwraps from the base circle, the velocity of the point Q is perpendicular to the cord, line AQ. This means the tangent to the involute profile is perpendicular to AQ and therefore AQ is the normal to the involute profile at the point Q. This leads to the fundamental principle which makes the involute an excellent gear tooth profile:

At any instant, the normal direction to the involute of a circle is the line of the cord generating the involute. Remember this, it's important.

Now suppose two involutes have been generated, each with its own base circle, and consider that those involutes are rigid bodies that extend from their base circles as gear teeth. Figure 3.43 shows these two gear teeth brought into contact. Notice that each of the involute profiles has an unwrapping cord as the normal to its surface at any point of engagement. Since these are rigid bodies, these normal directions are common to each other, as Figure 3.43 shows. This means the contacting involute profiles are being generated by a point on a cord common to both base circles that wraps around one of

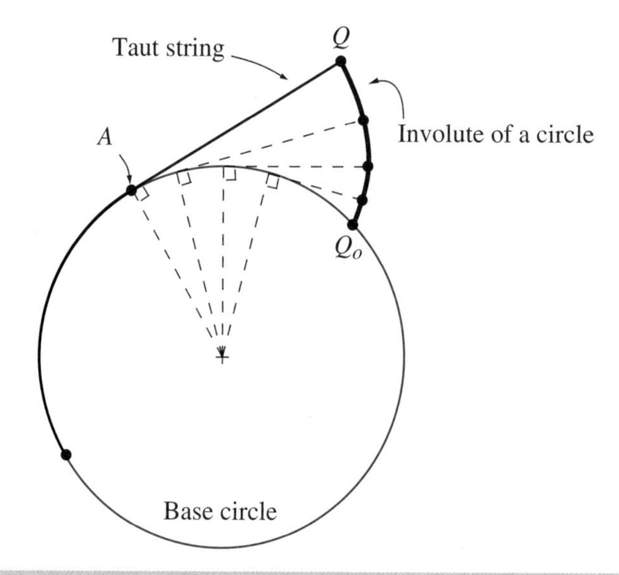

FIGURE 3.42 The involute of a circle

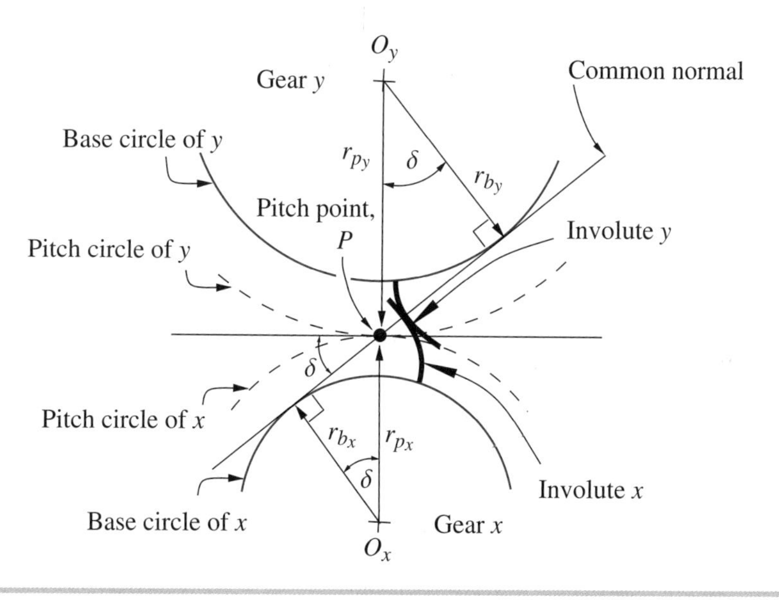

FIGURE 3.43 A pair of involute profile gear teeth in engagement

the base circles and unwraps from the other base circle. The common normal to the two involute gear tooth profiles at any instant in their engagement is always this hypothetical cord wrapped around the base circles. This causes P to remain stationary, satisfying the law of conjugate gear tooth action.

Let r_{p_x} represent the radius of the pitch circle of x, which is the distance from O_x to P. Let r_{p_y} represent the radius of the pitch circle of y, which is the distance from O_y to P. Then, from the definition of P, namely $\bar{v}_{p_x} = \bar{v}_{p_y}$, we know that

$$\omega_y \, r_{p_y} = -\omega_x \, r_{p_x},$$

or

$$\omega_y = -\omega_x \left(\frac{r_{p_x}}{r_{p_y}} \right). \tag{3.99}$$

The negative sign is introduced into Equation (3.99) to account for ω_x and ω_x being in opposite directions. Now consider the two right triangles in Figure 3.43 and note that

$$r_{p_x} \cos\delta = r_{b_x} \quad \text{and} \quad r_{p_y} \cos\delta = r_{b_y}; \tag{3.100}$$

that is,

$$\frac{r_{b_x}}{r_{b_y}} = \frac{r_{p_x}}{r_{p_y}}. \tag{3.101}$$

Substituting Equation (3.101) into Equation (3.99) gives

$$\omega_y = -\omega_x \left(\frac{r_{b_x}}{r_{b_y}} \right). \tag{3.102}$$

Comparing Equation (3.102) to Equation (3.99) shows that the rolling of the pitch circles causes a motion identical to that of a cord being wrapped around and unwrapped from the base circles. So the overall conclusion is as follows:

The movement caused by the slipping action of a pair of involute tooth profiles is identical to the movement caused by the rolling of their respective pitch circles, which is identical to the movement caused by a cord unwrapping from their respective base circles. This is the magic of an involute tooth profile — the three cases occur simultaneously.

3.7.3 Pressure Angle

In Figure 3.43 the angle δ is known as the *pressure angle*. Figure 3.44 shows a free body diagram of the two gear teeth, with a driving torque T_x on gear x and a load torque T_y on gear y. The gear force F is transmitted between the gear teeth along their common normal. Summing the moments about O_x yields

$$F \, r_{b_x} - T_x = 0 \quad \Longrightarrow \quad F = \frac{T_x}{r_{b_x}}, \tag{3.103}$$

and summing the moments about O_y gives

$$F \, r_{b_y} - T_y = 0 \quad \Longrightarrow \quad F = \frac{T_y}{r_{b_y}}. \tag{3.104}$$

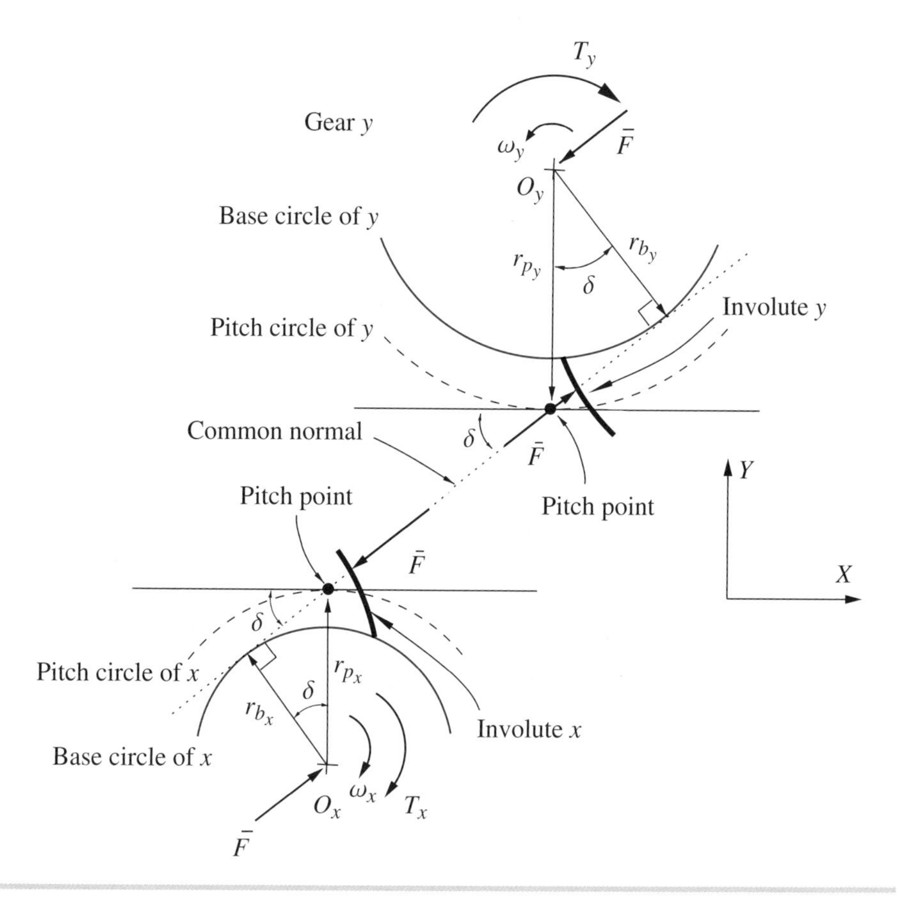

FIGURE 3.44 Free body diagram of involute profile gear teeth in engagement

Eliminating F between Equations (3.103) and (3.104) gives the output torque in terms of the input torque and the base circle radii:

$$T_y = T_x \frac{r_{b_y}}{r_{b_x}}, \tag{3.105}$$

which is independent of δ. Substituting Equations (3.101) into Equations (3.103) and (3.104) gives

$$F = \frac{T_x}{r_{p_x} \cos\delta} \quad \text{and} \quad F = \frac{T_y}{r_{p_y} \cos\delta}, \tag{3.106}$$

and eliminating F between the two expressions in Equation (3.106) gives

$$T_y = T_x \frac{r_{p_y}}{r_{p_x}}, \tag{3.107}$$

which is also independent of δ. Equations (3.105) and (3.107) are equivalent relations between the output and input torque. Either the base circle radii or the pitch circle radii can be used. Most often the pitch circle radii are used because the base circle radius of

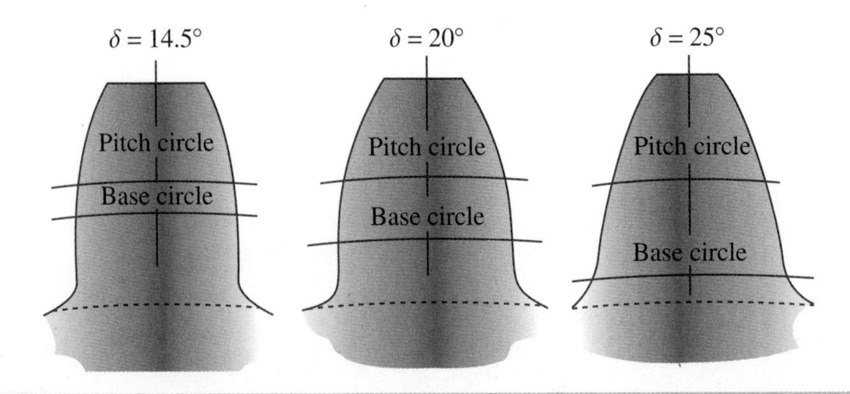

FIGURE 3.45 The standard pressure angles

a gear is typically unknown, whereas the pitch circle radii are published in catalogs and must be known in order to dimension the center-to-center distance between gears in a CAD model.

So what is the effect of pressure angle δ? The expression on the left of Equation (3.106) shows that as δ increases, F increases for a given driving torque T_x. Larger F increases the bearing loads at O_x and O_y, as well as the contact stresses between the teeth and tooth wear. It would be desirable to have $\delta = 0°$, in which case there would be no cord between the base circles, and the base circles would be directly rolling on one another. This is not a reasonable thing to do. Wrapped cords (i.e., chains on sprockets) and gear teeth provide a positive drive. Rolling wheels (i.e., rolling base circles) are friction drives and they can slip. They do not provide a positive drive. So some nonzero value of δ must be accepted for positive drives.

The value of δ is set by the ratio chosen for base and pitch circle radii, as seen in Figure 3.43, where $r_{p_i}\cos\delta = r_{b_i}$ $(i = x, y)$. Figure 3.45 illustrates the three standard values of pressure angle as specified by the American Gear Manufacturers Association, the AGMA. Notice that gear teeth with lower pressure angles have more of a vertical side, which is why they use the gear force more effectively in transmitting torque. However, they are also weaker since there is a greater amount of undercutting caused by the issue of clearances to avoid teeth interferences, and their base areas are smaller. The 14.5° pressure angle is no longer in use and is somewhat difficult to find.

Also, observe from Figure 3.43 that if the center to center distance between the gears were not located precisely as the sum of the pitch circle radii, there would no effect on the developed speed ratio. The only effect would be on the value of the pressure angle. This center-to-center dimension would always be toleranced with a positive value since any negative value would result in tooth jamming. Hence, any inaccuracy in the center-to-center distance results in an increased pressure angle and as well an increase in backlash, but it has no effect on the kinematic parameter—that is, the speed ratio!

3.7.4 Nomenclature

Figure 3.46 shows a gear that consists of involute teeth. It also illustrates some standard nomenclature.

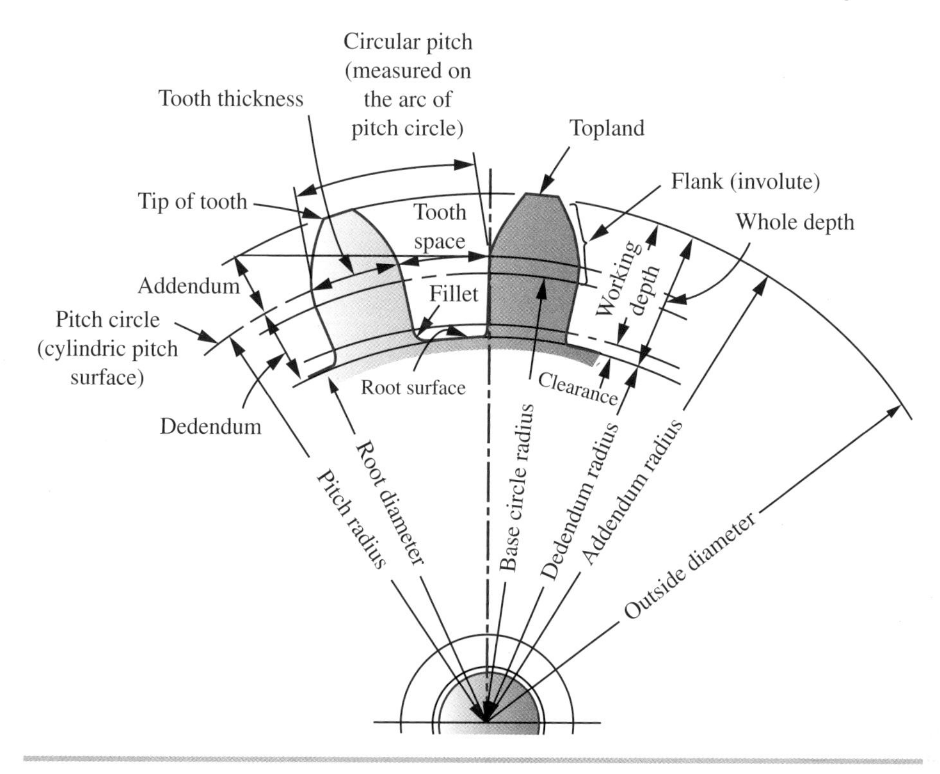

FIGURE 3.46 Involute gear tooth nomenclature

Source: Based on Erdman, Sandor, and Kota, "Mechanism Design Analysis and Synthesis," Vol.1, 4th edition. Prentice Hall Figure 7.5.

DEFINITIONS:

Pitch Circle *The equivalent rolling circles of a pair of gears.*

Circular Pitch, P_c *The distance along the arc of the pitch circle, measured in inches, between corresponding points on adjacent gear teeth. This is a direct measure of tooth size, which affects tooth strength.*

Addendum *The height to which a tooth extends beyond the pitch circle (radial distance from pitch circle to addendum circle).*

Dedendum *The height to which a tooth extends below the pitch circle (radial distance from pitch circle to dedendum circle).*

Clearance *The amount by which the dedendum exceeds the addendum (which it must).*

Working Depth *The depth of engagement of two gears, which would be the sum of their addenda.*

Whole Depth *The total depth of a tooth space—that is, the sum of the addendum and dedendum.*

Tooth Thickness *The distance across the gear tooth along the pitch circle.*

Tooth Space *The distance between adjacent teeth along the pitch circle.*

Backlash *The amount by which the width of the tooth space exceeds the tooth thickness of the engaging tooth; see Figure 3.47.*

Backlash can be removed with an antibacklash gear. One of the more common forms of an antibacklash gear splits the gear into two gears of equal thickness, each half as thick as the original gear. One half of the gear is fixed to the shaft, while the other half of the gear is allowed to turn on the shaft but is connected to the other gear by a preloaded spring. The tension in the spring rotates the free half gear so that it contacts one side of the mating gear's tooth, while the half gear fixed to the shaft contacts the other side of the mating gear's tooth. This eliminates backlash up to a level of torque transmission that would cause the springs to stretch and reopen the backlash.

Diametral Pitch, P_d *The number of teeth on a gear divided by the diameter of the pitch circle in inches.*

This is the standard for measuring tooth size in the US Customary and British Gravitational systems of units. Figure 3.48 shows actual tooth sizes, along with their diametral pitches.

Module, m *The diameter of the pitch circle divided by the number of teeth.*

This is standard for measuring tooth size in the Metric and European Engineering systems of units.

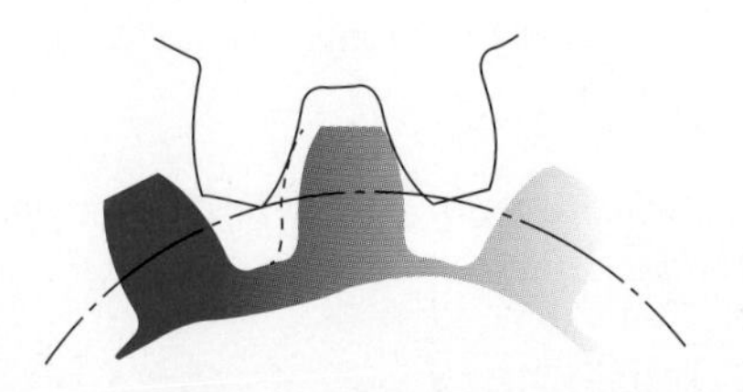

FIGURE 3.47 Backlash in gear teeth

Source: Based on Wilson and Sadler, Kinematics and Dynamics of Machinery, 3rd edition, Prentice Hall.

FIGURE 3.48 Actual gear tooth sizes

Source: Based on Norton, Design of Machinery: An Introduction to the Synthesis and Analysis of Mechanisms and Machines, 3rd edition, McGraw Hill.
NB: N+S261orton acknowledges Colman Co. of Loves Park, Illinois for the image.

P_c and P_d are not independent. Let r_{p_i} represent the radius of the ith gear's pitch circle, and let N_i represent the number of teeth on that gear. Then, from the definition of the circular pitch,

$$P_{c_i} = \frac{2\pi r_{p_i}}{N_i}, \tag{3.108}$$

and from the definition of the diametral pitch,

$$P_{d_i} = \frac{N_i}{2r_{p_i}}. \tag{3.109}$$

Also, observe from Equations (3.108) and (3.109) that

$$P_{c_i} P_{d_i} = \pi. \tag{3.110}$$

Thus, as a tooth gets larger—that is, as P_{c_i} increases, P_{d_i} decreases, as seen in Figure 3.48. $P_d = \frac{1}{m}$ and Equations (3.108)–(3.110) change accordingly.

In order for two gears i and j to mesh, it is necessary that they have the same size teeth, which requires $P_{d_i} = P_{d_j}$, so

$$\frac{N_i}{2r_{p_i}} = \frac{N_j}{2r_{p_j}}, \tag{3.111}$$

from which

$$\frac{N_i}{N_j} = \frac{r_{p_i}}{r_{p_j}}. \tag{3.112}$$

In any calculations, these ratios are interchangeable. Generally speaking, results are developed in terms of r_{p_i} and r_{p_j} but are presented in terms of N_i and N_j.

CHAPTER 4

Kinematic Analysis Part III: Kinematic Coefficients

This chapter introduces kinematic coefficients. We will apply kinematic coefficients to the velocity and acceleration analysis of mechanisms, which are steps 7 and 8 in the definition of kinematic analysis and the vector loop method given in Section 2.1. We will also apply kinematic coefficients to finding limit positions and dead positions of mechanisms.

First let us consider a time-based approach to velocity and acceleration analysis, an approach that does not exploit kinematic coefficients directly. We do this with an example of a four bar mechanism.

4.1 TIME-BASED VELOCITY AND ACCELERATION ANALYSIS OF THE FOUR BAR MECHANISM

Consider the four bar mechanism and its vector loop in Figure 2.3. The X and Y components of its vector loop equation, Equation (2.4) and Equation (2.5), are repeated here:

$$r_2\cos\theta_2 + r_3\cos\theta_3 - r_4\cos\theta_4 - r_1 = 0 \tag{4.1}$$

$$r_2\sin\theta_2 + r_3\sin\theta_3 - r_4\sin\theta_4 = 0. \tag{4.2}$$

Given the link lengths and a value for the input θ_2, first solve the position Equation (4.1) and Equation (4.2) for the unknowns θ_3 and θ_4, either in closed form, as in Section 2.3, or iteratively using Newton's Method, as in Section 2.4. This completes steps 1 through 6 in Section 2.1.

Suppose that link 2 is the input. In the velocity problem, step 7 in Section 2.1, we are given $\dot{\theta}_2$ and we must solve for $\dot{\theta}_3$ and $\dot{\theta}_4$. To do this, differentiate Equations (4.1) and (4.2) with respect to time and make use of the chain rule:

$$-r_2\sin\theta_2\,\dot{\theta}_2 - r_3\sin\theta_3\,\dot{\theta}_3 + r_4\sin\theta_4\,\dot{\theta}_4 = 0 \tag{4.3}$$

$$r_2\cos\theta_2\,\dot{\theta}_2 + r_3\cos\theta_3\,\dot{\theta}_3 - r_4\cos\theta_4\,\dot{\theta}_4 = 0. \tag{4.4}$$

Equations (4.3) and (4.4) are two linear equations in the two unknowns $\dot{\theta}_3$ and $\dot{\theta}_4$. Rewrite them in matrix form,

$$\begin{bmatrix} -r_3\sin\theta_3 & r_4\sin\theta_4 \\ r_3\cos\theta_3 & -r_4\cos\theta_4 \end{bmatrix} \begin{bmatrix} \dot{\theta}_3 \\ \dot{\theta}_4 \end{bmatrix} = \begin{bmatrix} r_2\sin\theta_2\,\dot{\theta}_2 \\ -r_2\cos\theta_2\,\dot{\theta}_2 \end{bmatrix}, \tag{4.5}$$

and solve Equation (4.5) for the two unknowns $\dot{\theta}_3$ and $\dot{\theta}_4$ using Cramer's Rule. Unlike the nonlinear position problem, the velocity problem is a simpler linear problem.

After both the position problem and the velocity problem have been solved, the acceleration problem can be addressed. In the acceleration problem, which is step 8, in Section 2.1, a value of $\ddot{\theta}_2$ is also given, and $\ddot{\theta}_3$ and $\ddot{\theta}_4$ must be determined. To do this, differentiate Equation (4.3) and Equation (4.4) with respect to time, making careful use of both the chain rule and the product rule.

$$-r_2\sin\theta_2\,\ddot{\theta}_2 - r_2\cos\theta_2\,\dot{\theta}_2^2 - r_3\sin\theta_3\,\ddot{\theta}_3 - r_3\cos\theta_3\,\dot{\theta}_3^2 + r_4\sin\theta_4\,\ddot{\theta}_4 + r_4\cos\theta_4\,\dot{\theta}_4^2 = 0 \tag{4.6}$$

$$r_2\cos\theta_2\,\ddot{\theta}_2 - r_2\sin\theta_2\,\dot{\theta}_2^2 + r_3\cos\theta_3\,\ddot{\theta}_3 - r_3\sin\theta_3\,\dot{\theta}_3^2 - r_4\cos\theta_4\,\ddot{\theta}_4 + r_4\sin\theta_4\,\dot{\theta}_4^2 = 0 \tag{4.7}$$

Equations (4.6) and (4.7) are two linear equations in the two unknowns $\ddot{\theta}_3$ and $\ddot{\theta}_4$. Rewrite them in matrix form,

$$\begin{bmatrix} -r_3\sin\theta_3 & r_4\sin\theta_4 \\ r_3\cos\theta_3 & -r_4\cos\theta_4 \end{bmatrix} \begin{bmatrix} \ddot{\theta}_3 \\ \ddot{\theta}_4 \end{bmatrix} = \begin{bmatrix} r_2\sin\theta_2\,\ddot{\theta}_2 + r_2\cos\theta_2\,\dot{\theta}_2^2 + r_3\cos\theta_3\,\dot{\theta}_3^2 - r_4\cos\theta_4\,\dot{\theta}_4^2 \\ -r_2\cos\theta_2\,\ddot{\theta}_2 + r_2\sin\theta_2\,\dot{\theta}_2^2 + r_3\sin\theta_3\,\dot{\theta}_3^2 - r_4\sin\theta_4\,\dot{\theta}_4^2 \end{bmatrix}, \tag{4.8}$$

and solve Equation (4.8) for the two unknowns using Cramer's Rule. Like the velocity problem, the acceleration problem is linear.

Although this method of time differentiation is straightforward enough, it does not reveal the purely geometric nature of the velocity and acceleration problems. Most important, it is very cumbersome when one is trying to solve the forward dynamics problem and develop dynamic simulations. For this reason, the concept of *kinematic coefficients* is introduced.

4.2 KINEMATIC COEFFICIENTS

This discussion of kinematic coefficients is limited to planar mechanisms with one degree of freedom. These are mechanisms in which there are n scalar position equations (these include scalar components of all vector loop equations, all geometric constraints, and all rolling contact equations) in $n+1$ unknown variables. Of the $n+1$ unknown variables in the mechanism, one of them will be the *input variable*, which will be denoted as S_i. The remaining n unknown variables will be the *output variables*. Denote an output variable as S_k, where $1 \leq k \leq n$. S_i and each S_k can be a vector direction or a vector magnitude.

The term *kinematic coefficient* will be used for the derivatives of an output variable S_k with respect to the input variable S_i.

DEFINITIONS:

First-Order Kinematic Coefficient *The first derivative of S_k with respect to S_i:*

$$\underbrace{\frac{dS_k}{dS_i}}_{\textit{First-order kinematic coefficient}}$$

Second-Order Kinematic Coefficient *The second derivative of S_k with respect to S_i:*

$$\underbrace{\frac{d^2S_k}{dS_i^2}}_{\textit{Second-order kinematic coefficient}}$$

4.2.1 Notation Used for the Kinematic Coefficients

In the forward dynamics problem (discussed in Chapter 7), there is a need for a specific notation for the first- and second-order kinematic coefficients. The notation depends on whether the output variable is a vector direction or a vector magnitude. If S_k is a vector direction θ_k, the first- and second-order kinematic coefficients are denoted as:

$$h_k = \frac{d\theta_k}{dS_i} \quad \text{and} \quad h_k' = \frac{d^2\theta_k}{dS_i^2}, \tag{4.9}$$

and if S_k is a vector length r_k, the first- and second-order kinematic coefficients are denoted as:

$$f_k = \frac{dr_k}{dS_i} \quad \text{and} \quad f_k' = \frac{d^2r_k}{dS_i^2}. \tag{4.10}$$

4.2.2 Units Associated with the Kinematic Coefficients

If S_i and S_k are vector directions θ_i and θ_k, then

$$h_k = \frac{d\theta_k}{d\theta_i} \text{ is dimensionless.} \qquad h_k' = \frac{d^2\theta_k}{d\theta_i^2} \text{ is dimensionless.}$$

If S_i and S_k are vector magnitudes r_i and r_k, then

$$f_k = \frac{dr_k}{dr_i} \text{ is dimensionless.} \qquad f_k' = \frac{d^2r_k}{dr_i^2} \text{ has the units } \textit{length}^{-1}.$$

If S_i is a vector direction θ_i and S_k is a vector magnitude r_k, then

$$f_k = \frac{dr_k}{d\theta_i} \text{ has the units } \textit{length}. \qquad f_k' = \frac{d^2r_k}{d\theta_i^2} \text{ has the units } \textit{length}.$$

If S_i is a vector magnitude r_i and S_k is a vector direction θ_k, then

$$h_k = \frac{d\theta_k}{dr_i} \text{ has the units } \textit{length}^{-1}. \qquad h_k' = \frac{d^2\theta_k}{dr_i^2} \text{ has the units } \textit{length}^{-2}.$$

4.2.3 Physical Meaning of the Kinematic Coefficients

Output variable S_k is *explicitly* a function of the input variable, S_i, through the position equations. S_k is *implicitly* a function of time *only through the input variable* S_i *and the time dependence of* S_i. This implicit relationship is seen through the chain rule.

By the chain rule for differentiation,

$$\frac{dS_k}{dt} = \frac{dS_k}{dS_i}\left(\frac{dS_i}{dt}\right).$$

Defining a "dot" as denoting a time derivative,

$$\dot{S}_k = \frac{dS_k}{dS_i}\dot{S}_i. \tag{4.11}$$

Recognize dS_k/dS_i is the first-order kinematic coefficient. From Equation (4.11),

$$\frac{dS_k}{dS_i} = \left(\frac{\dot{S}_k}{\dot{S}_i}\right), \tag{4.12}$$

which shows *the physical meaning of the first-order kinematic coefficient. It is the ratio of the output variable's velocity to the input variable's velocity*. That is, it is a velocity ratio.

Differentiating Equation (4.11) again with respect to time and carefully applying both chain and product rule for differentiation give

$$\frac{d^2S_k}{dt^2} = \frac{dS_k}{dS_i}\left(\frac{d^2S_i}{dt^2}\right) + \frac{d^2S_k}{dS_i^2}\left(\frac{dS_i}{dt}\right)^2,$$

that is,

$$\ddot{S}_k = \frac{dS_k}{dS_i}\ddot{S}_i + \frac{d^2S_k}{dS_i^2}\dot{S}_i^2. \tag{4.13}$$

From Equation (4.13) we recognize that the second-order kinematic coefficient is *not* the acceleration ratio $\ddot{S}_k/\ddot{S}_i$. The physical meaning of the second-order kinematic coefficient comes directly from its definition,

$$\frac{d^2S_k}{dS_i^2} = \frac{d}{dS_i}\left(\frac{dS_k}{dS_i}\right),$$

which means that *the second-order kinematic coefficient is the rate of change of the first-order kinematic coefficient with respect to the input variable.* Mechanisms with constant first-order kinematic coefficients have zero second-order kinematic coefficients. These are systems with a constant velocity ratios, such as geartrains and transmissions.

4.2.4 Use of Kinematic Coefficients in Velocity and Acceleration Analysis

Equations (4.11) and (4.13) illustrate how the kinematic coefficients are used in velocity and acceleration analysis. If S_k is an angle, θ_k, then from Equations (4.11) and (4.13),

$$\dot{\theta}_k = h_k\, \dot{S}_i \quad \text{and} \quad \ddot{\theta}_k = h_k\, \ddot{S}_i + h_k'\, \dot{S}_i^2, \tag{4.14}$$

and if S_k is a vector magnitude, r_k, then from Equations (4.11) and (4.13),

$$\dot{r}_k = f_k\, \dot{S}_i \quad \text{and} \quad \ddot{r}_k = f_k\, \ddot{S}_i + f_k'\, \dot{S}_i^2. \tag{4.15}$$

The following example illustrates finding the first-order and second-order kinematic coefficients of a mechanism and using them to compute the output velocities and accelerations.

▶ EXAMPLE 4.1

The mechanism shown in Figure 4.1 is an inverted crank-slider mechanism. The vector loop equation,

$$\bar{r}_1 + \bar{r}_2 - \bar{r}_3 - \bar{r}_4 = \bar{0},$$

has the X and Y components

$$r_1 + r_2\cos\theta_2 - r_3\cos\theta_3 - r_4\cos\theta_4 = 0 \tag{4.16}$$

$$r_2\sin\theta_2 - r_3\sin\theta_3 - r_4\sin\theta_4 = 0 \tag{4.17}$$

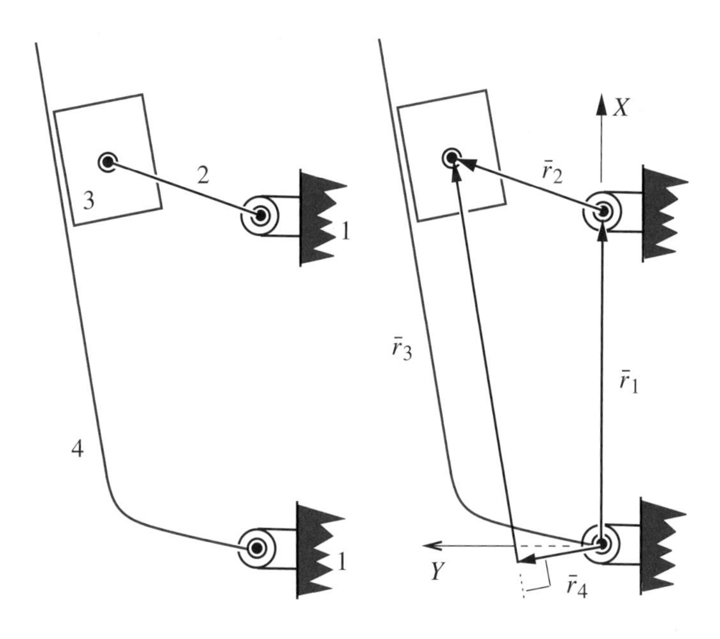

FIGURE 4.1 An inverted crank-slider mechanism

and the geometric constraint

$$\theta_4 - \frac{\pi}{2} - \theta_3 = 0. \tag{4.18}$$

Equations (4.16), (4.17), and (4.18) are the position equations with

scalar knowns: $r_1, \theta_1 = 0°, r_2, r_4$.

and

scalar unknowns: $\theta_2, \theta_3, r_3, \theta_4$.

Take θ_2 to be the input variable S_i, and the remaining three unknown variables are each an S_k. Suppose that the value of input variable θ_2 has been specified and the position equations have been solved for the output variables, θ_3, r_3, and θ_4. Proceed as follows:

1. Derive equations that can be used to compute the first- and second-order kinematic coefficients.
2. Write the equations with which to compute the first- and second-order time derivatives of the output variables.

Solution

The first-order kinematic coefficients are

$$h_3 = \frac{d\theta_3}{d\theta_2}, \quad f_3 = \frac{dr_3}{d\theta_2}, \quad \text{and } h_4 = \frac{d\theta_4}{d\theta_2}.$$

To find them, differentiate the position equations with respect to the input variable, and be careful to apply the chain and product rules of differentiation:

$$-r_2\sin\theta_2 + r_3\sin\theta_3 h_3 - f_3\cos\theta_3 + r_4\sin\theta_4 h_4 = 0 \tag{4.19}$$

$$r_2\cos\theta_2 - r_3\cos\theta_3 h_3 - f_3\sin\theta_3 - r_4\cos\theta_4 h_4 = 0 \tag{4.20}$$

$$h_4 - h_3 = 0. \tag{4.21}$$

These three equations are linear in the first-order kinematic coefficients and can be written in matrix form as

$$\underbrace{\begin{bmatrix} r_3\sin\theta_3 & -\cos\theta_3 & r_4\sin\theta_4 \\ -r_3\cos\theta_3 & -\sin\theta_3 & -r_4\cos\theta_4 \\ -1 & 0 & 1 \end{bmatrix}}_{J} \begin{bmatrix} h_3 \\ f_3 \\ h_4 \end{bmatrix} = \begin{bmatrix} r_2\sin\theta_2 \\ -r_2\cos\theta_2 \\ 0 \end{bmatrix}. \tag{4.22}$$

The 3×3 matrix of coefficients in Equation (4.22) is known as the *Jacobian matrix*, J. It is the same Jacobian matrix defined in Section 2.4, that you would have used if you had solved the position equations using Newton's Method. So there would be no need to recompute the Jacobian in a computer program because you already have it from your last iteration in Newton's Method. Equation (4.22) can be solved for the first-order kinematic coefficients, after which the second-order kinematic coefficients may be determined.

The second-order kinematic coefficients are

$$h_3' = \frac{d^2\theta_3}{d\theta_2^2}, \quad f_3' = \frac{d^2r_3}{d\theta_2^2}, \quad \text{and } h_4' = \frac{d^2\theta_4}{d\theta_2^2}.$$

To find them, differentiate Equations (4.19), (4.20), and (4.21) with respect to the input variable, carefully applying the chain and product rules of differentiation:

$$-r_2\cos\theta_2 + f_3\sin\theta_3 h_3 + r_3\cos\theta_3 h_3^2 + r_3\sin\theta_3 h_3' - f_3'\cos\theta_3 + f_3\sin\theta_3 h_3 + r_4\cos\theta_4 h_4^2 + r_4\sin\theta_4 h_4' = 0 \tag{4.23}$$

$$-r_2\sin\theta_2 - f_3\cos\theta_3 h_3 + r_3\sin\theta_3 h_3^2 - r_3\cos\theta_3 h_3' - f_3'\sin\theta_3 - f_3\cos\theta_3 h_3 + r_4\sin\theta_4 h_4^2 - r_4\cos\theta_4 h_4' = 0 \tag{4.24}$$

$$h_4' - h_3' = 0. \tag{4.25}$$

These three equations are linear in the three second-order kinematic coefficients and can be written in matrix form as

$$\underbrace{\begin{bmatrix} r_3\sin\theta_3 & -\cos\theta_3 & r_4\sin\theta_4 \\ -r_3\cos\theta_3 & -\sin\theta_3 & -r_4\cos\theta_4 \\ -1 & 0 & 1 \end{bmatrix}}_{J} \begin{bmatrix} h_3' \\ f_3' \\ h_4' \end{bmatrix} = \begin{bmatrix} r_2\cos\theta_2 - 2f_3h_3\sin\theta_3 - r_3\cos\theta_3 h_3^2 - r_4\cos\theta_4 h_4^2 \\ r_2\sin\theta_2 + 2f_3h_3\cos\theta_3 - r_3\sin\theta_3 h_3^2 - r_4\sin\theta_4 h_4^2 \\ 0 \end{bmatrix}. \tag{4.26}$$

The same Jacobian matrix from Equation (4.22) appears here. Equation (4.26) can be solved for the second-order kinematic coefficients. Notice the terms multiplied by $2f_3h_3$ on the r.h.s. of Equation (4.26). These are the Coriolis acceleration terms. They arose naturally in the second-order kinematic coefficients.

With the first- and second-order kinematic coefficients known, the time derivatives of the output variables can be computed if the values of the input velocity and acceleration, $\dot{\theta}_2$ and $\ddot{\theta}_2$, respectively, are given. From Equation (4.11),

$$\dot{\theta}_3 = h_3\dot{\theta}_2, \quad \dot{r}_3 = f_3\dot{\theta}_2, \quad \dot{\theta}_4 = h_4\dot{\theta}_2,$$

and from Equation (4.13),

$$\ddot{\theta}_3 = h_3'\dot{\theta}_2^2 + h_3\ddot{\theta}_2, \quad \ddot{r}_3 = f_3'\dot{\theta}_2^2 + f_3\ddot{\theta}_2, \quad \ddot{\theta}_4 = h_4'\dot{\theta}_2^2 + h_4\ddot{\theta}_2.$$

4.2.5 Checking the Kinematic Coefficients

If your position solution has been checked graphically as described in Section 2.4.1, and you are certain it is correct, then a first-order kinematic coefficient is the slope of the plot of the corresponding output variable versus the input variable. Choose several values of the input variable, and check to see whether the slope of this plot matches the corresponding first-order kinematic coefficients. If so, you can be confident that you are computing that first-order kinematic coefficient correctly. Perform this check for all your first-order kinematic coefficients.

To check your value of a second-order kinematic coefficient, plot the first-order kinematic coefficient against θ_2 and see whether the slope of that curve at several values of values of θ_2 matches your computed second-order kinematic coefficient. Perform this check for all your second-order kinematic coefficients.

4.3 FINDING DEAD POSITIONS USING KINEMATIC COEFFICIENTS

Recall from Equation (4.11) that an output velocity is the product of its first-order kinematic coefficient times the input velocity. When the kinematic coefficients are computed, the determinant of the system's Jacobian matrix will appear in each denominator. When the determinant goes to zero, all the kinematic coefficients go to infinity, which means that for any finite value of input velocity, all the output velocities become infinite. This is physically impossible. The output velocities must remain finite, and with the kinematic coefficients being infinite, we see from Equation (4.11) that $\dot{S}_i$ goes to zero. This means the input must stop moving—hence the name *dead position*. Dead positions of a mechanism occur when the determinant of the mechanism's Jacobian matrix goes to zero.

> Dead positions of a mechanism correspond to when $|J| = 0$.

▶ EXAMPLE 4.2

Return to Example 4.1 where we found the kinematic coefficients of the inverted crank-slider. Find the dead positions of this mechanism in terms of the scalar unknowns in the position equations.

Solution

Equation (4.22) gives the Jacobian matrix of this mechanism. The determinant of J is

$$|J| = \begin{vmatrix} r_3\sin\theta_3 & -\cos\theta_3 & r_4\sin\theta_4 \\ -r_3\cos\theta_3 & -\sin\theta_3 & -r_4\cos\theta_4 \\ -1 & 0 & 1 \end{vmatrix}.$$

Expanding the determinant along the bottom row gives

$$|J| = -\underbrace{\underbrace{\begin{vmatrix} -\cos\theta_3 & r_4\sin\theta_4 \\ -\sin\theta_3 & -r_4\cos\theta_4 \end{vmatrix}}_{r_4(\cos\theta_3\cos\theta_4 + \sin\theta_3\sin\theta_4)}}_{r_4\cos(\theta_3-\theta_4)} + \underbrace{\underbrace{\begin{vmatrix} r_3\sin\theta_3 & -\cos\theta_3 \\ -r_3\cos\theta_3 & -\sin\theta_3 \end{vmatrix}}_{r_3(-\sin^2\theta_3 - \cos^2\theta_3)}}_{-r_3},$$

$$|J| = -r_4\cos(\theta_3 - \theta_4) - r_3.$$

But recall from the geometric constraint in Equation (4.18) that the difference $(\theta_3 - \theta_4)$ equals $-90°$, whose cosine is zero, which simplifies the determinant to

$$|J| = -r_3. \tag{4.27}$$

So dead positions of the inverted crank-slider with θ_2 as the input occur when the joint variable, which is the vector magnitude r_3, goes to zero. The inverted crank-slider in Figure 4.1 cannot realize this dead position. But if we change r_2, the dimension of link 2, we get the mechanism

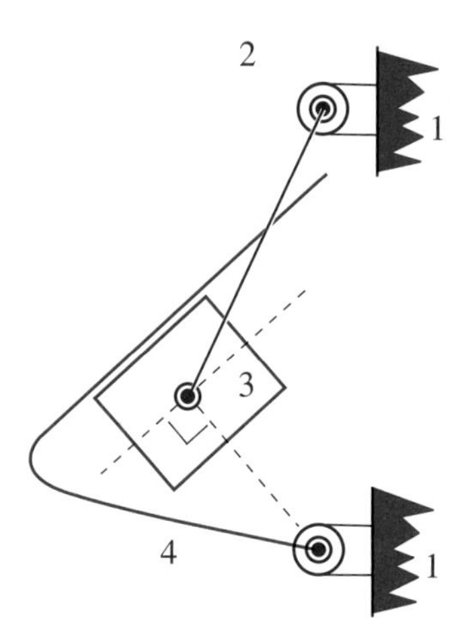

FIGURE 4.2 Dead positions of an inverted rocker-slider mechanism with θ_2 as the input

in Figure 4.2, which is shown in a dead position ($r_3 = 0$). It would be incorrect to call the input link a crank since it is no longer capable of continuous rotations. The input is now a rocker and this is now an inverted rocker-slider mechanism. Figure 4.2 shows that in a static force analysis, the force between bodies 3 and 4 would be along the vector $\bar{r}_4$ of the vector loop in Figure 4.1. The force would not create any moment about the ground pivot of 4, which is required to cause 4 to rotate—hence this is a dead position.

▶ EXAMPLE 4.3

Return to Example 2.1, where we derived the position equations of the four bar mechanism shown in Figure 2.2 using the vector loop shown in Figure 2.3.

1. Find the dead positions of the four bar mechanism in terms of the scalar unknowns in the position equations.
2. State the geometric condition that corresponds to a dead position of this mechanism.

Solution

Of the three scalar unknowns θ_2, θ_3, and θ_4 cited in Example 2.1, take θ_2 to be S_i and take θ_3 and θ_4 each as output variables S_k. Differentiating the position Equation (2.4) and Equation (2.5) with respect to θ_2 gives the following matrix equation for the first-order kinematic coefficients h_3 and h_4:

$$\underbrace{\begin{bmatrix} -r_3\sin\theta_3 & r_4\sin\theta_4 \\ r_3\cos\theta_3 & -r_4\cos\theta_4 \end{bmatrix}}_{J} \begin{bmatrix} h_3 \\ h_4 \end{bmatrix} = \begin{bmatrix} r_2\sin\theta_2 \\ -r_2\cos\theta_2 \end{bmatrix}, \tag{4.28}$$

where the 2×2 matrix above is the Jacobian matrix. $|J|$ is given by

$$|J| = r_3r_4(\sin\theta_3\cos\theta_4 - \cos\theta_3\sin\theta_4) = r_3r_4\sin(\theta_3 - \theta_4) = -r_3r_4\sin(\theta_4 - \theta_3). \tag{4.29}$$

Dead positions correspond to when $|J| = 0$ — that is, when $(\theta_4 - \theta_3) = 0$. This defines dead positions in terms of the scalar unknowns θ_3 and θ_4 and answers the first part of the question.

To answer the second part of the question, note that this condition corresponds to when the angle μ defined in Equation (2.2) and shown in Figure 2.3 has a value of 0 or π — that is, when vectors $\bar{r}_3$ and $\bar{r}_4$ in Figure 2.3 are colinear. Geometrically, this means that dead positions of the four bar mechanism occur when link 3 (the coupler) and link 4 (the output) are in line. This answers the second part of the question. The angle μ in the four bar mechanism is known as the *transmission angle*, and when $\mu = (0°, 180°)$ the four bar mechanism is in a dead position.

4.4 FINDING LIMIT POSITIONS USING KINEMATIC COEFFICIENTS

When a particular output variable of a mechanism reaches a limit position, it must be reversing its motion. And while reversing its motion, its instantaneous velocity must be zero. Recall from Equation (4.11) that the velocity of an output variable is the product of its first-order kinematic coefficient and the velocity of the input. Thus, in order that the velocity of the output variable be zero, its first-order kinematic coefficient must be zero. So, in order for a particular output variable to be in a limit position, the associated first-order kinematic coefficient must be zero.

▶ EXAMPLE 4.4

Return to Example 4.1. With 2 as the input:

1. Find the limit positions of 4 in terms of the scalar unknowns in the position equations.
2. State the corresponding geometric condition.

Solution

When 4 reaches a limit position it reverses, its direction of rotation, which means that at that instant $\dot{\theta}_4 = 0$. Thus the kinematic coefficient h_4 is zero. Enforcing this condition reveals the limit positions of 4.

Solving Equation (4.22) for h_4 using Cramer's Rule yields

$$h_4 = \frac{\begin{vmatrix} r_3\sin\theta_3 & -\cos\theta_3 & r_2\sin\theta_2 \\ -r_3\cos\theta_3 & -\sin\theta_3 & -r_2\cos\theta_2 \\ -1 & 0 & 0 \end{vmatrix}}{|J|}.$$

Expanding the determinant in the numerator and substituting for $|J|$ from Equation (4.29) give

$$h_4 = \frac{-r_2\cos(\theta_3 - \theta_2)}{-r_3} = \frac{r_2}{r_3}\cos(\theta_3 - \theta_2). \quad (4.30)$$

From Equation (4.31), h_4 is zero when

$$\cos(\theta_3 - \theta_2) = 0 \Longrightarrow (\theta_3 - \theta_2) = \pm 90°.$$

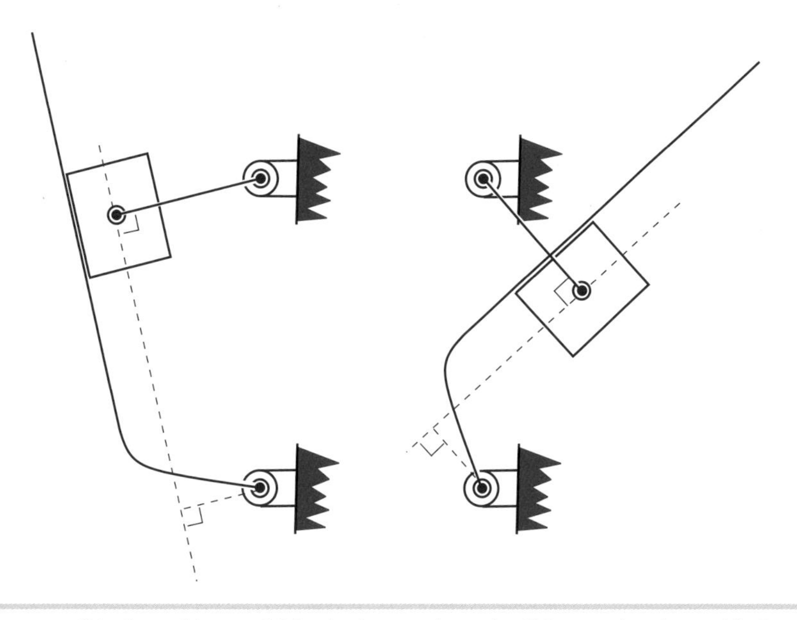

FIGURE 4.3 Limit positions of 4 in the inverted crank-slider mechanism with θ_2 as the input

This answers the first part. Geometrically, this condition means that vectors $\bar{r}_2$ and $\bar{r}_3$ in Figure 4.1 would be orthogonal, in which case link 2 is perpendicular to link 4, as shown in Figure 4.3. This answers the second part.

▶ EXAMPLE 4.5

Return to Example 4.3. With 2 as the input, continue as follows:

1. Find the limit positions of 4 and the limit positions in the rotation of link 3 in terms of the scalar unknowns in the position equations.
2. State the corresponding geometric conditions.

Solution

When 4 reaches a limit position it reverses its direction of rotation, which means that at that instant $\dot{\theta}_4 = 0$. Thus the kinematic coefficient h_4 is zero. Enforcing this condition reveals the limit positions of 4.

Solving Equation (4.28) for h_4 using Cramer's Rule yields

$$h_4 = \frac{\begin{vmatrix} -r_3\sin\theta_3 & r_2\sin\theta_2 \\ r_3\cos\theta_3 & -r_2\cos\theta_2 \end{vmatrix}}{|J|}.$$

Expanding the determinant in the numerator and substituting for $|J|$ from Equation (4.27) give

$$h_4 = -\frac{r_2}{r_4}\frac{\sin(\theta_3 - \theta_2)}{\sin(\theta_4 - \theta_3)} \tag{4.31}$$

From Equation (4.31), h_4 is zero when

$$\sin(\theta_3 - \theta_2) = 0 \Longrightarrow (\theta_3 - \theta_2) = 0°, 180°. \tag{4.32}$$

When the rotation of 3 reaches a limit, $h_3 = 0$. Solving Equation (4.28) for h_3 using Cramer's Rule yields

$$h_3 = \frac{\begin{vmatrix} r_2\sin\theta_2 & r_4\sin\theta_4 \\ -r_2\cos\theta_2 & -r_4\cos\theta_4 \end{vmatrix}}{|J|}.$$

Expanding the determinant in the numerator and substituting for $|J|$ from Equation (4.27) give

$$h_3 = \frac{r_2}{r_3}\frac{\sin(\theta_2 - \theta_4)}{\sin(\theta_4 - \theta_3)}. \tag{4.33}$$

From Equation (4.33), h_3 is zero when

$$\sin(\theta_2 - \theta_4) = 0 \Longrightarrow (\theta_2 - \theta_4) = 0°, 180°. \tag{4.34}$$

This answers part 1 for the limit positions in the rotation of 3.

Geometrically, the condition in Equation (4.32) corresponds to when the vectors $\bar{r}_2$ and $\bar{r}_3$ are in-line. Consequently, limit positions of the output link 4 occur when the input link 2 and the coupler 3 are in-line. The left side of Figure 4.4 shows this geometric condition for the case when $(\theta_3 - \theta_2) = 0°$. Geometrically, the condition in Equation (4.34) corresponds to when the vectors $\bar{r}_2$ and $\bar{r}_4$ are parallel. Consequently, limit positions in the rotation of link 3 occur when the input and output links are parallel. In this configuration the angular velocity of 3 is zero. This is shown on the right side of Figure 4.4.

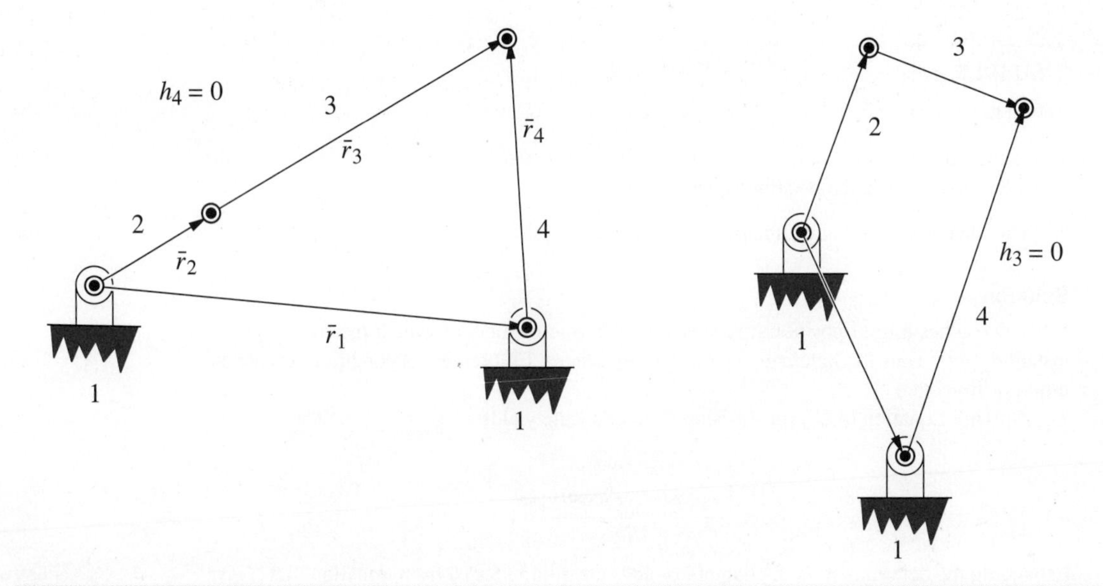

FIGURE 4.4 Configurations of four bar mechanisms where 4 and 3 are in limit positions, with 2 as the input

4.4.1 Time Ratio

When a mechanism has a crank input and a rocker output, it also has a *time ratio*. In a crank-rocker mechanism the crank will generally perform more rotation in moving the output rocker to one of its limit positions than it does in returning it to the other limit position. We can see in Figure 4.3 that if the crank were rotating clockwise, moving the rocker between the limit position on the left side to the limit position on the right side would require a greater rotation than moving the rocker from the limit on the right side back to the left side. This means that if the crank were rotating at a constant speed, the rocker would return from the right limit to left limit (called the return stroke) in less time than it took to advance from the left limit to the right limit (called the advance stroke). This is referred to as a quick return, and the time ratio Q is a direct measure of the quick-return nature of the mechanism.

Figure 4.5 shows the limit position overlayed. We see that the advance stroke requires a crank rotation of $180° + \alpha$ and the return stroke requires a crank rotation of $180° - \alpha$. Q is defined so that it is always greater than or equal to 1:

$$Q = \frac{180° + \alpha}{180° - \alpha}. \tag{4.35}$$

In a computer program, if we used the vector loop shown in Figure 4.1, we would note the values of θ_2 that correspond to $h_4 = 0$ and from this compute the numerator and denominator of Q in Equation (4.35). Figure 4.6 shows the angle α used to compute the time ratio for a crank-rocker four bar mechanism. Again, in a computer program, if we used the vector loop shown in Figure 2.3, we would note the values of θ_2 that correspond to $h_4 = 0$ and from this compute the numerator and denominator of Q in Equation (4.35).

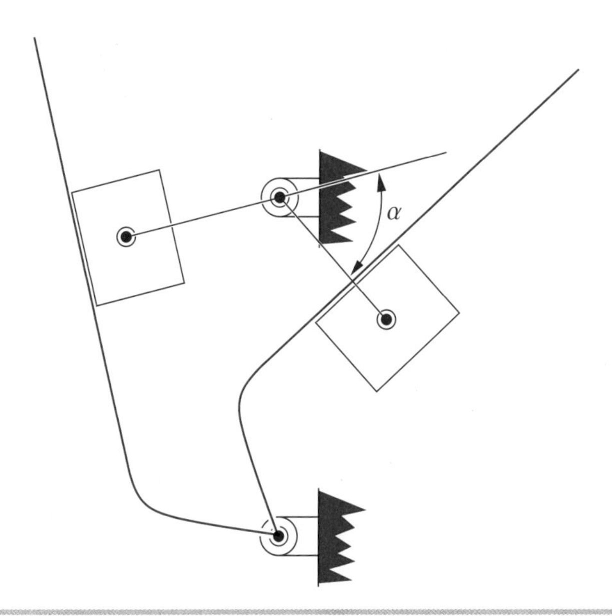

FIGURE 4.5 Time ratio for the inverted crank-slider mechanism

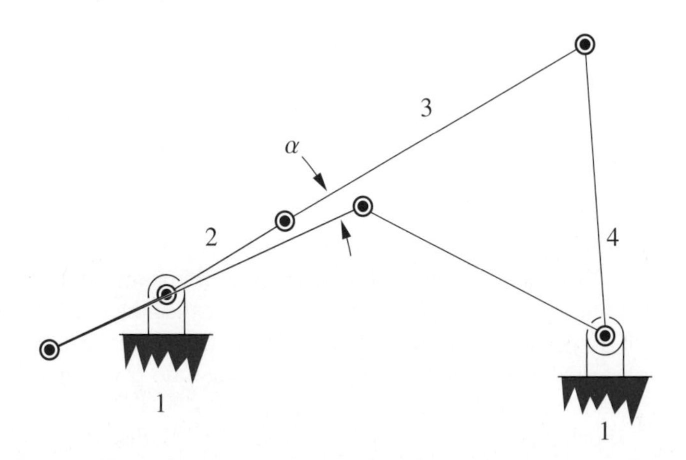

FIGURE 4.6 Time ratio for a crank-rocker four bar mechanism

Typically, the advance stroke would correspond to the stroke where the output of a machine is performing some type of work on the environment, and the return stroke would correspond to when the output of the machine is returning in order to perform another cycle of this work. A quick return is desirable since there is generally no work being done on the return stroke. However, too high a time ratio will lead to poor force transmission on the return stroke, because a high time ratio would push the design closer to a dead position.

4.5 KINEMATIC COEFFICIENTS OF POINTS OF INTEREST

In many situations the kinematic coefficients of particular points of interest must be determined. For example, when doing a dynamic force analysis, it is necessary to know the acceleration of the mass center, so the kinematic coefficients of the center of mass are needed.

DEFINITION

Basic Kinematic Coefficients *The derivatives of the unknown variables, S_k, in the position equations, with respect to the input variable, S_i.*

The kinematic coefficients we have studied up to this point are all basic kinematic coefficients. We make this distinction in kinematic coefficients because the basic kinematic coefficients must be known before one can determine the kinematic coefficients of points of interest. After finding the basic kinematic coefficients, one can find the kinematic coefficients of a point of interest by writing an expression for a vector locating the point of interest relative to the frame of reference used in the vector loop. The X and Y components of this vector can be differentiated with respect to the input to find the kinematic coefficients of the point. We will illustrate this procedure with a four bar mechanism.

▶ EXAMPLE 4.6

Find the kinematic coefficients of point C on the coupler of the four bar mechanism shown in Figure 4.7, where θ_2 is the input. Then give the expressions that would be used in a computer program to calculate the velocity and acceleration of C in terms of its kinematic coefficients and the velocity and acceleration of the input.

Solution

To find the kinematic coefficients of point C, we must first deal with the vector loop

$$\bar{r}_2 + \bar{r}_3 - \bar{r}_4 - \bar{r}_1 = \bar{0},$$

whose two scalar components (the position equations) contain the unknowns θ_2, θ_3, and θ_4. For a given value of the input θ_2 these position equations must be solved for θ_3 and θ_4. The position equations must then be differentiated with respect to θ_2, and the basic kinematic coefficients h_3, h_3', h_4, and h_4' must be computed.

After this, write a vector locating point C in the fixed frame, call it $\bar{r}_c$:

$$\bar{r}_c = \bar{r}_2 + \bar{r}_3'. \tag{4.36}$$

The vector $\bar{r}_3'$ locates the point C on the coupler relative to the reference point on 3 which is the pin between 3 and 2. It would also have been correct to use the combination

$$\bar{r}_c = \bar{r}_1 + \bar{r}_4 - \bar{r}_3 + \bar{r}_3',$$

but that option is more complicated. Point C could also have been referenced on 3 from the pin joint between 3 and 4.

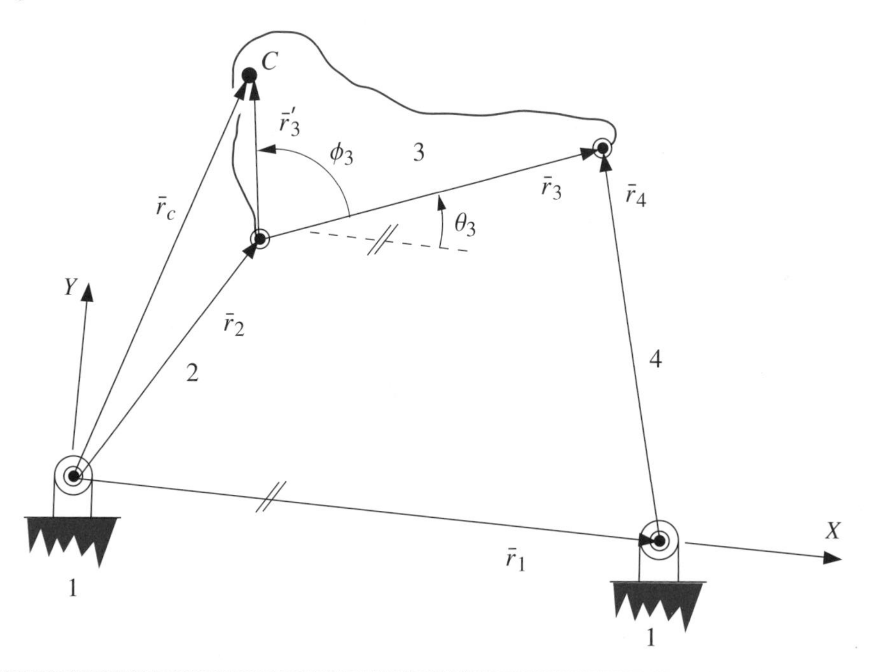

FIGURE 4.7 The kinematic coefficients of a coupler point

From Equation (4.36), the X and Y components of $\bar{r}_C$ can be written:

$$r_{C_x} = r_2\cos\theta_2 + r_3'\cos(\theta_3 + \phi_3) \quad (4.37)$$

$$r_{C_y} = r_2\sin\theta_2 + r_3'\sin(\theta_3 + \phi_3). \quad (4.38)$$

Differentiating Equations (4.37) and (4.38) with respect to θ_2 yields the equations that could be used in a program to calculate the desired kinematic coefficients of C in terms of the basic kinematic coefficients:

$$f_{C_x} = \frac{dr_{C_x}}{d\theta_2} = -r_2\sin\theta_2 - r_3'\sin(\theta_3 + \phi_3)h_3 \quad (4.39)$$

$$f_{C_y} = \frac{dr_{C_y}}{d\theta_2} = r_2\cos\theta_2 + r_3'\cos(\theta_3 + \phi_3)h_3 \quad (4.40)$$

$$f'_{C_x} = \frac{d^2r_{C_x}}{d\theta_2^2} = -r_2\cos\theta_2 - r_3'\cos(\theta_3 + \phi_3)h_3^2 - r_3'\sin(\theta_3 + \phi_3)h_3' \quad (4.41)$$

$$f'_{C_y} = \frac{d^2r_{C_x}}{d\theta_2^2} = -r_2\sin\theta_2 - r_3'\sin(\theta_3 + \phi_3)h_3^2 + r_3'\cos(\theta_3 + \phi_3)h_3'. \quad (4.42)$$

From Equation (4.11), the X and Y components of the velocity C are computed from

$$v_{C_x} = \dot{r}_{C_x} = f_{C_x}\dot{\theta}_2$$
$$v_{C_y} = \dot{r}_{C_y} = f_{C_y}\dot{\theta}_2,$$

and from Equation (4.13), the X and Y components of the acceleration of C are computed from

$$a_{C_x} = \ddot{r}_{C_x} = f'_{C_x}\dot{\theta}_2^2 + f_{C_x}\ddot{\theta}_2$$
$$a_{C_y} = \ddot{r}_{C_x} = f'_{C_y}\dot{\theta}_2^2 + f_{C_y}\ddot{\theta}_2.$$

4.6 KINEMATIC COEFFICIENTS OF GEARTRAINS

In Section 3.5 we learned that geartrains are mechanisms that consist only of rolling contacts (the pitch circles of gears) and the arms that connect the axes of rotation of these rolling elements. The position equations of these systems consist only of rolling contact equations that are linear relationships between the scalar unknowns. The scalar unknowns are all angles, so S_k and S_i are always θ_k and θ_i. Since the position equations are linear, geartrains have basic first-order kinematic coefficients h_k that are constants and second-order kinematic coefficients h_k' that are all zero.

If θ_k is the output rotation of a geartrain, and θ_i is the input rotation of the geartrain, then the output speed is ω_k and the input speed is ω_i, and per Equation (3.22),

$$n_s = \frac{\omega_k}{\omega_i}. \quad (4.43)$$

In terms of kinematic coefficients,

$$\omega_k = h_k\omega_i, \quad (4.44)$$

and we see that from Equations (4.43) and (4.44) that,

$$n_s = h_k. \quad (4.45)$$

R, n_t, n_s, and h_k are related. Comparing Equation (4.45) to Equations (3.21) and (3.24) show that,

$$R = -n_t = \frac{1}{n_s} = \frac{1}{h_k}. \tag{4.46}$$

▶ EXAMPLE 4.7

Consider the planetary geartrain in Example 3.8. Recall that 6 is the input, 5 is fixed, and 2 is the output.

1. Derive a symbolic expression for the first- and second-order kinematic coefficients of the output.
2. Write the equations used to compute the first and second time derivatives of the output.
3. Find the speed ratio and the gear ratio.

Solution

Expanding the corresponding rolling contact Equation (3.43) and Equation (3.44) to include assembly position and writing the equations in homogeneous form give

$$(\theta_{34} - \theta_{34_i}) - (\theta_6 - \theta_{6_i}) + \frac{\rho_2}{\rho_3}\left[\theta_2 - \theta_{2_i}) - (\theta_6 - \theta_{6_i})\right] = 0 \tag{4.47}$$

$$(\theta_{34} - \theta_{34_i}) - (\theta_6 - \theta_{6_i}) + \frac{\rho_5}{\rho_4}\left[\underbrace{(\theta_5 - \theta_{5_i})}_{=0} - (\theta_6 - \theta_{6_i})\right] = 0. \tag{4.48}$$

Gear 5 is fixed, so $(\theta_5 - \theta_{5_i})$ above is zero. θ_6 is the input variable S_i. Differentiating Equation (4.47) and Equation (4.48) with respect to the input θ_6 gives

$$h_{34} + \frac{\rho_2}{\rho_3} h_2 = 1 + \frac{\rho_2}{\rho_3} \tag{4.49}$$

$$h_{34} = 1 + \frac{\rho_5}{\rho_4}, \tag{4.50}$$

and putting the results in matrix form yields

$$\begin{bmatrix} 1 & \frac{\rho_2}{\rho_3} \\ 1 & 0 \end{bmatrix} \begin{bmatrix} h_{34} \\ h_2 \end{bmatrix} = \begin{bmatrix} 1 + \frac{\rho_2}{\rho_3} \\ 1 + \frac{\rho_5}{\rho_4} \end{bmatrix}. \tag{4.51}$$

Equation (4.51) can be solved for h_2 using Cramer's Rule:

$$h_2 = 1 - \frac{\rho_3 \rho_5}{\rho_2 \rho_4}. \tag{4.52}$$

This answers part 1. Just as an exercise, differentiate Equation (4.53) and Equation (4.54) a second time with respect to θ_6 to find second-order kinematic coefficients h'_2 and h'_{34}:

$$h'_{34} + \frac{\rho_2}{\rho_3} h'_2 = 0 \tag{4.53}$$

$$h'_{34} = 0. \tag{4.54}$$

Putting the results in matrix form yields

$$\begin{bmatrix} 1 & \frac{\rho_2}{\rho_3} \\ 1 & 0 \end{bmatrix} \begin{bmatrix} h'_{34} \\ h'_2 \end{bmatrix} = \begin{bmatrix} 0 \\ 0 \end{bmatrix}.$$

We see that $h'_{34} = 0$ and $h'_2 = 0$. The second-order kinematic coefficients of geartrains are always zero because the first-order kinematic coefficients are constant. That is, geartrains have constant velocity ratios.

From Equation (4.14),

$$\dot{\theta}_2 = h_2\,\dot{\theta}_6 \quad \text{and} \quad \ddot{\theta}_2 = h_2\,\ddot{\theta}_6 + \underbrace{h'_2}_{=0}\,\dot{\theta}_2^2 = h_2\,\ddot{\theta}_6.$$

This answers part 2. From Equation (4.45) (h_k is h_2),

$$n_s = h_2 = 1 - \frac{\rho_3\rho_5}{\rho_2\rho_4},$$

and from Equation (4.46),

$$R = \frac{1}{h_2} = \frac{1}{1 - \frac{\rho_3\rho_5}{\rho_2\rho_4}}.$$

This answers part 3 and agrees with the result in of Example 3.8.

4.7 PROBLEMS

Problem 4.1

Consider the four combinations of S_k and S_i as vector magnitudes or directions, and give the units of the first- and second-order kinematic coefficients.

Problem 4.2

In Example 3.2 as link 2 rotates ccw, does 3 move into or retract from 4? Use the kinematic coefficients to determine this.

Measure dimensions r_1, r_2, r_4, r_5, ρ_1, ρ_4 and joint variable r_3 from Figure 3.12 using the scale 1 in. = 5 in. Also measure joint variables θ_2, θ_3, θ_4, θ_5 from Figure 3.12 using a protractor. (The alternative is to program a computer to solve the position Equations (3.14)–(3.17) using Newton's Method.)

Problem 4.3

Continuing with Problem 3.9, if θ_2 is the input:

1. Derive a system of equations in matrix form that could be solved for the first-order kinematic coefficients h_3, h_4, f_4, and h_5.
2. Derive a system of equations in matrix form that could be solved for the second-order kinematic coefficients h'_3, h'_4, f'_4, and h'_5.
3. In terms of $\dot{\theta}_2$, $\ddot{\theta}_2$, h_3, and h'_3, give the equations used to compute the angular velocity and angular acceleration of 3.

Problem 4.4

Continue with Example 2.4 and use the vector loops shown in Figure 2.17. Use θ_2 as the input, and assume the position problem has been solved.

1. Derive a system of equations in matrix form that can be solved for the first-order kinematic coefficients.
2. Derive a system of equations in matrix form that can be solved for the second-order kinematic coefficients.
3. Call the pin joint connection between bodies 5 and 6 point A. In terms of the kinematic coefficients in steps 1 and 2 above, derive expressions for the first- and second-order kinematic coefficients of A.
4. In terms of the kinematic coefficients from part 3 above, write out the equations you would use to compute the velocity and acceleration of A.

Problem 4.5

The one-degree-of-freedom mechanism on the next page continues with Problem 2.15. The figure shows an appropriate vector loop.

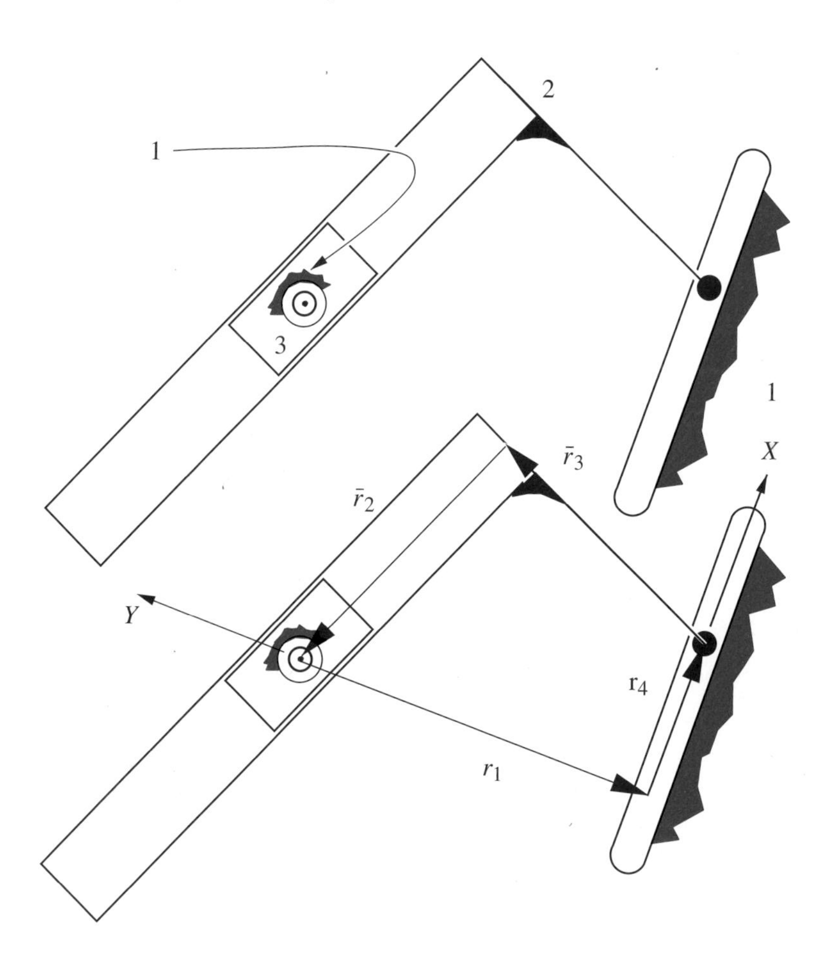

The vector loop for mechanism in Prob 2.15
The vector loop equation,

$$\bar{r}_1 + \bar{r}_3 + \bar{r}_2 + \bar{r}_4 = \bar{0},$$

The vector loop for mechanism in Problem 2.15.
has scalar components,

$$r_1 + r_3 \cos\theta_3 + r_2 \cos\theta_2 = 0$$
$$-r_4 + r_3 \sin\theta_3 + r_2 \sin\theta_2 = 0,$$

and a geometric constraint,

$$\theta_3 + (\pi/2) - \theta_2 = 0.$$

In this system of three position equations there are

scalar knowns: $r_4, \theta_4 = (-\pi/2), \theta_1 = 0, r_3$

and

scalar unknowns: $r_1, \theta_3, r_2, \theta_2$.

The system of three position equations in four unknowns means there is one degree of freedom. Take r_1 to be the input (S_i). Then:

1. Derive a system of equations in matrix form that can be solved for the basic first-order kinematic coefficients.
2. Derive a system of equations in matrix form that can be solved for the basic second-order kinematic coefficients.

Problem 4.6

Consider the transmission in homework Problem 3.13.

1. Derive expressions for the first- and second-order kinematic coefficients. Use θ_1 as the input variable S_i.
2. 5 is the output. Derive an expression for n_s in terms of the kinematic coefficients.
3. If ω_1 is 3000 rpm and α_1 is 80 rev/s^2, compute ω_5 and α_5 .
4. Compare n_s to R, which you derived in homework Problem 3.13.

Problem 4.7

Consider the transmission in homework Problem 3.15.

1. Derive a system of equations which you *could* solve for the first-order kinematic coefficients. No need to solve for them, just derive the system of equations and put them in matrix form. Use θ_6 as S_i.
2. If the input is 6 and the output is 9. derive an expression for n_s in terms of the kinematic coefficients.
3. In terms of the output's kinematic coefficients and the angular velocity and acceleration of the input, write the equations you would use to compute the output angular velocity and acceleration.

Problem 4.8

Consider the transmission in homework Problem 3.16. The output is the compound element consisting of 9 and 3 and the input is 2.

1. Derive an expression for the first-order kinematic coefficient of the output, h_{93}. (Suggestion - if you solve for h_{93} using Cramer's Rule, you will see that there is a column in both the determinant in the numerator and the determinant in the denominator which contains nothing but ρ_8 terms, If you expand both determinants down that column you will see the ρ_8 term cancels from the numerator and denominator.)
2. Derive an expression for n_s in terms of the kinematic coefficients.
3. If the input is a constant is 2000 rpm, compute the output angular velocity and acceleration.
4. Compare n_s to R, which you derived in homework Problem 3.16.

Problem 4.9

You derived the position equations for the windshield wiper mechanism below in Problem 3.7. A motor drives 2, and a wiper blade is attached to gear 5, which oscillates. If we use the vector loop shown in the bottom figure, we have the vector loop equation $\bar{r}_2 + \bar{r}_3 - \bar{r}_4 - \bar{r}_1 = \bar{0}$, which has the scalar components

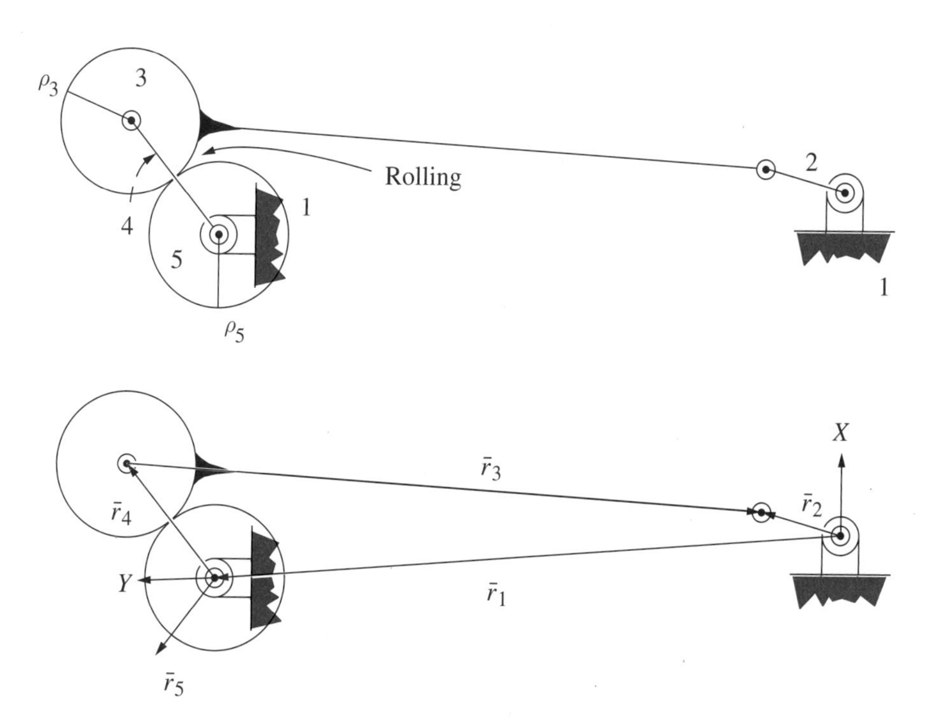

Windshield Wiper Mechanism from Problem 3.7

$$r_2 \cos\theta_2 + r_3 \cos\theta_3 - r_4 \cos\theta_4 = 0 \tag{4.55}$$

$$r_2 \sin\theta_2 + r_3 \sin\theta_3 - r_4 \sin\theta_4 - r_1 = 0. \tag{4.56}$$

There is also the rolling contact equation

$$\rho_5 \Delta\theta_{5/4} = -\rho_3 \Delta\theta_{3/4},$$

Where $\bar{r}_S$ is attached to gear S.
In this mechanism, $\rho_3 = \rho_5 = \rho$. Incorporating this and then fully expanding the rolling contact equation to include assembly configuration give the rolling contact equation in homogeneous form:

$$\rho\left[(\theta_5 - \theta_{5_i}) - 2(\theta_4 - \theta_{4_i}) + (\theta_3 - \theta_{3_i}) - (\theta_4 - \theta_{4_i})\right] = 0. \tag{4.57}$$

Equations (4.55)–(4.57) are system of three positions equations with

scalar knowns: $\rho, \theta_{5_i}, \theta_{4_i}, \theta_{3_i}, r_1, \theta_1 = \frac{\pi}{2}, r_2, r_3, r_4, r_5$

and

scalar unknowns: $\theta_2, \theta_3\ \theta_4, \theta_5,$

so this is a one degree of freedom mechanism. Determine:

1. The geometric condition that defines the dead positions.
2. The geometric condition that defines the limit positions in the oscillation of the wiper blade (i.e., link 5).

Problem 4.10

In the crank-slider mechanism shown in Figure 4.8, take θ_2 to be the input. Then:

1. Derive the geometric condition that defines the dead positions.
2. Derive the geometric condition that defines the limit positions in the rotation of the coupler (connecting rod).
3. Derive the geometric condition that defines the limit positions in the motion of the slider that is the output.
4. Compute the time ratio for the output.

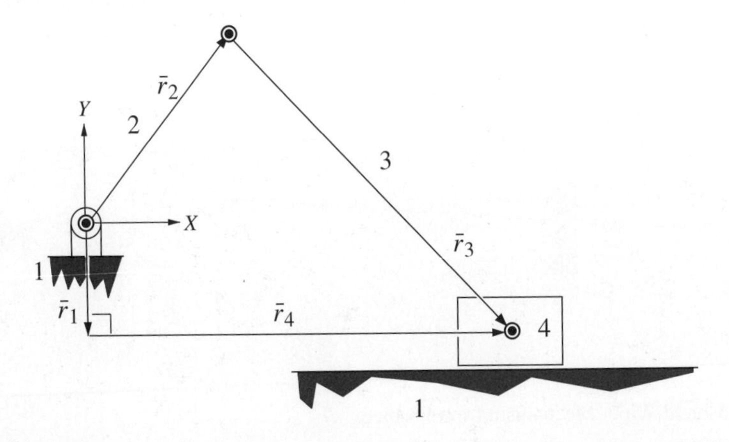

FIGURE 4.8 The offset crank-slider mechanism

4.8 PROGRAMMING PROBLEM

You may solve the following programming problem using any computer language you desire.

Programming Problem 4

For the open four bar shown in the programming problem 2 of Section 2.7, take θ_2 as S_i.

1. Derive symbolic expressions for the kinematic coefficients h_3 and h_4.
2. State the geometric conditions that correspond to $h_3 = 0$ and $h_4 = 0$.
3. Compare the geometric condition for $h_3 = 0$ to the automobile hood near its closed position in Figure 1.14. Why do you suppose they designed that four bar mechanism in that way?
4. State the geometric condition for $h_4 = \infty$. Compare this result to the door near its closed position in Figure 1.15. Why do you suppose they designed that four bar mechanism in that way?
5. Plot h_3 vs. θ_2, h_4 vs. θ_2, h'_3 vs. θ_2, and h'_4 vs. θ_2. Then check your first- and second-order kinematic coefficients graphically.
6. From these plots:
   a. Identify whether the mechanism is a crank-rocker or a crank-crank.
   b. If it is a crank-rocker, what values of θ_2 correspond to the limit positions of the output (4)? And what is the time ratio?

CHAPTER 5

Machine Dynamics Part I: The Inverse Dynamics Problem

This chapter considers the dynamic analysis of machines via Newton's Second Law. In machine dynamics, the forces acting on the links of a machine and the machine's motion are related. Inertia of the links plays a role in this relationship, as do the kinematic coefficients of the mechanism and the motion of the input. In dynamic analysis there are two possible cases. Either the forces are not completely known, or the motion is not completely known. Depending on which of these cases exists, one of two types of dynamics problems arises.

DEFINITIONS:

Inverse Dynamics Problem (IDP) *A problem in which the state of motion of a machine is known and the corresponding forces and torques must be determined.*

Forward Dynamics Problem (FDP) *A problem in which the forces and torques applied to a machine are known and the corresponding state of motion must be determined.*

Both the IDP and the FDP are basic tools used to design machines. In this chapter we discuss the IDP and Newton's Laws.

5.1 REVIEW OF PLANAR KINETICS

The material presented in this section is a review of material you studied in your sophomore-level mechanics class. The goal here is twofold:

1. To prepare you to model joint friction, which is the topic of Chapter 6.
2. To discuss three-dimensional aspects of planar motion force analysis.

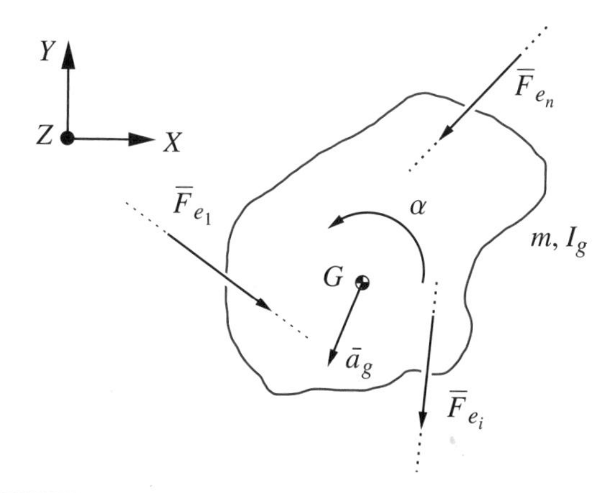

FIGURE 5.1 A system of externally applied forces acting on a rigid body
© Cengage Learning.

Figure 5.1 shows a rigid body moving in a plane under the action of a system of n externally applied forces, $\overline{F}_{e_i}$, $i = 1, 2, ..., n$. The body has a known center of mass at G, a mass m, and a rotational inertia I_g about G.

According to Newton's Second Law,

$$\Sigma\overline{F} = \Sigma\overline{F}_{e_i} = m\bar{a}_g, \tag{5.1}$$

where $\Sigma\overline{F}_{e_i}$ is the vector sum of external forces $\overline{F}_{e_i}$. This is referred to as force equilibrium. For planar motion, Equation (5.1) has only X and Y components; that is, if

$$\overline{F}_{e_i} = \begin{bmatrix} F_{e_{ix}} \\ F_{e_{iy}} \\ 0 \end{bmatrix} \quad \text{and} \quad \bar{a}_g = \begin{bmatrix} a_{gx} \\ a_{gy} \\ 0 \end{bmatrix},$$

then the nontrivial scalar components of Equation (5.1) are

$$\Sigma F_x = \Sigma F_{e_{ix}} = ma_{gx} \tag{5.2}$$

$$\Sigma F_y = \Sigma F_{e_{iy}} = ma_{gy}. \tag{5.3}$$

Likewise, for planar motion, Newton's Second Law states that

$$\Sigma\overline{M}_g = I_g\bar{\alpha}, \tag{5.4}$$

where $\Sigma\overline{M}_g$ is the vector sum of the moments about G of the externally applied forces $\overline{F}_{e_i}$ as well as any externally applied torques (couples or pure moments). This is referred to as moment equilibrium. Equation (5.4) has only a Z component and is a scalar equation. To see this, consider one of the forces $\overline{F}_{e_i}$ and the vector $\bar{r}_{(e/g)_i}$ that locates a point E on the line of action of $\overline{F}_{e_i}$ relative to G, as shown in Figure 5.2. The moment that force $\overline{F}_{e_i}$ creates about G, denoted as $\overline{M}_g$, is found from the vector cross product,

$$\overline{M}_g = \bar{r}_{(e/g)_i} \times \overline{F}_{e_i}. \tag{5.5}$$

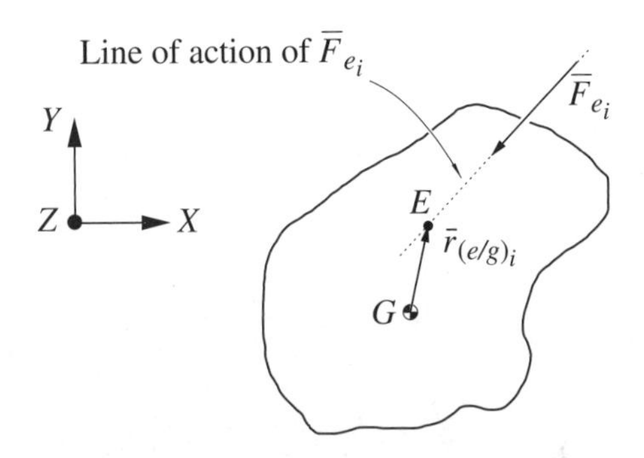

FIGURE 5.2 Moment of $\overline{F}_{e_i}$ about G

Let $\overline{r}_{(e/g)_i}$ have the components

$$\overline{r}_{(e/g)_i} = \begin{bmatrix} r_{(e/g)_{ix}} \\ r_{(e/g)_{iy}} \\ 0 \end{bmatrix},$$

and take the cross product in (5.5),

$$\overline{M}_g = \begin{bmatrix} 0 \\ 0 \\ M_g \end{bmatrix} = \begin{bmatrix} 0 \\ 0 \\ r_{(e/g)_{ix}} F_{e_{iy}} - r_{(e/g)_{iy}} F_{e_{ix}} \end{bmatrix},$$

which shows that $\overline{M}_g$ is a vector with only a Z component—that is, the scalar

$$M_g = r_{(e/g)_{ix}} F_{e_{iy}} - r_{(e/g)_{iy}} F_{e_{ix}}. \tag{5.6}$$

Summing M_g for each externally applied force F_{e_i} gives, from Equation (5.4),

$$\Sigma M_g = \Sigma\left(r_{(e/g)_{ix}} F_{e_{iy}} - r_{(e/g)_{iy}} F_{e_{ix}}\right) = I_g \alpha. \tag{5.7}$$

Equation (5.7) shows that Equation (5.4) is a scalar equation. Equations (5.2), (5.3), and (5.7) are the equilibrium equations for a rigid body undergoing planar motion.

5.1.1 Summing Moments about an Arbitrary Point

In many situations, when solving the IDP, summing moments about a point other than G results in a moment equilibrium condition that involves fewer unknowns. Such points are typically pin joint connections between links. When summing moments about such a point, the forces acting at the pin create no moments and do not appear in the moment equilibrium condition.

Consider an arbitrary point A on the moving body (or on a hypothetical extension of the body), as shown in Figure 5.3, and the moment of a force $\overline{F}_{e_i}$ about A, denoted as $\overline{M}_a$,

$$\overline{M}_a = \overline{r}_{(e/a)_i} \times \overline{F}_{e_i}, \tag{5.8}$$

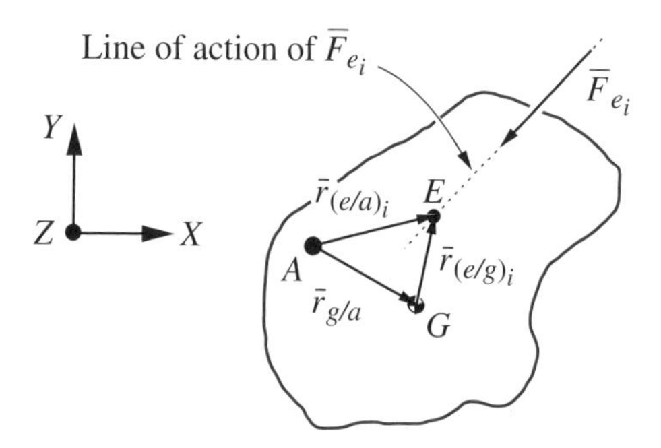

FIGURE 5.3 Moment of $\overline{F}_{e_i}$ about A

and note that although Equation (5.8) is a vector equation, $\overline{M}_a$ has only a Z component and is thus a scalar M_a, just as vector $\overline{M}_g$ was a scalar M_g in Equation (5.6).

From the vector loop in the figure, observe that

$$\bar{r}_{(e/a)_i} = \bar{r}_{g/a} + \bar{r}_{(e/g)_i}$$

so

$$M_a = (\bar{r}_{g/a} + \bar{r}_{(e/g)_i}) \times \overline{F}_{e_i} = \bar{r}_{g/a} \times \overline{F}_{e_i} + \bar{r}_{(e/g)_i} \times \overline{F}_{e_i}.$$

The sum of the moments about A of all the externally applied forces is given by the sum of all the moments M_a contributed by each force $\overline{F}_{e_i}$:

$$\Sigma M_a = \Sigma(\bar{r}_{g/a} \times \overline{F}_{e_i} + \bar{r}_{(e/g)_i} \times \overline{F}_{e_i}) = \Sigma(\bar{r}_{g/a} \times \overline{F}_{e_i}) + \Sigma(\bar{r}_{(e/g)_i} \times \overline{F}_{e_i}). \qquad (5.9)$$

For each force $\overline{F}_{e_i}$ the vector $\bar{r}_{g/a}$ is the same, so it can be factored out of the first summation on the r.h.s. of Equation (5.9):

$$\Sigma M_a = \bar{r}_{g/a} \times \Sigma\overline{F}_{e_i} + \Sigma(\bar{r}_{(e/g)_i} \times \overline{F}_{e_i}).$$

From Equation (5.1), we can replace $\Sigma\overline{F}_{e_i}$ with $m\bar{a}_g$. The second summation on the r.h.s. above is the sum of the moments about G, which from Equation (5.7) can be replaced by $I_g\alpha$, which gives

$$\Sigma M_a = I_g\alpha + \underbrace{\bar{r}_{g/a} \times m\bar{a}_g}_{Z \text{ component only}}. \qquad (5.10)$$

It is important to note that the cross product in Equation (5.10) has only a Z component, so Equation (5.10) is a scalar equation. Equation (5.10) is a generalized form of the moment equilibrium Equation (5.7). If A is taken to be at G ($\bar{r}_{g/a} = \bar{0}$), then Equation (5.10) reduces to Equation (5.7). Since the choice of A in Equation (5.10) is arbitrary, one can choose it wisely to eliminate unknowns from the moment equilibrium equation, thus simplifying the system of equations. This will be illustrated by examples.

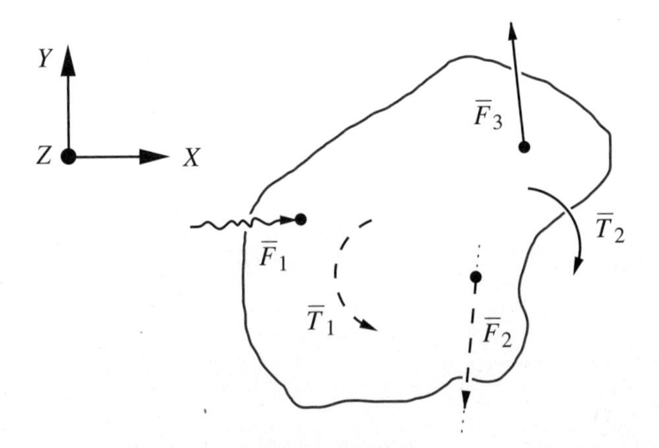

FIGURE 5.4 Graphical notation in free body diagrams

5.1.2 Notation in Free Body Diagrams

When drawing free body diagrams, it is recommended that you adopt a particular graphical notation. This notation considers vectors as having a magnitude and a direction, and it visually conveys which of these is unknown.

Refer to Figure 5.4. Here you see a force $\overline{F}_1$, which is drawn as a "squiggly" vector. This means that $\overline{F}_1$ has unknown magnitude and direction. This is equivalent to saying that $\overline{F}_1$ has unknown X and Y components, which would be denoted as F_{1x} and F_{1y}. Also in Figure 5.4 you see a force $\overline{F}_2$, which is shown as a dashed arrow. This means that $\overline{F}_2$ has the indicated known direction but has unknown magnitude, F_2. If, in the computation of F_2, it turns out that F_2 is negative, then you may reverse the assumed direction and make F_2 positive. Figure 5.4 also shows a vector $\overline{F}_3$, which is shown as a solid arrow. This indicates that $\overline{F}_3$ has known magnitude and direction—in other words, it has known X and Y components.

The figure also shows two torques. As a dashed circular arc, torque $\overline{T}_1$ is shown to be in the counterclockwise (ccw) direction with an unknown magnitude. If the magnitude of $\overline{T}_1$ is computed as being negative, then you may reverse its direction to clockwise (cw) and take its magnitude to be positive. Either way it will appear as a negative number in the Z component of the moment equilibrium equation. As a solid circular arc, $\overline{T}_2$ has a known magnitude and direction. Since its direction is cw, it will appear as a negative term when moments are summed. If its direction had been ccw, it would have been a positive term when moments were summed.

The following example outlines a step-by-step procedure we recommend that you follow when solving an inverse dynamics problem.

▶ EXAMPLE 5.1

Consider the offset crank-slider mechanism shown at the top of Figure 5.5. A known load torque $\overline{T}_2$ acts on link 2, and an unknown driving force $\overline{F}_4$ is applied to link 4. The line of action of $\overline{F}_4$ is offset from wrist pin B by a perpendicular distance that is the magnitude of vector $\overline{r}_5$.

The center of Figure 5.5 shows the vector loop for the mechanism. The dimensions r_1, r_2, and r_3 are knowns. Take θ_2 to be the input for the mechanism, and suppose that θ_2, $\dot{\theta}_2$, and

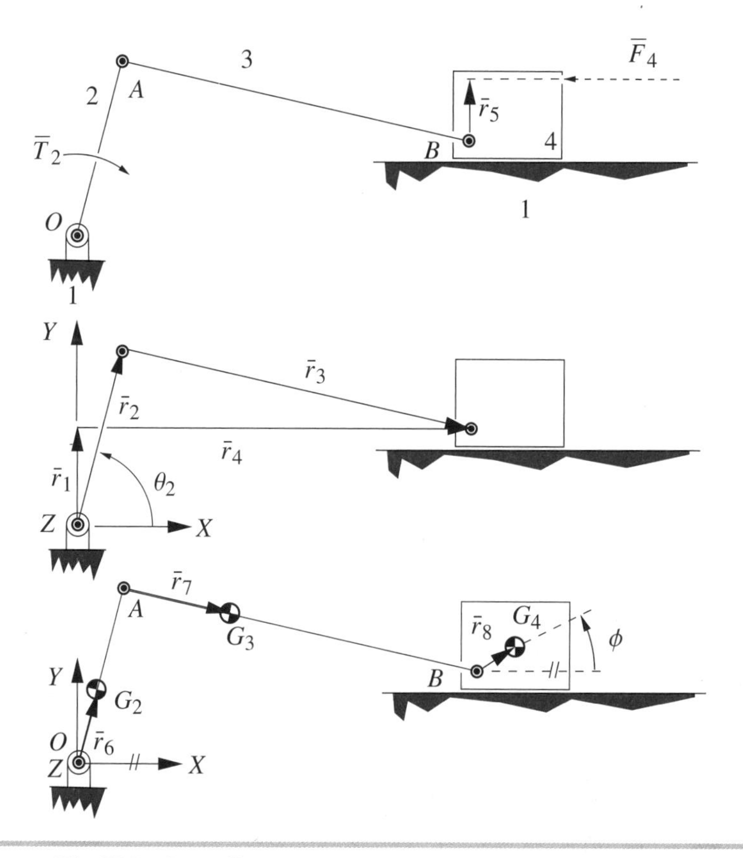

FIGURE 5.5 The IDP of an offset crank-slider

$\ddot{\theta}_2$ are known; that is, the state of motion is known. The outputs are θ_3 and r_4. The bottom of Figure 5.5 shows the location of the mass centers, defined by known vectors $\overline{r}_6$, $\overline{r}_7$, and $\overline{r}_8$ and angle ϕ. The inertias m_2, m_3, m_4, I_{g_2}, I_{g_3}, and I_{g_4} are known.

Outline the equations you would use in a computer program that would compute the magnitude of the driving force that overcomes the load torque and causes the desired state of motion. Neglect gravity and weight forces. (These known forces only add known terms to the equilibrium equations.)

Solution

Step 1: Perform the standard kinematic analysis using the vector loop shown. Using $S_i = \theta_2$:

- Compute the solution to the position problem, finding θ_3 and r_4.
- Compute the basic first-order kinematic coefficients h_3 and f_4.
- Compute the basic second-order kinematic coefficients h_3' and f_4'.

Using the vector loop in Figure 5.5, we have the vector loop equation

$$\overline{r}_1 + \overline{r}_4 - \overline{r}_3 - \overline{r}_2 = \overline{0},$$

with scalar components

$$\begin{aligned} r_4 - r_3\cos\theta_3 - r_2\cos\theta_2 &= 0 \\ r_1 - r_3\sin\theta_3 - r_2\sin\theta_2 &= 0, \end{aligned}$$

which contain

scalar knowns: $r_1, \theta_1 = \frac{\pi}{2}, \theta_4 = 0, r_3, r_2$

and

scalar unknowns: r_4, θ_3, θ_2.

Two scalar position equations in three scalar unknowns means this mechanism has one dof. Of the three unknowns, take S_i to be θ_2. Solve the position problem for θ_3 and r_4 by implementing Newton's Method from Section 2.4, using

$$\bar{x} = \begin{bmatrix} \theta_3 \\ r_4 \end{bmatrix}$$

$$\bar{f}(\bar{x}) = \begin{bmatrix} r_4 - r_3\cos\theta_3 - r_2\cos\theta_2 \\ r_1 - r_3\sin\theta_3 - r_2\sin\theta_2 \end{bmatrix}$$

$$A = \begin{bmatrix} r_3\sin\theta_3 & 1 \\ -r_3\cos\theta_2 & 0 \end{bmatrix}.$$

Step 2: Find the first- and second-order kinematic coefficients of the mass centers.

$$\bar{r}_{g2} = \bar{r}_6 \rightarrow \begin{matrix} r_{g2x} = r_6\cos\theta_2 \\ r_{g2y} = r_6\sin\theta_2 \end{matrix} \rightarrow \begin{matrix} f_{g2x} = -r_6\sin\theta_2 \\ f_{g2y} = r_6\cos\theta_2 \end{matrix} \rightarrow \begin{matrix} f'_{g2x} = -r_6\cos\theta_2 \\ f'_{g2y} = -r_6\sin\theta_2 \end{matrix}$$

$$\bar{r}_{g3} = \bar{r}_2 + \bar{r}_7 \rightarrow \begin{matrix} r_{g3_x} = r_2\cos\theta_2 + r_7\cos\theta_3 \\ r_{g3_y} = r_2\sin\theta_2 + r_7\sin\theta_3 \end{matrix} \rightarrow \begin{matrix} f_{g3_x} = -r_2\sin\theta_2 - r_7\sin\theta_3 h_3 \\ f_{g3_y} = r_2\cos\theta_2 + r_7\cos\theta_3 h_3 \end{matrix} \rightarrow$$

$$\begin{matrix} f'_{g3x} = -r_2\cos\theta_2 - r_7\cos\theta_3 h_3^2 - r_7\sin\theta_3 h'_3 \\ f'_{g3y} = -r_2\sin\theta_2 - r_7\sin\theta_3 h_3^2 + r_7\cos\theta_3 h'_3 \end{matrix}$$

$$\bar{r}_{g4} = \bar{r}_1 + \bar{r}_4 + \bar{r}_8 \rightarrow \begin{matrix} r_{g4x} = r_4 + r_8\cos\phi \\ r_{g4y} = r_1 + r_8\sin\phi \end{matrix} \rightarrow \begin{matrix} f_{g4x} = f_4 \\ f_{g4y} = 0 \end{matrix} \rightarrow \begin{matrix} f'_{g4x} = f'_4 \\ f'_{g4y} = 0 \end{matrix}$$

Step 3: Compute the accelerations of the mass centers and the angular accelerations of the links.

$$a_{g2_x} = f_{g2x}\ddot{\theta}_2 + f'_{g2x}\dot{\theta}_2^2$$
$$a_{g2_y} = f_{g2y}\ddot{\theta}_2 + f'_{g2y}\dot{\theta}_2^2$$

$$a_{g3_x} = f_{g3x}\ddot{\theta}_2 + f'_{g3x}\dot{\theta}_2^2$$
$$a_{g3_y} = f_{g3y}\ddot{\theta}_2 + f'_{g3y}\dot{\theta}_2^2$$

$$a_{g4_x} = f_{g4x}\ddot{\theta}_2 + f'_{g4x}\dot{\theta}_2^2$$
$$a_{g4_y} = f_{g4y}\ddot{\theta}_2 + f'_{g4y}\dot{\theta}_2^2 = 0$$

$$\alpha_2 = \ddot{\theta}_2$$
$$\alpha_3 = h_3\ddot{\theta}_2 + h'_3\dot{\theta}_2^2$$
$$\alpha_4 = 0$$

Step 4: Compute the known inertial terms associated with each link. In this example, compute $m_2a_{g2_x}, m_2a_{g2_y}, m_3a_{g3_x}, m_3a_{g3_y}, m_4a_{g4_x}, m_4a_{g4_y} = 0, I_{g2}\alpha_2, I_{g3}\alpha_3$, and $I_{g4}\alpha_4 = 0$.

Step 5: Sketch a free body diagram (FBD) of each link. Wherever two bodies were previously connected, you should include the forces of action and reaction that exist at that connection. Figure 5.6 shows the free body diagrams. For now, we will neglect friction in the joints.

The FBD of link 2 has an unknown force at O acting from 1 to 2, called $\overline{F}_{12}$. There is also an unknown force at A from 3 to 2. This is taken to be the reaction of the force from 2 to 3 and thus is labeled $-\overline{F}_{23}$. Link 2 also has the load torque $\overline{T}_2$ acting on it.

The FBD of link 3 has the force from 2 to 3, $\overline{F}_{23}$, acting at point A, and the reaction to the force from 3 to 4 acting at B, labeled as $-\overline{F}_{34}$.

The FBD of link 4 has the force from 3 to 4, $\overline{F}_{34}$, acting at B, and the driving force with unknown magnitude but known direction $\overline{F}_4$. The line of action of $\overline{F}_4$ is located by the known vector $\bar{r}_5$. The reaction force from 1 to 4 is acting in the direction of the common normal to 1 and 4, so its direction is known. The line of action of $\overline{F}_{14}$ is located relative to point B by the vector $\bar{r}_9$, which is perpendicular to $\overline{F}_{14}$. The magnitude of $\bar{r}_9$ is an unknown.

Even before writing the equilibrium conditions, of which there are three for each body (two components of force equilibrium—Equations (5.2) and (5.3)—and one component of moment equilibrium, which is the Z component of Equation (5.10); recall that this equation has *only* a Z component), we can check now to see whether what we have drawn is correct by considering the number of unknowns,

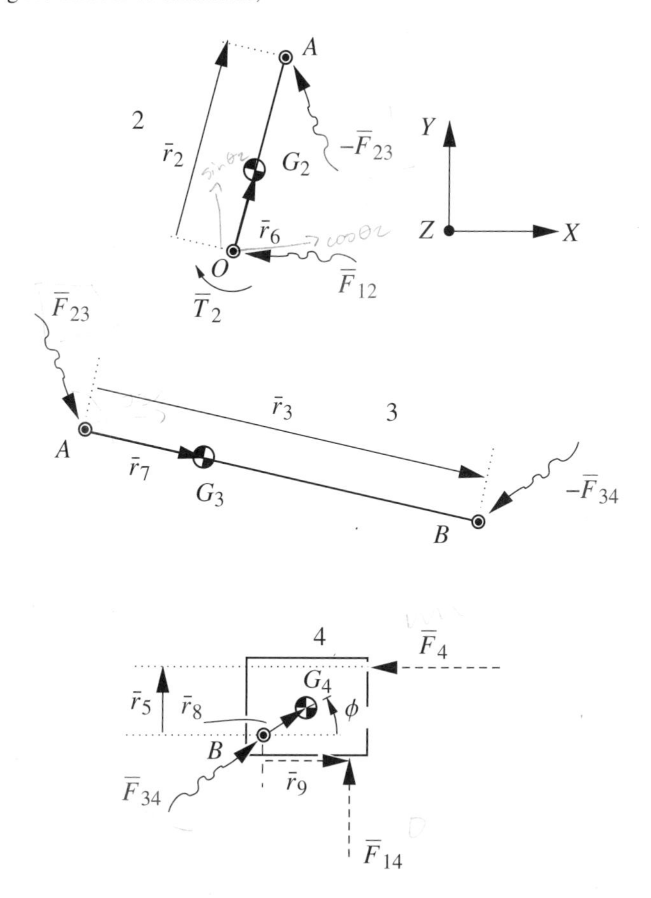

FIGURE 5.6 Free body diagrams for the IDP of an offset crank-slider

scalar unknowns: $F_{12x}, F_{12y}, F_{23x}, F_{23y}, F_{34x}, F_{34y}, F_4, r_9, F_{14},$

which is nine unknowns. There are three bodies with three equilibrium conditions per body, for a total of nine equilibrium equations. The number of unknowns and the number of equations are the same, so there is a solution and what we have done is correct. It is now a matter of writing the equilibrium conditions and solving them for the unknowns.

Step 6: Write the force and moment equilibrium equations, Equations (5.2), (5.3), and (5.10), for each link,

Link 2::
$\Sigma\overline{F} = \overline{F}_{12} - \overline{F}_{23} = m_2\overline{a}_{g2}$
which has scalar components
$\Sigma F_x = F_{12_x} - F_{23_x} = m_2 a_{g2_x}$
$\Sigma F_y = F_{12_y} - F_{23_y} = m_2 a_{g2_y}$

$$\Sigma\overline{M}_o = \overline{r}_2 \times (-\overline{F}_{23}) + \overline{T}_2 = I_{g2}\begin{bmatrix} 0 \\ 0 \\ \alpha_2 \end{bmatrix} + \overline{r}_6 \times m_2\overline{a}_{g2}, \text{ where } \overline{r}_6 = r_6\begin{bmatrix} \cos\theta_2 \\ \sin\theta_2 \\ 0 \end{bmatrix}.$$

This equation has only a Z component
$$\Sigma M_o = \left(r_2\cos\theta_2(-F_{23_y}) - r_2\sin\theta_2(-F_{23_x})\right) - T_2 = m_2 r_6(\cos\theta_2 a_{g2_y} - \sin\theta_2 a_{g2_x}) + I_{g2}\alpha_2.$$

Link 3::
$\Sigma\overline{F} = \overline{F}_{23} - \overline{F}_{34} = m_3\overline{a}_{g3}$
which has scalar components
$\Sigma F_x = F_{23_x} - F_{34_x} = m_3 a_{g3_x}$
$\Sigma F_y = F_{23_y} - F_{34_y} = m_3 a_{g3_y}$

$$\Sigma\overline{M}_a = \overline{r}_3 \times (-\overline{F}_{34}) = I_{g3}\begin{bmatrix} 0 \\ 0 \\ \alpha_3 \end{bmatrix} + \overline{r}_7 \times m_3\overline{a}_{g3}, \text{ where } \overline{r}_7 = r_7\begin{bmatrix} \cos\theta_3 \\ \sin\theta_3 \\ 0 \end{bmatrix}.$$

This equation has only a Z component
$$\Sigma M_a = \left(r_3\cos\theta_3(-F_{34_y}) - r_3\sin\theta_3(-F_{34_x})\right) = m_3 r_7(\cos\theta_3 a_{g3_y} - \sin\theta_3 a_{g3_x}) + I_{g3}\alpha_3.$$

Link 4::
$\Sigma\overline{F} = \overline{F}_{34} + \overline{F}_4 + \overline{F}_{14} = m_4\overline{a}_{g4}$
which has scalar components
$\Sigma F_x = F_{34_x} - F_4 = m_4 a_{g4_x}$
$\Sigma F_y = F_{34_y} + F_{14} = m_4 a_{g4_y} = 0$

$$\Sigma\overline{M}_b = \overbrace{\overline{r}_5 \times \overline{F}_4}^{\text{only has Z comp. } r_5F_4} + \overbrace{\overline{r}_9 \times \overline{F}_{14}}^{\text{only has Z comp. } r_9F_{14}} = r_8 \times m_4\overline{a}_{g4},$$

This equation has only a Z component
$\Sigma M_b = r_5F_4 + r_9F_{14} = -m_4 r_8 \sin\phi a_{g4_x}.$

The system of equations is linear, except for the term (r_9F_{14}) that appears in ΣM_b. Replace the unknown r_9 with an unknown that is the product (r_9F_{14}). This linearizes the system. After this linear system is solved, the values of F_{14} and (r_9F_{14}) can be used to compute r_9:

$$r_9 = \frac{(r_9F_{14})}{F_{14}}.$$

The linear system can be written in matrix form as

$$\begin{bmatrix} 1 & 0 & -1 & 0 & 0 & 0 & 0 & 0 & 0 \\ 0 & 1 & 0 & -1 & 0 & 0 & 0 & 0 & 0 \\ 0 & 0 & r_2\sin\theta_2 & -r_2\cos\theta_2 & 0 & 0 & 0 & 0 & 0 \\ 0 & 0 & 1 & 0 & -1 & 0 & 0 & 0 & 0 \\ 0 & 0 & 0 & 1 & 0 & -1 & 0 & 0 & 0 \\ 0 & 0 & 0 & 0 & r_3\sin\theta_3 & -r_3\cos\theta_3 & 0 & 0 & 0 \\ 0 & 0 & 0 & 0 & 1 & 0 & -1 & 0 & 0 \\ 0 & 0 & 0 & 0 & 0 & 1 & 0 & 1 & 0 \\ 0 & 0 & 0 & 0 & 0 & 0 & r_5 & 0 & 1 \end{bmatrix} \begin{bmatrix} F_{12x} \\ F_{12y} \\ F_{23x} \\ F_{23y} \\ F_{34x} \\ F_{34y} \\ F_4 \\ F_{14} \\ (r_9F_{14}) \end{bmatrix} =$$

$$\begin{bmatrix} m_2a_{g2_x} \\ m_2a_{g2_y} \\ m_2r_6\left(\cos\theta_2a_{g2_y} - \sin\theta_2a_{g2_x}\right) + I_{g2}\alpha_2 + T_2 = 0 \\ m_3a_{g3_x} \\ m_3a_{g3_y} \\ m_3r_7\left(\cos\theta_3a_{g3_y} - \sin\theta_3a_{g3_x}\right) + I_{g3}\alpha_3 = 0 \\ m_4a_{g4_x} \\ 0 \\ -m_4r_8\sin\phi a_{g4_x} \end{bmatrix}.$$

Simplifications in Moment Equations

Choosing a reference point for each moment equation wisely simplified the three moment equilibrium equations in this example, because they involved a minimum number of unknowns. For link 2, ΣM_{g2} would have involved unknowns F_{12x} and F_{12y}, whereas ΣM_o did not. For link 3, ΣM_{g3} would have involved unknowns F_{23x} and F_{23y}, whereas ΣM_a did not. For link 4, ΣM_{g4} would have involved unknowns F_{34x} and F_{34y}, whereas ΣM_b did not. This is the advantage of using Equation (5.10) instead of Equation (5.7).

The following example includes numerical computations. You should follow the units in the computations, and when you make numerical computations, you should show all your units just as the example does.

▶ EXAMPLE 5.2

Consider the inverted crank-slider mechanism shown in Figure 5.7. As a machine, the input crank, link 2, is driven in the cw direction by the cw torque $\overline{T}_2$. The output rocker, link 4, has a ccw load torque $\overline{T}_4$. The load torque is known,

$$T_4 = 150.0 \text{ ft} \cdot \text{lb}_\text{f} \text{ ccw} = 150.0 \text{ ft} \cdot \text{lb}_\text{f}, \tag{5.11}$$

and the magnitude of the driving torque is unknown.

The input is the rotation of link 2. Instantaneously, $\theta_2 = 30°$. The machine is driven at a constant speed of 60 rpm cw, so instantaneously,

$$\dot{\theta}_2 = -60 \text{ rpm} = -2\pi \text{ rad/s} \quad \text{and} \quad \ddot{\theta}_2 = 0. \tag{5.12}$$

The inertia properties are listed in Figure 5.8. Determine the value of $\overline{T}_2$ that will overcome the load in Equation (5.11) and produce the motion in Equation (5.12). This is an IDP. The solution

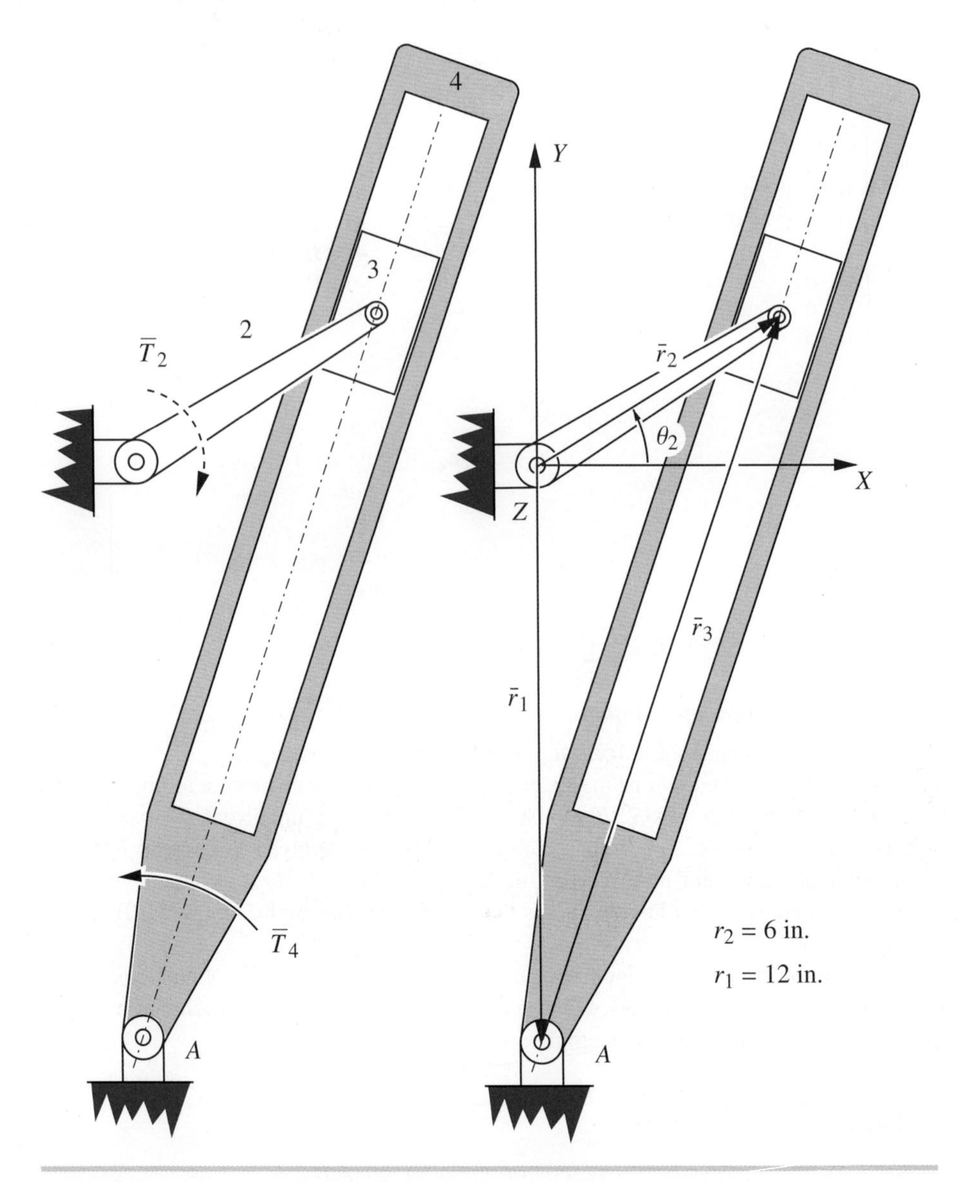

FIGURE 5.7 An inverted crank-slider mechanism and its vector loop

in this example will include all the numerical details, such as using the gravitational constant, g_c. You should be in the same habit in your own work. g_c is involved in most computations in mechanics, no matter which of the following systems of units is in use.

United States Customary System: (lb_m, lb_f, ft, s) $\longrightarrow$ $g_c = \dfrac{32.174\,\text{ft}\cdot\text{lb}_m}{\text{lb}_f\cdot\text{s}^2}$

British Gravitational System: (slug, lb_f, ft, s) $\longrightarrow$ $g_c = \dfrac{\text{ft}\cdot\text{slug}}{\text{lb}_f\cdot\text{s}^2}$

European Engineering System: (kg_m, kg_f, m, s) $\longrightarrow$ $g_c = \dfrac{9.81\,\text{kg}_m\cdot\text{m}}{\text{kg}_f\cdot\text{s}^2}$

Système Internationale: (kg, N, m, s) $\longrightarrow$ $g_c = \dfrac{\text{kg}\cdot\text{m}}{\text{N}\cdot\text{s}^2}$

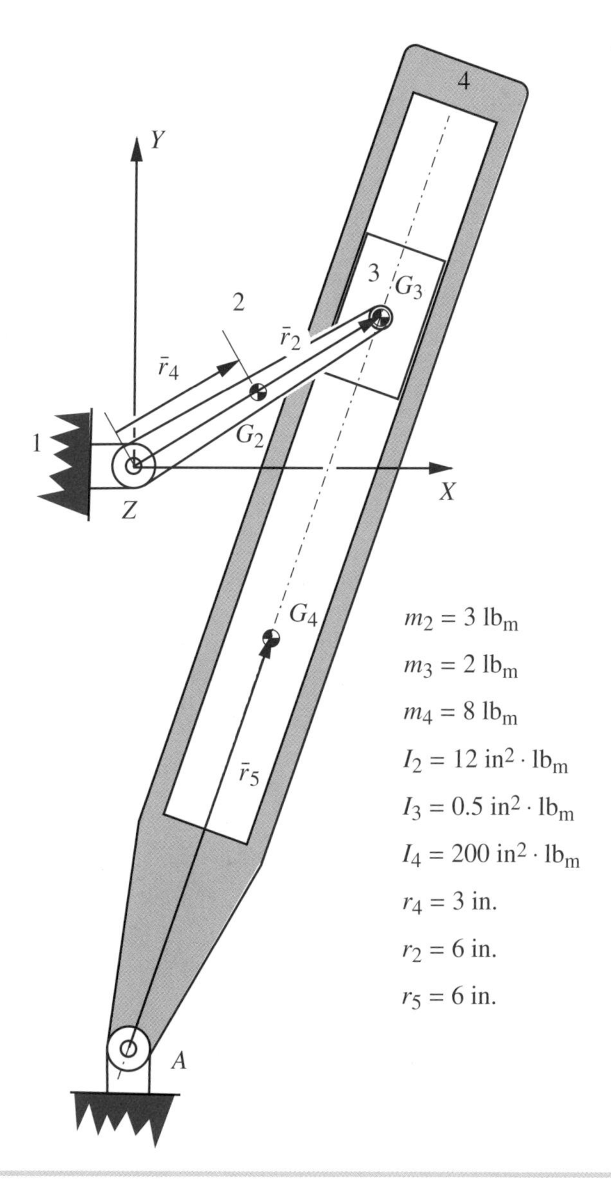

FIGURE 5.8 Inertia properties of the moving links

In any computer programs you write, you should work in units of the British Gravitational System or the Système Internationale. These are referred to as "consistent" sets of units because g_c has a magnitude of 1. Failing to use a consistent sets of units is asking for trouble because you will have to follow the magnitude of g_c through all of your equations. Always input numbers in a consistent set of units and recognize that the numbers coming out of the program are also in these units. So that you can appreciate the role of g_c in these computations, this numerical example will not be worked in a consistent set of units. Please note the abundant use of g_c. If you were to work in this system of units in a computer program, at every point where you forgot to apply g_c your results would be off by a factor of 32.174, and as you will see, this occurs many times. So be certain to compute in a consistent set of units whenever you make computations.

Solution

The solution follows the six steps outlined in Example 5.1. The solution presents all the numerical results. In general, unless you are specifically asked for numerical results, you need only outline all the equations, as in Example 5.1.

Step 1: Find the position solution and the basic first- and second-order kinematic coefficients. In Appendix I you will find the standard kinematic analysis using the vector loop method that you have learned. The vector loop and the link dimensions are shown in Figure 5.7. The value of the input is $\theta_2 = 30° = \frac{\pi}{6}$ radians. The result of the kinematic analysis is the position solution,

$$\theta_3 = 70.893°, \quad r_3 = 15.875 \text{ in.},$$

and the basic first- and second-order kinematic coefficients are

$$h_3 = 0.286, \quad f_3 = 3.928 \text{ in.}$$
$$h_3' = 0.106, \quad f_3' = -3.240 \text{ in.}$$

Step 2: Find the first- and second-order kinematic coefficients of the mass centers. In Appendix I these are computed as

$$f_{g2_x} = -1.500 \text{ in.}, \quad f'_{g2_x} = -2.598 \text{ in.}$$
$$f_{g2_y} = 2.598 \text{ in.}, \quad f'_{g2_y} = -1.500 \text{ in.}$$
$$f_{g3_x} = -3.000 \text{ in.}, \quad f'_{g3_x} = -5.196 \text{ in.}$$
$$f_{g3_y} = 5.196 \text{ in.}, \quad f'_{g3_y} = -3.000 \text{ in.}$$
$$f_{g4_x} = -2.430 \text{ in.}, \quad f'_{g4_x} = -1.142 \text{ in.}$$
$$f_{g4_y} = 0.842 \text{ in.}, \quad f'_{g4_y} = -0.382 \text{ in.}$$

Step 3: Compute the acceleration of the mass centers and the angular acceleration of the links.

$$a_{g2_x} = f_{g2_x}\ddot{\theta}_2 + f'_{g2_x}\dot{\theta}_2^2 = (-1.500 \text{ in.})(0) + -2.598 \text{ in.}(-2\pi \text{ rad/s})^2 = -102.57 \text{ in./s}^2$$
$$a_{g2_y} = f_{g2_y}\ddot{\theta}_2 + f'_{g2_y}\dot{\theta}_2^2 = (2.598 \text{ in.})(0) + -1.500 \text{ in.}\ (-2\pi \text{ rad/s})^2 = -59.22 \text{ in./s}^2$$
$$a_{g3_x} = f_{g3_x}\ddot{\theta}_2 + f'_{g3_x}\dot{\theta}_2^2 = (-3.000 \text{ in.})(0) + -5.196 \text{ in.}\ (-2\pi \text{ rad/s})^2 = -205.14 \text{ in./s}^2$$
$$a_{g3_y} = f_{g3_y}\ddot{\theta}_2 + f'_{g3_y}\dot{\theta}_2^2 = (5.196 \text{ in.})(0) + -3.000 \text{ in.}\ (-2\pi \text{ rad/s})^2 = -118.44 \text{ in./s}^2$$
$$a_{g4_x} = f_{g4_x}\ddot{\theta}_2 + f'_{g4_x}\dot{\theta}_2^2 = (-2.4230 \text{ in.})(0) + -1.142 \text{ in.}\ (-2\pi \text{ rad/s})^2 = -45.10 \text{ in./s}^2$$
$$a_{g4_y} = f_{g4_y}\ddot{\theta}_2 + f'_{g4_y}\dot{\theta}_2^2 = 0 = (0.842 \text{ in.})(0) + -0.382 \text{ in.}\ (-2\pi \text{ rad/s})^2 = -15.07 \text{ in./s}^2$$

$$\alpha_2 = \ddot{\theta}_2 = 0 \text{ rad/s}^2$$
$$\alpha_3 = h_3\ddot{\theta}_2 + h_3'\dot{\theta}_2^2 = 0.286(0) + 0.106(-2\pi \text{ rad/s})^2 = 4.19 \text{ rad/s}^2$$
$$\alpha_4 = h_3\ddot{\theta}_2 + h_3'\dot{\theta}_2^2 = 0.286(0) + 0.106(-2\pi \text{ rad/s})^2 = 4.19 \text{ rad/s}^2$$

Step 4: Compute the inertial terms associated with each link. Watch your units in these calculations.

$$m_2 a_{g2_x} = 3\ \text{lb}_\text{m}\left(-102.57\ \frac{\text{in.}}{\text{s}^2}\right)\left(\frac{\text{ft}}{12\ \text{in.}}\right)\left(\frac{\text{lb}_\text{f}\cdot\text{s}^2}{32.174\ \text{ft}\cdot\text{lb}_\text{m}}\right) = -0.80\ \text{lb}_\text{f}$$

$$m_2 a_{g2_y} = 3\ \text{lb}_\text{m}\left(-59.22\ \frac{\text{in.}}{\text{s}^2}\right)\left(\frac{\text{ft}}{12\ \text{in.}}\right)\left(\frac{\text{lb}_\text{f}\cdot\text{s}^2}{32.174\ \text{ft}\cdot\text{lb}_\text{m}}\right) = -0.46\ \text{lb}_\text{f}$$

$$m_3 a_{g3_x} = 2\ \text{lb}_\text{m}\left(-205.14\ \frac{\text{in.}}{\text{s}^2}\right)\left(\frac{\text{ft}}{12\ \text{in.}}\right)\left(\frac{\text{lb}_\text{f}\cdot\text{s}^2}{32.174\ \text{ft}\cdot\text{lb}_\text{m}}\right) = -1.06\ \text{lb}_\text{f}$$

$$m_3 a_{g3_y} = 2\ \text{lb}_\text{m}\left(-118.44\ \frac{\text{in.}}{\text{s}^2}\right)\left(\frac{\text{ft}}{12\ \text{in.}}\right)\left(\frac{\text{lb}_\text{f}\cdot\text{s}^2}{32.174\ \text{ft}\cdot\text{lb}_\text{m}}\right) = -0.61\ \text{lb}_\text{f}$$

$$m_4 a_{g4_x} = 8\ \text{lb}_\text{m}\left(-45.10\ \frac{\text{in.}}{\text{s}^2}\right)\left(\frac{\text{ft}}{12\ \text{in.}}\right)\left(\frac{\text{lb}_\text{f}\cdot\text{s}^2}{32.174\ \text{ft}\cdot\text{lb}_\text{m}}\right) = -0.93\ \text{lb}_\text{f}$$

$$m_4 a_{g4_y} = 8\ \text{lb}_\text{m}\left(-15.07\ \frac{\text{in.}}{\text{s}^2}\right)\left(\frac{\text{ft}}{12\ \text{in.}}\right)\left(\frac{\text{lb}_\text{f}\cdot\text{s}^2}{32.174\ \text{ft}\cdot\text{lb}_\text{m}}\right) = -0.31\ \text{lb}_\text{f}$$

$$I_{g2}\alpha_2 = 12.0\ \text{in}^2\cdot\text{lb}_\text{m}(0) = 0$$

$$I_{g3}\alpha_3 = 0.5\ \text{in}^2\cdot\text{lb}_\text{m}\left(4.19\ \frac{\text{rad}}{\text{s}^2}\right)\left(\frac{\text{ft}^2}{144\ \text{in}^2}\right)\left(\frac{\text{lb}_\text{f}\cdot\text{s}^2}{32.174\ \text{ft}\cdot\text{lb}_\text{m}}\right) = .00045\ \text{ft}\cdot\text{lb}_\text{f}$$

$$I_{g4}\alpha_4 = 200.0\ \text{in}^2\cdot\text{lb}_\text{m}\left(4.19\ \frac{\text{rad}}{\text{s}^2}\right)\left(\frac{\text{ft}^2}{144\ \text{in}^2}\right)\left(\frac{\text{lb}_\text{f}\cdot\text{s}^2}{32.174\ \text{ft}\cdot\text{lb}_\text{m}}\right) = 0.18\ \text{ft}\cdot\text{lb}_\text{f}$$

Step 5: Draw a free body diagram of each link as in Figure 5.9. To simplify the moment equilibrium equation for link 2, sum moments about point O. This eliminates the unknown components of $\overline{F}_{12}$ from the equation. For link 3, sum moments about G_3. This eliminates the unknown components of $\overline{F}_{43}$ from the equation. For link 4, sum moments about A. This eliminates the unknown components of $\overline{F}_{14}$ from the equation.

The system of three bodies has a total of nine equilibrium conditions. There are nine unknowns in the free body diagrams—F_{12_x}, F_{12_y}, F_{32_x}, F_{32_y}, F_{14_x}, F_{14_y}, F_{43}, $(r_6 F_{43})$, and T_2—so our free body diagrams are correct.

Step 6: Write the equations for force and moment equilibrium, and solve them for the unknowns.

Link 2::
$\Sigma\overline{F} - \overline{F}_{12} - \overline{F}_{32} = m_2\overline{a}_{g2}$
which has scalar components
$\Sigma F_x = F_{12_x} + F_{32_x} = m_2 a_{g2_x}$
$\Sigma F_y = F_{12_y} + F_{32_y} = m_2 a_{g2_y}$

$$\Sigma\overline{M}_o = \overline{r}_2\times\overline{F}_{32} + \begin{bmatrix}0\\0\\-T_2\end{bmatrix} = I_{g2}\begin{bmatrix}0\\0\\\alpha_2\end{bmatrix} + \overline{r}_4\times m_2\overline{a}_{g2},\ \text{where}\ \overline{r}_4 = r_4\begin{bmatrix}\cos\theta_2\\\sin\theta_2\\0\end{bmatrix}.$$

$$\Sigma M_o = -T_2 + r_2\cos\theta_2 F_{32_y} - r_2\sin\theta_2 F_{32_x} = I_{g2}\alpha_2 + m_2 r_4(\cos\theta_2 a_{g2_y} - \sin\theta_2 a_{g2_x}).$$

Link 3::
$\Sigma\overline{F} = -\overline{F}_{32} + \overline{F}_{43} = m_3\overline{a}_{g3}$

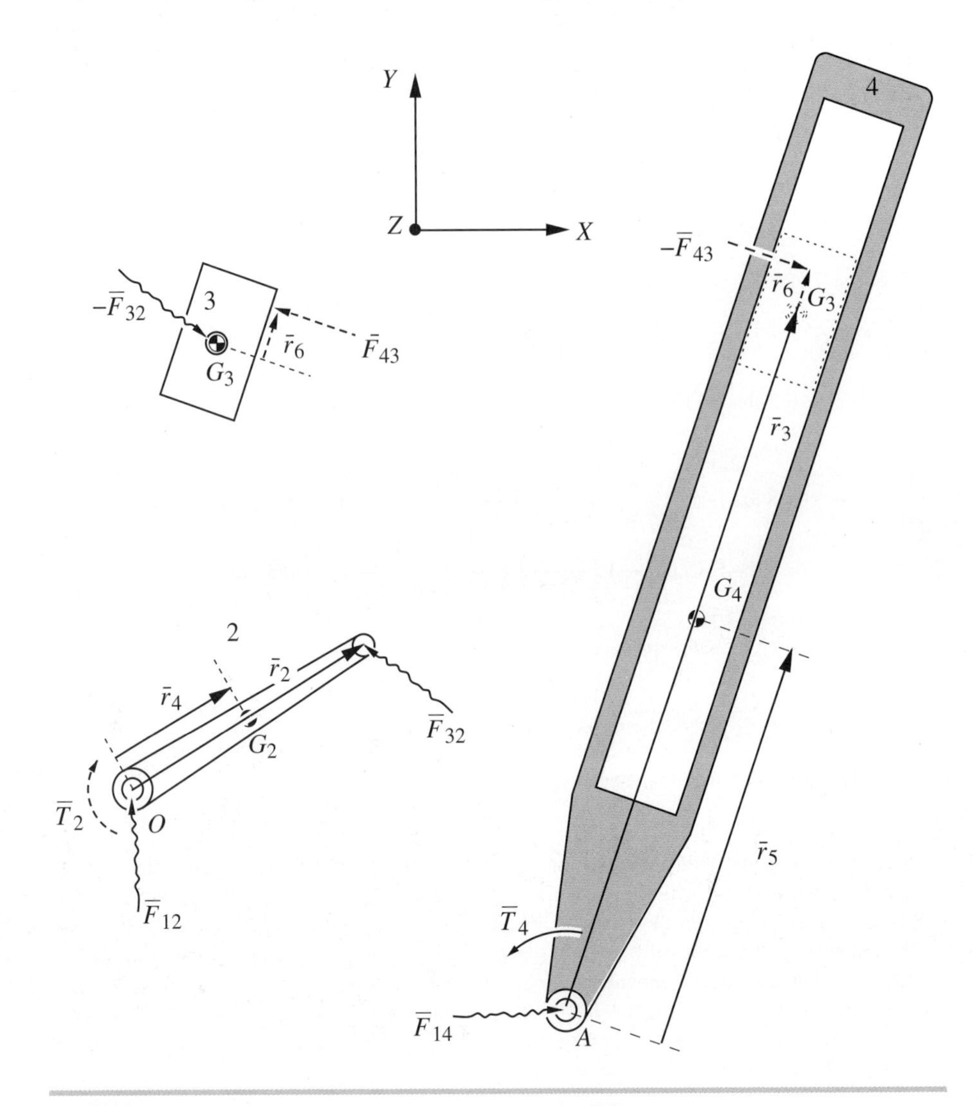

FIGURE 5.9 Free body diagrams of inverted crank-slider

which has scalar components

$\Sigma F_x = -F_{32_x} - F_{43}\sin\theta_3 = m_3 a_{g3_x}$

$\Sigma F_y = -F_{32_y} + F_{43}\cos\theta_3 = m_3 a_{g3_y}$

$$\Sigma \overline{M}_{g3} = \overbrace{\overline{r}_6 \times \overline{F}_{43}}^{\text{has only } Z \text{ comp. } r_6 F_{43}} = I_{g3}\begin{bmatrix} 0 \\ 0 \\ \alpha_3 \end{bmatrix}$$

which has only the Z component

$\Sigma M_{g_3} = (r_6 F_{43}) = I_{g_3}\alpha_3.$

Link 4::

$\Sigma \overline{F} = \overline{F}_{14} - \overline{F}_{43} = m_4 \overline{a}_{g4}$

which has scalar components

$$\Sigma F_x = F_{14_x} + F_{43}\sin\theta_3 = m_4 a_{g4_x}$$
$$\Sigma F_y = F_{14_y} - F_{43}\cos\theta_3 = m_4 a_{g4_y}$$

$$\Sigma \overline{M}_a = \begin{bmatrix} 0 \\ 0 \\ T_4 \end{bmatrix} + \overbrace{(r_3 + \overline{r}_6) \times (-\overline{F}_{34})}^{\text{has only } Z \text{ comp. } -(r_3+r_6)F_{34}} = I_{g4}\begin{bmatrix} 0 \\ 0 \\ \alpha_4 \end{bmatrix} + \overline{r}_5 \times m_4 \overline{a}_{g4}$$

which has only the Z component

$$\Sigma M_a = T_4 - r_3 F_{43} - (r_6 F_{43}) = m_4 r_5(\cos\theta_3 a_{g4_y} - \sin\theta_3 a_{g4_x}) + I_{g4}\alpha_4.$$

Putting these into a matrix form gives

$$\begin{bmatrix} 1 & 0 & 1 & 0 & 0 & 0 & 0 & 0 & 0 \\ 0 & 1 & 0 & 1 & 0 & 0 & 0 & 0 & 0 \\ 0 & 0 & -r_2\sin\theta_2 & r_2\cos\theta_2 & 0 & 0 & 0 & 0 & -1 \\ 0 & 0 & -1 & 0 & 0 & 0 & -\sin\theta_3 & 0 & 0 \\ 0 & 0 & 0 & -1 & 0 & 0 & \cos\theta_3 & 0 & 0 \\ 0 & 0 & 0 & 0 & 0 & 0 & 0 & 1 & 0 \\ 0 & 0 & 0 & 0 & 1 & 0 & \sin\theta_3 & 0 & 0 \\ 0 & 0 & 0 & 0 & 0 & 1 & -\cos\theta_3 & 0 & 0 \\ 0 & 0 & 0 & 0 & 0 & 0 & -r_3 & -1 & 0 \end{bmatrix} \begin{bmatrix} F_{12x} \\ F_{12y} \\ F_{32x} \\ F_{32y} \\ F_{14x} \\ F_{14y} \\ F_{43} \\ (r_6F_{43}) \\ T_2 \end{bmatrix} =$$

$$\begin{bmatrix} m_2 a_{g2_x} \\ m_2 a_{g2_y} \\ I_{g2}\alpha_2 + m_2 r_4\left(\cos\theta_2 a_{g2_y} - \sin\theta_2 a_{g2_x}\right) \\ m_3 a_{g3_x} \\ m_3 a_{g3_y} \\ I_3\alpha_3 \\ m_4 a_{g4_x} \\ m_4 a_{g4_y} \\ -T_4 + I_{g4}\alpha_4 + m_4 r_5\left(\cos\theta_3 a_{g4_y} - \sin\theta_3 a_{g4_x}\right) \end{bmatrix}. \quad (5.13)$$

Equation (5.13) can be solved for T_2 by using Cramer's Rule or by writing a Matlab script or by writing a FORTRAN code. Summing moments as we did simplifies the system of Equation (5.13) to the point where they can also be solved for T_2 by back substitution. Specifically, solve the sixth row for (r_6F_{43}),

$$(r_6F_{43}) = 0.00045 \text{ ft} \cdot \text{lb}_\text{f},$$

then solve the ninth row for F_{43},

$$F_{43} = 112.81 \text{ lb}_\text{f},$$

then solve the fourth and fifth rows for F_{32x} and F_{32y},

$$F_{32x} = -105.53 \text{ lb}_\text{f}, \quad F_{32y} = 37.54 \text{ lb}_\text{f},$$

and finally, solve the third row for T_2,

$$T_2 = -42.64 \text{ in.} \cdot \text{lb}_\text{f}, \quad (5.14)$$

which is the desired solution to the IDP. Thus T_2 has a magnitude of 42.62 in · $\mathrm{lb_f}$ and acts in the cw direction.

For the example in Section 5.2.6 in the upcoming discussion of the three-dimensional aspects of planar force analysis, we will need to know $\overline{F}_{12}$. From the first and second row of Equation (5.13),

$$F_{12_x} = 104.73\,\mathrm{lb_f} \quad \text{and} \quad F_{12_y} = -38.00\,\mathrm{lb_f}.$$

Effects of Gravity

The effect of gravity on the IDP is to add known forces to the problem. Let gravity act in the $-Y$ direction. Figure 5.10 shows the new free body diagrams,

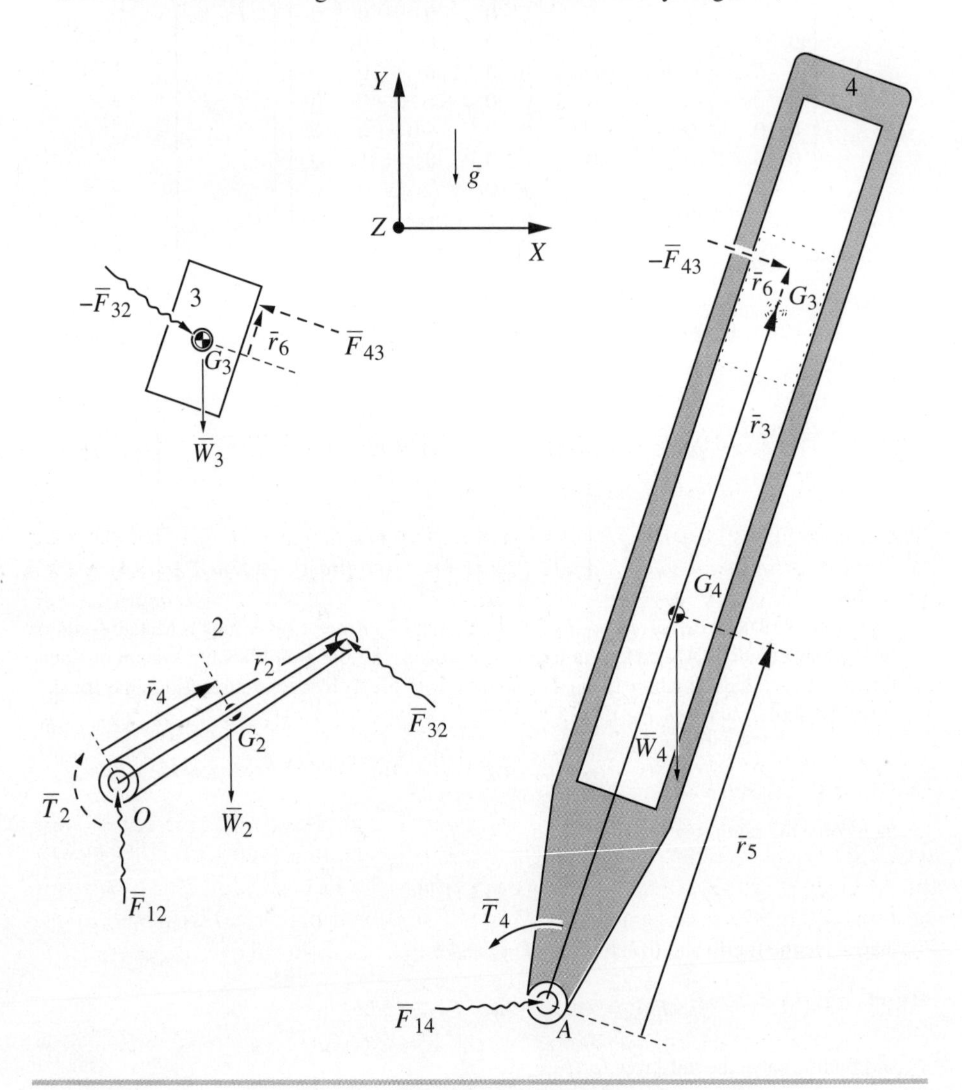

FIGURE 5.10 Free body diagrams of inverted crank-slider with weight forces

and (5.13) becomes

$$\begin{bmatrix} 1 & 0 & 1 & 0 & 0 & 0 & 0 & 0 & 0 \\ 0 & 1 & 0 & 1 & 0 & 0 & 0 & 0 & 0 \\ 0 & 0 & -r_2\sin\theta_2 & r_2\cos\theta_2 & 0 & 0 & 0 & 0 & -1 \\ 0 & 0 & -1 & 0 & 0 & 0 & -\sin\theta_3 & 0 & 0 \\ 0 & 0 & 0 & -1 & 0 & 0 & \cos\theta_3 & 0 & 0 \\ 0 & 0 & 0 & 0 & 0 & 0 & 0 & 1 & 0 \\ 0 & 0 & 0 & 0 & 1 & 0 & \sin\theta_3 & 0 & 0 \\ 0 & 0 & 0 & 0 & 0 & 1 & -\cos\theta_3 & 0 & 0 \\ 0 & 0 & 0 & 0 & 0 & 0 & -r_3 & -1 & 0 \end{bmatrix} \begin{bmatrix} F_{12x} \\ F_{12y} \\ F_{32x} \\ F_{32y} \\ F_{14x} \\ F_{14y} \\ F_{43} \\ (r_6 F_{43}) \\ T_2 \end{bmatrix} =$$

$$\begin{bmatrix} m_2 a_{g_{2x}} \\ m_2 a_{g_{2y}} + W_2 \\ I_{g_2}\alpha_2 + m_2 r_4\left(\cos\theta_2 a_{g_{2y}} - \sin\theta_2 a_{g_{2x}}\right) + r_4\cos\theta_2 W_2 \\ m_3 a_{g_{3x}} \\ m_3 a_{g_{3y}} + W_3 \\ I_{g_3}\alpha_3 \\ m_4 a_{g_{4x}} \\ m_4 a_{g_{4y}} + W_4 \\ -T_4 + I_{g_4}\alpha_4 + m_4 r_5\left(\cos\theta_3 a_{g_{4y}} - \sin\theta_3 a_{g_{4x}}\right) + r_5\cos\theta_3 W_4 \end{bmatrix}. \quad (5.15)$$

Notice the additional knowns on the r.h.s. of Equation (5.15) in comparison to Equation (5.13).

5.2 THREE-DIMENSIONAL ASPECTS IN THE FORCE ANALYSIS OF PLANAR MACHINES

This text discusses planar mechanisms and machines. In a planar mechanism, the mass center of every link has only *X* and *Y* components of velocity and acceleration (the *XY* plane being parallel to the plane of motion) and every link has only a *Z* component of angular velocity or angular acceleration. Thus the force analysis is also two-dimensional (or planar), as we have seen in the examples up until now.

What makes the force analysis of these devices three-dimensional is that the links do not move in the *same XY* plane. This is necessary in order to avoid interferences and collisions between the links. The links and their mass centers move in *parallel planes* that are offset from each other in the *Z* direction.

Figure 5.11 illustrates this for a four bar mechanism. On the left side of the figure you see the planar four bar mechanism by viewing onto the plane of motion, the *XY* plane. On the right side of the figure the view is onto the *YZ* plane, and in that edge view you can see that the mass centers, G_2, G_3, and G_4, move in planes that are parallel to the *XY* plane, resulting in a *partial three-dimensionality*.

DEFINITION:

Partially Three-Dimensional Planar Mechanism *A planar mechanism whose links have mass centers that move in parallel planes.*

As before, since all mass centers have accelerations with only *X* and *Y* components, and there are no external loads applied in the *Z* direction, the joints transmit reaction

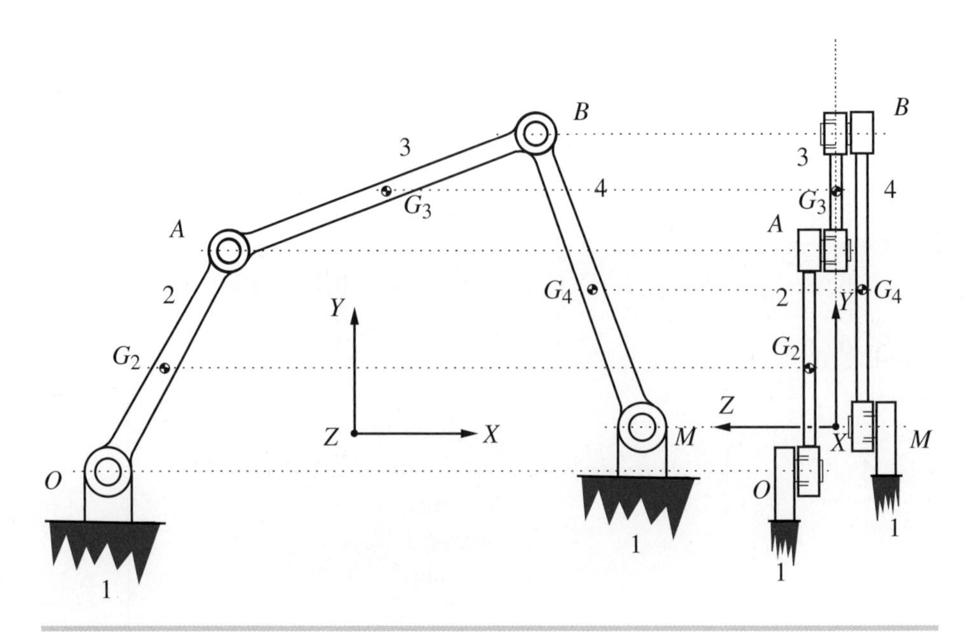

FIGURE 5.11 "Out of plane" mass centers of a four bar mechanism

forces with only X and Y components. Also as before, the joints cannot transmit any components of reaction moments in the Z direction, because they rotate freely.

However, the result of the partial three-dimensionality is that the joints will now transmit reaction moments in the X and Y directions. Let us investigate this. Our goal is to develop a procedure by which we can take the results of our planar force analysis and map them into the results for the actual system that has a partial three-dimensionality. This is important because the designer must determine the forces acting at each of *two* bearings that support a shaft that rotates with the link or support a link that rotates on a fixed shaft. The designer is interested in these bearing forces (loads) in order to be able to select the proper size bearing for a given situation.

But first, we present a review of three-dimensional rigid body kinetics, a subject you should have studied in your sophomore mechanics course.

5.2.1 Spatial Kinetics of a Rigid Body; The Newton–Euler Equations

Consider the rigid body shown in Figure 5.12, which is undergoing a spatial motion. Attached to the body is coordinate system xyz, which moves with the body and has its origin at G. The XYZ system is fixed. Let the body (that is, coordinate system xyz) have an angular velocity $\overline{\omega}$ and an angular acceleration $\overline{\alpha}$ ($\overline{\alpha} = \frac{d\,\overline{\omega}}{dt}$), where

$$\overline{\omega} = \begin{bmatrix} \omega_x \\ \omega_y \\ \omega_z \end{bmatrix} \quad \text{and} \quad \overline{\alpha} = \begin{bmatrix} \alpha_x \\ \alpha_y \\ \alpha_z \end{bmatrix}.$$

According to Newton's Laws, force equilibrium requires

$$\Sigma\overline{F} = \Sigma\overline{F}_{e_i} = m\bar{a}_g, \tag{5.16}$$

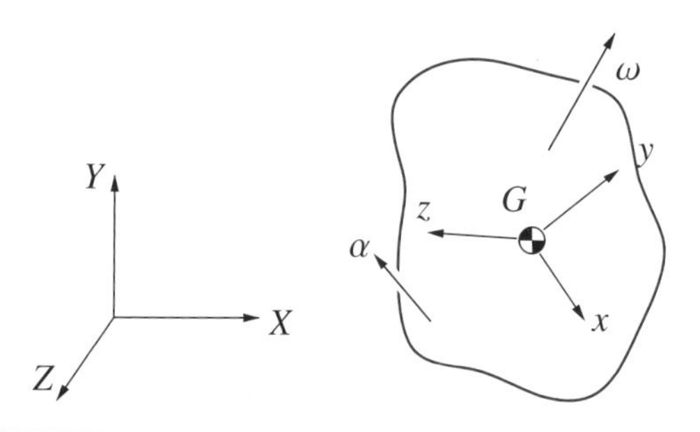

FIGURE 5.12 Spatial motion of a rigid body

where $\overline{F}_{e_i}$ $(i = 1, 2, 3, \ldots, n)$ is one of a system of n externally applied forces.

Moment equilibrium is more complex and involves the Inertia Matrix I, where I is defined as

$$I = \begin{bmatrix} I_{xx} & -I_{xy} & -I_{xz} \\ -I_{yx} & I_{yy} & -I_{yz} \\ -I_{zx} & -I_{zy} & I_{zz} \end{bmatrix}.$$

The elements of I are known as products of inertia, and they are found by evaluating integrals over the entire mass (volume) of the body. These are triple integrals denoted by the symbol $\int_m$. If dm is a differential mass element in the body, and the vector $\bar{r} = [x, y, z]^T$ is a vector locating dm in the xyz coordinate system, then the products of inertia are defined as

$$\begin{aligned} I_{xx} &= \int_m (y^2 + z^2)\, dm \quad & I_{xy} &= \int_m (xy)\, dm \quad & I_{xz} &= \int_m (xz)\, dm \\ I_{yx} &= \int_m (yx)\, dm \quad & I_{yy} &= \int_m (x^2 + z^2)\, dm \quad & I_{yz} &= \int_m (yz)\, dm \\ I_{zx} &= \int_m (zx)\, dm \quad & I_{zy} &= \int_m (zy)\, dm \quad & I_{zz} &= \int_m (x^2 + y^2)\, dm. \end{aligned} \tag{5.17}$$

Clearly $I_{xy} = I_{yx}$, $I_{xz} = I_{zx}$, and $I_{yz} = I_{zy}$. According to the Newton-Euler Equations, moment equilibrium requires

$$\Sigma \overline{M}_g = I\,\overline{\alpha} + \overline{\omega} \times \left(I\,\overline{\omega}\right). \tag{5.18}$$

Equations (5.16) and (5.18) are known as the Newton–Euler Equations. They are the general equations of spatial rigid body motion.

5.2.2 The Newton–Euler Equations Reduced for a Planar Motion

Consider what happens to the Newton–Euler Equations in the case of a rigid body undergoing planar motion. The xy plane of the moving coordinate system is taken parallel to the plane of motion. The fixed XY axes define what is thought of as the plane of motion. The moving and fixed coordinate systems will be offset from each other in the z-Z direction, and you will see that this offset is the source of the three-dimensionality in the force analysis.

For a planar motion, $\bar{a}_g$ has only X and Y components, that is,

$$\bar{a}_g = \begin{bmatrix} a_{g_x} \\ a_{g_y} \\ 0 \end{bmatrix},$$

so the r.h.s. of Equation (5.16) becomes

$$\Sigma\overline{F} = \Sigma\overline{F}_{e_i} = \begin{bmatrix} ma_{g_x} \\ ma_{g_y} \\ 0 \end{bmatrix}. \tag{5.19}$$

For a planar motion, $\overline{\omega}$ and $\overline{\alpha}$ have only have Z components, that is,

$$\overline{\omega} = \begin{bmatrix} 0 \\ 0 \\ \omega \end{bmatrix} \quad \text{and} \quad \overline{\alpha} = \begin{bmatrix} 0 \\ 0 \\ \alpha \end{bmatrix}.$$

Restrict the body to have a plane of symmetry parallel to the *xy* plane of the body-fixed *xyz* system, as shown in Figure 5.13. Due to the symmetry, the body's mass center would lie on this plane. You may think this is very restrictive, but it is not. Consider that most machine parts are fabricated on a milling machine using profiling and/or pocketing operations. This results in parts that have the required plane of symmetry. As seen in the view on the right of Figure 5.13, for every point in the body with particular x, y, and z coordinates, there is a corresponding point reflected across the *xy* plane of symmetry, which has the same x and y coordinates the but opposite z coordinate. The result of this is that

$$I_{xz} = \int_m (xz)\, dm = 0 \quad \text{and} \quad I_{yz} = \int_m (yz)\, dm = 0.$$

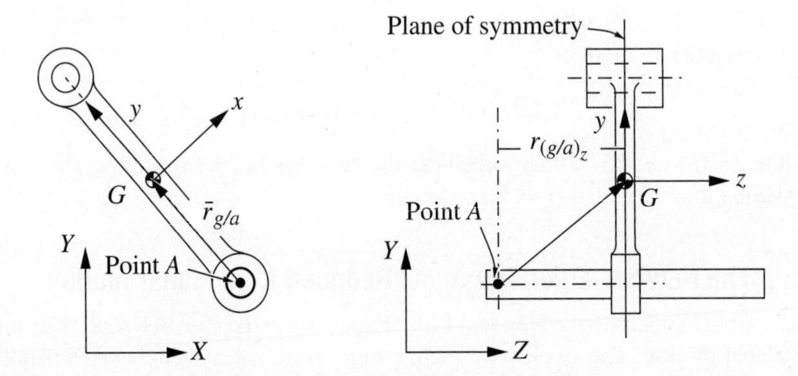

FIGURE 5.13 A moving link with plane of symmetry parallel to the plane of motion (the *XY* plane)

Writing Euler's Equation (5.18) and incorporating these simplifications gives

$$\Sigma\overline{M}_g = \begin{bmatrix} I_{xx} & -I_{xy} & 0 \\ -I_{yx} & I_{yy} & 0 \\ 0 & 0 & I_{zz} \end{bmatrix} \begin{bmatrix} 0 \\ 0 \\ \alpha \end{bmatrix} + \begin{bmatrix} 0 \\ 0 \\ \omega \end{bmatrix} \times \left(\begin{bmatrix} I_{xx} & -I_{xy} & 0 \\ -I_{yx} & I_{yy} & 0 \\ 0 & 0 & I_{zz} \end{bmatrix} \begin{bmatrix} 0 \\ 0 \\ \omega \end{bmatrix} \right),$$

and expanding the matrix products gives

$$\Sigma\overline{M}_g = \underbrace{\begin{bmatrix} 0 \\ 0 \\ I_{zz}\alpha \end{bmatrix}}_{I_{zz}=I_g} + \underbrace{\begin{bmatrix} 0 \\ 0 \\ \omega \end{bmatrix} \times \begin{bmatrix} 0 \\ 0 \\ I_{zz}\omega \end{bmatrix}}_{\text{equals zero}} = \begin{bmatrix} 0 \\ 0 \\ I_g\alpha \end{bmatrix}. \tag{5.20}$$

Equations (5.19) and (5.20) are the Newton–Euler Equations reduced for a planar motion. They are identical to the force and moment equilibrium equations of a planar motion—Equations (5.2), (5.3), and (5.7) in Section 5.1—but there are some subtle differences that we will see in the next section when we consider the form of the equation of moment equilibrium, when moments are referenced to a point other than the center of mass.

5.2.3 Summing Moments about an Arbitrary Point for a Partially Three-Dimensional Planar Rigid Body Motion

Consider an arbitrary point A on the body and the moment created by an external force $\overline{F}_{e_i}$ about the point A, as is shown in Figure 5.13:

$$\overline{M}_a = \bar{r}_{(e/a)_i} \times \overline{F}_{e_i} = \left(\bar{r}_{(e/g)_i} + \bar{r}_{g/a}\right) \times \overline{F}_{e_i} = (\bar{r}_{(e/g)_i} \times \overline{F}_{e_i}) + (\bar{r}_{g/a} \times \overline{F}_{e_i}).$$

The sum of the moments about point A of all the externally applied forces is given by the sum of all the moments $\overline{M}_a$ contributed by each force $\overline{F}_{e_i}$,

$$\Sigma\overline{M}_a = \Sigma(\bar{r}_{(e/g)_i} \times \overline{F}_{e_i}) + \Sigma(\bar{r}_{g/a} \times \overline{F}_{e_i}).$$

The first summation on the r.h.s. of this equation is ΣM_g, whose three components are given in Equation (5.20). In the second summation, the term $\bar{r}_{g/a}$ is a constant that can be factored out of the summation, so the above equation becomes

$$\Sigma\overline{M}_a = \begin{bmatrix} 0 \\ 0 \\ I_g\alpha \end{bmatrix} + \bar{r}_{g/a} \times \Sigma\overline{F}_{e_i}.$$

The term $\Sigma\overline{F}_{e_i}$ on the r.h.s. of this equation is replaced by $m\bar{a}_g$ from Equations (5.16) and (5.19):

$$\Sigma\overline{M}_a = \begin{bmatrix} 0 \\ 0 \\ I_g\alpha \end{bmatrix} + \bar{r}_{g/a} \times \begin{bmatrix} ma_{g_x} \\ ma_{g_y} \\ 0 \end{bmatrix}. \tag{5.21}$$

It is possible to choose our reference frames so that one of the moving mass centers will lie in the XY plane, but not all of the moving mass centers. This makes the problem different from the purely two-dimensional case in that for a particular body, the vector $\bar{r}_{g/a}$ can now have a Z component,

$$\bar{r}_{g/a} = \begin{bmatrix} r_{(g/a)x} \\ r_{(g/a)y} \\ r_{(g/a)z} \end{bmatrix},$$

as seen on the r.h.s. of Figure 5.13. In this case Equation (5.21) is no longer a scalar equation with only a Z component, as it was in Equation (5.10) for the purely two-dimensional problem. Now, due to a Z component in $\bar{r}_{g/a}$, there will be X and Y components in the moment equilibrium condition that arise from the cross product on the r.h.s. of Equation (5.21). The result is

$$\Sigma\overline{M}_a = \begin{bmatrix} \Sigma M_{a_x} \\ \Sigma M_{a_y} \\ \Sigma M_{a_z} \end{bmatrix} = \begin{bmatrix} -mr_{(g/a)_z} a_{gy} \\ mr_{(g/a)_z} a_{gx} \\ I_g\alpha + m\left(r_{(g/a)_x} a_{gy} - r_{(g/a)_y} a_{gx}\right) \end{bmatrix}. \tag{5.22}$$

To summarize, for a partially three-dimensional planar motion, force equilibrium in Equation (5.16) requires

$$\Sigma F_x = \Sigma F_{e_{ix}} = ma_{gx} \tag{5.23}$$

$$\Sigma F_y = \Sigma F_{e_{iy}} = ma_{gx} \tag{5.24}$$

(the Z component of Equation (5.16) is trivial), and moment equilibrium in Equation (5.22) requires

$$\Sigma M_{a_x} = -mr_{(g/a)_z} a_{gy} \tag{5.25}$$

$$\Sigma M_{a_y} = mr_{(g/a)_z} a_{gx} \tag{5.26}$$

$$\Sigma M_{a_z} = I_g\alpha + m\left(r_{(g/a)_x} a_{gy} - r_{(g/a)_y} a_{gx}\right), \tag{5.27}$$

where A is any arbitrary point on the body. Thus there are five conditions of force and moment equilibrium in a partially three-dimensional motion.

Composite Bodies

Many machine parts do not have the xy plane of symmetry required for (5.22) to be valid. However, virtually all machine parts are composites of shapes, where each of the individual shapes making up the composite shape has this symmetry. The crank in Figure 5.14 is an example of such a composite shape. The image on the left is the planar view of the link. When we look at it edgewise, in the middle view, it is apparent the link has some three-dimensionality and does not have the xy plane as a plane of symmetry. However, one can see that the link in the middle is a composite of three shapes, shaded differently to distinguish them. The edge view on the right shows the individual shapes and their respective xyz systems. Notice that each of these three individual shapes has the required xy plane of symmetry. In this case Equations (5.25) and (5.26), which are supplementing the planar analysis, must be modified as follows:

$$\Sigma M_{a_x} = -\Sigma(mr_{(g/a)_z} a_{gy}) \tag{5.28}$$

$$\Sigma M_{a_y} = \Sigma(mr_{(g/a)_z} a_{gx}), \tag{5.29}$$

where the right hand sides are evaluated by summing the terms for each symmetrical piece of the composite shape.

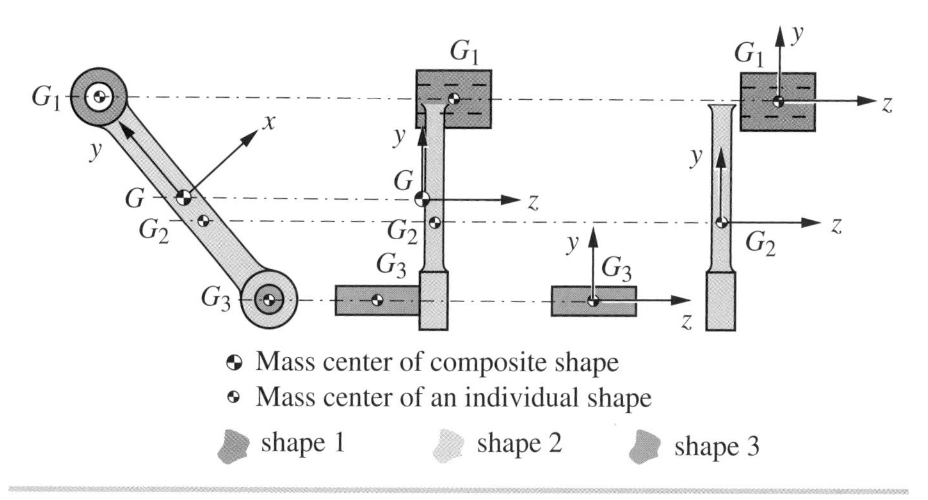

FIGURE 5.14 A body that is a composite of shapes each of which has the xy plane of symmetry

Comparison to Purely Planar Case

In the purely planar case there were two conditions of force equilibrium, Equations (5.2) and (5.3), and one condition of moment equilibrium, the Z component of Equation (5.10). These three conditions also appear in the conditions for force and moment equilibrium in the partially three-dimensional case of planar motion, in Equations (5.23), (5.24), and (5.27). What this means is that the results of the force analysis for the planar case are not useless. In fact, the purely planar case has considered the solution to Equation (5.23), (5.24), and (5.27). When the partially three-dimensional aspect of the force analysis is considered, Equations (5.28) and (5.29) are available to solve for the additional unknowns. This will be illustrated by an example, following the discussion in Section 5.2.4, of the fully three-dimensional aspects of planar mechanisms.

5.2.4 Discussion of the Four Bar Linkage

Consider the four bar mechanism shown in Figure 5.11. For the purposes of this discussion, suppose that the driving torque, $\overline{T}_2$, acts on link 2 and is unknown. Suppose the load torque, $\overline{T}_4$, acts on link 4 and is known. Also suppose that the state of motion, θ_2, $\dot{\theta}_2$, and $\ddot{\theta}_2$, is known. As you will see, moment loads now exist at the joints due to the offset planes of motion of the mass centers of links 2, 3, and 4. We want to solve for $\overline{T}_2$ and the forces *and moments* being transmitted by the joints. This is an IDP.

The free body diagrams of links 2, 3, and 4 are shown in Figure 5.15. The forces at the joints $\overline{F}_{12}$, $\overline{F}_{23}$, $\overline{F}_{43}$, and $\overline{F}_{14}$, have only X and Y components,

$$\overline{F}_{12} = \begin{bmatrix} F_{12_x} \\ F_{12_y} \\ 0 \end{bmatrix}, \quad \overline{F}_{23} = \begin{bmatrix} F_{23_x} \\ F_{23_y} \\ 0 \end{bmatrix}, \quad \overline{F}_{43} = \begin{bmatrix} F_{43_x} \\ F_{43_y} \\ 0 \end{bmatrix}, \quad \overline{F}_{14} = \begin{bmatrix} F_{14_x} \\ F_{14_y} \\ 0 \end{bmatrix}; \tag{5.30}$$

the moments at the joints $\overline{M}_{12}$, $\overline{M}_{23}$, $\overline{M}_{43}$, and $\overline{M}_{14}$, have only X and Y components;

$$\overline{M}_{12} = \begin{bmatrix} M_{12_x} \\ M_{12_y} \\ 0 \end{bmatrix}, \quad \overline{M}_{23} = \begin{bmatrix} M_{23_x} \\ M_{23_y} \\ 0 \end{bmatrix}, \quad \overline{M}_{43} = \begin{bmatrix} M_{43_x} \\ M_{43_y} \\ 0 \end{bmatrix}, \quad \overline{M}_{14} = \begin{bmatrix} M_{14_x} \\ M_{14_y} \\ 0 \end{bmatrix}, \tag{5.31}$$

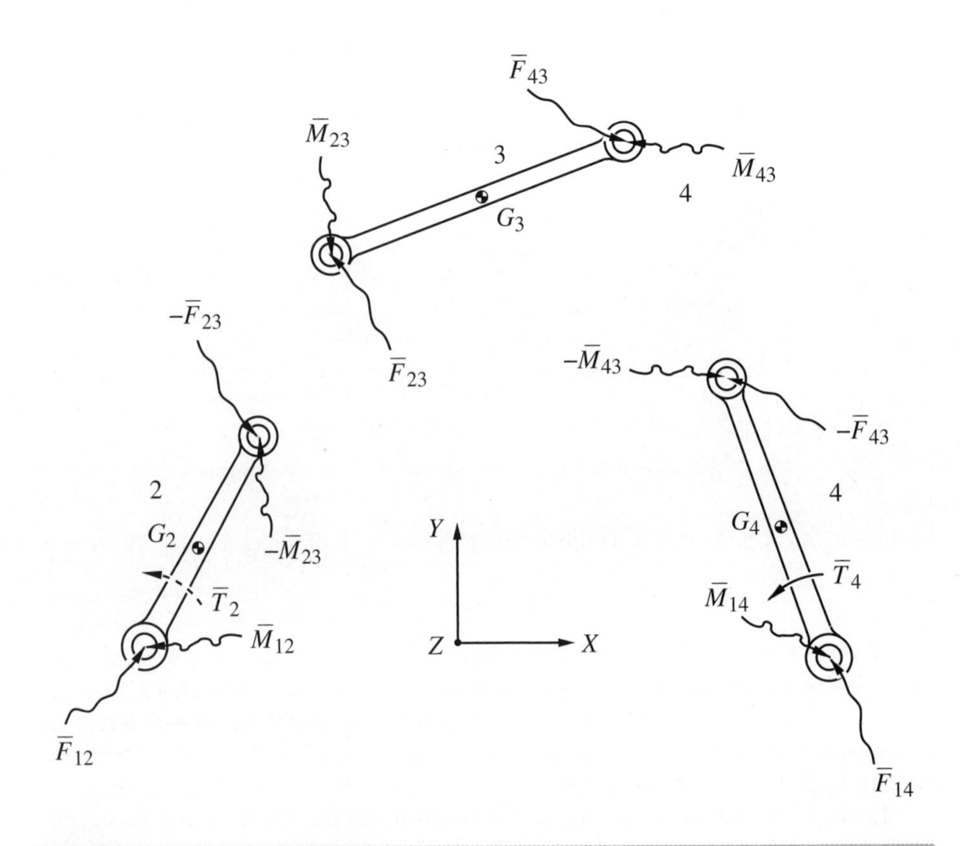

FIGURE 5.15 Free body diagrams for the three-dimensional force analysis of a four bar mechanism

and the load torque and driving torque have only a Z component,

$$\overline{T}_4 = \begin{bmatrix} 0 \\ 0 \\ T_4 \end{bmatrix}, \quad \overline{T}_2 = \begin{bmatrix} 0 \\ 0 \\ T_2 \end{bmatrix}. \tag{5.32}$$

There are 8 unknown components of forces in Equation (5.30), 8 unknown components of moments in Equation (5.31), and 1 unknown component of torque, T_2, in Equation (5.32), for a total of 17 unknowns. In Section 5.2.3 we saw that there are 5 scalar equilibrium equations for each link. Thus we have a total of 15 equations. Something is wrong. There are two more unknowns than there are equations. This problem is statically indeterminate.

An explanation for this is as follows. In planar four bar mechanisms, the axes of the four joints must be in the Z direction. The fact is that due to manufacturing tolerances, they will not be parallel. There will be some degree of misalignment, even though it may be very small. To solve this problem, we must know what the angular misalignments between the joints are. Just as in statically indeterminate problems in solid mechanics, we need to have information about deformations in order to solve the problem.

The problem is that these misalignments cannot be precisely predicted. They are unknown. The misalignments have a *range* of possible values dictated by the tolerances.

To alleviate the statically indeterminate problem, one of the two floating joints can be replaced with a ball-and-socket joint, a.k.a. a ball joint. The ball joint cannot transmit any moment loads, so incorporating it in place of a pin joint will eliminate the two components of reaction moment that existed in the original pin joint. This brings us to a system of 15 equations in 15 unknowns. But the problem is even a bit more complicated.

Consider that the four bar mechanism is completely three-dimensional and that the planes of motion of the mass centers are no longer parallel. This is the actual case. In this case, each of the four joints will have 3 unknown components of reaction force and 2 unknown components of reaction moments, for a total of 20 unknowns. Including the unknown driving torque gives 21 unknowns. There are 3 force equilibrium and 3 moment equilibrium equations for each body, giving a total of 18 equations. Now there is an excess of 3 unknowns, and the system is even more statically indeterminate.

To eliminate 3 unknowns, once again, one of the floating joints can be replaced by a ball joint, eliminating 2 unknown components of reaction moment. The other floating joint can be replaced by a universal joint, eliminating 1 unknown component of reaction moment. This makes the problem determinate, with 18 equations in 18 unknowns.

In practice, instead of using a coupler link with a ball joint at one end and a universal joint at the other, a ball joint is used at both ends. It is simpler and less expensive to use ball joints at both ends. This results in the system having two degrees of freedom. The second degree of freedom is a trivial "spin" of the coupler about the line between the centers of its ball-and-socket joints. This mechanism is the well-known revolute-spherical-spherical-revolute (RSSR) spatial four bar mechanism, shown in Figure 5.16. When the two revolute joints are parallel ($\gamma = 0$ in Figure 5.16), the RSSR degenerates into the planar four bar mechanism.

A common form of the ball joint is called a rod end, shown in Figure 5.17. Early into World War II, the Brits shot down a German aircraft, and when they

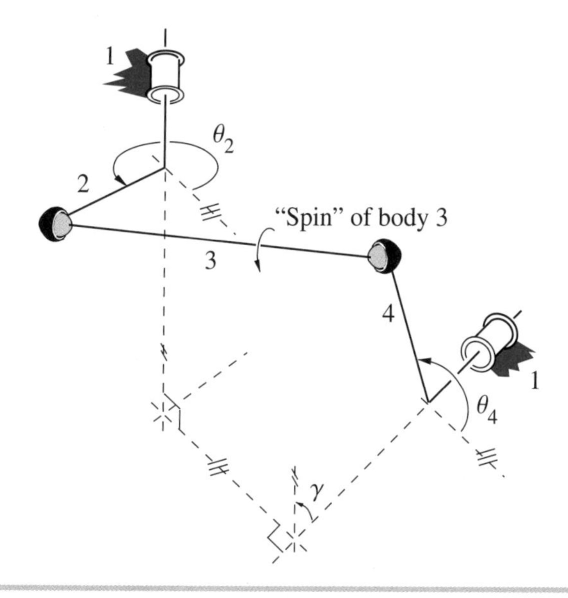

FIGURE 5.16 The RSSR spatial four bar mechanism

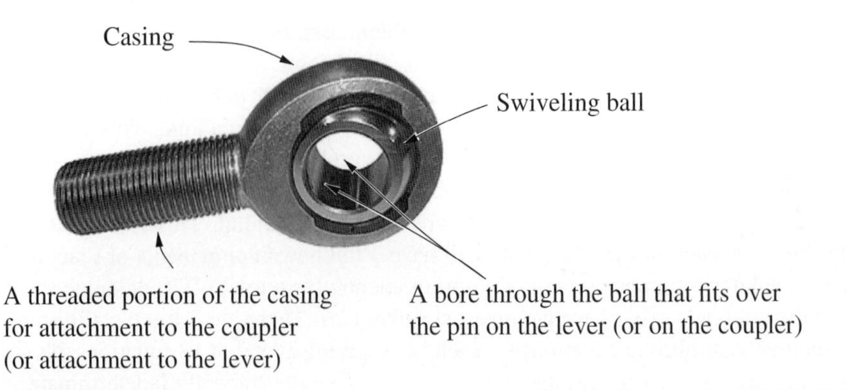

FIGURE 5.17 A rod end

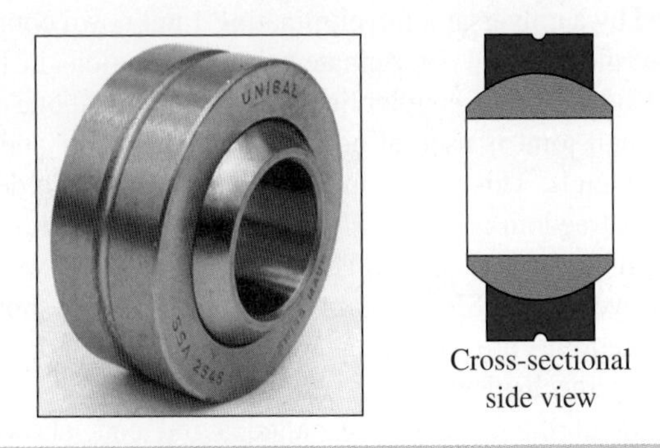

FIGURE 5.18 An insertable ball joint

disassembled it (reverse-engineered it), they found these ball joints. In the United States the H.G. Heim Co. was given exclusive rights to manufacture such joints, and in England the same rights went to the Rose Bearings Ltd. Hence they became known as a heim joint here and as a rose joint in England, although they probably should have been called by their original name, "kugelgelenk." The rod end is frequently found in the control linkages for automobile steering or aircraft rudder movements. The threaded portion (shown as male but also available as female) makes for easy attachment/detachment.

Another form of the ball joint is shown in Figure 5.18. This ball joint is inserted into a bore, with a light press fit, and many times is also retained by internal retaining rings. You can imagine that the pin joints at points *A* and *B* on the l.h.s. of Figure 5.11 are side views of insertable ball joints. Figure 5.19 shows the edge view of the four bar mechanism on the r.h.s. of Figure 5.11, with the pin joints at points *A* and *B* replaced by insertable ball joints.

To summarize this discussion of the three-dimensional aspects of the bearing forces in a planar four bar mechanism:

1. If all four joints being used are pin joints, the force analysis is statically indeterminate.
2. The force analysis will be statically determinate if ball joints are used in place of the floating pin joints.

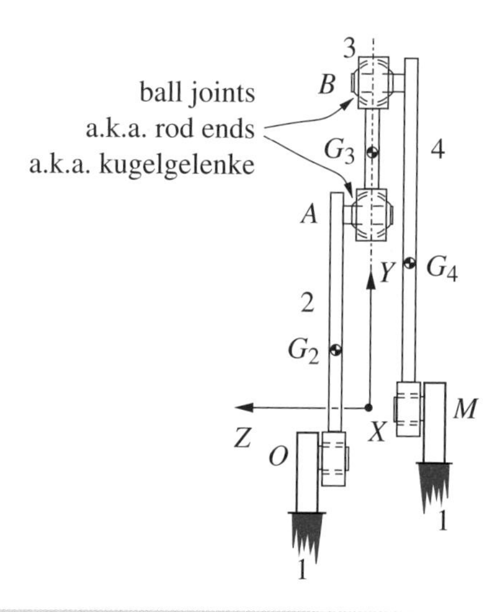

FIGURE 5.19 Edge view of a four bar mechanism using rod ends as floating joints © Cengage Learning.

In situations where extremely high levels of force and torque are being transmitted (this means nothing without some quantification), you may find the floating joints as pin joints—i.e., the statically indeterminate case. In that case, for the purpose of force analysis, the designer will have to assume that the reaction force between the coupler and each lever lies in the plane of motion of the coupler's mass center, which would have been the case if the floating joints were ball joints and the problem were statically determinate.

5.2.5 Equivalence of a Reaction Force and Moment to Two Reaction Forces

Depending on the type of bearing used at a joint, it may not be possible to support a reaction moment at the joint. Many bearings are not intended to withstand a moment load. Ball bearings, for example, are designed to withstand only radial or axial (thrust) forces. In these cases, to accommodate the reaction moment, the reaction force and moment at the joint are replaced by reaction forces that would be acting on two bearings at the joint. A general rule of thumb is that every shaft must be supported by two bearings.

The left side of Figure 5.20 shows a pin joint that transmits a moment and a force, each with X and Y components, like the joint between links 1 and 2 in the free body diagrams of Figure 5.15. The right side of Figure 5.20 shows an edge view where you see the shaft is supported by two bearings. You can see that four scalars on the left side—M_{12_x}, M_{12_y}, F_{12_x}, and F_{12_y}—are replaced by four scalars—$F_{12_{lx}}$, $F_{12_{ly}}$, $F_{12_{rx}}$, and $F_{12_{ry}}$—which are the X and Y components of the forces that are acting on the two bearings.

For the two cases to be equivalent requires force equivalence,

$$F_{12_x} = F_{12_{lx}} + F_{12_{rx}} \tag{5.33}$$

$$F_{12_y} = F_{12_{ly}} + F_{12_{ry}}, \tag{5.34}$$

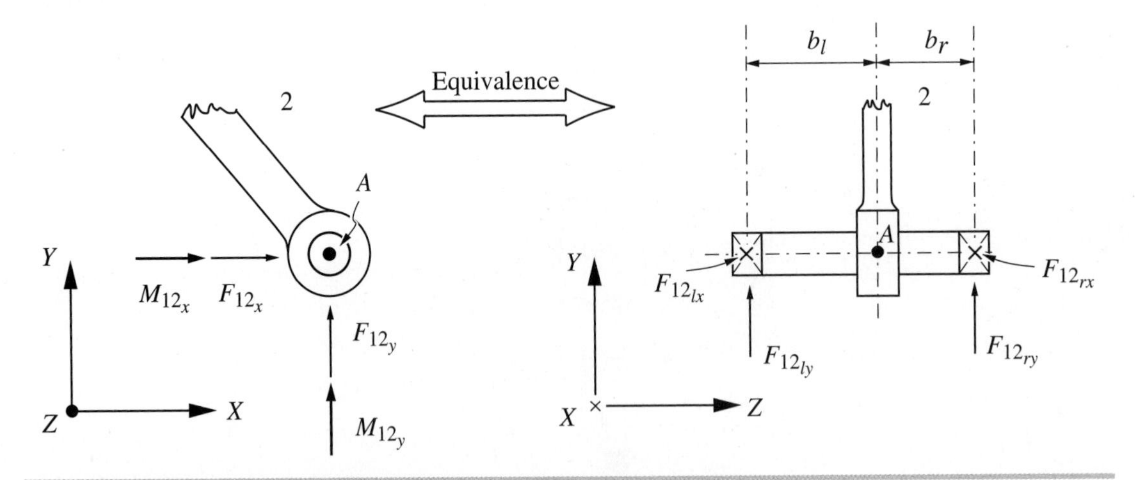

FIGURE 5.20 Equivalence of a force and a moment to two forces
© Cengage Learning.

and moment equivalence (referencing moments to point A),

$$M_{12_x} = F_{12_{ly}} b_l - F_{12_{ry}} b_r \tag{5.35}$$

$$M_{12_y} = -F_{12_{lx}} b_l + F_{12_{rx}} b_r. \tag{5.36}$$

We can write Equations (5.33) through (5.36) in matrix form,

$$\begin{bmatrix} 1 & 0 & 1 & 0 \\ 0 & 1 & 0 & 1 \\ 0 & b_l & 0 & -b_r \\ -b_l & 0 & b_r & 0 \end{bmatrix} \begin{bmatrix} F_{12_{lx}} \\ F_{12_{ly}} \\ F_{12_{rx}} \\ F_{12_{ry}} \end{bmatrix} = \begin{bmatrix} F_{12_x} \\ F_{12_y} \\ M_{12_x} \\ M_{12_y} \end{bmatrix}, \tag{5.37}$$

from which we see the two-force case can always be mapped into the force and moment case, and as long as $b_l \neq -b_r$ (that is, the two bearings are spaced axially), the matrix can be inverted and the force-and-moment case can be mapped into the two-force case.

The conclusion here is that when we consider three-dimensional aspects in the force analysis of a planar mechanism, we can always consider a joint as either

1. Transmitting two forces that are spaced axially (such as two ball bearings or the two edges of a bushing), or
2. A single force and a moment.

The first situation is more realistic, because most bearings will not support the moment load in the second case. (There are some extremely expensive "four point contact" ball bearings that will support a moment load.) Thus most shafts will have two supporting bearings, and the forces acting at those bearings need to be determined so that the bearings of the proper size may be selected. The following section gives an example of this.

5.2.6 Three-Dimensional Aspects of Planar Force Analysis in Section 5.2

Figure 5.21 shows the planar machine analyzed in Example 5.2 on the left. On the right is an edge view onto the *ZY* plane that shows the three-dimensionality of the machine. The *XY* plane is taken as the plane of motion of G_3. Our goal is to describe the three-dimensional aspects of the force analysis using the results from the planar analysis of Example 5.2, along with the equations for the three-dimensional aspects of planar force analysis developed in Section 5.2.3 This planar machine, like all planar machines, will suffer from the same statically indeterminate conditions as the four bar mechanisms in

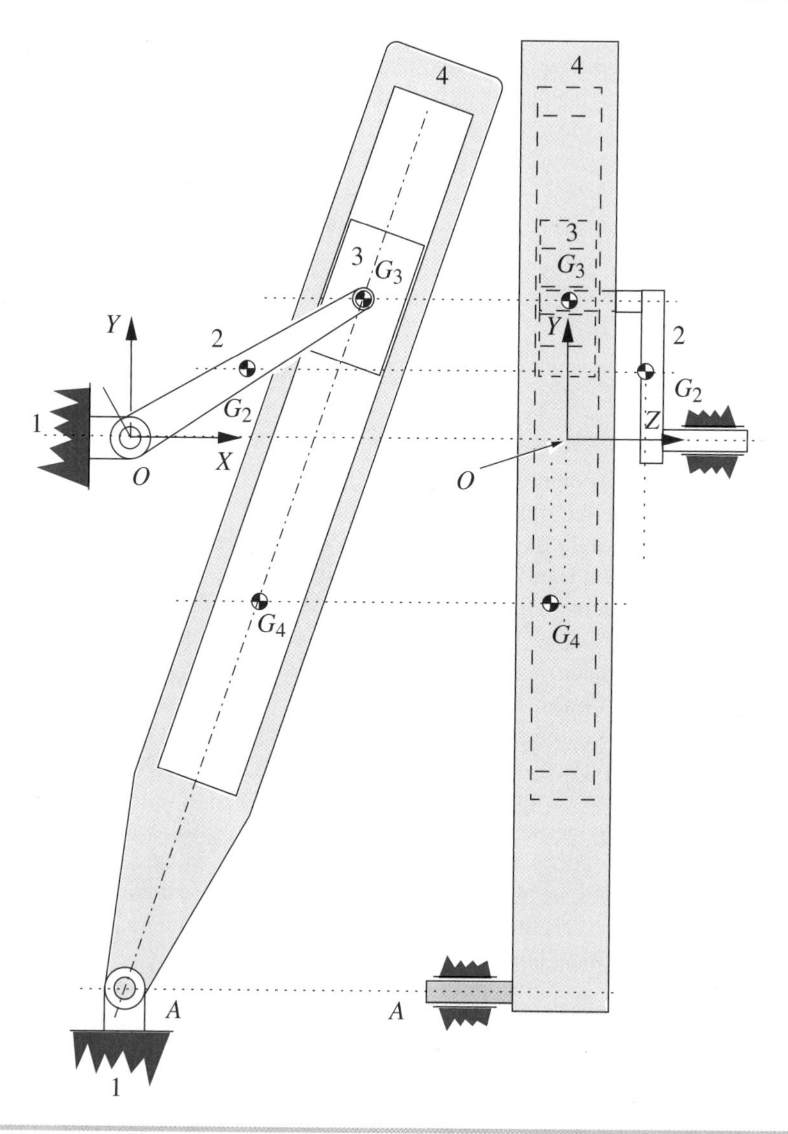

FIGURE 5.21 Three-dimensional aspects of the machine in Section 5.2

Section 5.2.4, because in order for the mechanism to be movable, it must meet some very exacting conditions:

1. All points should move on parallel planes.
2. All rotations must be about axes perpendicular to the parallel planes.

But in real systems, such exact conditions cannot be met. There are always tolerances. So, in fact, a planar mechanism is a fantasy. As a consequence of the manufacturing imprecision that exists in all cases, all mechanisms are actually spatial (three-dimensional).

Figure 5.22 shows the free body diagrams for the case when the motion is considered three-dimensional. The unknowns associated with the forces and torques in the free body diagram are listed below.

$\overline{F}_{12}$: 3 unknown components	$\overline{M}_{12}$: 2 unknown components
$\overline{F}_{32}$: 3 unknown components	$\overline{M}_{32}$: 2 unknown components
$\overline{F}_{43}$: 2 unknown components	$\overline{M}_{43}$: 3 unknown components
$\overline{F}_{14}$: 3 unknown components	$\overline{M}_{14}$: 2 unknown components
$\overline{T}_{2}$: 1 unknown component	

Thus the total number of unknowns is 21. There are 6 equilibrium conditions for each of the 3 bodies, for a total of 18 equations, so there are 3 more unknowns than there are equations. The problem is statically indeterminate. Again, we must relieve the problem of three of the unknown components in the list above.

To do so, consider the pin joint between links 2 and 3. The reaction moment at their contact, $\overline{M}_{32}$, has only 2 components (there is no component of reaction moment about the axis of the pin joint). If we replace this pin joint with a ball joint, these 2 components of reaction moment are removed and $\overline{M}_{23} = \overline{0}$. This leaves us with only one more component of force or moment to eliminate. One possibility is to replace the sliding (a.k.a. prismatic) joint between links 3 and 4 with a cylindrical joint. The cylindrical joint allows translation along the axis of the joint and rotation about the axis of the joint. The reaction moment $\overline{M}_{43}$ will now have only 2 components that are perpendicular to the direction of sliding. The third component has been eliminated, and the problem is determinate with 18 equations in 18 unknowns.

Replacing sliding joints with cylindrical joints is frequently done in planar mechanisms specifically to allow the mechanism to be movable and also for ease of manufacturing. Examples are seen in hydraulic cylinders, internal combustion engines, and reciprocating pumps. The pistons (the sliders) in these examples are cylindrical, and they move inside cylindrical bores.

When we replace the pin joint between 2 and 3 with a ball joint (a.k.a. spherical joint) and the sliding joint between 3 and 4 with a cylindrical joint, the resulting mechanism is the one-degree-of-freedom mechanism known as the revolute-spherical-cylindrical-revolute (RSCR) spatial four bar mechanism. Other reaction forces and moments could have been relieved in place of the ones chosen here. This particular substitution works well when the two joints to ground must have a fixed axis of rotation by which to transmit power to the surroundings.

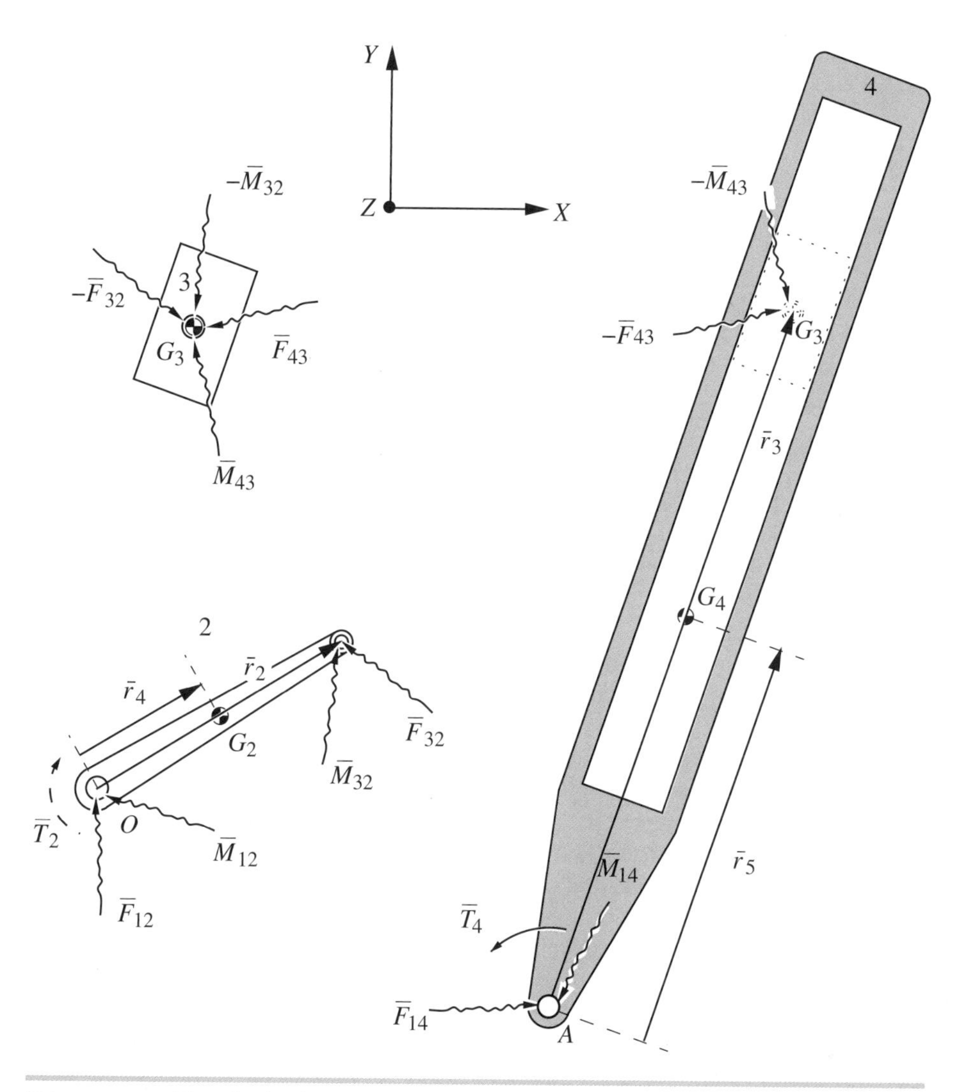

FIGURE 5.22 Free body diagrams of the machine in Section 5.2, considering three-dimensional effects

Figure 5.23 shows a spherical bearing on the left, which can be inserted in the cylindrically shaped slider shown on the right. The drawing is a conceptual one. It does not show the details of how the bearing is mounted and supported.

An important point that is shown in the figure is that *ideally*, $\overline{F}_{32}$ lies in the plane of motion of G_3. This is the case when the centerline of the spherical bearing lies in this plane and when the planes of motion of the mass centers are all parallel. These are the same "ideal" conditions that need to be met in order for the mechanism to be movable. It is not possible to exactly meet them, but we have no choice but to make this assumption for the force analysis, because these misalignments are unknowns. In the actual situation, $\overline{F}_{32}$ will have a component in the plane of motion of G_3, but an axial

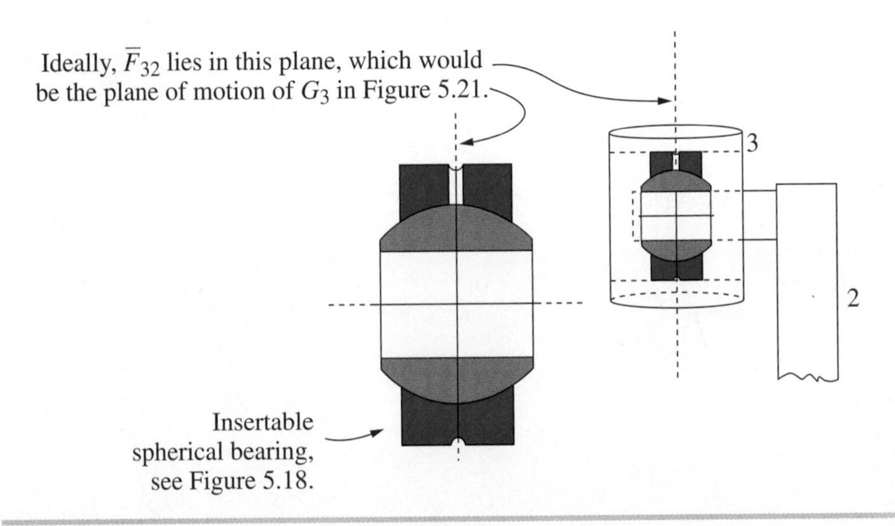
Ideally, $\overline{F}_{32}$ lies in this plane, which would be the plane of motion of G_3 in Figure 5.21.
3
2
Insertable spherical bearing, see Figure 5.18.

FIGURE 5.23 A cylindrically shaped slider with an inserted ball joint

component in the direction of the shaft between 2 and 3 will also exist. We could make a worst case analysis if tolerances were specified.

As in the four bar mechanism, these changes to the joints are not always made to the mechanism. There are two possible reasons why. First, the mechanism may be of low precision, and so, to keep costs lower, a pin joint between 2 and 3 may continue to be used in place of the more expensive ball joint. In this case there must be clearance in the joints to allow the mechanism to accommodate the tolerances and not bind up. Second, the loads being transmitted may be quite high, and the force between 2 and 3 may need the added strength of a pin joint. If this is a high precision machine, tolerances should be maintained and ball bearing mountings should incorporate some compliance. If this is a low precision machine, clearances in the joints will accommodate low tolerances.

Determination of the Bearing Forces

Recall the numerical example of an IDP for the inverted crank-slider mechanism in Example 5.2. The force analysis there was two-dimensional. Our goal here is to determine the bearing forces on the crank of that mechanism, and this will require considering the partially three-dimensional motion of the mechanism.

Figure 5.24 shows a planar view of the crank and the forces acting on the crank that result from the planar force analysis in Example 5.2. Figure 5.25 shows the planar view of the crank on the l.h.s. The crank is a composite of shapes that has the required plane of symmetry discussed in Section 5.2.3. The individual shapes are shaded differently and numbered in Roman numerals. The r.h.s. shows an edge view of the crank, a view onto the ZY plane. Figure 5.25 is a free body diagram of the crank that reflects the partially three-dimensional aspects of the motion. The force $\overline{F}_{32}$ is assumed to lie in the plane of motion of G_3, which does not coincide with the plane of motion of G_2. This induces a reaction moment in the bearing between 1 and 2, $\overline{M}_{12}$, which has X and Y components. As seen in Section 5.2.5, this reaction moment and the reaction force $\overline{F}_{12}$ are equivalent to a system of two forces. These are the forces $\overline{F}_{12_l}$ and $\overline{F}_{12_r}$ which act on the two bearings supporting the shaft. There are now four unknowns, which are the X and Y

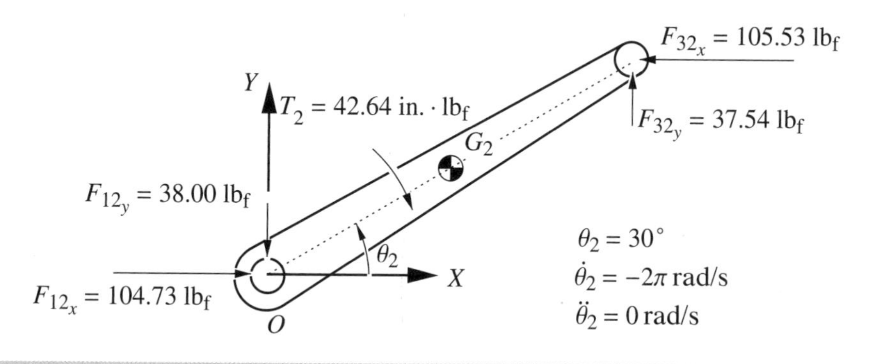

FIGURE 5.24 Results of planar force analysis for the crank in Figure 5.9

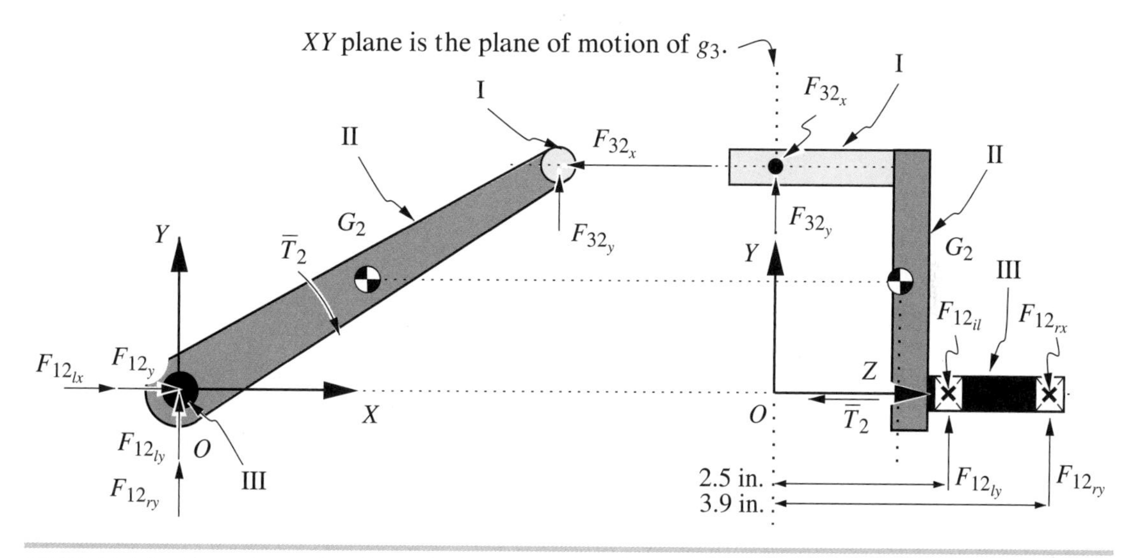

FIGURE 5.25 Three-dimensionality of the crank in Figure 5.24 and free body diagram showing bearing forces

components of $\overline{F}_{12_l}$ and $\overline{F}_{12_r}$ To solve for them, we use Equations (5.2) and (5.3),

$$\Sigma F_x \longrightarrow F_{12_{lx}} + F_{12_{rx}} = F_{12_x} = 104.73 \text{ lb}_\text{f} \tag{5.38}$$

$$\Sigma F_y \longrightarrow F_{12_{ly}} + F_{12_{ry}} = F_{12_y} = -38.00 \text{ lb}_\text{f}, \tag{5.39}$$

and Equations (5.28) and (5.29), using point O as the reference for the moments:

$$\Sigma M_{o_x} \longrightarrow -(2.5 \text{ in.})F_{12_{ly}} - (3.9 \text{ in.})F_{12_{ry}} = -\Sigma(mr_{(g/o)_z} a_{g_y}) \tag{5.40}$$

$$\Sigma M_{o_y} \longrightarrow (2.5 \text{ in.})F_{12_{lx}} + (3.9 \text{ in.})F_{12_{rx}} = \Sigma(mr_{(g/o)_z} a_{g_x}). \tag{5.41}$$

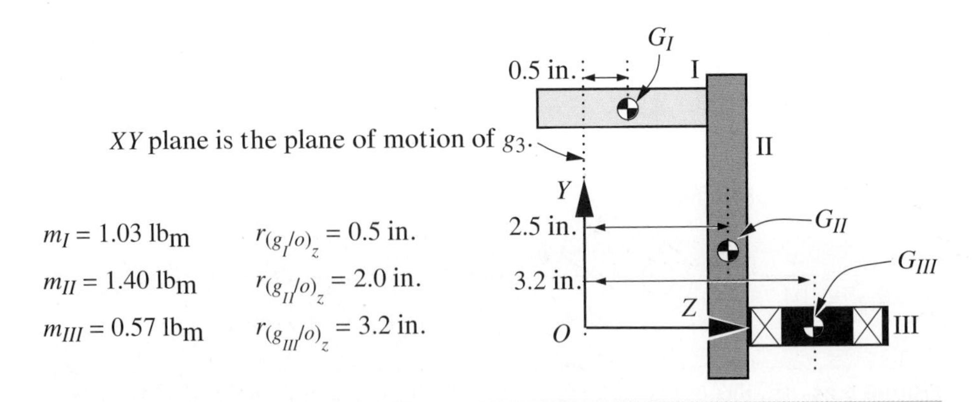

FIGURE 5.26 Dimensions and inertia properties of the shapes within the composite crank

Appendix II calculates the accelerations of the mass centers of the shapes within the composite that are needed to compute the summations on the right hand side of Equations (5.40) and (5.41).

$$a_{g_{Ix}} = -205.14 \text{ in./s}^2, \quad a_{g_{Iy}} = -118.44 \text{ in./s}^2$$
$$a_{g_{IIx}} = -68.38 \text{ in./s}^2, \quad a_{g_{IIy}} = -39.48 \text{ in./s}^2$$
$$a_{g_{IIIx}} = 0 \text{ in./s}^2, \quad a_{g_{IIIy}} = 0 \text{ in./s}^2$$

Figure 5.26 shows the inertia properties of the composite shapes that are needed to compute the summations on the right hand side of Equations (5.40) and (5.41).

Thus

$$-\Sigma(mr_{(g/o)_z}a_{g_y}) = -\Big((1.03)(0.5)(-118.44) + (1.40)(2.0)(-39.48) + (0.57)(3.2)(0)\Big)\left(\frac{\text{in}^2 \cdot \text{lb}_\text{m}}{\text{s}^2}\right)$$
$$= 171.54\left(\frac{\text{in}^2 \cdot \text{lb}_\text{m}}{\text{s}^2}\right)\left(\frac{\text{ft}}{12 \text{ in.}}\right)^2\left(\frac{\text{lb}_\text{f} \cdot \text{s}^2}{32.174 \text{ ft} \cdot \text{lb}_\text{m}}\right) = 0.014 \text{ ft} \cdot \text{lb}_\text{f}$$

$$\Sigma(mr_{(g/o)_z}a_{g_x}) = \Big((1.03)(0.5)(-205.14) + (1.40)(2.0)(-68.38) + (0.57)(3.2)(0)\Big)\left(\frac{\text{in}^2 \cdot \text{lb}_\text{m}}{\text{s}^2}\right)$$
$$= -297.11\left(\frac{\text{in}^2 \cdot \text{lb}_\text{m}}{\text{s}^2}\right)\left(\frac{\text{ft}}{12 \text{ in.}}\right)^2\left(\frac{\text{lb}_\text{f} \cdot \text{s}^2}{32.174 \text{ ft} \cdot \text{lb}_\text{m}}\right) = -0.024 \text{ ft} \cdot \text{lb}_\text{f}, \tag{5.42}$$

and the r.h.s. of Equation (5.40) and Equation (5.41) are computed. Writing Equation (5.38)–Equation (5.41) in a matrix yields

$$\begin{bmatrix} 1 & 0 & 1 & 0 \\ 0 & 1 & 0 & 1 \\ 0 & -\frac{2.5}{12} & 0 & -\frac{3.9}{12} \\ \frac{2.5}{12} & 0 & \frac{3.9}{12} & 0 \end{bmatrix} \begin{bmatrix} F12_{lx} \\ F12_{ly} \\ F12_{rx} \\ F12_{ry} \end{bmatrix} = \begin{bmatrix} 104.73 \\ -38.00 \\ 0.014 \\ -0.024 \end{bmatrix} \text{lb}_\text{f},$$

and solving gives

$$\overline{F}_{12_l} = \begin{bmatrix} F_{12_{lx}} \\ F_{12_{ly}} \end{bmatrix} = \begin{bmatrix} 291.95 \\ -105.74 \end{bmatrix} \text{lb}_\text{f}$$

$$\overline{F}_{12_r} = \begin{bmatrix} F_{12_{rx}} \\ F_{12_{ry}} \end{bmatrix} = \begin{bmatrix} -187.22 \\ 67.74 \end{bmatrix} \text{lb}_\text{f}.$$

This is not really the best example, because in this problem the inertia effects are quite small and the r.h.s. of Equation (5.40) and Equation (5.41) are nearly zero, so the effect of the mass centers moving in parallel but offset planes is not very pronounced. Nonetheless, it correctly illustrates the procedure.

5.3 STATIC FORCE ANALYSIS AND INERTIA FORCE ANALYSIS

In the IDP the known quantities can be divided into two categories. The first category is *externally applied forces and torques*. These include all driving forces and torques, all load forces and torques, and all weight forces. The second category is *inertia loads*. These include the products of mass and acceleration—that is, the "*ma*" terms and the "$I_g\alpha$" terms. In most cases, some of the terms in one of these categories will be an order of magnitude greater than the largest terms in the other category. The larger terms will dominate the problem, and when this occurs, the category of terms that are smaller may be neglected—in other words, set to zero. This leads to two different types of analysis:

1. Static force analysis
2. Inertia force analysis.

In static force analysis, the external loads dominate and the inertia terms can be neglected. This is equivalent to setting all inertias (all m's and I_g's) to zero. This does *not* mean that the links' weights are set to zero. Machines whose force analysis falls into this category are slow-moving machines with heavy links that transmit large forces to the environment. Examples include the bridge-raising machine in Figure 8.6. Cranes, front loaders, crushing or compacting machines, and slow-moving freight elevators are other examples.

In inertia force analysis, the inertia terms dominate. Machines in this category are high-speed machines with light links, including high-speed reciprocating pumps, high-speed indexing or packaging machines, and most high-speed rotational machinery such as turbines. In this case, weight forces and perhaps load torques or forces may be neglected. Of course, the driving force or torque is not neglected. In fact, it is probably the unknown being sought.

Static force analysis is computationally simpler than inertia force analysis because of the kinematic analysis associated with computing the accelerations of the mass centers, the angular accelerations, and the inertia terms. Steps 2, 3, and 4 in Example 5.1 are not needed. In situations where neither category of terms dominates, both categories should be included in the analysis.

5.4 FORCE ANALYSIS OF ROLLING CONTACTS

There are two types of rolling contacts. They are known as friction drives and positive drives. Friction drives include rolling wheels, and belts and pulleys. Friction drives require a static friction force at the point of contact in order to maintain the no-slip condition. The static friction force has a maximum value that is the product of the normal force and the static coefficient of friction. When this maximum static friction force is exceeded, slipping occurs and there is no longer a rolling contact. Figure 5.27 shows the two possible situations when links 2 and 3 are rigid bodies that are rolling wheels within a friction drive. The figure illustrates that if 2 and 3 are rigid, the situations are unrealistic because the length of the arm, 1, will not precisely match the sum of the radii of 2 and 3. Either you will have the situation on the left, where there will be a gap between 2 and 3 at their point of contact, or you will have the situation on the right, where 2 and 3 overlap, which cannot be assembled. *In a friction drive there must be some compliance in the system.*

Figure 5.28 shows two realistic possibilities. On the left, 2 and 3 have a deformable (compliant) surface that compensates for any error in the length of the arm. On the right

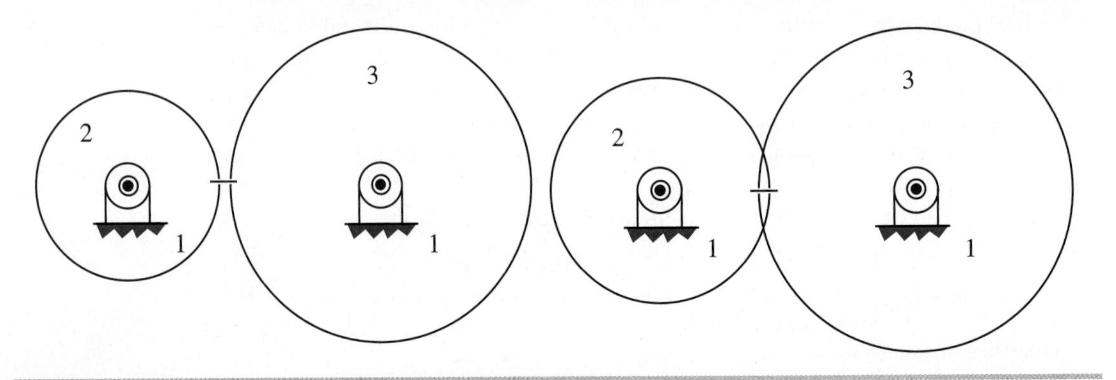

FIGURE 5.27 Rolling wheels with a friction drive

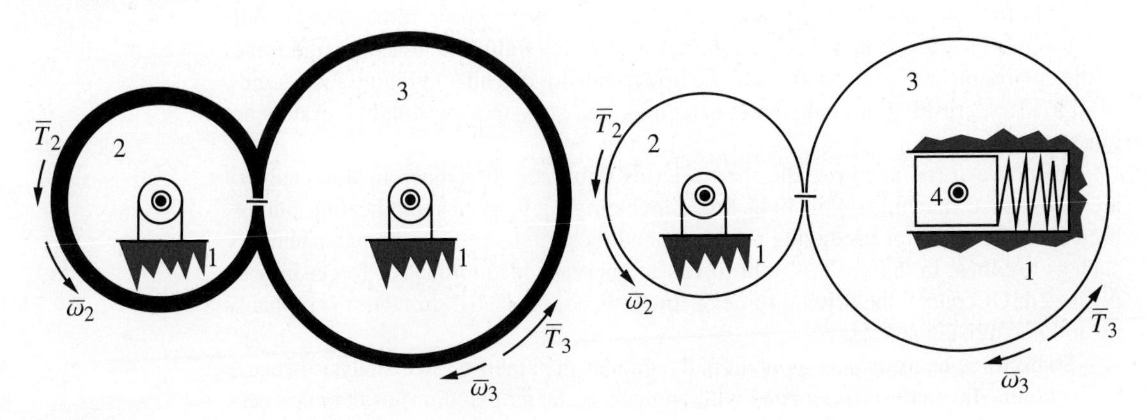

FIGURE 5.28 Compliance in a friction drive

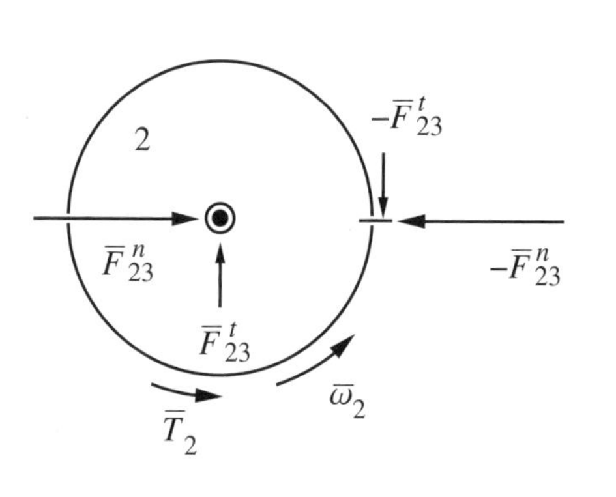

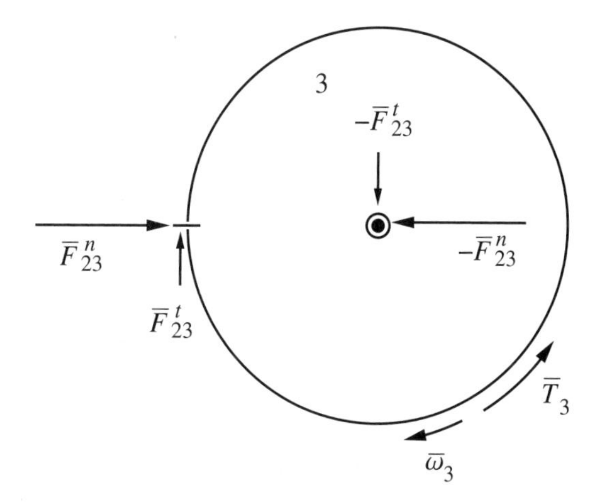

FIGURE 5.29 Free body diagram of a friction drive

side, link 3 is pinned to slider 4, which has a spring pushing it to the left, thus forcing 3 against 2. In either case there is now a preload between 2 and 3, which creates a sufficiently large normal force so that there is a no-slip condition. In order to perform a force analysis of a rolling contact with friction drive, this preload must be known.

Figure 5.29 shows a free body diagram for a static force analysis of either of the friction drives in Figure 5.28. Gravity is also being neglected. $\overline{F}^t_{23}$ is the friction force. It is tangent to the wheel at the point of contact. Force $\overline{F}^n_{23}$ is the preload. It is in the radial direction at the point of contact. The maximum magnitude of the friction force F^t_{23} is related to F^n_{23} by the coefficient of static friction, μ_{static}:

$$F^t_{23} \leq \mu_{\text{static}} F^n_{23}.$$

The other type of friction drive is belts and pulleys. They also require a preload. You will study these in a machine elements course.

A rolling contact with positive drive is achieved either with gears that have involute tooth profiles or by chains and sprockets. Here we will consider gears with involute tooth profiles. (See Appendix 1 of Chapter 3 for development of the involute tooth profile.)

The top of Figure 5.30 shows a pair of meshed gear teeth that belong to 2 and 3 and have involute profiles. The gear teeth are instantaneously in contact at coincident points C_2 and C_3 that belong to 2 and 3. The contact between 2 and 3 is a slipping contact, and according to Gruebler's Criterion, the mechanism has one degree of freedom. As the teeth engage and disengage, the points C_2 and C_3 move along their tooth profiles. If these are involute tooth profiles, then as C_2 and C_3 move along them, there are three very important properties:

1. The direction of the common normal $n - n$ does not change, so ϕ is constant and the direction of the contact force between 2 and 3 (which acts along line $n - n$) is constant.

2. There is a virtual point of rolling contact, P, known as the pitch point, that does not move.

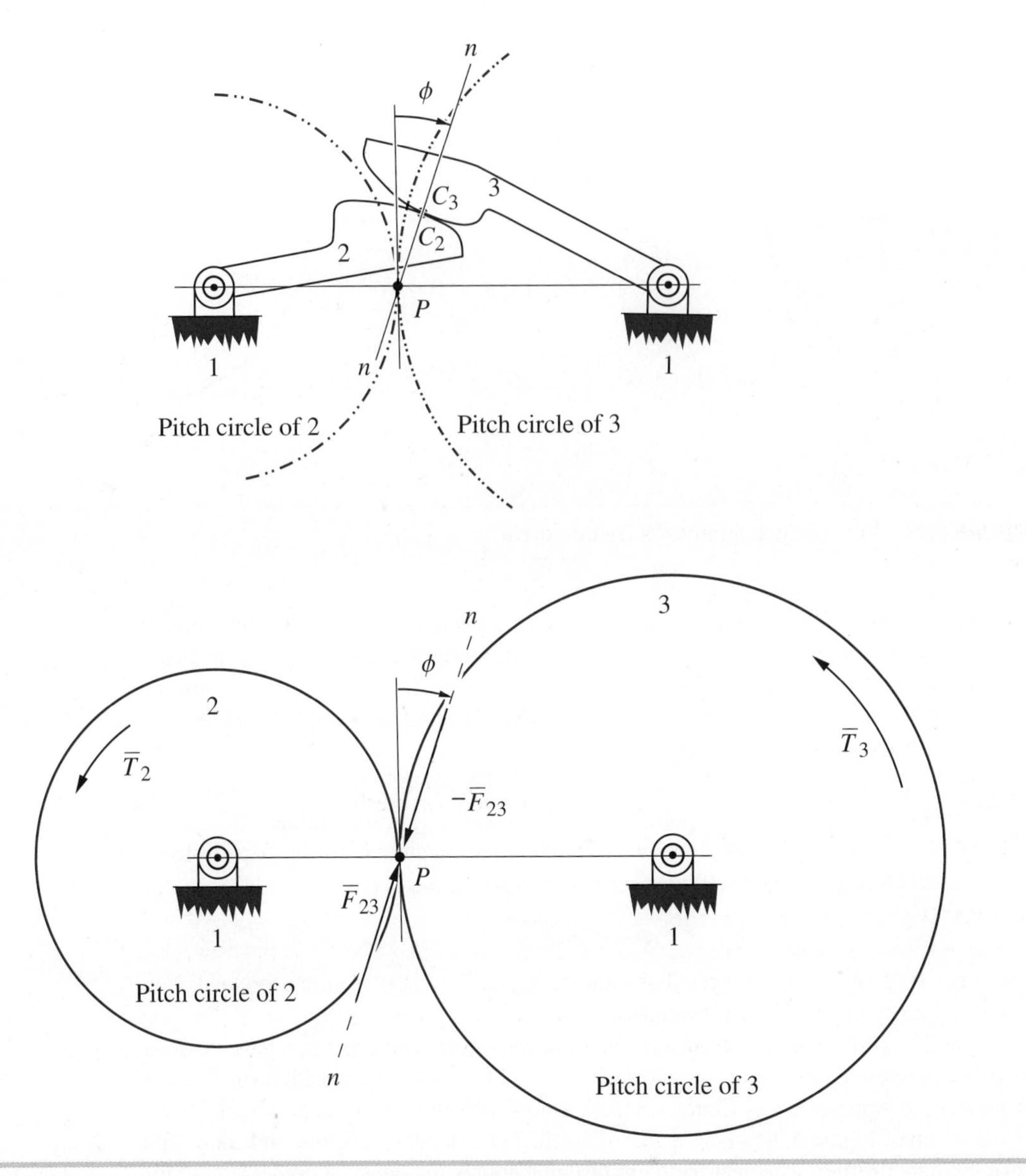

FIGURE 5.30 The gear force between a pair of gear teeth in mesh

3. The common normal always passes through the pitch point, so the contact force between 2 and 3 also always passes through P.

P defines the radius of the pitch circles of 2 and 3. The angle ϕ is known as the pressure angle. Gears have standard pressure angles of 14.5°, 20°, and 25°. You do not mix gears with different pressure angles because if you do so, these properties of an involute tooth profile will not exist. As a result of these three properties, the contact force between the gears can be represented as shown at the bottom of Figure 5.30.

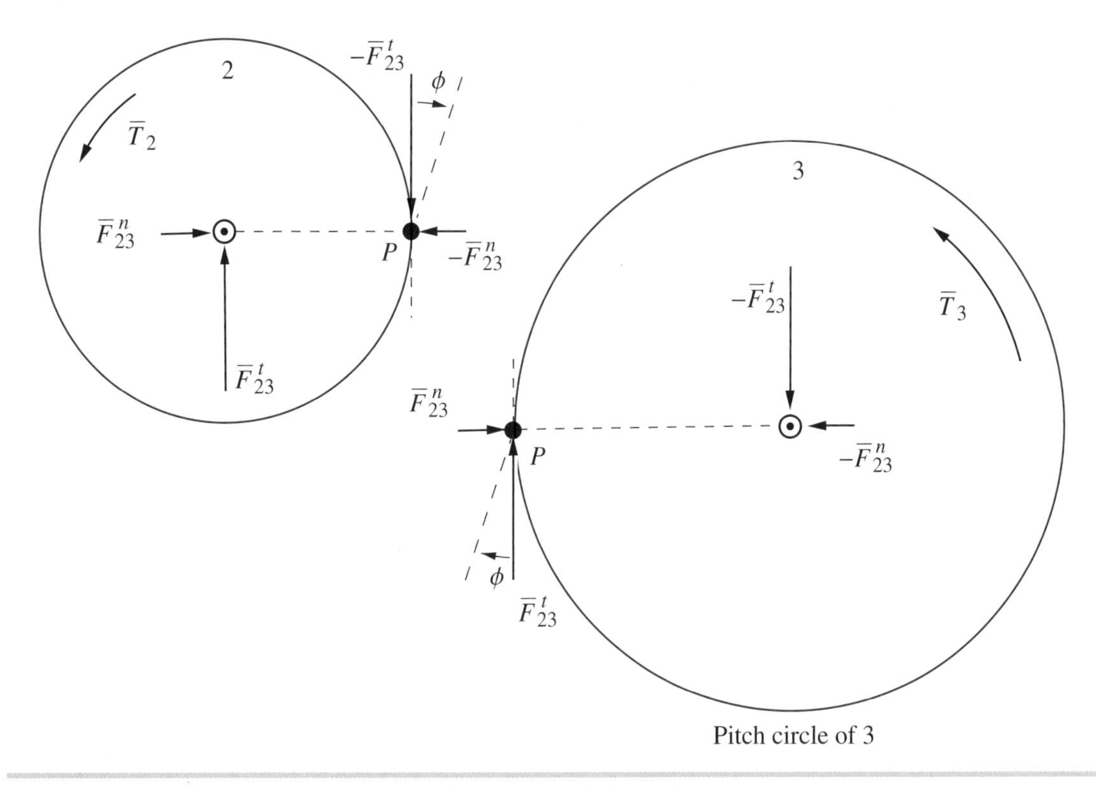

FIGURE 5.31 The gear force between a pair of gear teeth in mesh

Figure 5.31 shows the free body diagram of these two gears in a static force analysis that neglects gravitational effects. The gear force $\overline{F}_{23}$ has been decomposed into a tangential component $\overline{F}^t_{23}$ that acts tangential to the pitch circles at P, and a normal component $\overline{F}^n_{23}$ that acts radially at P. These components of the gear force are related by the pressure angle, In the force analysis of a positive drive rolling contact, the normal component of gear force is always computed from the tangential component, using Equation (5.43).

$$\tan\phi = \frac{F^n_{32}}{F^t_{32}} \quad \longrightarrow F^n_{32} = F^t_{32}\tan\phi \tag{5.43}$$

As a side note, there is an interesting duality between friction drives and positive drives. In friction drives, the normal component of force develops the tangential component of force. In positive drives, the tangential component of force develops the normal component of force, through the pressure angle and Equation (5.43).

▶ EXAMPLE 5.3

Consider the drive mechanism of an automotive rear windshield wiper that we saw earlier in Figure 1.29 and Problems 3.7 and 4.9. The motor drives the crank, link 2, and the windshield wiper is connected to gear 5. The load torque $\overline{T}_5$ acting on gear 5 can be estimated, so we assume it given. We need to find the driving torque $\overline{T}_2$ that acts on 2, as shown in Figure 5.32.

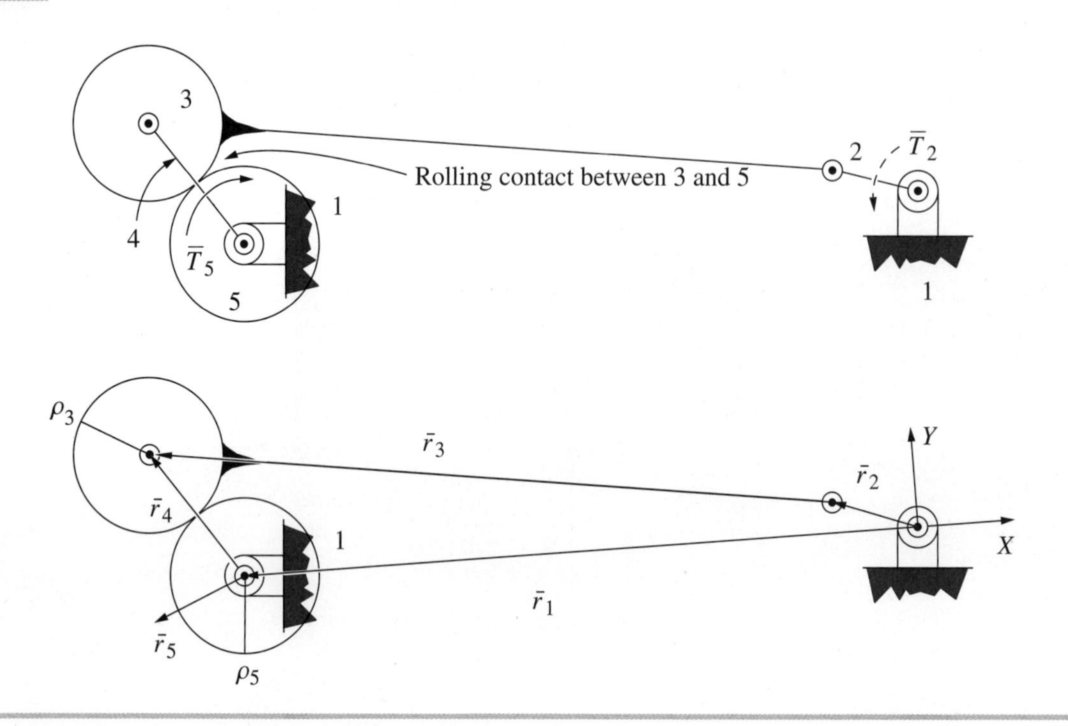

FIGURE 5.32 Vector loop for the windshield wiper drive mechanism with a positive drive rolling contact

Let the rolling contact between 3 and 5 be a positive drive using gears with a pressure angle of 20°. Draw the free body diagrams, and deduce whether they are correct by counting the unknowns.

Solution

First the position problem must be solved. You did this in Problem 3.7. We will repeat the solution here, briefly. There is the vector loop equation,

$$\bar{r}_2 + \bar{r}_3 - \bar{r}_4 - \bar{r}_1 = \bar{0}, \tag{5.44}$$

which has scalar components

$$r_2\cos\theta_2 + r_3\cos\theta_3 - r_4\cos\theta_4 + r_1 = 0 \tag{5.45}$$

$$r_2\sin\theta_2 + r_3\sin\theta_3 - r_4\sin\theta_4 = 0. \tag{5.46}$$

(Do you know why $\bar{r}_1$ is negative in Equation (5.44) and r_1 is positive in Equation (5.45)?) There is also a rolling contact equation,

$$\rho_3\Delta\theta_{3/4} = -\rho_5\Delta\theta_{5/4},$$

which we expand and introduce the assembly configuration to give

$$\rho_3(\theta_3 - \theta_{3_i}) + \rho_5(\theta_5 - \theta_{5_i}) - (\rho_3 + \rho_5)(\theta_4 - \theta_{4_i}) = 0. \tag{5.47}$$

The three position Equations (5.45), (5.46), and (5.47) contain

scalar knowns: $\rho_3, \rho_5, \theta_{3_i}, \theta_{4_i}, \theta_{5_i}, r_1, r_2, r_3, r_4, \theta_1 = \pi$

and

scalar unknowns: $\theta_2, \theta_3, \theta_4, \theta_5$.

Since there are four unknowns in the three position equations, the mechanism has one degree of freedom. The input θ_2 would be known, and the remaining three position unknowns can be found using Newton's Method, which we will not outline.

With the position problem solved, we can address the force analysis. We will perform a static force analysis and neglect gravity. The coordinate system for the force analysis is the same as used in the vector loop in Figure 5.32. Recognize that 4 is a two-force member, so its direction is known. Call this force $\overline{F}_4$ and draw the free body diagrams. The two components of the gear force, $\overline{F}^n_{35}$ and $\overline{F}^t_{35}$, have known directions. All other forces have unknown magnitudes and directions. The free body diagrams are shown in Figure 5.33.

Unknowns in the free body diagrams are the magnitude of $\overline{F}_4$, magnitude of $\overline{F}^n_{35}$, magnitude of $\overline{F}^t_{35}$, magnitude and direction of $\overline{F}_{15}$, magnitude and direction of $\overline{F}_{12}$, magnitude and direction of $\overline{F}_{32}$, and magnitude of $\overline{T}_2$. $\overline{T}_5$ is a known load. This adds up to ten unknowns. There are two scalar force equilibrium conditions and one scalar moment equilibrium condition for each of links 2, 3, and 5, which amounts to nine scalar equilibrium conditions. There are no scalar equilibrium conditions for link 4, because we can recognize it as a two-force member from its three equilibrium conditions. We are missing one equation, and that is the equivalent of Equation (5.43) applied to this problem:

$$F^n_{35} = F^t_{35}\tan\phi, \tag{5.48}$$

where ϕ is the known as pressure angle. So we have ten equations in ten unknowns, and we can solve for the unknowns.

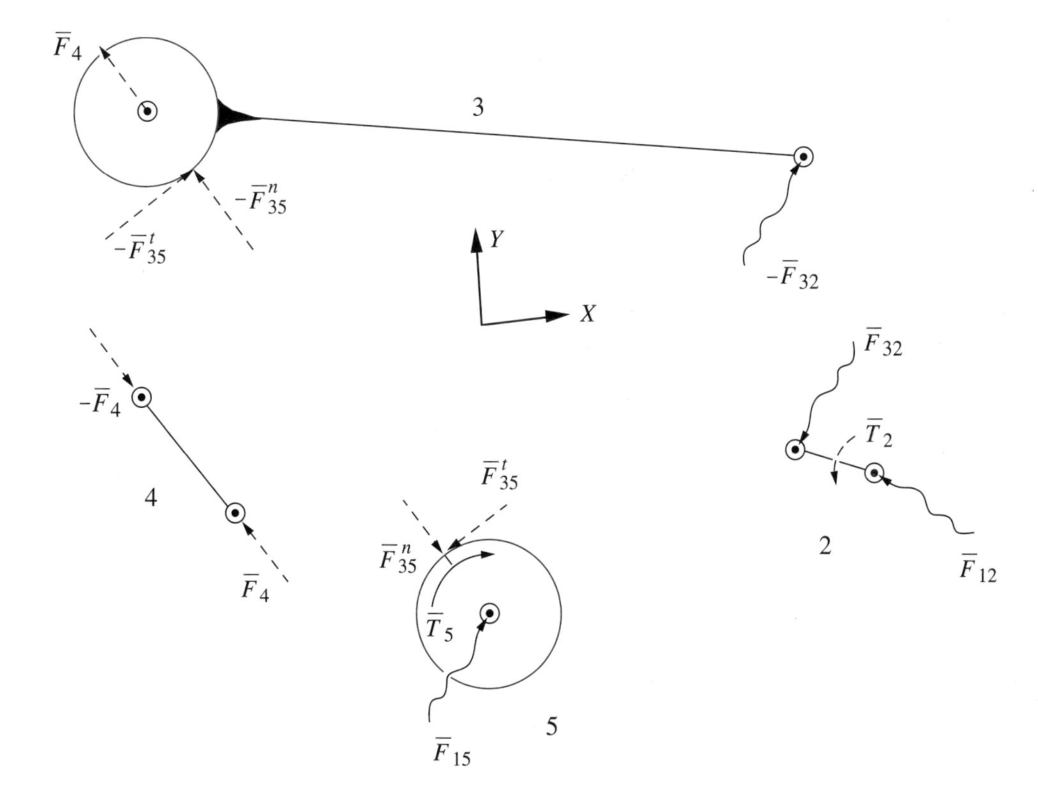

FIGURE 5.33 Free body diagrams for the windshield wiper drive mechanism with a positive drive rolling contact

▶ EXAMPLE 5.4

If in Example 5.3 the rolling contact had been a friction drive, what would the force and moment equilibrium situation have been?

Solution

The free body diagrams would have been identical, and the count of unknowns would also have been the same. Equation (5.48), however, would no longer apply. If this were a friction drive, the force in the arm, $\overline{F}_4$, would be a known preload. This would reduce the system to nine unknowns and nine equations. Realistically, a friction drive would not be used in this mechanism, and in general, friction drives are not as common as positive drives.

5.5 PROBLEMS

Problem 5.1

The input to the machine represented in Figure 5.34 is θ_2. The values of $\theta_2, \dot{\theta}_2$, and $\ddot{\theta}_2$ are known—that is, the "state of motion" is known. In this machine, the load torque, $\overline{T}_5$, is known. This is a two-loop mechanism. The figure shows appropriate vector loops.

There are two vector loop equations,

$$\bar{r}_2 + \bar{r}_3 + \bar{r}_4 - \bar{r}_5 + \bar{r}_1 = \bar{0}$$

$$\bar{r}_8 - \bar{r}_7 - \bar{r}_6 - \bar{r}_5 + \bar{r}_1 = \bar{0},$$

and three geometric constraints,

$$\theta_8 + \gamma - \theta_2 = 0$$

$$\theta_6 + (\pi - \psi) - \theta_4 = 0$$

$$\text{and } \theta_6 + \frac{\pi}{2} - \theta_7 = 0,$$

with

scalar knowns: $r_1, \theta_1 = \pi, r_2, r_3, r_4, r_5, r_6, r_8, \phi, \psi$

and

scalar unknowns: $\theta_2, \theta_3, \theta_4, \theta_5, \theta_6, r_7, \theta_7, \theta_8$.

With seven equations in eight unknowns, there is one degree of freedom. Take θ_2 as S_i.

Figure 5.35 defines the locations of the mass centers. Vectors $\bar{r}_9, \bar{r}_{10}, \bar{r}_{11}$, and $\bar{r}_{12}$ and angles ϕ_2 and ϕ_4 are known.

The inertias $m_2, I_{g_2}; m_3, I_{g_3}; m_4, I_{g_4}$; and m_5, I_{g_5} are known.

Assume that steps 1–4 in Example 5.1 are completed. You should complete steps 5 and 6 here and derive a system of equations in matrix form that could be solved for the magnitude of the driving torque, $\overline{T}_2$, and all the bearing forces.

Figure 5.36 is for your free body diagrams.

For body 2, sum moments about point A.
For link 3, sum moments about point B.
For link 4, sum moments about point E.
For link 5, sum moments about point F.
Show your coordinate system in the free body diagrams.

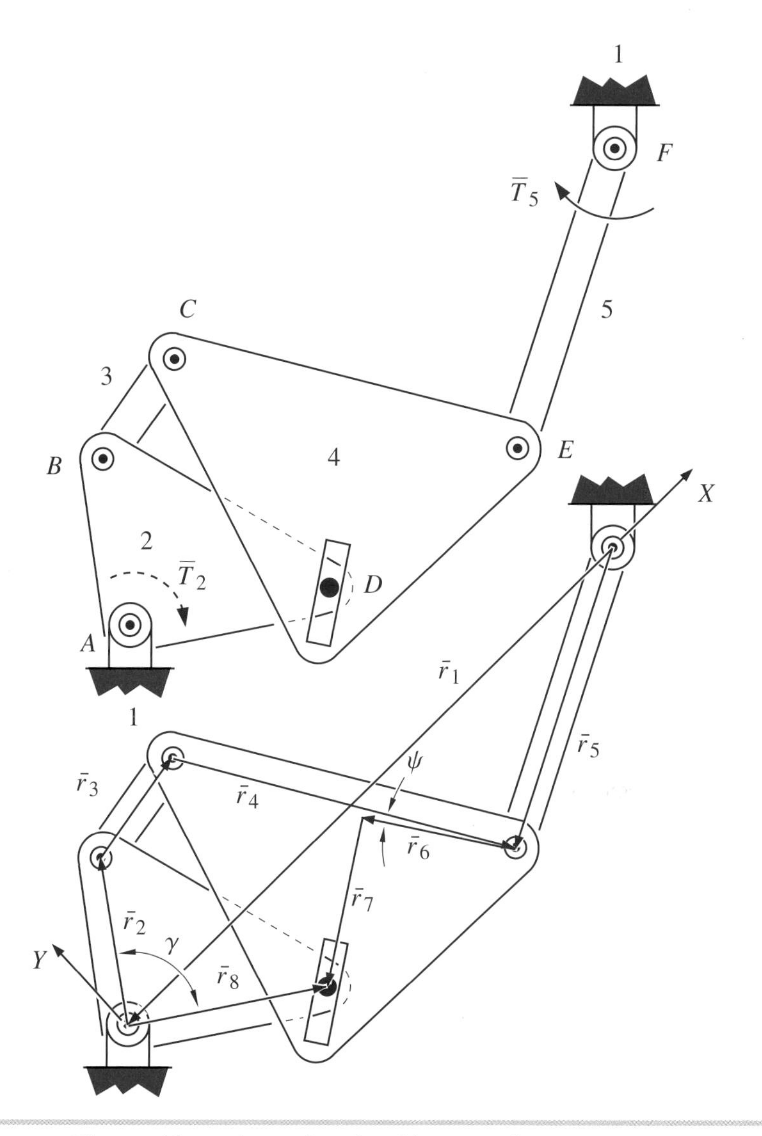

FIGURE 5.34 The machine to be analyzed and its vector loop

As a part of your solution, for each moment equation specify which unknowns were eliminated from that equation by summing moments about the designated point, compared with summing moments about the mass center.

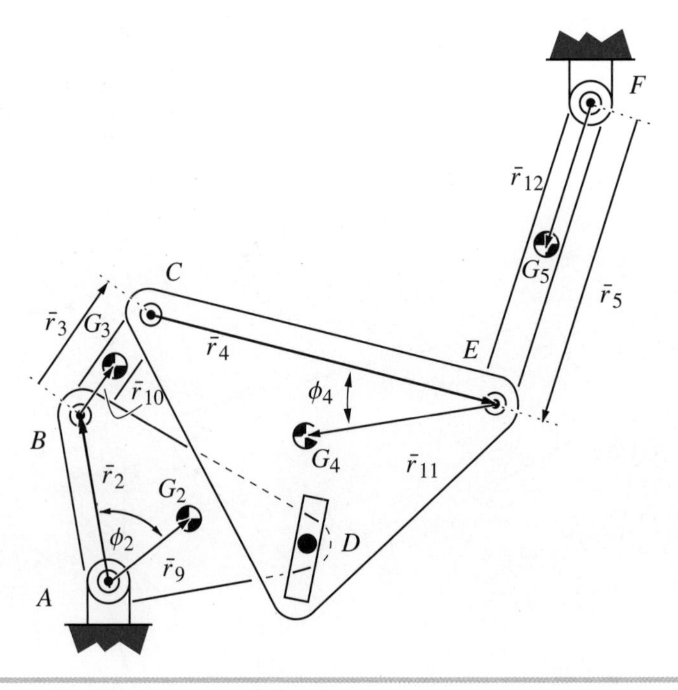

FIGURE 5.35 Definition of inertia properties

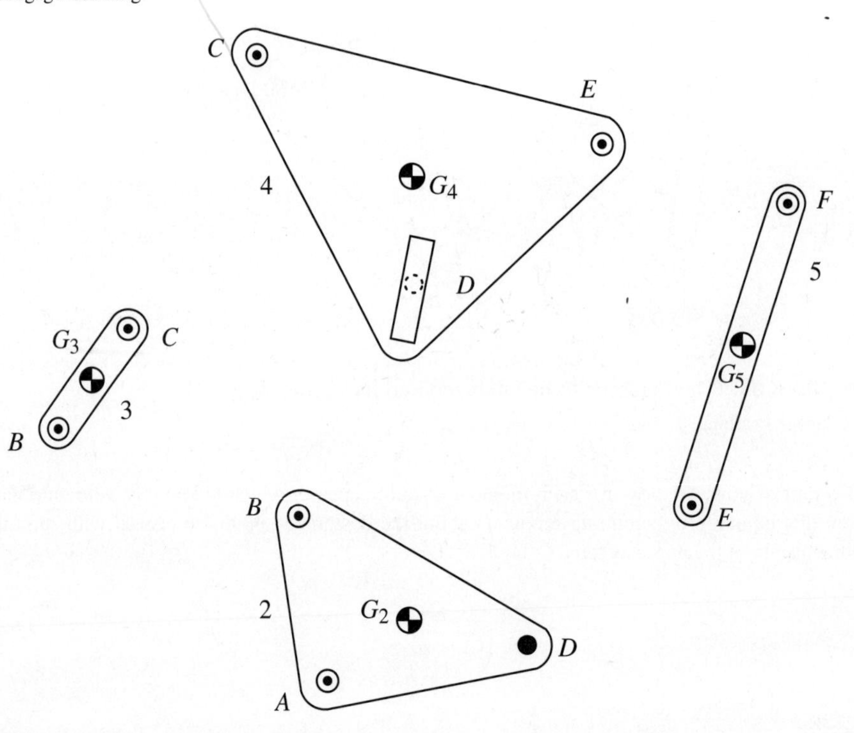

FIGURE 5.36 Free body diagrams

Problem 5.2

The Figure 5.37 shows a "walkable hoist." As link 2 is rotated counterclockwise by the driving torque $\overline{T}_2$, its center moves up and to the right. This causes link 4 to move down and to the right. As it does so, a cable attached to link 4 hoists a known load, $\overline{F}_4$.

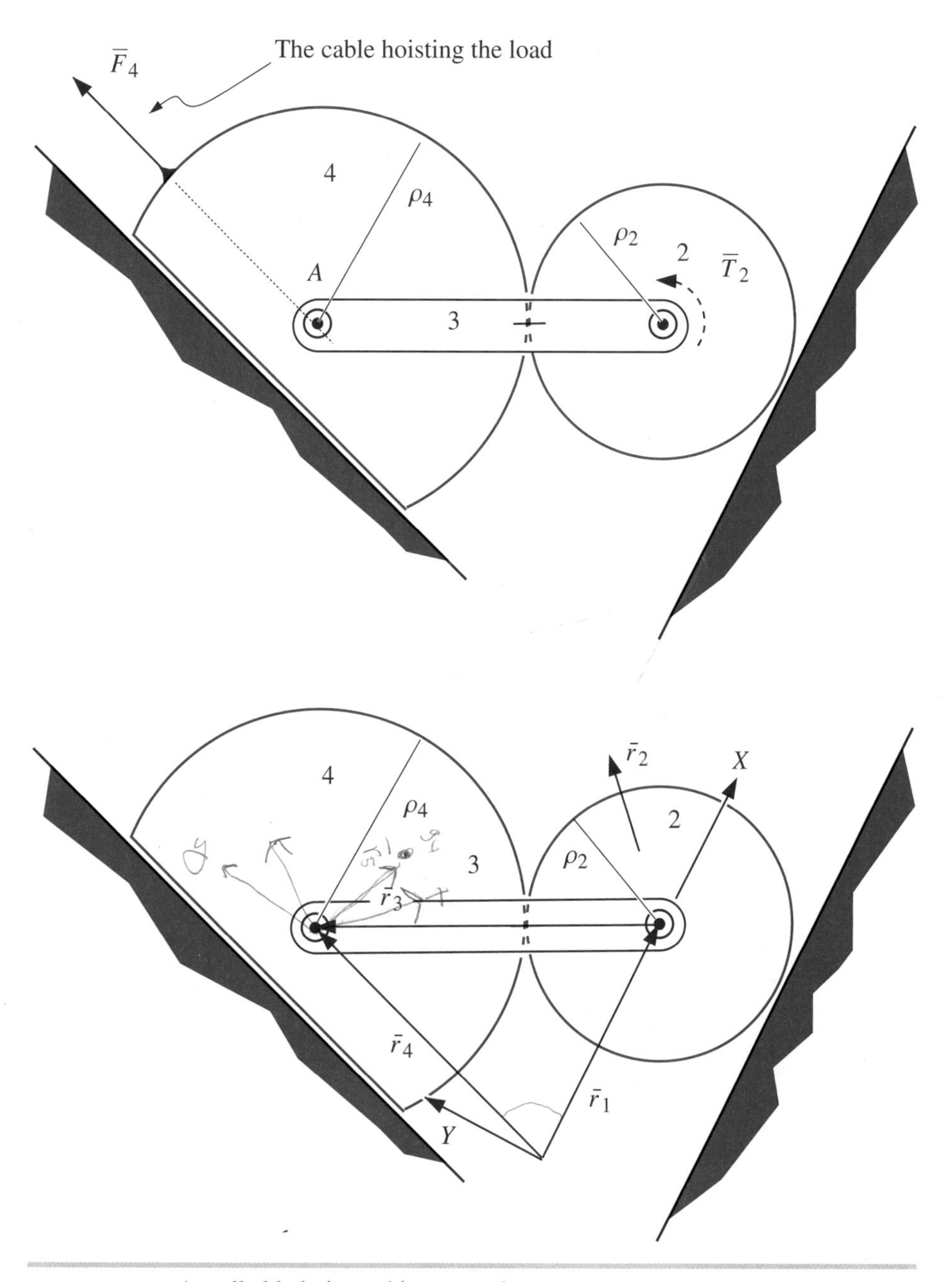

FIGURE 5.37 A walkable hoist and its vector loop

The load is to be hoisted at a desired constant rate. This means the "state of motion" of the machine is defined by desired values of r_4, $\dot{r}_4$, and $\ddot{r}_4$, and since the hoisting is to be at a constant rate, $\ddot{r}_4 = 0$.

The figure also shows an appropriate vector loop.

Figure 5.38 defines the location of the mass centers. Vectors $\bar{r}_5$ and $\bar{r}_6$ are known. The values of m_2, m_3, m_4, I_{g_2}, I_{g_3}, and I_{g_4} are known.

Assume that steps 1–4 in Example 5.1 are completed. You should complete steps 5 and 6 here and derive a system of equations in matrix form that could be solved for the magnitude of the driving torque, $\overline{T}_2$, and all the bearing forces.

Figure 5.39 is for your free body diagrams. Since your kinematic analysis uses the vector loop coordinate system, the X and Y components of the mass center accelerations are expressed in that system, and therefore that system defines the directions of the X and Y components of all the forces in your free body diagrams.

For link 2, sum moments about point G_2.
For link 3, sum moments about point A.
For link 4, sum moments about point A.
Show your coordinate system in the free body diagram.

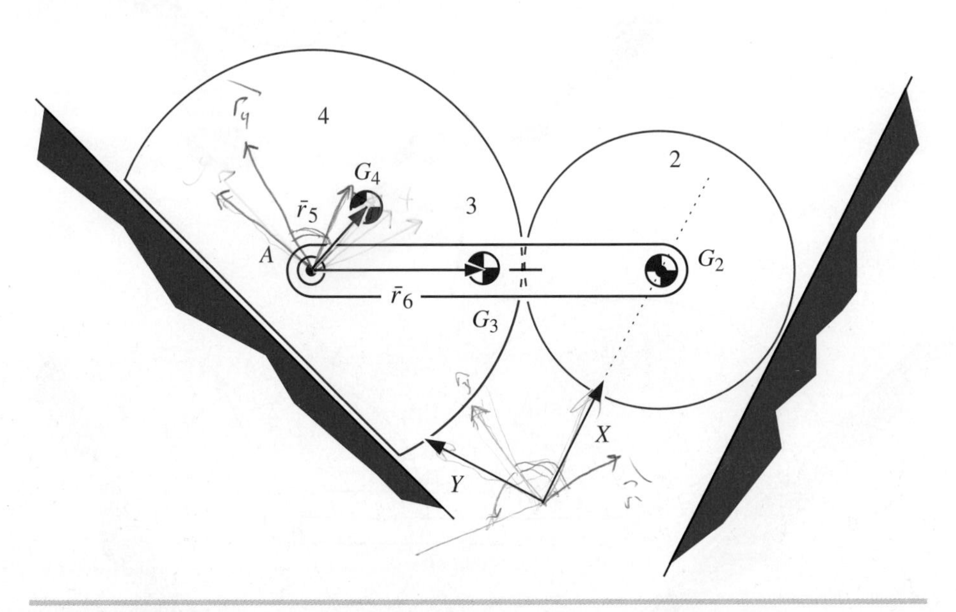

FIGURE 5.38 Definition of inertia properties

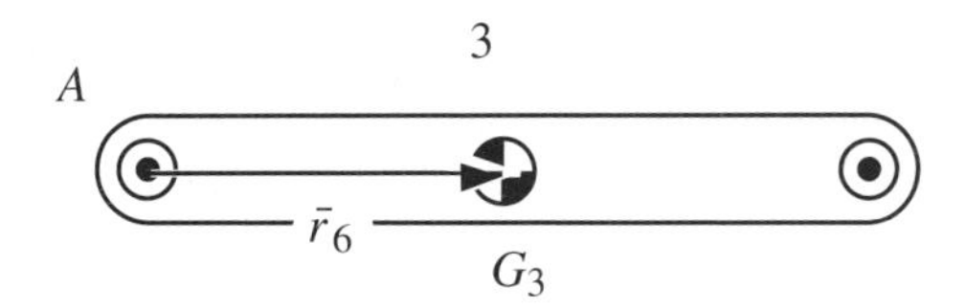

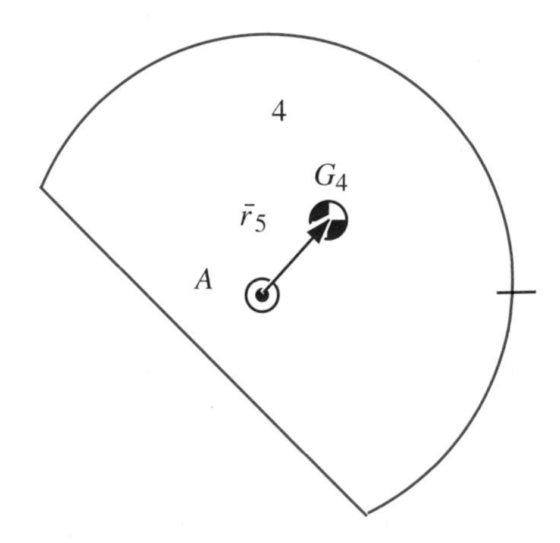

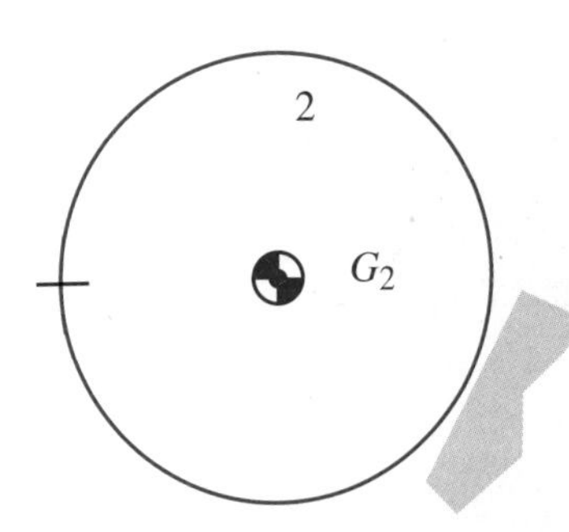

FIGURE 5.39 Free body diagrams

Problem 5.3

In the machine represented in Figure 5.40, link 2 is rotated clockwise by the driving torque, $\overline{T}_2$. This causes the center of link 4 to move to the right. This motion of 4 is resisted by a known load $\overline{F}_4$. The figure also shows an appropriate vector loop and the direction of the gravity field relative to the coordinate system for the vector loop method.

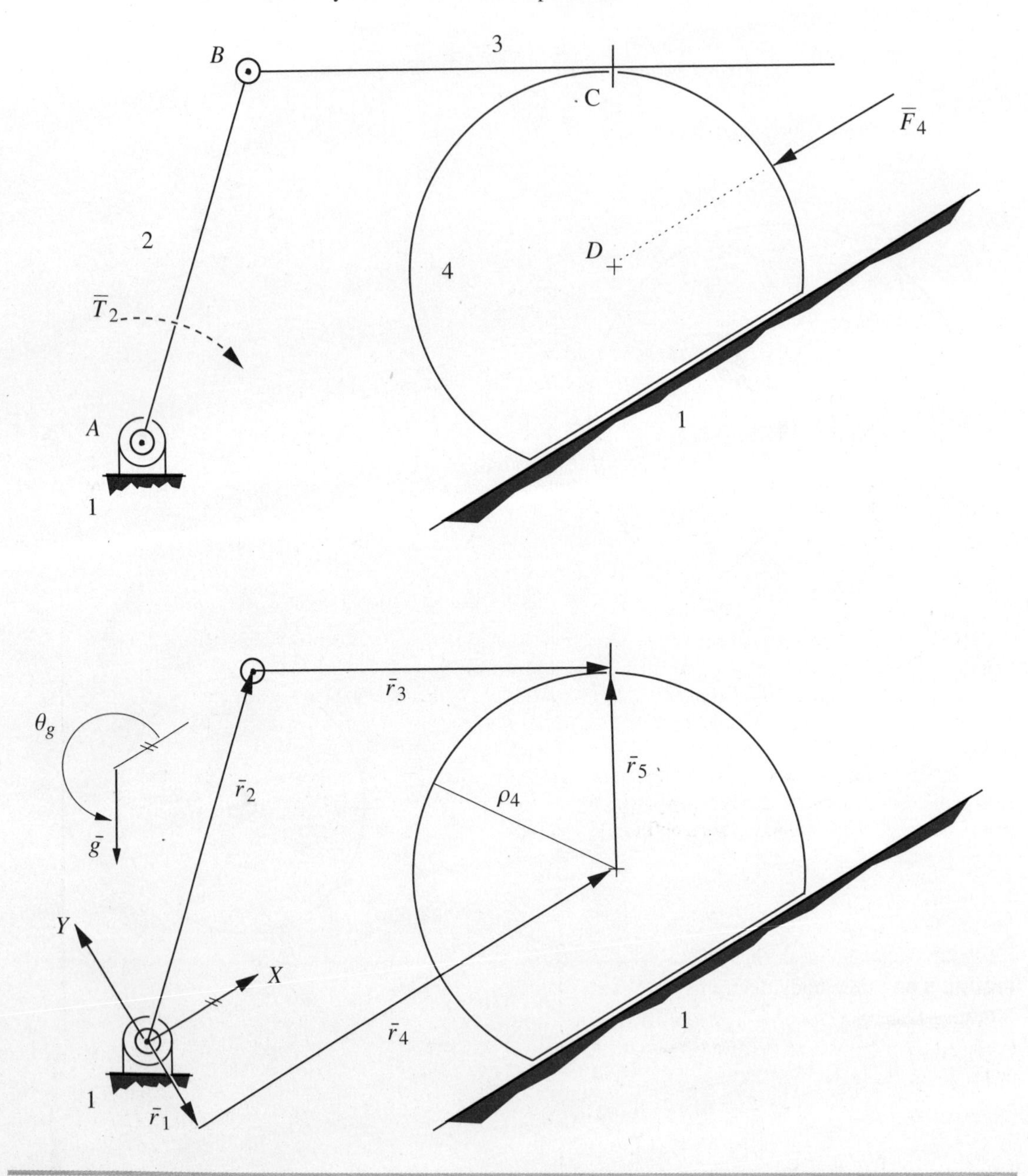

FIGURE 5.40 A machine transmitting power

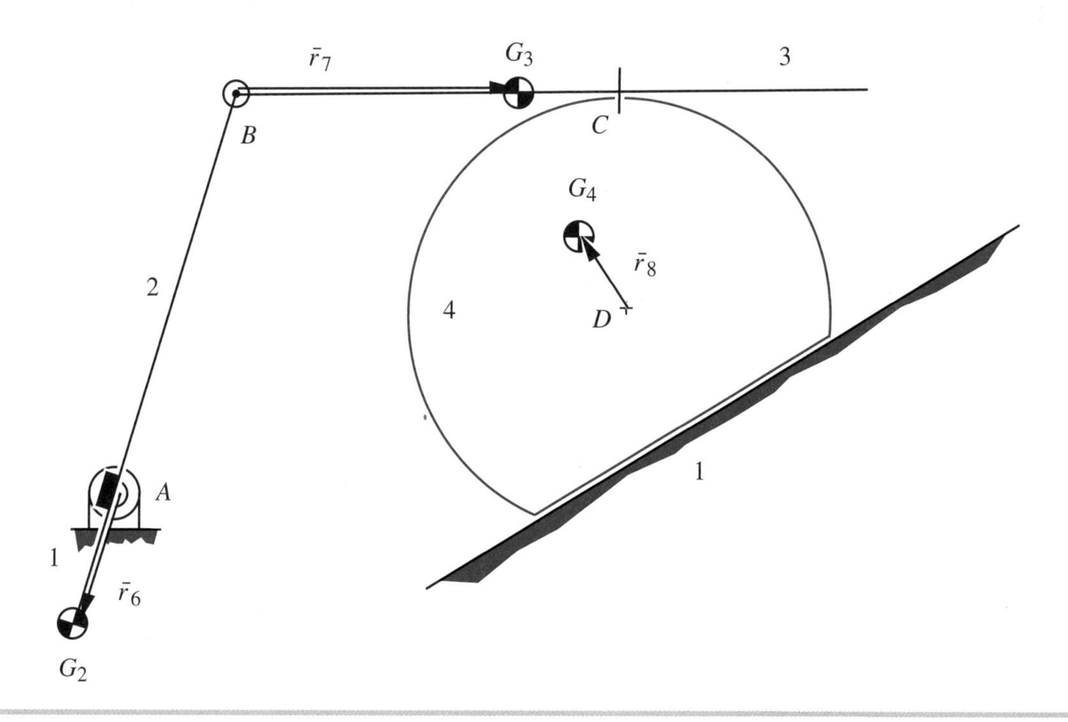

FIGURE 5.41 Definition of inertia properties

Figure 5.41 defines the location of the mass centers. Vectors $\bar{r}_6$, $\bar{r}_7$, and $\bar{r}_8$ are known. The inertias m_2, m_3, m_4, I_{g_2}, I_{g_3}, and I_{g_4} are known.

Assume that steps 1–4 in Example 5.1 are completed. You should complete steps 5 and 6 here and derive a system of equations in matrix form that could be solved for the magnitude of the driving torque, $\overline{T}_2$, and all the bearing forces.

Figure 5.42 is for your free body diagrams. Since your kinematic analysis uses the vector loop coordinate system, the X and Y components of the mass center accelerations are expressed in that system, and therefore that system defines the directions of the X and Y components of all the forces in your free body diagrams.

For link 2, sum moments about point A.
For link 3, sum moments about point B.
For link 4, sum moments about point D.

As a part of your solution, for each moment equation specify which unknowns were eliminated from that equation by summing moments about the designated point, compared with summing moments about the mass center. Show your coordinate system in your free body diagrams.

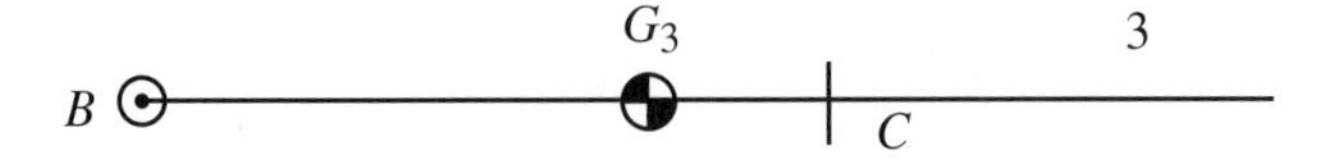

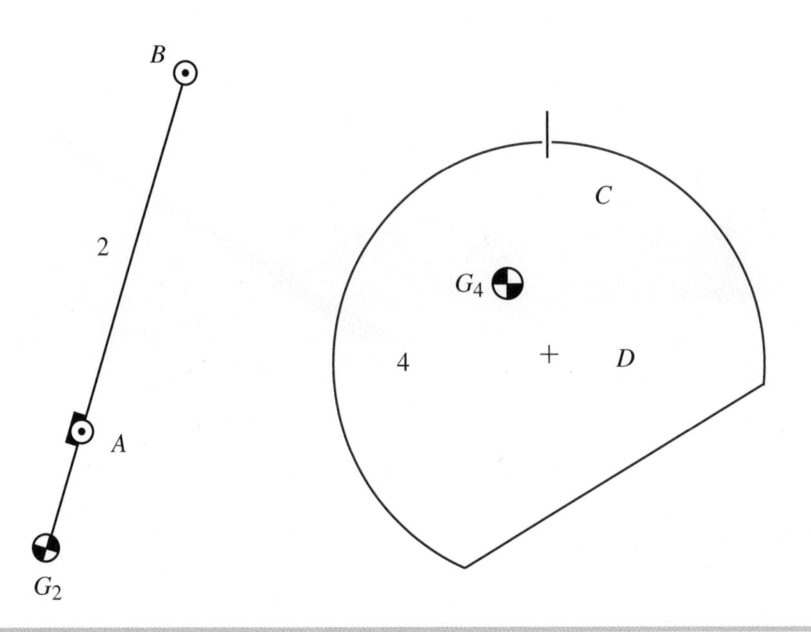

FIGURE 5.42 Free body diagrams

Problem 5.4

The machine represented in Figure 5.43 is a scissor jack. It is commonly used to lift an automobile in order to change a flat tire. $\overline{W}$ is the known weight of the automobile. $\overline{F}$ is the driving force. This force is usually developed through a screw that is turned by the operator of the jack.

This is a two-loop mechanism. The right side shows two appropriate vector loops. You may assume that $r_2 = r_4 = R$ and $r_6 = r_7 = r$. (You should realize that this would be true only to a certain tolerance.)

The system has one dof. Suppose that $S_i = r_1$. The weight being lifted is large and it is being lifted very slowly. This means that it is reasonable to neglect inertia effects. Likewise, the weights of the links are a fraction of a percent of the weight $\overline{W}$, and they can also be neglected.

Derive all equations necessary in order to compute F and the bearing forces in terms of the prescribed W, link lengths, and state of motion. Present all systems of equations in matrix form. Figure 5.44 is for your free body diagrams. For your moment equations, please sum moments as follows:

For link 2, sum moments about the point B.
For link 3, sum moments about the point C.
For link 4, sum moments about the point A.
Show your coordinate system in your free body diagrams.

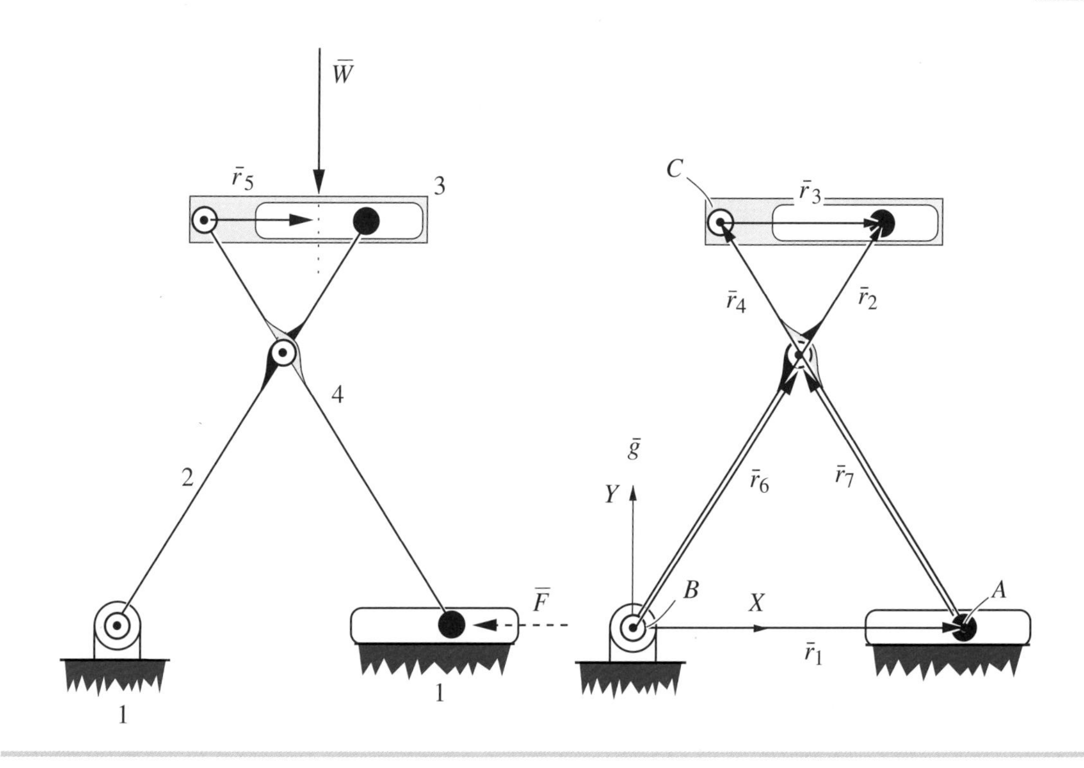

FIGURE 5.43 A scissor jack

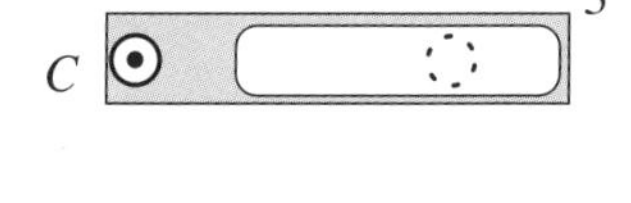

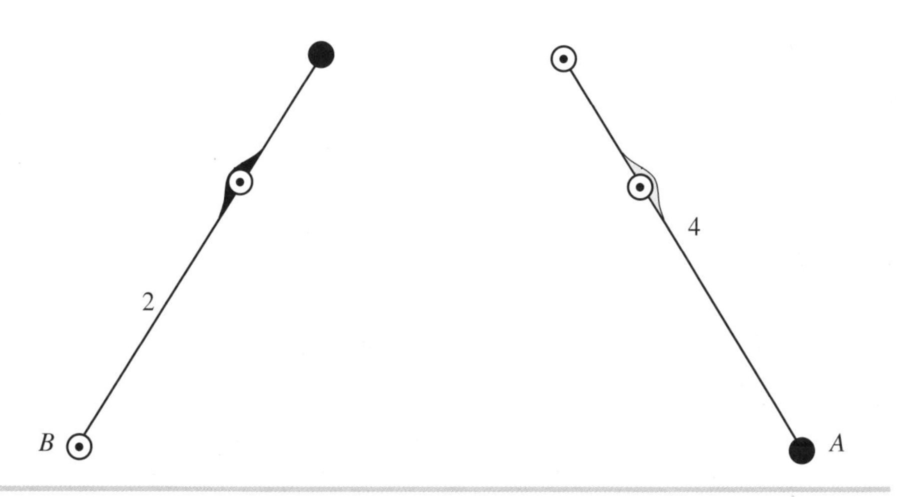

FIGURE 5.44 Free body diagrams

Problem 5.5

The indexing machine represented in Figure 5.45 has a known load force $\overline{F}$. The input crank, 2, is rotating at a known constant speed. The inertia properties of the links are known, and the locations of the mass centers are defined in Figure 5.46. The vectors $\bar{r}_9$ and $\bar{r}_{10}$ locate the centerlines of the bearings on the left and right, relative to the origin of the vector loop's coordinate system.

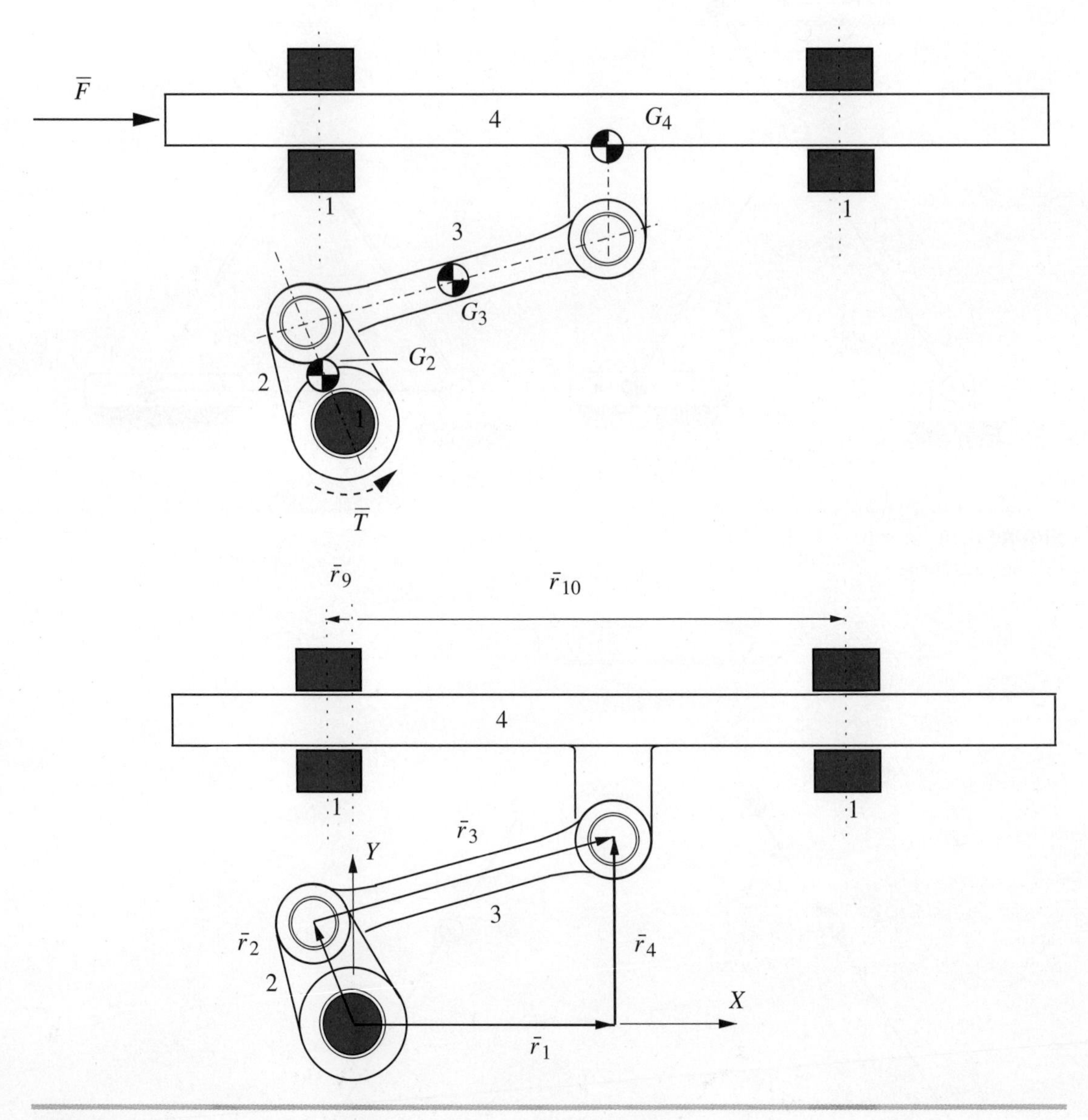

FIGURE 5.45 An indexing machine and an appropriate vector loop

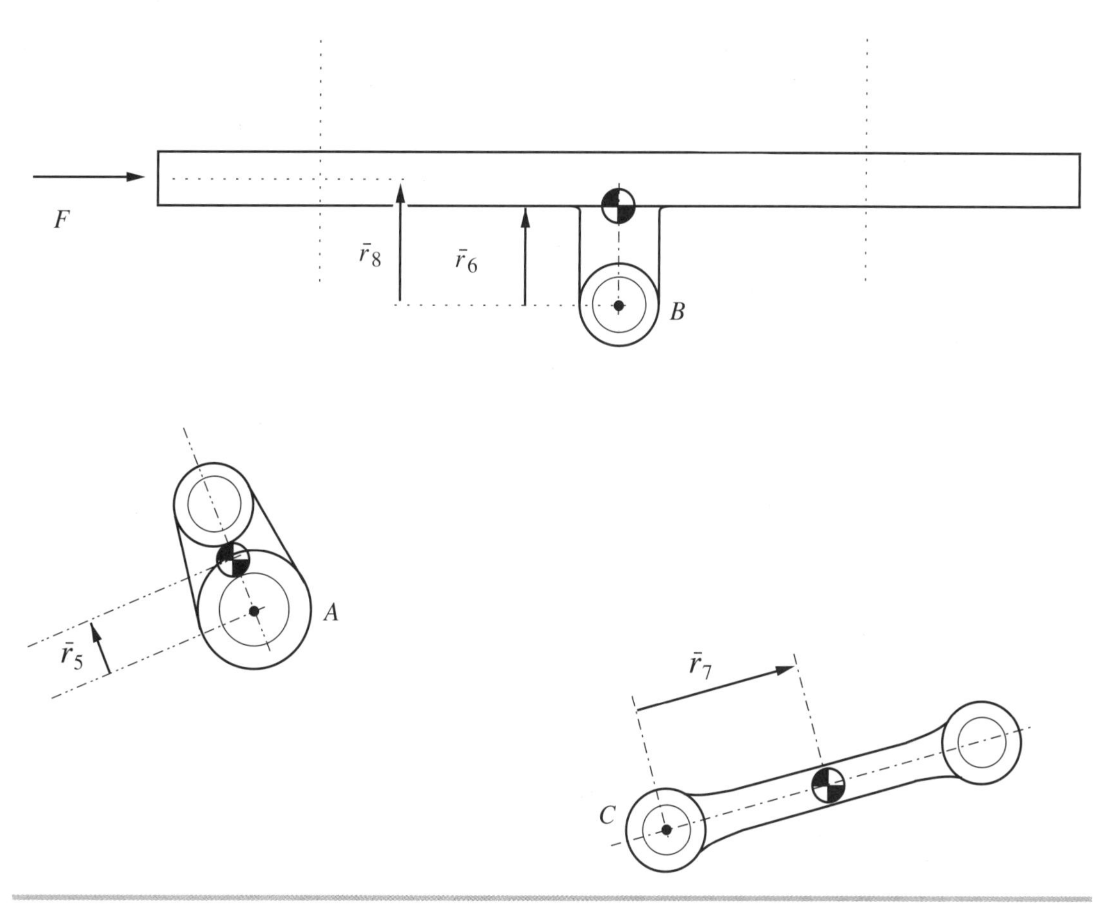

FIGURE 5.46 Free body diagrams

Derive all the equations that you would use in a computer program to solve for the driving torque $\overline{T}$ and all the bearing forces. Give all the details, including steps 1–6 in Example 5.1. Figure 5.46 is for your free body diagrams.

For link 2, sum moments about point *A*.
For link 3, sum moments about point *B*.
For link 4, sum moments about point *C*.
Show your coordinate system in your free body diagrams.

Problem 5.6

In the machine represented in Figure 5.47, $\overline{P}$ is a known force on link 8. $\overline{Q}$ is an unknown torque (couple) on link 2. The machine is slow-moving, so inertia effects are neglected. P is very large, so weight forces are also negligible.

Outline the equations you would implement in a computer code that would solve for Q and the bearing forces. You are advised to take advantage of all two-force members. (That's good advice in general.) For your moment equilibrium, use points A, B, and C.

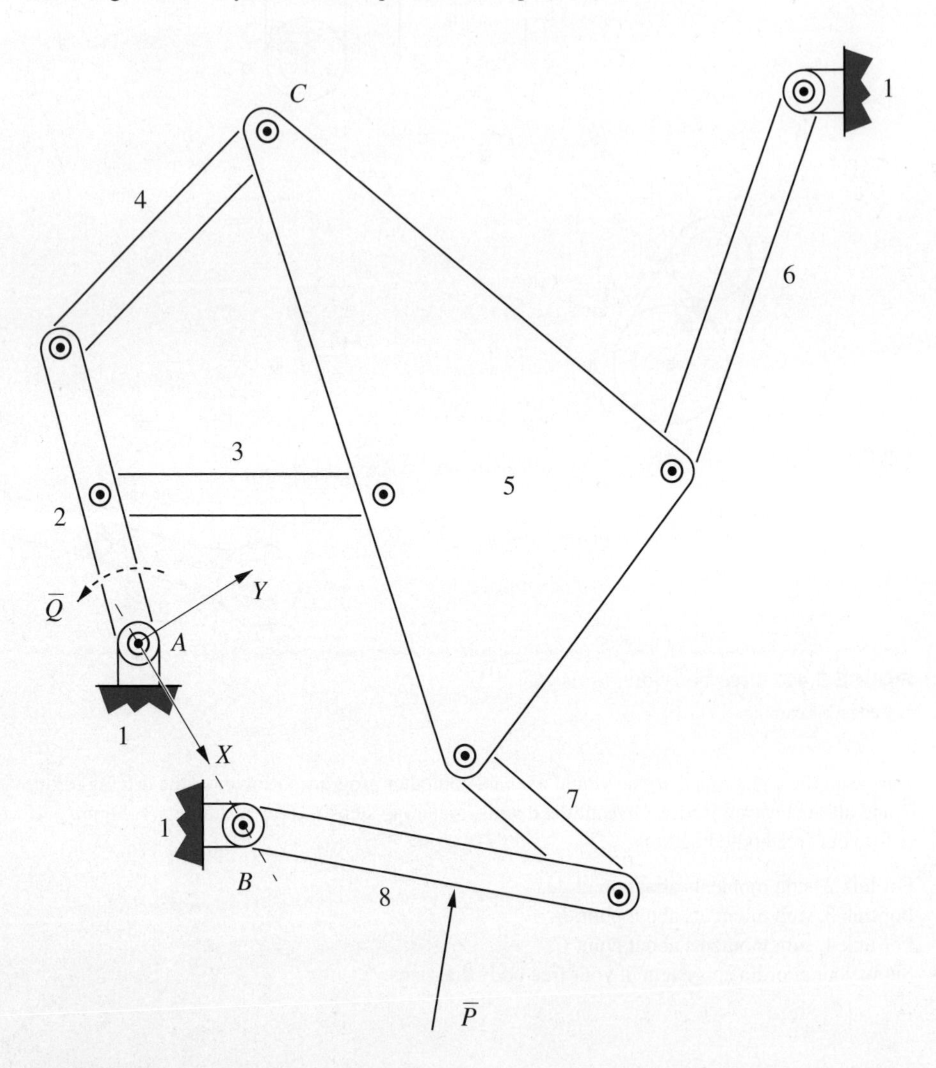

FIGURE 5.47 An eight bar mechanism

Figure 5.48 is for your free body diagrams.

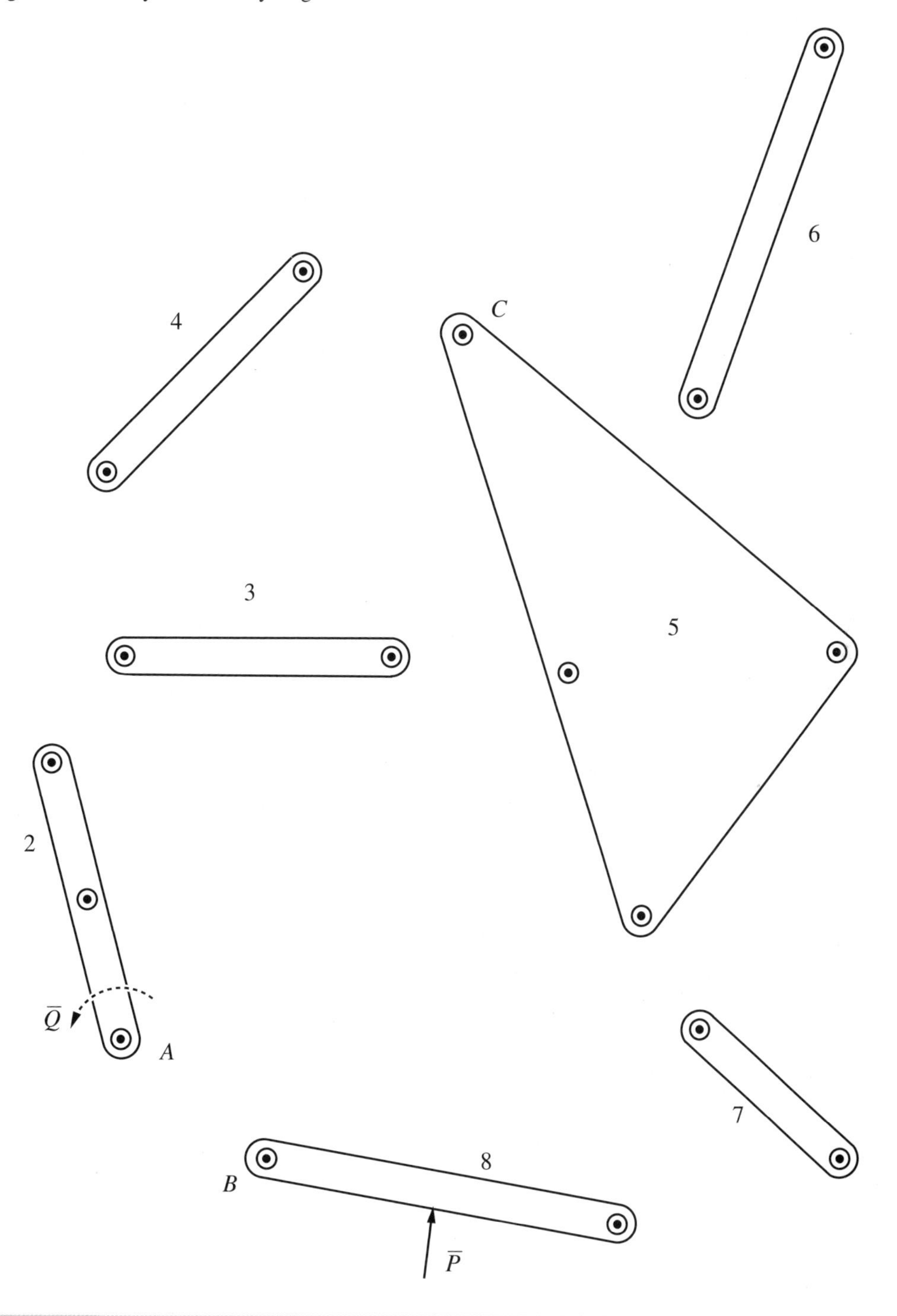

FIGURE 5.48 Free body diagrams

Problem 5.7

In the machine represented in Figure 5.49, the load $\overline{P}$ applied to link 4 is known. The driving torque $\overline{Q}$ applied to link 2 is unknown. Figure 5.50 shows locations of mass centers. The inertia properties of all links are known. Assume that steps 1–4 in Example 5.1 are completed. You should complete steps 5 and 6 here and derive a system of equations in matrix form that could be solved for the magnitude of the driving torque, and all the bearing forces.

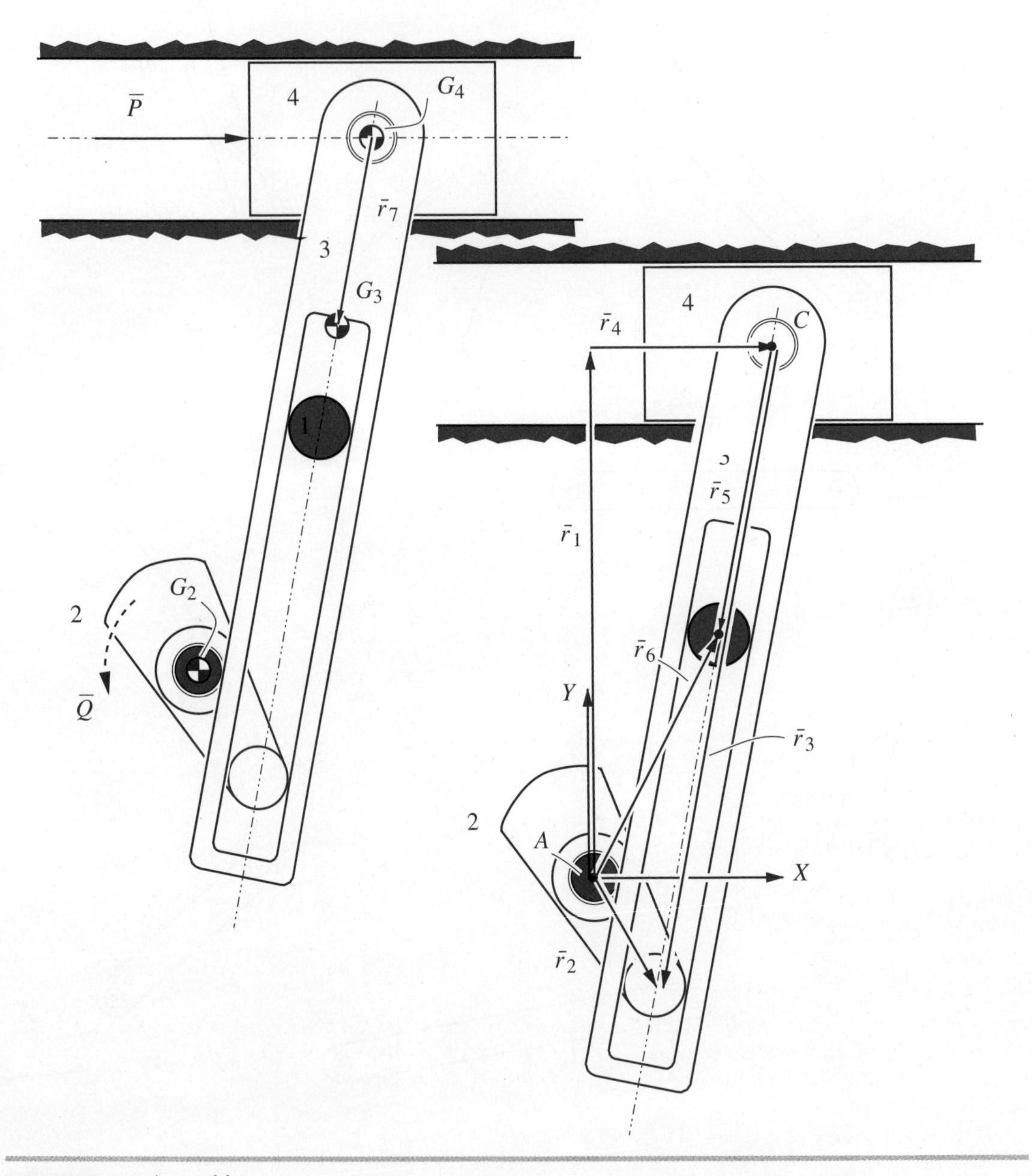

FIGURE 5.49 A machine

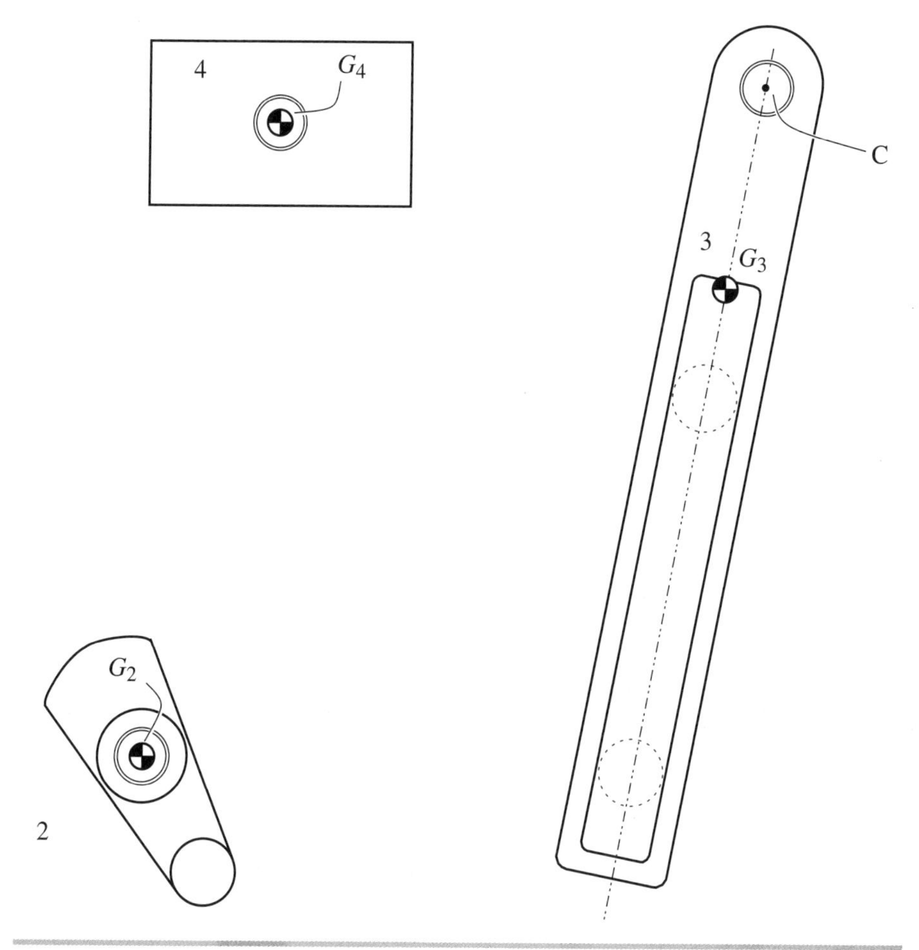

FIGURE 5.50 Free body diagrams

For link 2, sum moments about G_2.
For link 3, sum moments about C.
For link 4, sum moments about G_4.

(Vector $\bar{r}_3$ would actually be directly on top of $\overline{5}$, but for visual clarity it has been drawn as slightly offset.) Show your coordinate system in your free body diagrams.

Problem 5.8

In the machine represented in Figure 5.51, the angular position, velocity, and acceleration of link 2 are known. The inertia properties of all links are known. The load torque $\overline{T}_5$ is known, and the magnitude of the driving torque $\overline{T}_2$ is to be determined. An appropriate vector loop is shown on the right.

Assume that steps 1–4 in Example 5.1 are completed. You should complete steps 5 and 6 here and derive a system of equations in matrix form that could be solved for the magnitude of the driving torque, $\overline{T}_2$, and all the bearing forces.

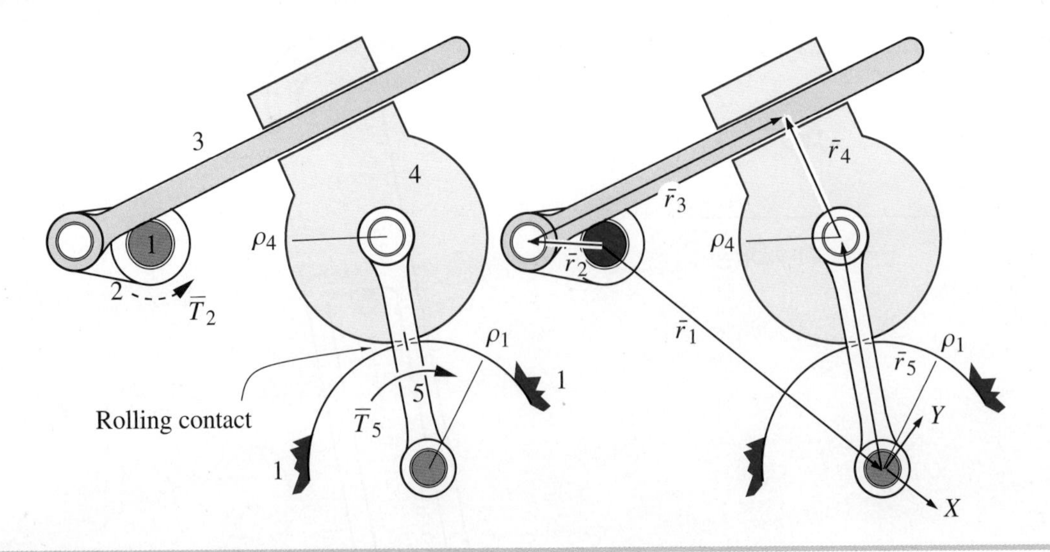

FIGURE 5.51 A machine

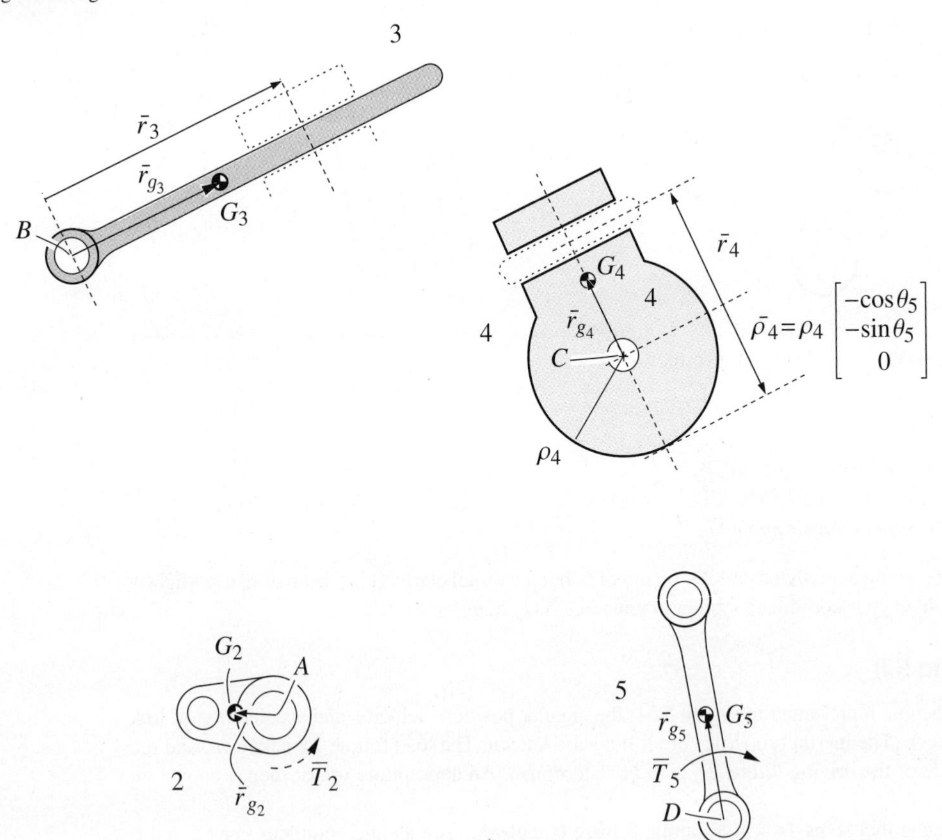

FIGURE 5.52 Free body diagrams

Vectors $\bar{r}_{g2}$, $\bar{r}_{g3}$, $\bar{r}_{g4}$, and $\bar{r}_{g5}$ are in the direction of vectors $\bar{r}_2$, $\bar{r}_3$, $\bar{r}_4$, and $\bar{r}_5$, respectively. For links 2, 3, 4, and 5, sum moments about points A, B, C, and D, respectively. Show your coordinate system in your free body diagrams.

Problem 5.9

Consider the machine represented in Figure 5.53. The weight of link 4 dominates all other weights, which means that only this weight needs to be included in your analysis. Gravity acts in the direction shown. The load $\overline{F}_6$ is known and it acts on 6. The driver $\overline{T}_2$ acts on 2 and is unknown. Complete steps 1–6 in Example 5.1, and outline a procedure for computing $\overline{T}_2$. Figure 5.54 is for your free body diagrams. Show your coordinate system there.

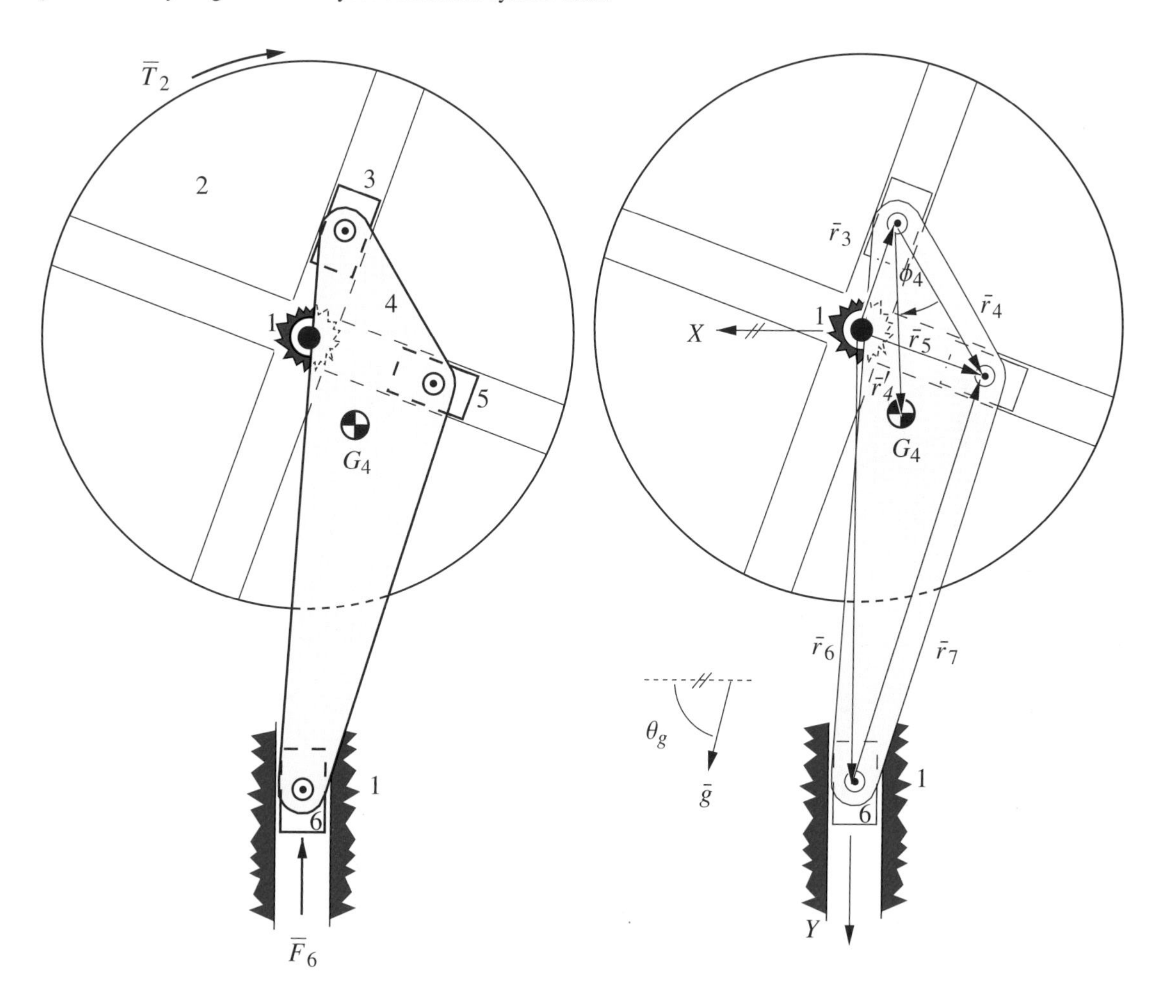

FIGURE 5.53 A machine and its vector loop

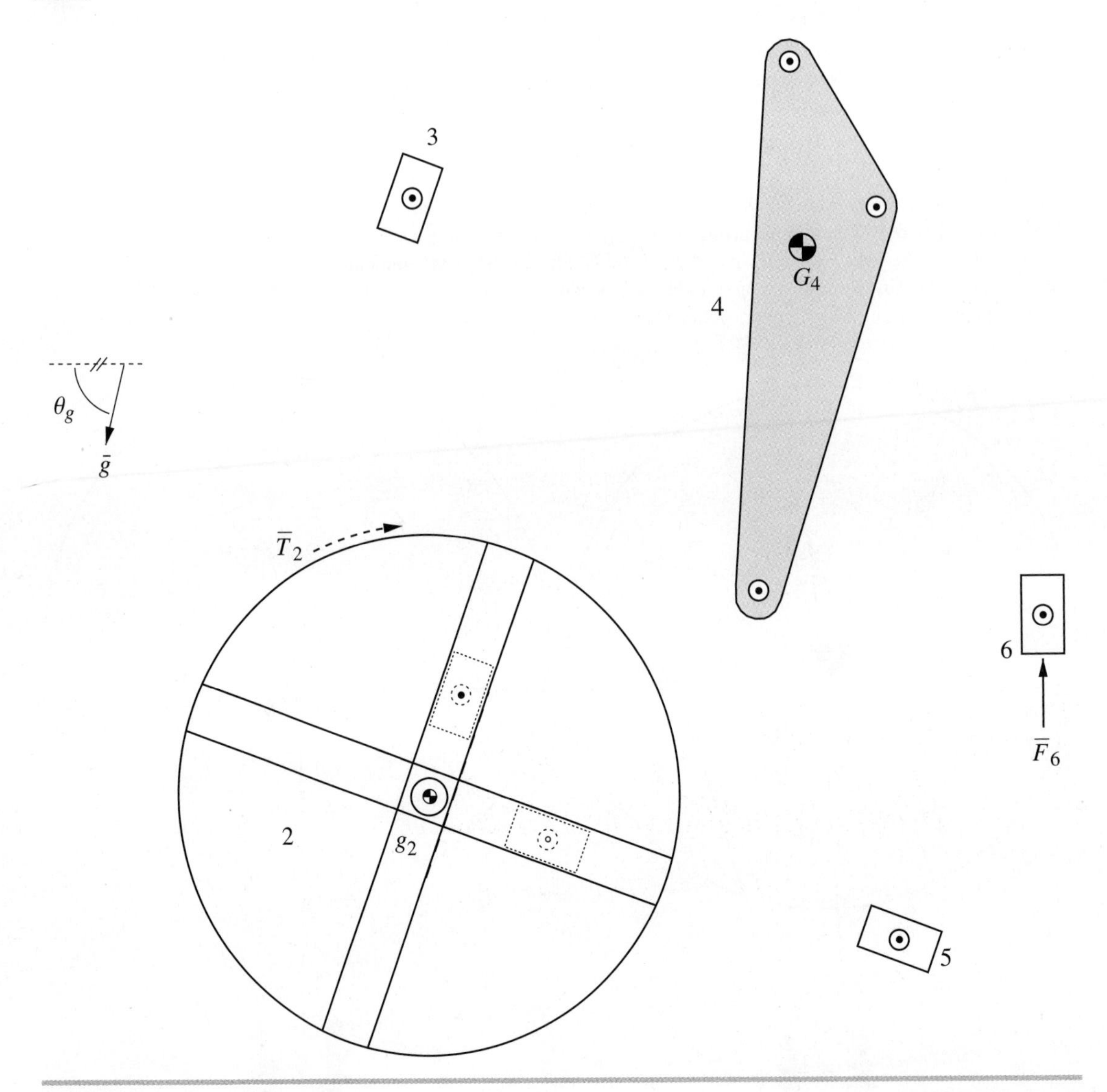

FIGURE 5.54 Free body diagrams

Problem 5.10

The machine represented in Figures 5.55, 5.56, and 5.57 moves slowly. The load $\overline{F}_3$ is known and is an order of magnitude larger than the weight of any link. Complete steps 1–6 in Example 5.1, and outline a procedure for computing $\overline{T}_2$. Show your coordinate system in your free body diagram.

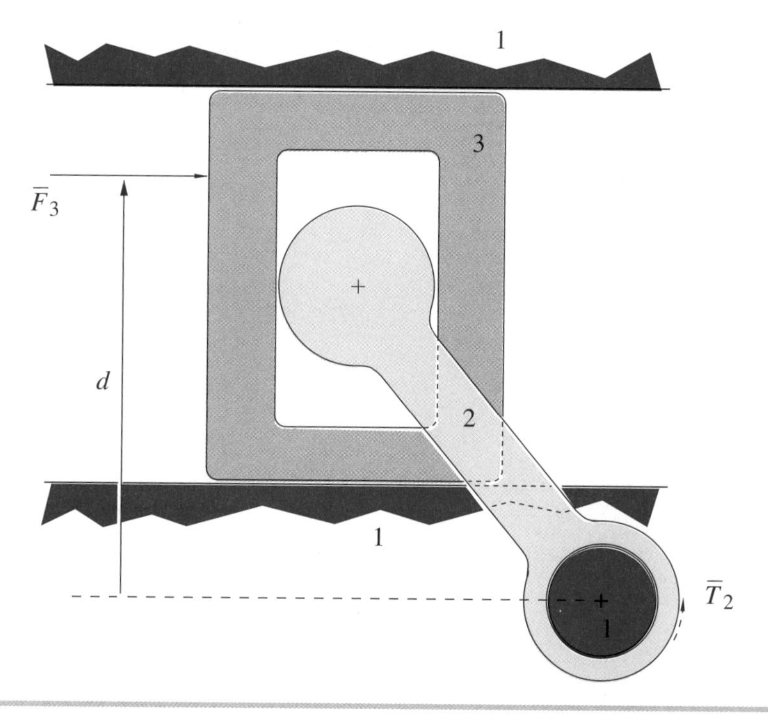

FIGURE 5.55 A machine

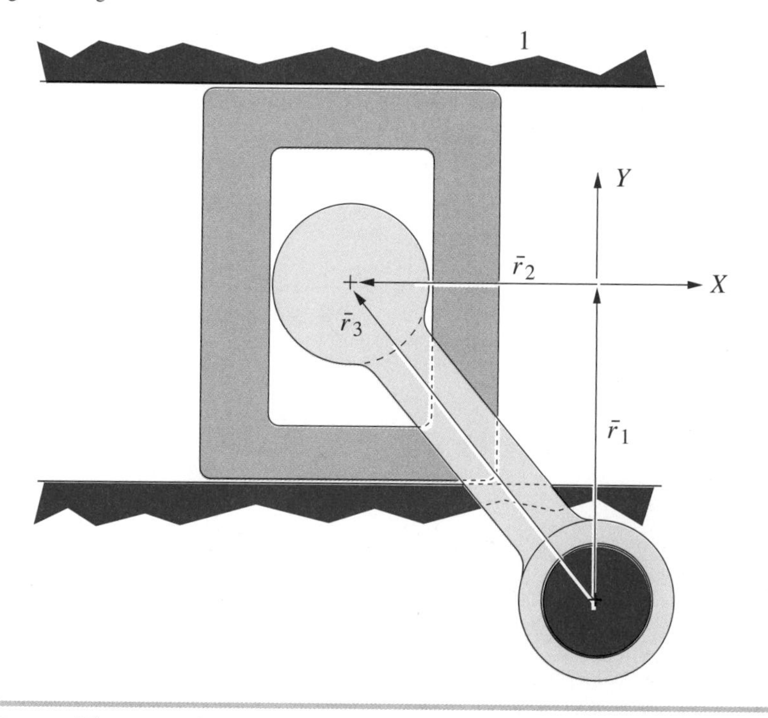

FIGURE 5.56 The vector loop

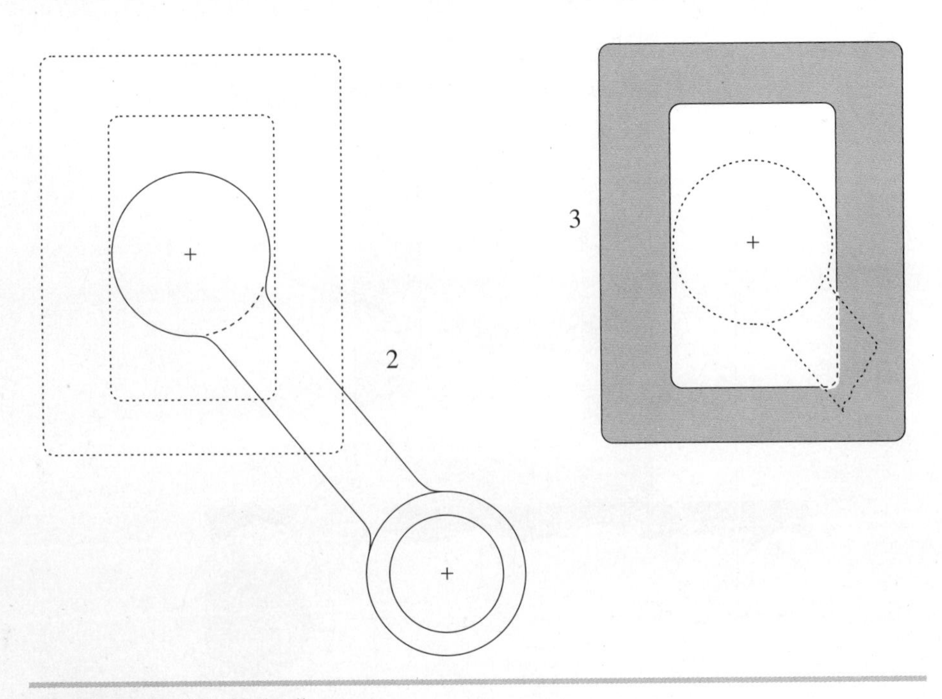

FIGURE 5.57 Free body diagrams

Problem 5.11

This problem continues with the problem stated in Problem 4.5. Suppose that the load $\overline{T}_3$ is known, and we want to find the driver $\overline{F}_2$ and all the forces at the joints. Neglect the effects of inertia and gravity. Show your coordinate system.

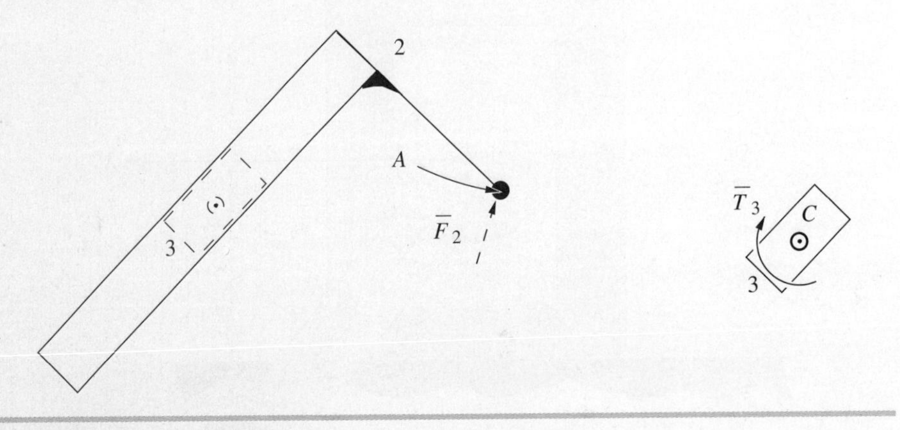

Figure Free body diagrams

Problem 5.12

Repeat Example 3.7, but use a force analysis instead of a kinematic analysis to determine the ratio. The planetary geartrain is redrawn in Figure 5.58. 4 is fixed, the load torque is applied to link 5, and the driving torque acts on 2.

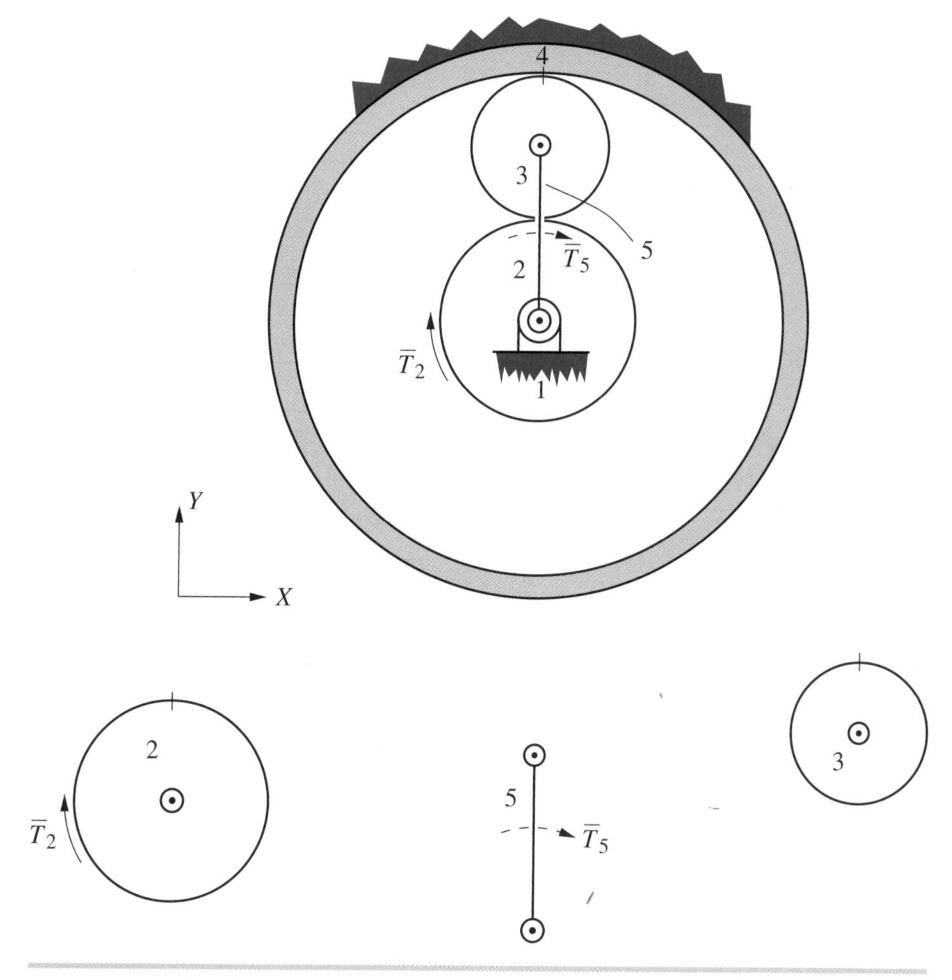

FIGURE 5.58 Free body diagrams

A planetary geartrain built with a single planet gear, as shown in Figure 5.58, would have an imbalance. To balance the system, muultiple planet gears, such as those shown in Figure 5.59, are typically used. How do you suppose the gear forces would be computed in this case?

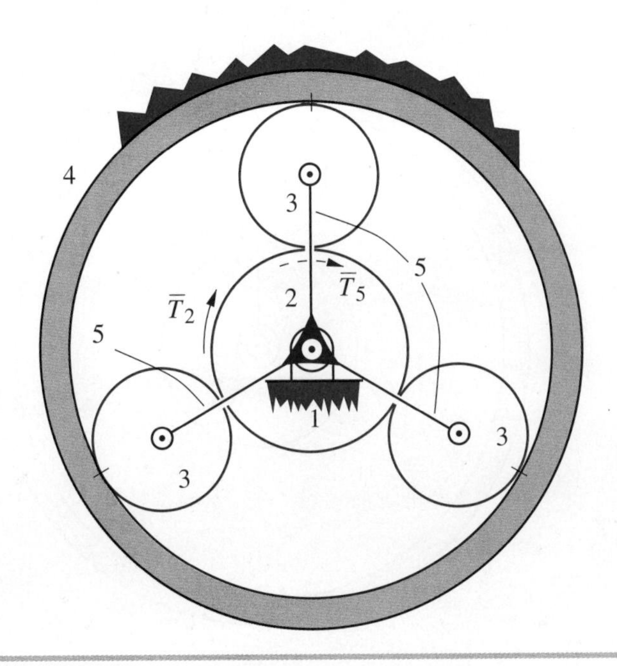

FIGURE 5.59 Planetary geartrain with multiple planet gears

5.6 APPENDIX I: KINEMATIC ANALYSIS FOR EXAMPLES IN SECTION 5.1 (EXAMPLE 5.2) AND SECTION 7.2

Step 1: Solve the position problem, and find the basic first- and second-order kinematic coefficients. Referring to Figure 5.7, the vector loop equation is $\bar{r}_2 - \bar{r}_3 - \bar{r}_1 = \bar{0}$. This has the X and Y components

$$r_2\cos\theta_2 - r_3\cos\theta_3 = 0 \tag{5.49}$$

$$r_2\sin\theta_2 - r_3\sin\theta_3 + r_1 = 0 \qquad \text{(note the sign of } r_1\text{),} \tag{5.50}$$

with

scalar knowns: $r_2, r_1, \theta_1 = \frac{3\pi}{2}$

and

scalar unknowns: θ_2, θ_3, r_3.

Take S_i to be θ_2, which has a value of $\pi/6$. Write a Newton's Method code, and compute

$$\theta_3 = 70.893° \quad \text{and} \quad r_3 = 15.875 \text{ in.}$$

Differentiate Equations (5.49) and (5.50) with respect to θ_2 once and twice to find the basic first- and second-order kinematic coefficients, respectively. First differentiation:

$$r_3\sin\theta_3 h_3 \;-\; \cos\theta_3 f_3 = r_2\sin\theta_2$$
$$-r_3\cos\theta_3 h_3 \;-\; \sin\theta_3 f_3 = -r_2\cos\theta_2$$

Computing gives

$$h_3 = 0.286 \quad \text{and} \quad f_3 = 3.928 \text{ in.}$$

Second differentiation:

$$r_3\sin\theta_3 h_3' - \cos\theta_3 f_3' = -r_3\cos\theta_3 h_3^2 - 2f_3h_3\sin\theta_3 + r_2\cos\theta_2$$
$$-r_3\cos\theta_3 h_3' - \sin\theta_3 f_3' = -r_3\sin\theta_3 h_3^2 + 2f_3h_3\cos\theta_3 + r_2\sin\theta_2$$

Computing gives

$$h_3' = 0.106 \quad \text{and} \quad f_3' = -3.240 \text{ in.}$$

Step 2: Find the first- and second-order kinematic coefficients of the moving mass centers.

Body 2, $\bar{r}_{g_2} = \bar{r}_4$:

$$r_{g2_x} = r_4\cos\theta_2 \rightarrow f_{g2_x} = -r_4\sin\theta_2 \rightarrow f'_{g2_x} = -r_4\cos\theta_2$$
$$r_{g2_y} = r_4\sin\theta_2 \rightarrow f_{g2_y} = r_4\cos\theta_2 \rightarrow f'_{g2_y} = -r_4\sin\theta_2$$

Computing gives

$$f_{g2_x} = -1.500 \text{ in}, \quad f'_{g2_x} = -2.598 \text{ in.}$$
$$f_{g2_y} = 2.598 \text{ in}, \quad f'_{g2_y} = -1.500 \text{ in.}$$

Body 3, $\bar{r}_{g_3} = \bar{r}_2$:

$$r_{g3_x} = r_2\cos\theta_2 \rightarrow f_{g3_x} = -r_2\sin\theta_2 \rightarrow f'_{g3_x} = -r_2\cos\theta_2$$
$$r_{g3_y} = r_2\sin\theta_2 \rightarrow f_{g3_y} = r_2\cos\theta_2 \rightarrow f'_{g3_y} = -r_2\sin\theta_2$$

Computing gives

$$f_{g3_x} = -3.000 \text{ in.}, \quad f'_{g3_x} = -5.196 \text{ in.}$$
$$f_{g3_y} = 5.196 \text{ in.}, \quad f'_{g3_y} = -3.000 \text{ in.}$$

Body 4, $\bar{r}_{g_4} = \bar{r}_1 + \bar{r}_5$:

$$r_{g4_x} = r_5\cos\theta_3 \rightarrow f_{g4_x} = -r_5\sin\theta_3 h_3 \rightarrow f'_{g4_x} = -r_5\cos\theta_3 h_3^2 - r_5\sin\theta_3 h_3'$$
$$r_{g4_y} = r_1 + r_5\sin\theta_3 \rightarrow f_{g4_y} = r_5\cos\theta_3 h_3 \rightarrow f'_{g4_y} = -r_5\sin\theta_3 h_3^2 + r_5\cos\theta_3 h_3'$$

Computing gives

$$f_{g4_x} = -2.430 \text{ in.}, \quad f'_{g4_x} = -1.142 \text{ in.}$$
$$f_{g4_y} = 0.842 \text{ in.}, \quad f'_{g4_y} = -0.382 \text{ in.}$$

5.7 APPENDIX II: COMPUTING THE ACCELERATIONS OF THE MASS CENTERS OF THE COMPOSITE SHAPES IN SECTION 5.2.6

Figure 5.60 shows the locations of the mass centers of the composite shapes that make up the crank. It also gives the masses of the shapes. Since the crank has a fixed pivot, the accelerations of the mass centers of the shapes will be proportional to the acceleration of the overall mass center, so there is no need for any additional kinematic analysis. The fixed pivot is what makes the scaling possible. Think about it. $a_{g_{2x}}$ and $a_{g_{2y}}$ were computed in Example 5.2.

$$a_{g_{2x}} = -102.57 \text{ in./s}^2, \qquad a_{g_{2y}} = -59.22 \text{ in./s}^2$$

$$a_{g_{Ix}} = \tfrac{r_2}{r_4}\, a_{g_{2x}} = \tfrac{6}{3}(-102.57 \text{ in./s}^2) = -205.14 \text{ in./s}^2$$
$$a_{g_{Iy}} = \tfrac{r_2}{r_4}\, a_{g_{2y}} = \tfrac{6}{3}(-59.22 \text{ in./s}^2) = -118.44 \text{ in./s}^2$$

$$a_{g_{IIx}} = \tfrac{r_6}{r_4}\, a_{g_{2x}} = \tfrac{2}{3}(-102.57 \text{ in./s}^2) = -68.38 \text{ in./s}^2$$
$$a_{g_{IIy}} = \tfrac{r_6}{r_4}\, a_{g_{2y}} = \tfrac{2}{3}(-59.22 \text{ in./s}^2) = -39.48 \text{ in./s}^2$$

$$a_{g_{IIIx}} = 0 \text{ in./s}^2$$
$$a_{g_{IIIy}} = 0 \text{ in./s}^2$$

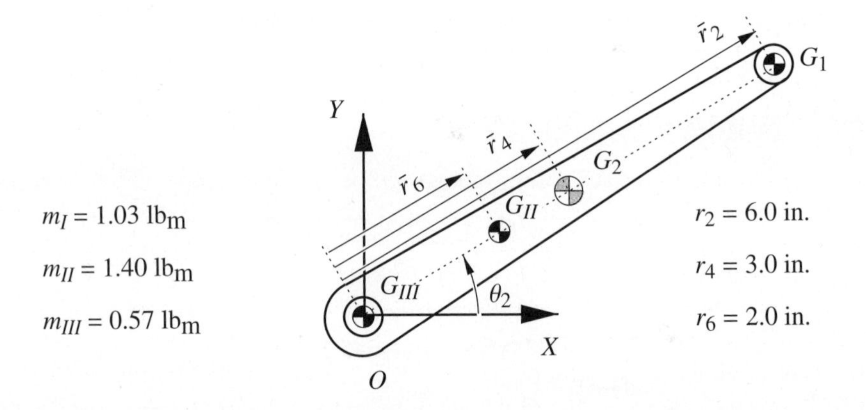

FIGURE 5.60 Computing accelerations of the composite shapes' mass centers

CHAPTER 6

Machine Dynamics Part II: Joint Friction

Chapter 5 discussed the IDP and neglected friction. In moderate- to high-speed machinery with good bearings and/or well lubricated joints, neglecting friction gives satisfactory results for purposes of design. For machines that operate at lower speeds, which may have poorer bearings, the effects of joint friction can be quite significant. The purpose of this chapter is to show you how to include the effects of Coulomb friction in the joints in your solution to the IDP. As you will see, joint friction makes the problem nonlinear. Consequently, an IDP that includes joint friction requires an iterative solution.

6.1 FRICTION IN A PIN JOINT

Figure 6.1 illustrates several details regarding the forces within a pin joint. Figure 6.1(a) shows a pin joint between bodies x and y and indicates the direction of the angular velocity of y relative to x, $\omega_{y/x}$. This relative velocity indicates the direction that y rotates relative to x—that is, the direction in which y rubs on x.

Figure 6.1(b) shows how the normal force from y to x is distributed over a portion of the joint as a pressure. This pressure distribution is never fully known. Assuming the bodies are rigid, the differential normal forces that make up the pressure distribution must all pass through the center of the joint; that is, they create no moment about the center of the joint. The net normal force from y to x, $\overline{F}^n_{yx}$, shown in Figures 6.1(b) and (c), also creates no moment about the center. It is therefore typically assumed that $\overline{F}^n_{yx}$ acts as a point load on the surface of the joint and is directed through the center of the joint, as shown in Figures 6.1(b) and (c).

If we define R as the radius of the pin and μ as the kinetic coefficient of friction in the joint, then $\overline{F}^n_{yx}$ induces a kinetic friction force, $\overline{f}_{yx}$, which acts at the point of contact and in the direction that y rubs on x. The magnitudes of these forces, denoted as F^n_{yx} and f_{yx}, are related through the coefficient of friction:

$$f_{yx} = \mu F^n_{yx}.$$

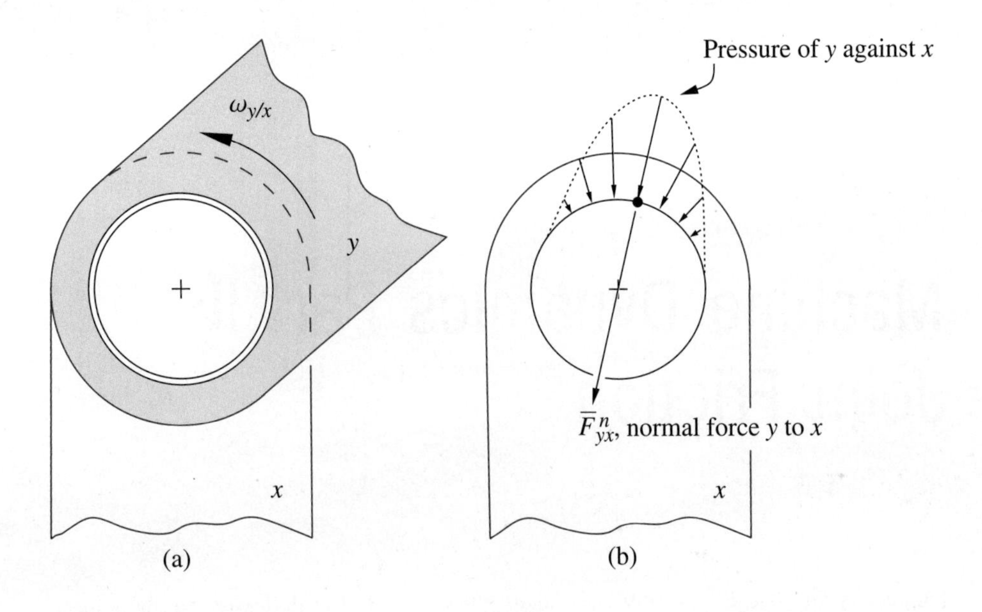

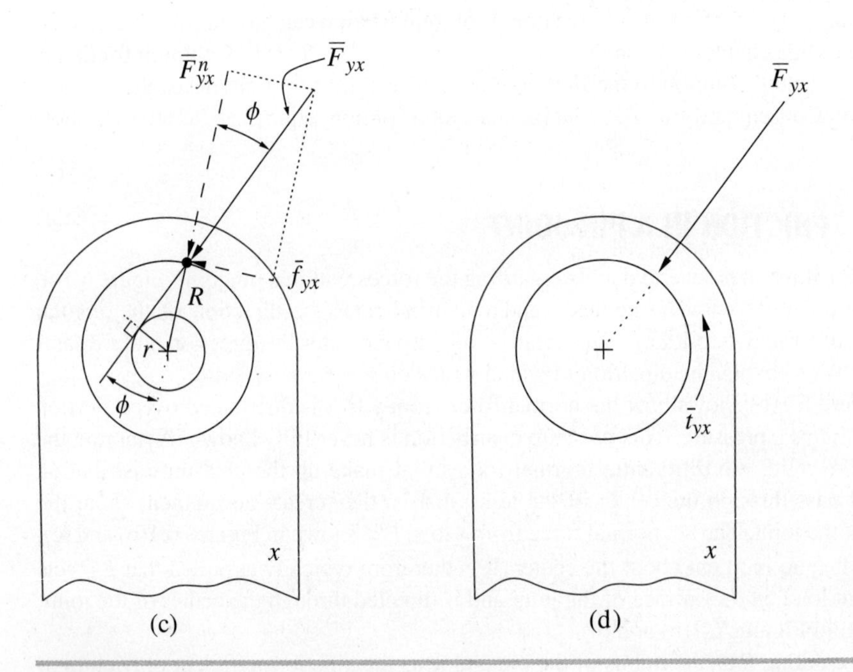

FIGURE 6.1 A pin joint with Coulomb friction

The friction angle ϕ in Figure 6.1(c) is given by

$$\tan\phi = \frac{f_{yx}}{F^n_{yx}} = \mu.$$

Typically $\mu < 0.2$, so ϕ is a small angle; μ is exaggerated in Figure 6.1(c). Thus we can approximate

$$\tan\phi \approx \sin\phi$$

In Figure 6.1(c) we see that the net force $\overline{F}_{yx}$ is the vector sum of the two orthogonal vectors $\overline{F}^n_{yx}$ and $\overline{f}_{yx}$, and so the magnitude F_{yx} is given by

$$F_{yx} = \sqrt{(F^n_{yx})^2 + f^2_{yx}} = F^n_{yx}\sqrt{1+\mu^2}.$$

Since $\mu < 0.2$, the term $\sqrt{1+\mu^2}$ is approximately equal to 1, and thus

$$F_{yx} \approx F^n_{yx}.$$

Also note from Figure 6.1(c) that $\overline{F}_{yx}$ is tangent to a circle of radius r given by

$$r = R\sin\phi \approx R\tan\phi \approx R\mu.$$

Hence we say that the force $\overline{F}_{yx}$ is tangent to a "friction circle" whose radius is μR. Furthermore, it is tangent to this circle on the side where the moment created by $\overline{F}_{yx}$ is in the direction of $\omega_{y/x}$; i.e. the direction that y rubs on x. We account for this with a "direction indicator" D_{yx} such that

$$D_{yx} = 1 \quad \text{when} \quad \omega_{y/x} \text{ is ccw}$$
$$\text{and } D_{yx} = -1 \quad \text{when} \quad \omega_{y/x} \text{ is cw.} \tag{6.1}$$

Last, in Figure 6.1(d) the force $\overline{F}_{yx}$ is translated so that it passes through the center of the joint, and a "friction torque" $\overline{t}_{yx}$ is added where,

$$\overline{t}_{yx} = t_{yx}\hat{k}, \quad \text{and} \quad t_{yx} = r|F_{yx}|D_{yx} = \mu R|F_{yx}|D_{yx} \tag{6.2}$$

where the direction indicator will produce a negative magnitude t_{yx} when the friction is in the clockwise (cw) direction.

Note that Equation (6.2) for the friction torque involves the *magnitude* of $\overline{F}_{yx}$, which is nonlinear in terms of the unknown X and Y components of $\overline{F}_{yx}$. This is what causes the IDP to be nonlinear when there is friction in a pin joint.

6.1.1 Computing the Direction Indicator D_{yx} for a Pin Joint

According to the definition of D_{yx} in Equation (6.1), the direction indicator is given by

$$D_{yx} = \frac{\omega_{y/x}}{|\omega_{y/x}|} = \frac{\omega_y - \omega_x}{|\omega_y - \omega_x|}.$$

In terms of kinematic coefficients,

$$\omega_y = h_y\dot{S}_i \quad \text{and} \quad \omega_x = h_x\dot{S}_i,$$

which gives D_{yx} as

$$D_{yx} = \frac{h_y - h_x}{|h_y - h_x|} \frac{\dot{S}_i}{|\dot{S}_i|}.$$

If we define a direction indicator for the input, D_i, as

$$D_i = \frac{\dot{S}_i}{|\dot{S}_i|}, \tag{6.3}$$

then D_{yx} becomes

$$D_{yx} = \frac{h_y - h_x}{|h_y - h_x|} D_i. \tag{6.4}$$

Note that

$$\begin{aligned} D_i &= 1 \quad \text{when} \quad \dot{S}_i > 0 \\ \text{and} \quad D_{yx} &= -1 \quad \text{when} \quad \dot{S}_i < 0. \end{aligned} \tag{6.5}$$

It makes good sense that D_{yx} would depend on D_i, because if the direction of the input is reversed, the direction of rubbing reverses and this reverses the friction torque.

Programming

The presence of $|F_{yx}|$ in the friction torque makes the problem nonlinear and there is no closed-form solution. The diameter of the friction circle is relatively small, and so this nonlinearity is "weak." To solve the system of equations we use a Method of Successive Iterations. It works as shown in Figure 6.2.

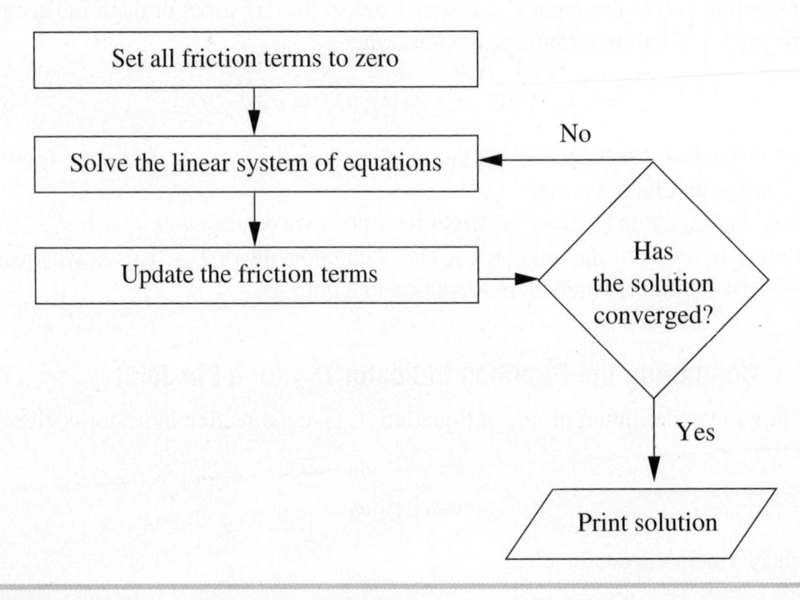

FIGURE 6.2 Method of Successive Iterations

6.2 FRICTION IN A PIN-IN-A-SLOT JOINT

Figure 6.3 shows a body x that has a circular pin with center at C and a body y that contains a straight slot. The pin on x moves in the slot on y, and there is Coulomb friction present in the joint. If the pin and slot are rigid bodies, they must have a common normal direction as shown. This leads to a common tangent direction as well.

Consider the forces acting on the pin shown in Figure 6.4. It is assumed that there is clearance in the joint, so y contacts x at either point D or point E, depending on the direction of the contact force $\overline{F}^n_{yx}$, which acts in the normal direction. There will also be a kinetic friction force on the pin, $\overline{f}_{yx}$, which acts in the tangential direction in which y rubs against x. There are four possible cases for the forces acting on the pin shown in Figure 6.4. The top two figures show the case when y contacts x at E. On the left image y rubs against x toward the left, and in the right image y rubs against x toward the right. The bottom two figures show the case when y contacts x at D and the two possible directions of y rubbing on x. Our goal is to develop a set of equations along with a logic that we can incorporate into a computer program to describe all four of the above possibilities.

To do this we introduce a $\hat{u}_t - \hat{u}_n$ coordinate system whose axes define the positive tangent and normal directions of the path. It is important to realize that the two directions are not independent, because we know from the definition of normal and tangential coordinates that

$$\hat{u}_t \times \hat{u}_n = \hat{k} \quad (\hat{k} \text{ comes out of page}). \tag{6.6}$$

The significance of Equation (6.6) is that once the positive direction of either $\hat{u}_t$ or $\hat{u}_n$ is set, Equation (6.6) gives the positive direction of the other unit vector.

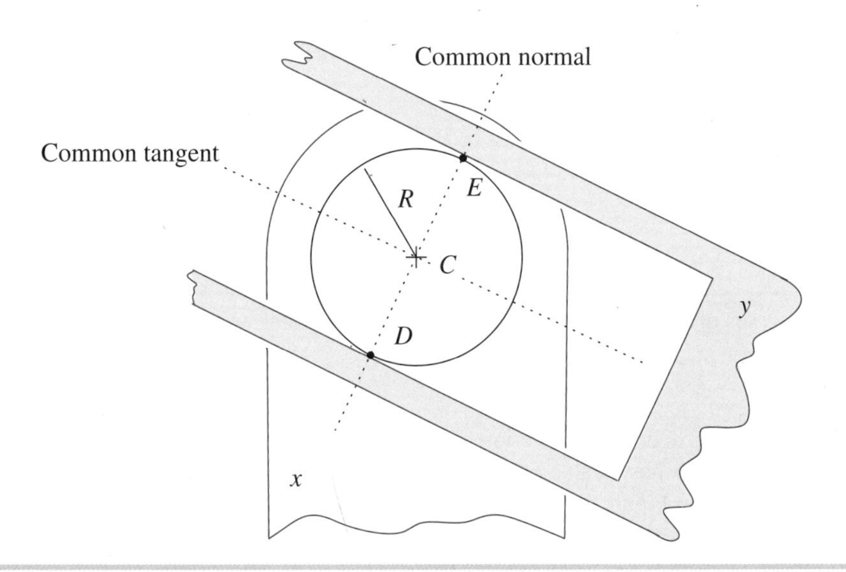

FIGURE 6.3 Coulomb friction in a pin-in-a-slot joint

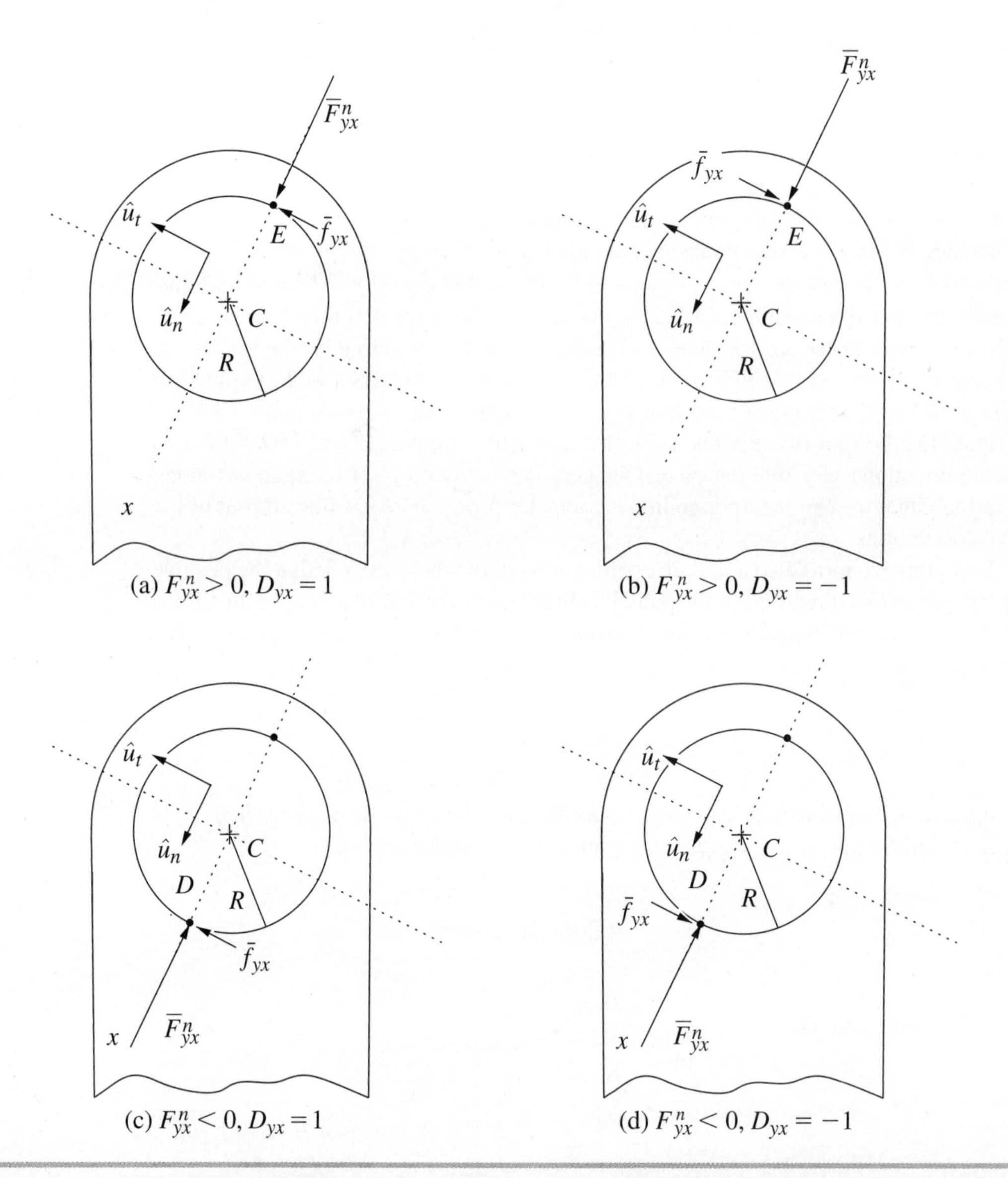

FIGURE 6.4 Four possible cases of forces acting on the pin of a slot in a joint with Coulomb friction

We also need to define a direction indicator D_{yx} that has a value of 1 when y rubs on x in the $\hat{u}_t$ direction has a value of and –1 when y rubs on x in the $-\hat{u}_t$ direction. In each of the four cases in Figure 6.4, the sign of F^n_{yx} and the value of D_{yx} are indicated.

The top two images show the case when contact is at E ($F^n_{yx} > 0$), and the bottom two images show the case when contact is at D ($F^n_{yx} < 0$). The left two images show the case when y rubs on x in the $\hat{u}_t$ direction ($D_{yx} = 1$), and the right two images show the case when y rubs on x in the $-\hat{u}_t$ direction ($D_{yx} = -1$).

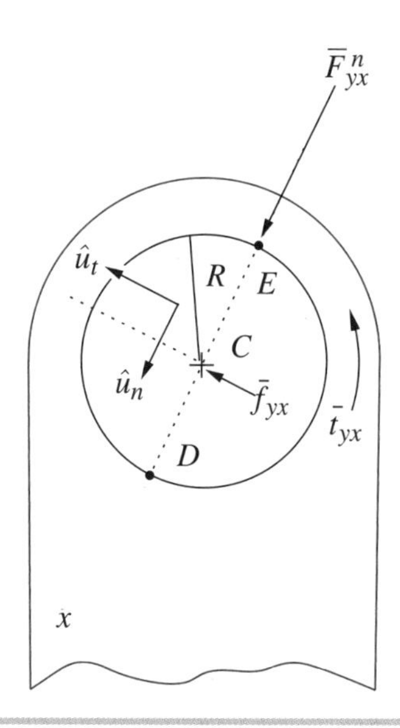

FIGURE 6.5 The general representation of the forces acting on the pin of a slot in a joint with Coulomb friction

Using the direction indicator, all four cases in Figure 6.4 can be represented by the generic situation shown in Figure 6.5, where $\overline{f}_{yx}$ has been translated and made to pass through the center of the joint and a friction torque $\overline{t}_{yx}$ has been added to account for the moment that the original $\overline{f}_{yx}$ created about C.

In terms of the $\hat{u}_t - \hat{u}_n$ system, we can define the normal force $\overline{F}^n_{yx}$ and kinetic friction $\overline{f}_{yx}$ as

$$\overline{F}^n_{yx} = F^n_{yx}\hat{u}_n \tag{6.7}$$

$$\overline{f}_{yx} = f_{yx}\hat{u}_t, \quad \text{where} \quad f_{yx} = \mu \left|F^n_{yx}\right| D_{yx}, \tag{6.8}$$

$$\text{and} \quad \overline{t}_{yx} = t_{yx}\hat{k}, \quad \text{where} \quad t_{yx} = \mu R F^n_{yx} D_{yx}. \tag{6.9}$$

To verify the correctness of Equation (6.7), Equation (6.8), and Equation (6.9), consider the signs of F^n_{yx} and D_{yx} in each case, and compare them to the signs of f_{yx} and t_{yx} in the four possible situations in Figure 6.4(a)–(d). Comparing the signs of f_{yx} and t_{yx} predicted by Equation (6.8) and Equation (6.9) in Table 6.1 to the directions observed from Figures 6.4(a)–(d) shows that they are in agreement.

6.2.1 Computing the Direction Indicator D_{yx} for a Pin-in-a-Slot Joint

If we define a unit vector in the direction that y rubs on x as $\hat{u}_{yx}$, then D_{yx} will be given by

$$D_{yx} = \hat{u}_{yx} \cdot \hat{u}_t.$$

	Figure 6.4(a)	Figure 6.4(b)	Figure 6.4(c)	Figure 6.4(d)
Sign of F^n_{yx} in Figure 6.4	positive	positive	negative	negative
Value of D_{yx} in Figure 6.4	1	–1	1	–1
Sign of f_{yx} in Eq. (6.8)	positive	negative	positive	negative
Sign of t_{yx} in Eq. (6.9)	positive	negative	negative	positive

TABLE 6.1 Verifying Correctness of Signs Produced by Equations (6.8) and (6.9)

Because of the relative rotation between x and y, the direction in which y rubs on x will depend on whether the contact is at E (corresponds to $F^n_{yx} > 0$) or D (corresponds to $F^n_{yx} < 0$). We need to consider these two cases separately.

Case 1: $F^n_{yx} > 0 \rightarrow$ contact at E

In Figure 6.6 imagine a pair of instantaneously coincident points at point E: E_y on body y and E_x on body x. Also imagine a pair of instantaneously coincident points at point C: C_y on body y and C_x on body x. The rubbing velocity of y against x is then given by the relative velocity $\overline{v}_{e_y/e_x}$, and $\hat{u}_{yx}$ is

$$\hat{u}_{yx} = \frac{\overline{v}_{e_y/e_x}}{\left|\overline{v}_{e_y/e_x}\right|}. \tag{6.10}$$

The vector loop for the mechanism would have included a vector $\overline{r}_y$ that locates the pin in the slot and rotates with body y, and a vector $\overline{r}_x$ that would rotate with body x as shown in Figure 6.6. If S_i is the input to the mechanism, then we would have basic

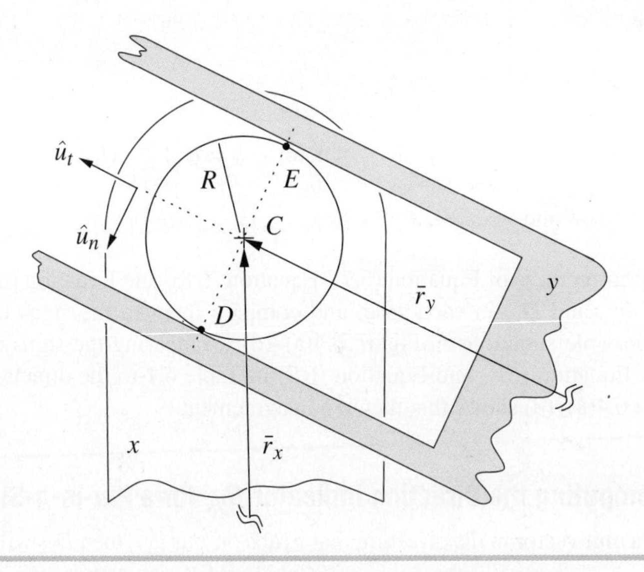

FIGURE 6.6 Formulating direction indicators for a pin-in-a-slot joint

first-order kinematic coefficients, f_y, h_y, and h_x. We need to formulate $\overline{v}_{e_y/e_x}$ in terms of these kinematic coefficients and $\dot{S}_i$. Here we go.

By definition of a relative velocity,

$$\overline{v}_{e_y/e_x} = \overline{v}_{e_y} - \overline{v}_{e_x},$$

and we can express the terms on the r.h.s. as

$$\overline{v}_{e_y} = \overline{v}_{c_y} + \overline{v}_{e_y/c_y} \quad \text{and} \quad \overline{v}_{e_x} = \overline{v}_{c_x} + \overline{v}_{e_x/c_x},$$

which gives

$$\begin{aligned}\overline{v}_{e_y/e_x} &= \left(\overline{v}_{c_y} + \overline{v}_{e_y/c_y}\right) - \left(\overline{v}_{c_x} + \overline{v}_{e_x/c_x}\right)\\ &= \left(\overline{v}_{c_y} - \overline{v}_{c_x}\right) + \overline{v}_{e_y/c_y} - \overline{v}_{e_x/c_x}\\ &= \overline{v}_{c_y/c_x} + \overline{v}_{e_y/c_y} - \overline{v}_{e_x/c_x}.\end{aligned}$$

Consider each of the three terms on the r.h.s. of this equation. The vector $\overline{r}_y$ locates the pin (point C_x) relative to the slot. Take the $\hat{u}_t$ direction to be the direction in which C_x moves relative to C_y as the length of $\overline{r}_y$ increases. Therefore

$$\overline{v}_{c_x/c_y} = \frac{d\,r_y}{dt}\,\hat{u}_t = f_y\dot{S}_i\hat{u}_t,$$

so

$$\overline{v}_{c_y/c_x} = -\overline{v}_{c_x/c_y} = -f_y\dot{S}_i\hat{u}_t.$$

The term $\overline{v}_{e_y/c_y}$ is the relative velocity of two points on body y and is given by

$$\overline{v}_{e_y/c_y} = \overline{\omega}_y \times \overline{r}_{e_y/c_y} = \left(h_y\dot{S}_i\hat{k}\right) \times \left(-R(\overline{u}_n)\right) = Rh_y\dot{S}_i\hat{u}_t.$$

Likewise, the term $\overline{v}_{e_x/c_x}$ is the relative velocity of two points on body x and is given by

$$\overline{v}_{e_x/c_x} = \overline{\omega}_x \times \overline{r}_{e_x/c_x} = \left(h_x\dot{S}_i\hat{k}\right) \times \left(-R(\overline{u}_n)\right) = Rh_x\dot{S}_i\hat{u}_t,$$

so

$$\begin{aligned}\overline{v}_{e_y/e_x} &= -f_y\dot{S}_i\hat{u}_t + Rh_y\dot{S}_i\hat{u}_t - Rh_x\dot{S}_i\hat{u}_t\\ &= \left(-f_y + R(h_y - h_x)\right)\dot{S}_i\hat{u}_t\end{aligned}$$

and

$$\begin{aligned}\hat{u}_{yx} &= \frac{\overline{v}_{e_y/e_x}}{\left|\overline{v}_{e_y/e_x}\right|}\\ &= \frac{-f_y + R(h_y - h_x)}{\left|-f_y + R(h_y - h_x)\right|}\,\frac{\dot{S}_i}{\left|\dot{S}_i\right|}\,\hat{u}_t,\end{aligned}$$

and finally,

$$D_{yx} = \hat{u}_{yx}\cdot\hat{u}_t = \frac{-f_y + R(h_y - h_x)}{\left|-f_y + R(h_y - h_x)\right|}\,D_i. \tag{6.11}$$

Case 2: $F^n_{yx} < 0 \longrightarrow$ contact at D

When the contact is at D, the rubbing velocity of y against x is $\overline{v}_{d_y/d_x}$, so

$$\hat{u}_{yx} = \frac{\overline{v}_{d_y/d_x}}{\left|\overline{v}_{d_y/d_x}\right|}. \tag{6.12}$$

Expressing $\overline{v}_{d_y/d_x}$ in terms of kinematic coefficients is the same process as in Case 1. The difference is that in Case 1 $\overline{r}_{e_y/c_y} = \overline{r}_{e_x/c_x} = -R\overline{u}_n$, and in Case 2 there will be the term $\overline{r}_{d_y/c_y} = \overline{r}_{d_x/c_x} = R\overline{u}_n$. The result is

$$D_{yx} = \frac{-f_y - R(h_y - h_x)}{\left|-f_y - R(h_y - h_x)\right|} D_i. \tag{6.13}$$

A computer program implementing Equation (6.11) and Equation (6.13) for computing the direction indicator for a pin-in-a-slot joint would involve the following "if" statement:

$$\begin{aligned} &\text{if} \quad F^n_{yx} > 0, \text{ then } D_{yx} = \frac{-f_y + R(h_y - h_x)}{\left|-f_y + R(h_y - h_x)\right|} D_i \\ &\text{else} \\ &\text{if} \quad F^n_{yx} < 0, \text{ then } D_{yx} = \frac{-f_y - R(h_y - h_x)}{\left|-f_y - R(h_y - h_x)\right|} D_i. \end{aligned} \tag{6.14}$$

Sign of the Direction Indicator for a Pin-in-a-Slot Joint

If in Figure 6.6 the direction of $\overline{r}_y$ had been reversed, it would have had no effect on the sign of the direction indicator because an increase in the length of $\overline{r}_y$ would still have moved C_x relative to C_y in the $\hat{u}_t$ direction. This would have been a somewhat awkward assignment of $\overline{r}_y$, but not an incorrect assignment.

However, if you had chosen the opposite direction for $\hat{u}_t$ so that an increase in length of $\overline{r}_y$ moves C_x relative to C_y in the negative $\hat{u}_t$ direction, then the sign of f_y in both Equation (6.11) and Equation (6.13) would have changed. You are advised to make your assignment of $\overline{r}_y$ and $\hat{u}_t$ consistent with that in Figure 6.6 so that Equation (6.11) and Equation (6.13) can be applied directly.

Programming

A problem that involves friction in a pin-in-a-slot joint contains the same nonlinearity as the pin joint friction problem due to the presence of $\left|F^n_{yx}\right|$ in Equation (6.8). A numerical solution can be programmed using the Method of Successive Iterations, because this nonlinearity is weak.

In addition to this, there will need to be a test on the sign of F^n_{yx} in order to determine which of the two expressions for the direction indicator in Equation (6.14) should be used.

6.3 FRICTION IN A STRAIGHT SLIDING JOINT

Figure 6.7 shows a straight sliding joint between body x, the slider, and body y, which contains the path. The path has a tangent direction and a normal direction. There must be a reference point on the slider, point C, which would establish the connection from x to another link in the mechanism. Typically this would be the center of a pin joint.

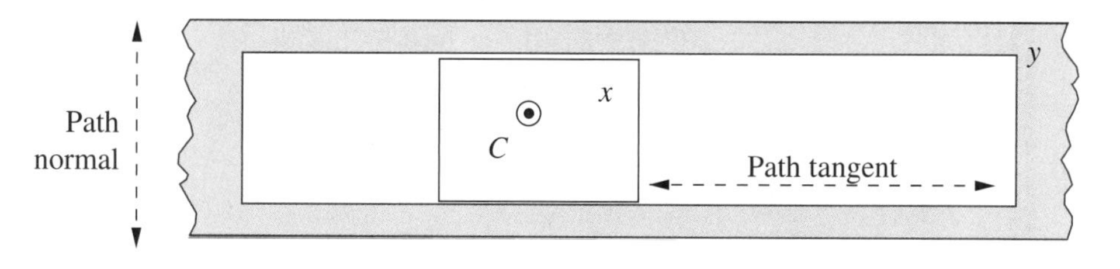

FIGURE 6.7 Coulomb friction in a straight sliding joint

Consider the forces acting on the slider, and assume there is some clearance in the joint. There are four possibilities for the normal force from y to x, $\overline{N}_{yx}$:

1. One-sided contact at the top
2. One-sided contact at the bottom
3. ccw cocking so contact will be at the bottom left and top right corners
4. cw cocking so contact will be at the top left and bottom right corners

In each of these cases the motion of y relative to x can be to the left or to the right, so there are two possible directions for the friction forces for each possible case of normal forces. The total of eight cases is illustrated in Figures 6.8(a)–(h).

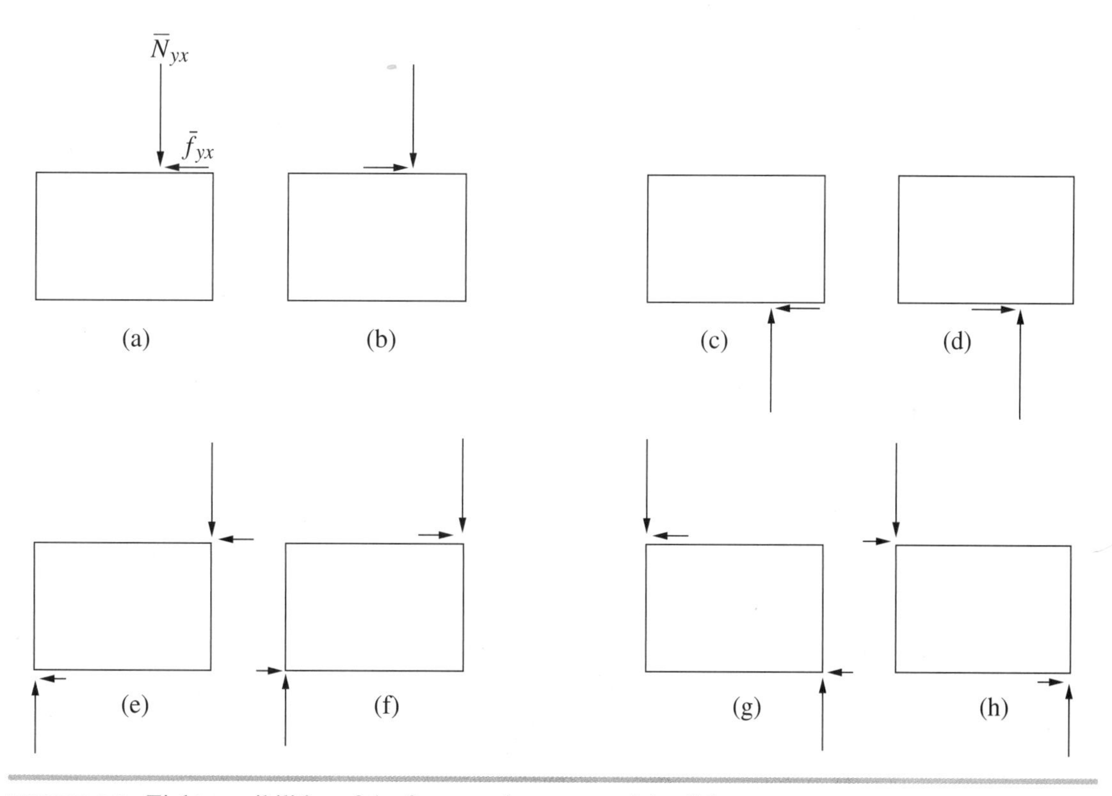

FIGURE 6.8 Eight possibilities of the forces acting on a straight slider

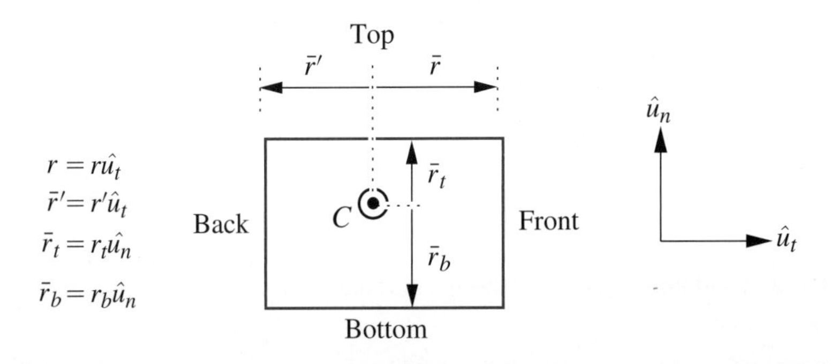

FIGURE 6.9 Definition of the $\hat{u}_t - \hat{u}_n$ system and the dimensions of a straight slider

Our challenge is to find a generic representation for all of these cases and a logic for implementing this representation in a computer program.

Figure 6.9 shows a generic slider and defines a $\hat{u}_t - \hat{u}_n$ system whose axes are aligned with the slider's path tangent and path normal, respectively. The two directions are not independent, because we know from the definitions of normal and tangential coordinates that again Equation (6.6) applies. The significance of Equation (6.6) is that once the positive direction of either $\hat{u}_t$ or $\hat{u}_n$ is set, Equation (6.6) gives the positive direction of the other unit vector. Since $\hat{k}$ is always coming out of the page.

In terms of the positive directions of $\hat{u}_t$ and $\hat{u}_n$, we establish a convention to define the dimensions of the slider block. Specifically, we think of $\hat{u}_t$ as pointing in the direction from the back surface to the front surface of the slider, and we think of $\hat{u}_n$ as pointing from the bottom surface to the top surface of the slider. Then, relative to C and in terms of the $\hat{u}_t - \hat{u}_n$ directions, the dimensions of the block are defined as follows:

- r, the distance from C to the front surface of the block measured in the $\hat{u}_t$ direction
- r', the distance from C to the back surface of the block measured in the $\hat{u}_t$ direction
- r_t, the distance from C to the top surface of the block measured in the $\hat{u}_n$ direction
- r_b, the distance from C to the bottom surface of the block measured in the $\hat{u}_n$ direction.

These dimensions are positive or negative, depending on whether they are in the positive or negative direction of their associated unit vector. For the case shown in Figure 6.9,

$$r > 0, \quad r' < 0, \quad r_t > 0, \quad r_b < 0.$$

Figure 6.10 shows a slider whose dimensions would all be positive. It is rather awkward and you are unlikely to encounter it. One can sense that the slider would be prone to cocking and jamming. You are more likely to see mixed signs, such as in Figure 6.9.

An important observation that will help us develop a generic representation of the forces acting on the slider is that the cases of one-sided contacts in Figures 6.8(a)–(d) can be equivalently represented by a pair of forces that act at the ends of the slider, as in the case of Figures 6.8(e)–(h). To see this, consider Figure 6.11. On the left is a slider with a one-sided contact on the bottom (as in Figure 6.8(c)). The normal force is F_{yx}, and it acts at a point located a distance r_f from the reference point C. On the

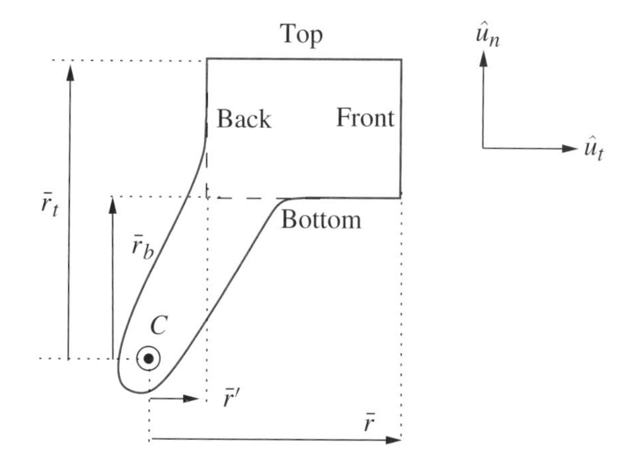

FIGURE 6.10 A slider with all positive dimensions

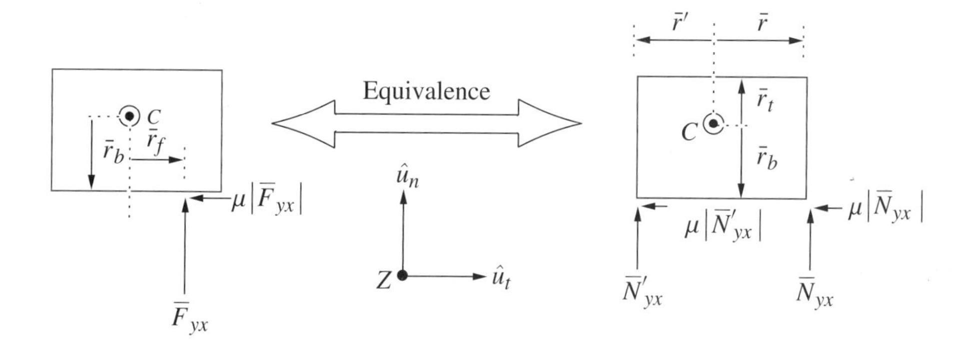

FIGURE 6.11 Equivalence of one-sided contact and end point contact in a straight slider

right is the alleged equivalent pair of normal forces acting on the ends of the slider and their associated kinetic friction forces. The dimensions of the slider (r', r, r_t, and r_b) are known, as defined in Figure 6.9. (Note that $r_b < 0$ and $r' < 0$ in Figure 6.11.) In order for there to be equivalence, the sum of the forces in the $\hat{u}_n$ direction and the $\hat{u}_t$ direction and the sum of the moments in the Z direction should be the same:

$$\Sigma F_{\hat{u}_n} \longrightarrow F_{yx} = N'_{yx} + N_{yx} \tag{6.15}$$

$$\Sigma F_{\hat{u}_t} \longrightarrow -\mu F_{yx} = -\mu N'_{yx} - \mu N_{yx} \tag{6.16}$$

$$\Sigma M_z \longrightarrow r_f F_{yx} + r_b \mu F_{yx} = r' N'_{yx} + r N_{yx} + r_b \mu N'_{yx} + r_b \mu N_{yx}. \tag{6.17}$$

Equation (6.16) is simply Equation (6.15) multiplied by ($-\mu$) and is not an independent equation. (Keep in mind that for one-sided contact F_{yx}, N_{yx} and N'_{yx} have the same sign.) Equation (6.15) also shows that in Equation (6.17), the term $r_b \mu F_{yx}$ on the l.h.s. cancels with the terms $r_b \mu N'_{yx}$ + $r_b \mu N_{yx}$ on the r.h.s. So Equation (6.15), Equation (6.16), and Equation (6.17) reduce to two linearly independent equations,

$$F_{yx} = N'_{yx} + N_{yx} \tag{6.18}$$

$$r_f = \frac{rN_{yx} + r'N'_{yx}}{F_{yx}}. \tag{6.19}$$

This system of two equations in two unknowns maps the solution of the two-point-contact case into the solution of the equivalent one-sided-contact case. This could be repeated for the case when the friction force is reversed and/or when the one-sided contact is on the top surface, and the results would be identical. The conclusion is that the one-sided contact can be represented by an equivalent pair of forces that act at the ends of the slider. Observe that if this pair of forces are in the same direction there is a one-sided contact, and if this pair of forces are in opposite directions there is cocking.

With the equivalence established, we propose the situation shown in Figure 6.12 as a general representation of the forces acting on a slider. The slider is modeled as having normal forces $\overline{N}_{yx}$ and $\overline{N}'_{yx}$ acting at the front and back surfaces. The kinetic friction forces are summed into a single force $\overline{f}_{yx}$ and translated to the reference point C on the slider. Friction torques $\overline{t}_{yx}$ and $\overline{t}'_{yx}$ are added to account for the moments that the original friction forces created about C. A direction indicator D_{yx} is defined as

$$D_{yx} = \hat{u}_{yx} \cdot \hat{u}_t,$$

where $\hat{u}_{yx}$ is a unit vector in the direction that y rubs on x—that is, a unit vector in the direction of $\overline{v}_{c_y/c_x}$:

$$\hat{u}_{yx} = \frac{\overline{v}_{c_y/c_x}}{\left|\overline{v}_{c_y/c_x}\right|}.$$

With these definitions,

$$\overline{N}_{yx} = N_{yx}\hat{u}_n \quad \text{and} \quad \overline{N}'_{yx} = N'_{yx}\hat{u}_n \tag{6.20}$$

$$\overline{f}_{yx} = f_{yx}\,\hat{u}_t, \quad \text{where} \quad f_{yx} = \mu\left(\left|N_{yx}\right| + \left|N'_{yx}\right|\right) D_{yx}. \tag{6.21}$$

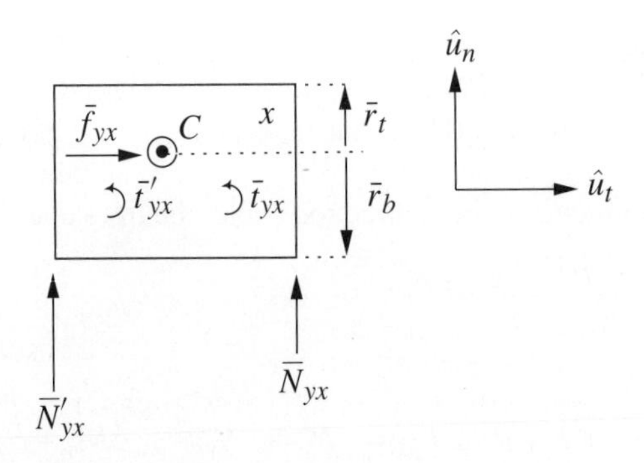

FIGURE 6.12 A general representation of the forces acting on a straight slider with Coulomb friction

The friction torques are a bit more complex, because they depend on whether the normal force is acting at the top or the bottom corner of the slider:

$$\bar{t}_{yx} = t_{yx}\hat{k} \text{ where, } \quad \begin{array}{l} \text{if} \quad N_{yx} > 0, \text{ then } t_{yx} = -\mu r_b N_{yx} D_{yx} \\ \text{else} \\ \text{if} \quad N_{yx} < 0, \text{ then } t_{yx} = \mu r_t N_{yx} D_{yx}. \end{array} \tag{6.22}$$

and

$$\bar{t}'_{yx} = t'_{yx}\hat{k} \text{ where, } \quad \begin{array}{l} \text{if} \quad N'_{yx} > 0, \text{ then } t'_{yx} = -\mu r_b N'_{yx} D_{yx} \\ \text{else} \\ \text{if} \quad N'_{yx} < 0, \text{ then } t'_{yx} = \mu r_t N'_{yx} D_{yx}. \end{array} \tag{6.23}$$

The "if" statements above are a part of the computer program's logic.

6.3.1 Computing the Direction Indicator D_{yx} for a Straight Sliding Joint

Figure 6.13 shows a vector $\bar{r}_y$ that would have been assigned to the vector loop in order to locate where C_x is in the slot. The $\hat{u}_t$ direction is taken to be in the direction of $\bar{r}_y$. Thus an increase in the length of $\bar{r}_y$ would cause C_x to move in the $\hat{u}_t$ direction. It follows from this definition of $\bar{r}_y$ that

$$\bar{v}_{C_x/C_y} = \frac{dr_y}{dt}(\hat{u}_t) = f_y \dot{S}_i \hat{u}_t$$

and

$$\bar{v}_{C_y/C_x} = -\bar{v}_{C_x/C_y} = -f_y \dot{S}_i \hat{u}_t,$$

which gives us

$$\hat{u}_{yx} = \frac{\bar{v}_{C_y/C_x}}{|\bar{v}_{C_y/C_x}|} = \frac{-f_y}{|f_y|}\frac{\dot{S}_i}{|\dot{S}_i|}\hat{u}_t = \frac{-f_y}{|f_y|} D_i \hat{u}_t,$$

from which the direction indicator is

$$D_{yx} = \hat{u}_{yx} \cdot \hat{u}_t = \frac{-f_y}{|f_y|} D_i. \tag{6.24}$$

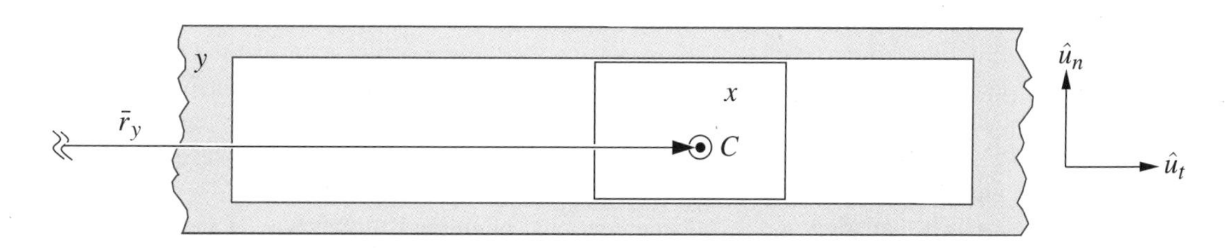

FIGURE 6.13 Formulating the direction indicator for a straight sliding joint

Sign of the Direction Indicator for a Straight Sliding Joint

If in Figure 6.13 the direction of $\bar{r}_y$ had been reversed, it would have had no effect on the sign of the direction indicator, because an increase in the length of $\bar{r}_y$ would still have moved C_x in the $\hat{u}_t$ direction. This would have been a somewhat awkward assignment of $\bar{r}_y$ but not an incorrect assignment. Just an awkward one.

However, if you had chosen the opposite direction for $\hat{u}_t$, then the sign of D_{yx} would have changed, because an increase in length of $\bar{r}_y$ would have moved C_x in the negative $\hat{u}_t$ direction. To make your life simple, you are advised to use a vector $\bar{r}_y$ and $\hat{u}_t$ in the direction of $\bar{r}_y$, as shown in Figure 6.13, so that Equation (6.24) can be applied directly.

Programming

A problem that involves friction in a straight sliding joint contains the same nonlinearity as the previous cases due to the presence of $|N_{yx}|$ and $|N'_{yx}|$ in Equation (6.21). A numerical solution can be programmed using the Method of Successive Iterations because this nonlinearity is weak.

In addition to this, tests on the signs of N_{yx} and N'_{yx} will be needed to determine which of the expressions in Equation (6.22) and Equation (6.23) should be used to compute the friction torques.

6.3.2 Cocking and Jamming in a Straight Sliding Joint

In a one-sided contact, the magnitude of the normal force is set by force equilibrium in the normal direction. The magnitude of $\bar{r}_f$ in Figure 6.11 is set by the magnitude of the moment that this normal force must equilibrate. As this moment increases, the normal force moves outward toward either the front or back surface (the magnitude of $\bar{r}_f$ increases). If the moment continues to increase, eventually the normal force moves all the way to either the front surface or back surface, and if the moment continues to grow, cocking ensues.

When cocking occurs, the normal forces at the ends form a couple to equilibrate the moment, and upon this is superimposed a force equilibrium in the normal direction. These are cases (e)–(h) in Figure 6.8. The "arm" of the couple is fixed (it is the distance from the back surface to the front surface), so the magnitude of the pair of forces associated with the couple increases with the moment and eventually will get high enough that the resulting kinetic friction could cause the slider to stop moving. This is jamming. The machine ceases to operate when jamming has occurred. Many times, the slider is then wedged in its bore, and the system cannot be repaired. The results of jamming are catastrophic in many cases.

Furthermore, when cocking occurs, the normal forces acting at the ends are point contacts, whereas with a one-sided contact the normal force is distributed as a pressure. Associated with the point contacts are contact stresses an order of magnitude greater. This results in greater surface fatigue and wear and can also cause surface "spalling" or pitting, all of which further contribute to the catastrophe of jamming.

The situation of cocking is precarious. Jamming is disastrous. Typically a machine is designed so that there is one-sided contact in the sliding joint, and cocking is avoided. To avoid cocking, the designer needs to provide sufficient length on the slider from the back surface to the front surface. One can also design the slider so that there is no moment load or angular acceleration to equilibrate, and the normal force then passes through the pin joint C in Figure 6.11.

▶ EXAMPLE 6.1

In the machine represented in Figure 6.14, the load force $\overline{F}_3$ and the state of motion (θ_2, $\overline{\omega}_2$, and $\overline{\alpha}_2$) are known. There is friction in all the joints. $\overline{T}_2$ is the driving (input) torque, so $\overline{\omega}_2$ would be in the same direction, clockwise.

The inertia properties of all links are known. Gravity has the direction shown. Outline the equations you would use in a computer program to solve the inverse dynamics problem, and compute the driving torque $\overline{T}_2$. Create a flowchart for the program.

Solution

In this example we will consider both the effects of gravity and inertia effects. In Section 5.3 we discussed how one of these sets of terms typically dominates the problem, in which case the other set of terms is assumed to be negligible. The purpose of this example is to outline a step-by-step procedure that you should follow when setting up the equations you would use in a computer program that solves this problem.

Step 1: Perform a kinematic analysis as follows.

- Solve the position problem.
- Determine the basic first-order kinematic coefficients.
- If you are including inertia effects, determine the basic second-order kinematic coefficients, then the first- and second-order kinematic coefficients of the mass centers, and then the accelerations of the mass centers.

The solution starts with the vector loop in Figure 6.15.

The vector loop equation

$$\overline{r}_1 + \overline{r}_3 - \overline{r}_2 = \overline{0}$$

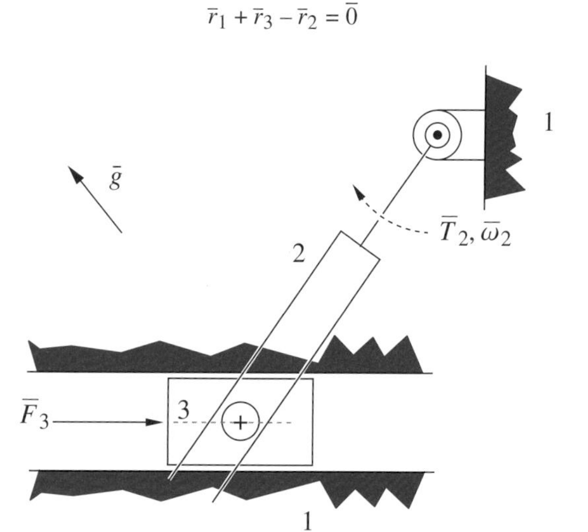

FIGURE 6.14 A machine with friction in the joints

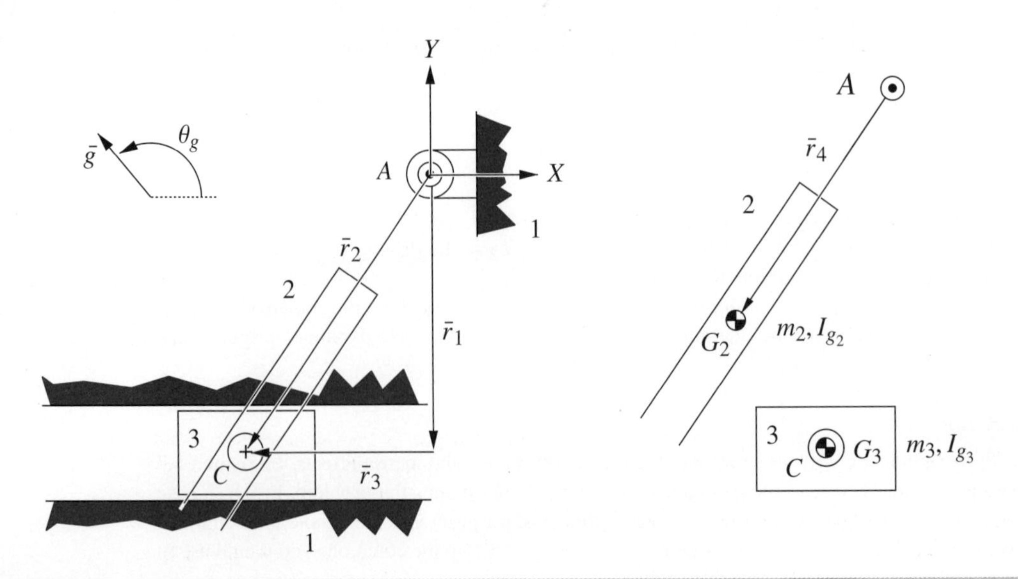

FIGURE 6.15 Vector loop for kinematic analysis and mass center definitions

has scalar components

$$-r_3 + r_2\cos\theta_2 = 0$$
$$-r_1 + r_2\sin\theta_2 = 0,$$

with

scalar knowns: $r_1, \theta_1 = -\pi/2, \theta_3 = \pi$

and

scalar unknowns: θ_2, r_2, and r_3,

so the mechanism has one degree of freedom. Take θ_2 as the input, and solve the position equations for r_2 and r_3. This simple mechanism has the closed-form solution

$$r_2 = r_1/\sin\theta_2 \quad \text{and} \quad r_3 = r_2\cos\theta_2. \tag{6.25}$$

If there had not been a closed-form solution, you would have had to outline a Newton's Method for the position solution, as in Example 5.1.

Differentiating the position equations with respect to input θ_2 gives

$$\begin{bmatrix} -1 & \cos\theta_2 \\ 0 & \sin\theta_2 \end{bmatrix} \begin{bmatrix} f_3 \\ f_2 \end{bmatrix} = \begin{bmatrix} r_2\sin\theta_2 \\ -r_2\cos\theta_2 \end{bmatrix}, \tag{6.26}$$

which can be solved for f_3 and f_2. Since we are including inertia effects, we also need the second-order kinematic coefficients. Differentiating the position equations again with respect to θ_2 gives

$$\begin{bmatrix} -1 & \cos\theta_2 \\ 0 & \sin\theta_2 \end{bmatrix} \begin{bmatrix} f_3' \\ f_2' \end{bmatrix} = \begin{bmatrix} r_2\cos\theta_2 + 2f_2\sin\theta_2 \\ r_2\sin\theta_2 - 2f_2\cos\theta_2 \end{bmatrix}, \tag{6.27}$$

which can be solved for f_3' and f_2'. Since we plan to include inertia effects, we also need the first- and second-order kinematic coefficients of the mass centers. G_2 is located on body 2 by the vector $\bar{r}_4$ in Figure 6.15, and G_3 is at point C. From Figure 6.15,

$$\bar{r}_{g2} = \bar{r}_4 \rightarrow \begin{matrix} r_{g2x} = r_4\cos\theta_2 \\ r_{g2y} = r_4\sin\theta_2 \end{matrix} \rightarrow \begin{matrix} f_{g2x} = -r_4\sin\theta_2 \\ f_{g2y} = r_4\cos\theta_2 \end{matrix} \rightarrow \begin{matrix} f'_{g2x} = -r_4\cos\theta_2 \\ f'_{g2y} = -r_4\sin\theta_2 \end{matrix} \quad (6.28)$$

$$\text{and} \quad \bar{r}_{g3} = \bar{r}_1 + \bar{r}_3 \rightarrow \begin{matrix} r_{g3x} = -r_3 \\ r_{g3y} = -r_1 \end{matrix} \rightarrow \begin{matrix} f_{g3x} = -f_3 \\ f_{g3y} = 0 \end{matrix} \rightarrow \begin{matrix} f'_{g3x} = -f'_3 \\ f'_{g3y} = 0 \end{matrix}. \quad (6.29)$$

Next we compute the accelerations of the mass centers:

$$a_{g2_x} = f_{g2x}\alpha_2 + f'_{g2x}\omega_2^2 \quad \text{and} \quad a_{g2_y} = f_{g2y}\alpha_2 + f'_{g2y}\omega_2^2 \quad (6.30)$$

$$a_{g3_x} = f_{g3x}\alpha_2 + f'_{g3x}\omega_2^2 \quad \text{and} \quad a_{g3_y} = f_{g3y}\alpha_2 + f'_{g3y}\omega_2^2 = 0 \quad (f_{g2_y} = f'_{g2_y} = 0). \quad (6.31)$$

This completes the kinematic analysis.

Step 2: Define the path-dependent coordinate systems needed to represent the forces at the joints, and express the forces and the dimensions of any sliding joints in terms of these coordinate systems. You are advised to make these definitions in accordance with Figure 6.13 for any sliding joints and Figure 6.6 for any pin-in-a-slot joints. In this way you can directly apply the equations for the respective direction indicators, Equation (6.24) and Equation (6.4). Otherwise, you will need to rederive the direction indicators on a case-by-case basis.

According to Section 6.3, we need unit vectors aligned with the path tangent and path normal of the sliding joint, $\hat{u}_{t_1}$ and $\hat{u}_{n_1}$, respectively, in order to define dimensions of the slider and the forces acting at the sliding joint. The directions of these unit vectors will always be related to the directions of vectors in the original vector loop, which are known since the position problem was solved in step 1. To maintain consistency with Figure 6.13, we align $\hat{u}_{t_1}$ with $\bar{r}_3$. According to Equation (6.6), $\hat{u}_{n_1}$ is in the direction of $\hat{u}_{t_1}$ rotated $\pi/2$ counterclockwise. $\hat{u}_{t_1}$ and $\hat{u}_{n_1}$ follow and are shown in Figure 6.16.

$$\hat{u}_{t_1} = \begin{bmatrix} -1 \\ 0 \end{bmatrix} \quad \text{and} \quad \hat{u}_{n_1} = \begin{bmatrix} 0 \\ -1 \end{bmatrix} \quad (6.32)$$

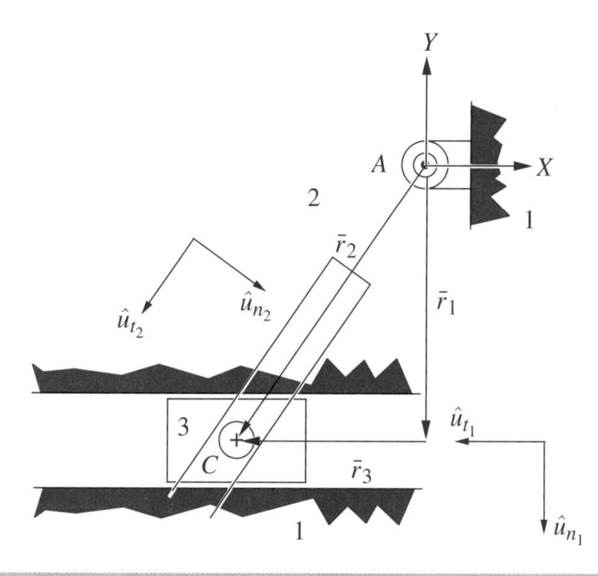

FIGURE 6.16 Definition of path-dependent coordinate systems

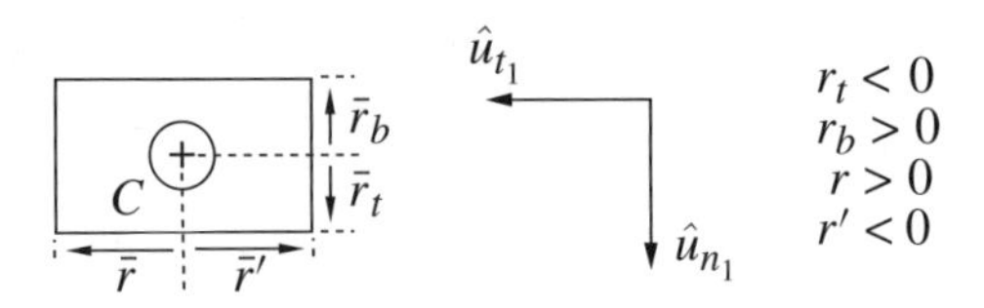

FIGURE 6.17 Signs on slider block dimensions

In terms of $\hat{u}_{t_1}$ and $\hat{u}_{n_1}$, the slider block's dimensions and their signs are as shown in Figure 6.17.

In terms of $\hat{u}_{t_1}$ and $\hat{u}_{n_1}$, the forces acting on the sliding block are given by

$$\overline{N}_{13} = N_{13}\hat{u}_{n_1} \quad \text{and} \quad \overline{N}'_{13} = N'_{13}\hat{u}_{n_1} \quad \text{and} \quad \bar{f}_{13} = f_{13}\hat{u}_{t_1}. \tag{6.33}$$

Substitute Equation (6.32) into Equation (6.33). Then

$$\overline{N}_{13} = N_{13}\begin{bmatrix} -1 \\ 0 \end{bmatrix} \quad \text{and} \quad \overline{N}'_{13} = N'_{13}\begin{bmatrix} -1 \\ 0 \end{bmatrix} \quad \text{and} \quad \bar{f}_{13} = f_{13}\begin{bmatrix} 0 \\ -1 \end{bmatrix}. \tag{6.34}$$

According to Section 6.2, to define the forces at the pin-in-a-slot joint, we need unit vectors $\hat{u}_{t_2}$ and $\hat{u}_{n_2}$, which are aligned with the path tangent and path normal of the slot. To maintain consistency with Figure 6.6, we align $\hat{u}_{t_2}$ with $\bar{r}_2$. Then $\hat{u}_n$ follows from Equation (6.6). These unit vectors are also shown in Figure 6.16.

$$\hat{u}_{t_2} = \begin{bmatrix} \cos\theta_2 \\ \sin\theta_2 \end{bmatrix} \quad \text{and} \quad \hat{u}_{n_2} = \begin{bmatrix} \cos(\theta_2 + \pi/2) \\ \sin(\theta_2 + \pi/2) \end{bmatrix}\begin{bmatrix} -\sin\theta_2 \\ \cos\theta_2 \end{bmatrix} \tag{6.35}$$

In terms of $\hat{u}_{t_2}$ and $\hat{u}_{n_2}$, the forces acting on the pin are given by

$$\overline{F}^n_{23} = F^n_{23}\hat{u}_{n_2} \quad \text{and} \quad \bar{f}_{23} = f_{23}\hat{u}_{t_2}. \tag{6.36}$$

Substitute Equation (6.35) into Equation (6.36). Then

$$\overline{F}^n_{23} = F^n_{23}\begin{bmatrix} -\sin\theta_2 \\ \cos\theta_2 \end{bmatrix} \quad \text{and} \quad \bar{f}_{23} = f_{23}\begin{bmatrix} \cos\theta_2 \\ \sin\theta_2 \end{bmatrix}. \tag{6.37}$$

Step 3: Draw the free body diagrams with the joint forces consistent with the path-dependent coordinates.

We will begin with the sliding block. Figure 6.18 shows the slider; the $\hat{u}_{t_1} - \hat{u}_{n_1}$ directions; associated forces $\overline{N}_{13}, \overline{N}'_{13}$, and $\bar{f}_{13}$; and the friction torques $\bar{t}_{13}$ and $\bar{t}'_{13}$. The free body diagram also shows the $\hat{u}_{t_2} - \hat{u}_{n_2}$ directions, associated forces $\overline{F}^n_{23}$ and $\bar{f}_{23}$, and the friction torque $\bar{t}_{23}$ acting on the pin. Following the convention defined in Section 5.1.2, these forces are all drawn as dashed vectors because their directions are known but their magnitudes are unknown. The free body diagram also includes the known forces $\overline{F}_3$ (the load) and $\overline{W}_3$ (the weight of 3). They are drawn as solid vectors because they are completely known.

Figure 6.19 is the free body diagram of the slotted member. It has the reactions of the forces and torques from the slot to the pin that we have denoted as $-\overline{F}^n_{23}, -\bar{f}_{23}$, and $-\bar{t}_{23}$. The free body diagram also includes the force at the pin joint, $\overline{F}_{12}$; the associated friction torque, $\bar{t}_{12}$; the driving torque $\overline{T}_2$; and the weight $\overline{W}_2$.

Following the convention in Section 5.1.2, $\overline{W}_2$ is solid because it is a known vector, and $\bar{t}_{12}, -\bar{t}_{23}, -\overline{F}^n_{23}, -\bar{f}_{23}$, and $\overline{T}_2$ are dashed because they have known directions and unknown magnitudes. $\overline{F}_{12}$ is drawn squiggly because it is completely unknown.

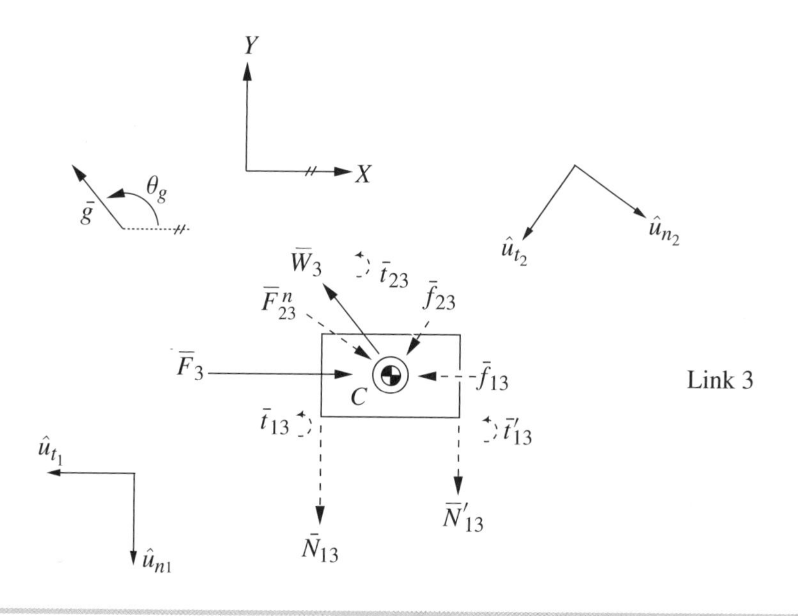

FIGURE 6.18 Free body diagram of slider block

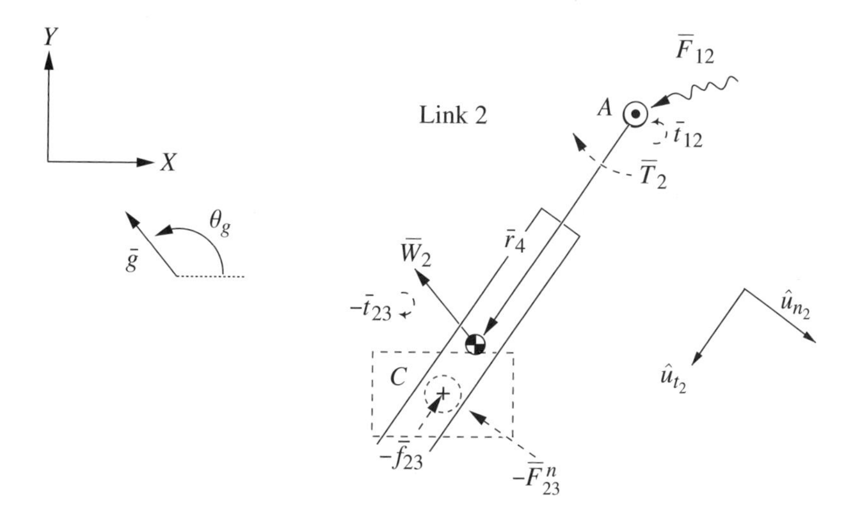

FIGURE 6.19 Free body diagram of the slotted link

Step 4: Write equations for the magnitudes of the friction terms and for the direction indicators. Consider the sliding joint first. From Equation (6.21),

$$f_{13} = \mu_{13}\left(|N_{13}| + |N'_{13}|\right) D_{13}, \tag{6.38}$$

where μ in Equation (6.21) is replaced by μ_{13} to distinguish it from the kinetic coefficient of friction in other joints. The friction torques are a bit more complex because they depend on whether the normal force is acting at the top or bottom corner of the slider. From Equation (6.22),

$$\begin{array}{ll} \text{if} & N_{13} > 0, \text{ then } t_{13} = -\mu_{13} r_b N_{13} D_{13} \\ \text{else} & \\ \text{if} & N_{13} < 0, \text{ then } t_{13} = \mu_{13} r_t N_{13} D_{13}, \end{array} \tag{6.39}$$

and from Equation (6.23),

$$\begin{array}{ll} \text{if} & N'_{13} > 0, \text{ then } t'_{13} = -\mu_{13} r_b N'_{13} D_{13} \\ \text{else} & \\ \text{if} & N'_{13} < 0, \text{ then } t'_{13} = \mu_{13} r_t N'_{13} D_{13}. \end{array} \tag{6.40}$$

From Equation (6.24), the direction indicator D_{13} is given by

$$D_{13} = \frac{-f_3}{|f_3|} D_i, \tag{6.41}$$

where

$$D_i = \frac{\dot{\theta}_2}{|\dot{\theta}_2|}. \tag{6.42}$$

The problem gives ω_2 as clockwise, so $\dot{\theta}_2 < 0$ and

$$D_i = -1. \tag{6.43}$$

Consider next the pin-in-a-slot joint. From Equation (6.8) and Equation (6.9),

$$f_{23} = \mu_{23} \left|F^n_{23}\right| D_{23} \tag{6.44}$$

$$\text{and} \quad t_{23} = (\mu R)_{23} F^n_{23} D_{23}, \tag{6.45}$$

where again μ and μR are written as μ_{23} and $(\mu R)_{23}$ in order to distinguish them from the friction properties of other joints. In the case of a pin-in-a-slot joint, the direction indicator is a bit more complicated because it depends on the direction of $\overrightarrow{F}^n_{23}$. From Equation (6.14),

$$\begin{array}{ll} \text{if} & F^n_{23} > 0, \text{ then } D_{23} = \dfrac{-f_2 + R(h_2 - h_3)}{|-f_2 + R(h_2 - h_3)|} D_i \\ \text{else} & \\ \text{if} & F^n_{23} < 0, \text{ then } D_{23} = \dfrac{-f_2 - R(h_2 - h_3)}{|-f_2 - R(h_2 - h_3)|} D_i, \end{array} \tag{6.46}$$

and D_i was given in Equation (6.43). Consider now the pin joint. From Equation (6.2),

$$t_{12} = (\mu R)_{12} \left|F_{12}\right| D_{12} = \mu R \sqrt{F^2_{12_x} + F^2_{12_y}}\, D_{12}, \tag{6.47}$$

where $(\mu R)_{12}$ has been used for the term μR in Equation (6.2) and

$$D_{12} = \frac{h_1 - h_2}{|h_1 - h_2|} D_i = \frac{-h_2}{|h_2|} D_i \quad (h_1 = 0), \tag{6.48}$$

and D_i was given in Equation (6.43).

Step 5: Write out the force and moment equilibrium conditions. Consider link 3. Moments will be taken about G_3, which is coincident with C.

Link 3:

Force equilibrium requires

$$\Sigma\overline{F} = \overline{N}_{13} + \overline{N}'_{13} + \overline{f}_{13} + \overline{F}^n_{23} + \overline{f}_{23} + \overline{F}_3 + \overline{W}_3 = m_3\overline{a}_{g_3}.$$

Note that

$$\overline{F}_3 = F_3\begin{bmatrix}1\\0\end{bmatrix} \quad \text{and} \quad \overline{W}_3 = W_3\begin{bmatrix}\cos\theta_g\\ \sin\theta_g\end{bmatrix},$$

and refer to Equation (6.34) and Equation (6.37) for the components of $\overline{N}_{13}$, $\overline{N}_{13}$, $\overline{f}_{13}$, $\overline{F}^n_{23}$, and $\overline{f}_{23}$ to get

$$\Sigma F_x = -N_{13} - N'_{13} - F^n_{23}\sin\theta_2 + f_{23}\cos\theta_2 + F_3 + W_3\cos\theta_g = m_3 a_{g_{3x}} \tag{6.49}$$

$$\Sigma F_y = -f_{13} + F^n_{23}\cos\theta_2 + f_{23}\sin\theta_2 + W_3\sin\theta_g = m_3 a_{g_{3y}}. \tag{6.50}$$

Moment equilibrium requires

$$\Sigma\overline{M}_{g_3} = r\hat{u}_{t_1} \times N_{13}\hat{u}_{n_1} + r'\hat{u}_{t_1} \times N'_{13}\hat{u}_{n_1} + \overline{t}_{13} + \overline{t}'_{13} + \overline{t}_{23} = I_{g_3}\overline{\alpha}_3$$

and this equation has only a Z component,

$$\Sigma M_{g_3} = N_{13}r + N'_{13}r' + t_{13} + t'_{13} + t_{23} = I_{g_3}\alpha_3 = 0 \qquad (\alpha_3 = 0). \tag{6.51}$$

Consider link 2.

Link 2:

Force equilibrium requires

$$\Sigma\overline{F} = -\overline{F}^n_{23} - \overline{f}_{23} + \overline{W}_2 + \overline{F}_{12} = m_2\overline{a}_{g_{2x}}.$$

Note that

$$\overline{W}_2 = W_2\begin{bmatrix}\cos\theta_g\\ \sin\theta_g\end{bmatrix},$$

and refer to Equation (6.37) for the components of $\overline{F}^n_{23}$ and $\overline{f}_{23}$ to get

$$\Sigma F_x = F^n_{23}\sin\theta_2 - f_{23}\cos\theta_2 + W_2\cos\theta_g + F_{12_x} = m_2 a_{g_{2x}} \tag{6.52}$$

$$\Sigma F_y = -F^n_{23}\cos\theta_2 - f_{23}\sin\theta_2 + W_2\sin\theta_g + F_{12_y} = m_2 a_{g_{2y}}. \tag{6.53}$$

Moments will be taken about A in order to eliminate the unknowns F_{12_x} and F_{12_y} from our system of equations. Moment equilibrium requires

$$\Sigma\overline{M}_a = \underbrace{\overline{r}_4 \times \overline{W}_2 + \overline{r}_2 \times (-\overline{F}^n_{23})}_{\text{Z component only}} - \overline{t}_{23} + \overline{t}_{12} = I_{g_2}\overline{\alpha}_2 + \underbrace{\overline{r}_4 \times m_2\overline{a}_{g_2}}_{\text{Z component only}}.$$

Recall our discussion of Equation (5.10), which showed that the above equation has only a Z component and is thus a scalar equation. Expanding the cross products gives

$$\begin{aligned}\Sigma M_a = r_4 W_2(\cos\theta_2\sin\theta_g - \sin\theta_2\cos\theta_g) - r_2F^n_{23} - t_{23} + t_{12}\\ = I_{g_2}\alpha_2 + r_4 m_2(\cos\theta_2 a_{g_{2y}} - \sin\theta_2 a_{g_{2x}}),\end{aligned}$$

which can be simplified by recognizing the double angle $(\theta_g - \theta_2)$ on the left hand side:

$$\Sigma M_a = r_4 W_2\sin(\theta_g - \theta_2) - r_2F^n_{23} - t_{23} + t_{12} = I_{g_2}\alpha_2 + r_4 m_2(\cos\theta_2 a_{g_{2y}} - \sin\theta_2 a_{g_{2x}}). \tag{6.54}$$

Before proceeding to describing how to solve these equations, we should count the number of equations and unknowns as a partial check of correctness. Our goal is to solve the six linear equations (6.49)–(6.54) for the unknown forces and torques, of which there are 12.

unknowns: $N_{13}, N'_{13}, f_{13}, F''_{23}, f_{23}, t_{13}, t'_{13}, t_{23}, F_{12_x}, F_{12_y}, t_{12}$, and T_2

The six additionally required equations are those that relate the friction forces and torques to the joint force: Equation (6.21), Equation (6.22), Equation (6.23), Equation (6.8), Equation (6.9), and Equation (6.47). These six equations involve direction indicators D_i, D_{12}, D_{13}, and D_{23} computed through Equation (6.43), Equation (6.48), Equation (6.41), and Equation (6.46) using the kinematic coefficients that were computed in step 1. The six additional equations are nonlinear. The nonlinearity is weak, and the equations can be solved using the Method of Successive Iterations.

The Method of Successive Iterations treats the friction terms as knowns. It begins by assuming the friction terms are all zero, and then Equation (6.49)–Equation (6.54) are solved for the joint forces. These joint forces are used to compute the friction terms through Equation (6.21), Equation (6.22), Equation (6.23), Equation (6.8), Equation (6.9), and Equation (6.47), which are then used to recompute the joint forces, and so on. The iterations continue until there is convergence. To assist in implementing the Method of Successive Iterations, Equation (6.49)–Equation (6.54) are written in matrix form with the friction terms treated as knowns and brought to the right side:

$$\begin{bmatrix} -1 & -1 & -\sin\theta_2 & 0 & 0 & 0 \\ 0 & 0 & \cos\theta_2 & 0 & 0 & 0 \\ r & r' & 0 & 0 & 0 & 0 \\ 0 & 0 & \sin\theta_2 & 1 & 0 & 0 \\ 0 & 0 & \cos\theta_2 & 0 & 1 & 0 \\ 0 & 0 & -r_2 & 0 & 0 & -1 \end{bmatrix} \begin{bmatrix} N_{13} \\ N'_{13} \\ F''_{23} \\ F_{12_x} \\ F_{12_y} \\ T_2 \end{bmatrix} = \begin{bmatrix} m_3 a_{g3_x} - f_{23}\cos\theta_2 - F_3 - W_3\cos\theta_g \\ m_3 a_{g3_y} + f_{13} - f_{23}\sin\theta_2 - W_3\sin\theta_g \\ -t_{13} - t'_{13} - t_{23} \\ m_2 a_{g2_x} + f_{23}\cos\theta_2 - W_2\cos\theta_g \\ m_2 a_{g2_y} + f_{23}\sin\theta_2 - W_2\sin\theta_g \\ I_{g2}\alpha_2 + r_4 m_2(\cos\theta_2 a_{g2_y} - \sin\theta_2 a_{g2_x}) - r_4 W_2 \sin(\theta_g - \theta_2) + t_{23} - t_{12} \end{bmatrix}. \quad (6.55)$$

Step 6: Create a flowchart for a program to compute T_2 that shows which terms are to be calculated and refers to the equations for making that calculation. This is shown in Figure 6.20.

Remember that, as pointed out in Example 5.2, in a computer program when you input the data in the first block, you should be certain that all quantities have units belonging to either the British Gravitational System or the International System. Do not use units belonging to the USCS or European Engineering System.

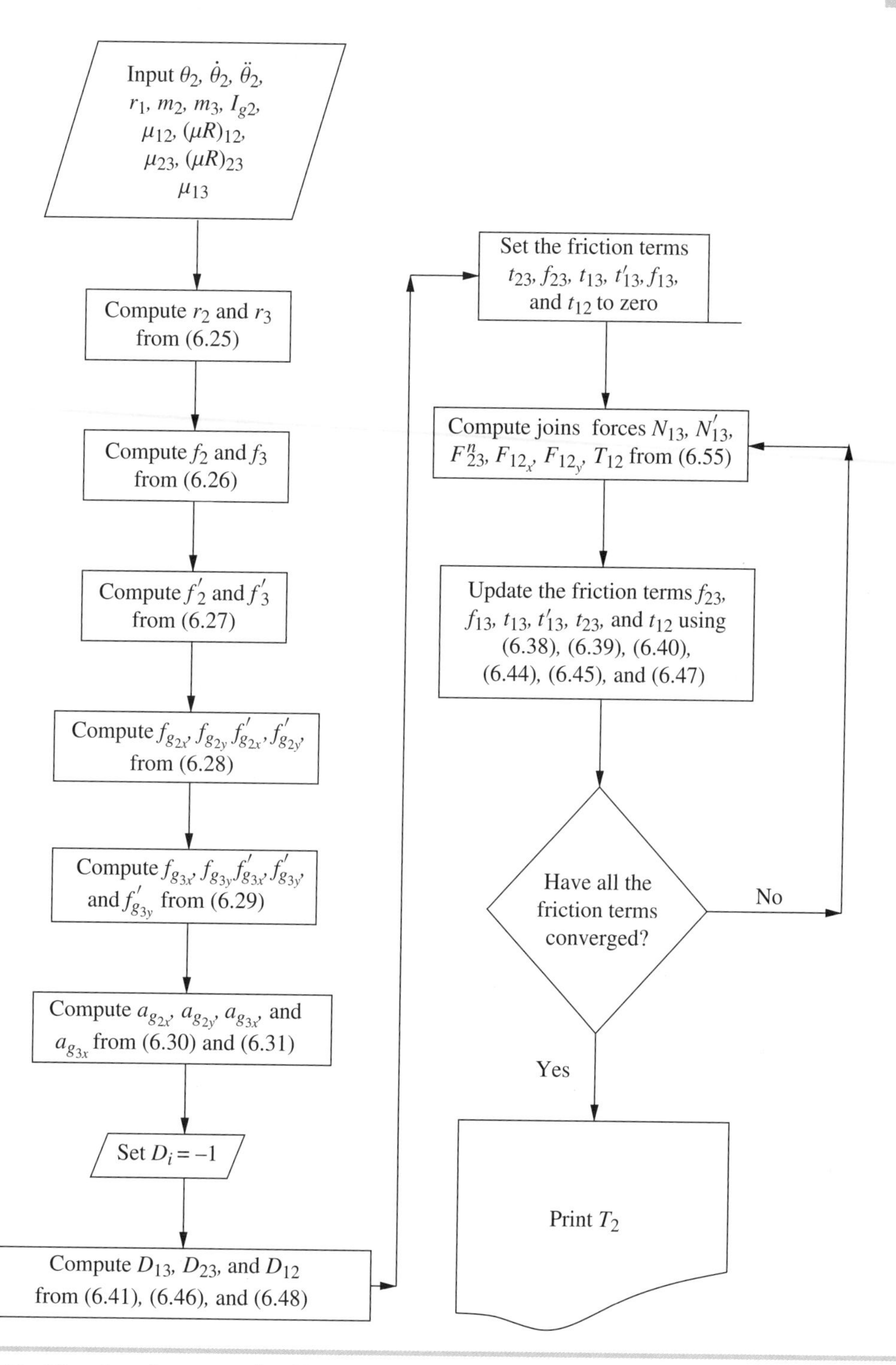

FIGURE 6.20 Flowchart for computing T_2

6.4 PROBLEMS

Problem 6.1

If in Equation (6.4) the input variable S_i had been θ_y, what would the expression for D_{yx} simplify to?

Problem 6.2

Repeat Problem 6.1 except that now S_i is θ_x.

Problem 6.3

If in Equations (6.11) and (6.13) the input variable S_i had been θ_y, what would the expressions for D_{yx} simplify to?

Problem 6.4

Repeat Problem 6.3 except that now S_i is θ_x.

Problem 6.5

Repeat Problem 5.7 and include friction in all the joints. The mechanism is redrawn in Figure 6.21 Outline all necessary equations, and show a flowchart for a computer program that calculates the result. (Vector $\bar{r}_3$ would actually be directly on top of $\bar{r}_5$, but for visual clarity it has been drawn as slightly offset.) Use Figure 6.22 for your free body diagrams.

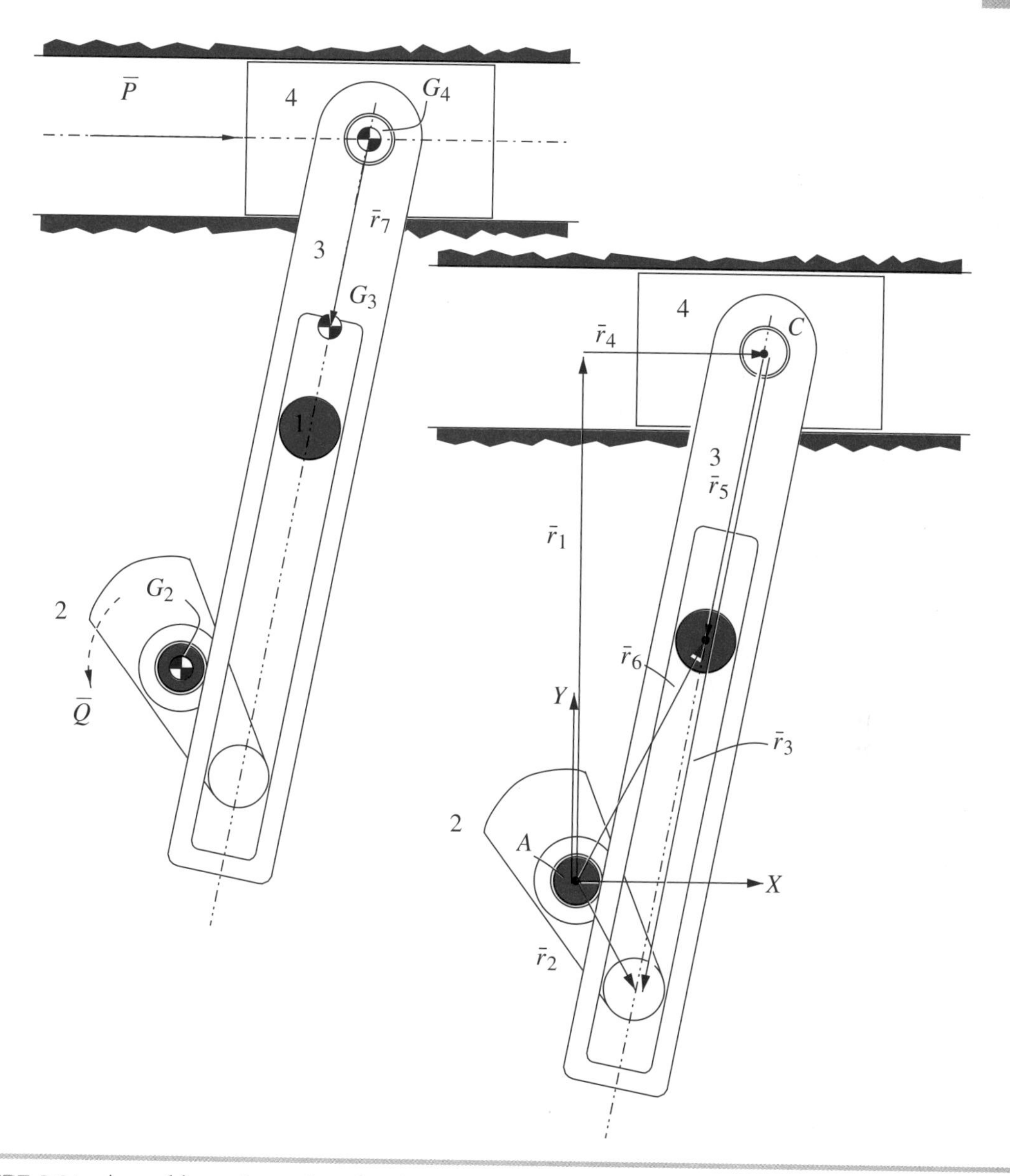

FIGURE 6.21 A machine to be analyzed and its vector loop (Problem 6.5)

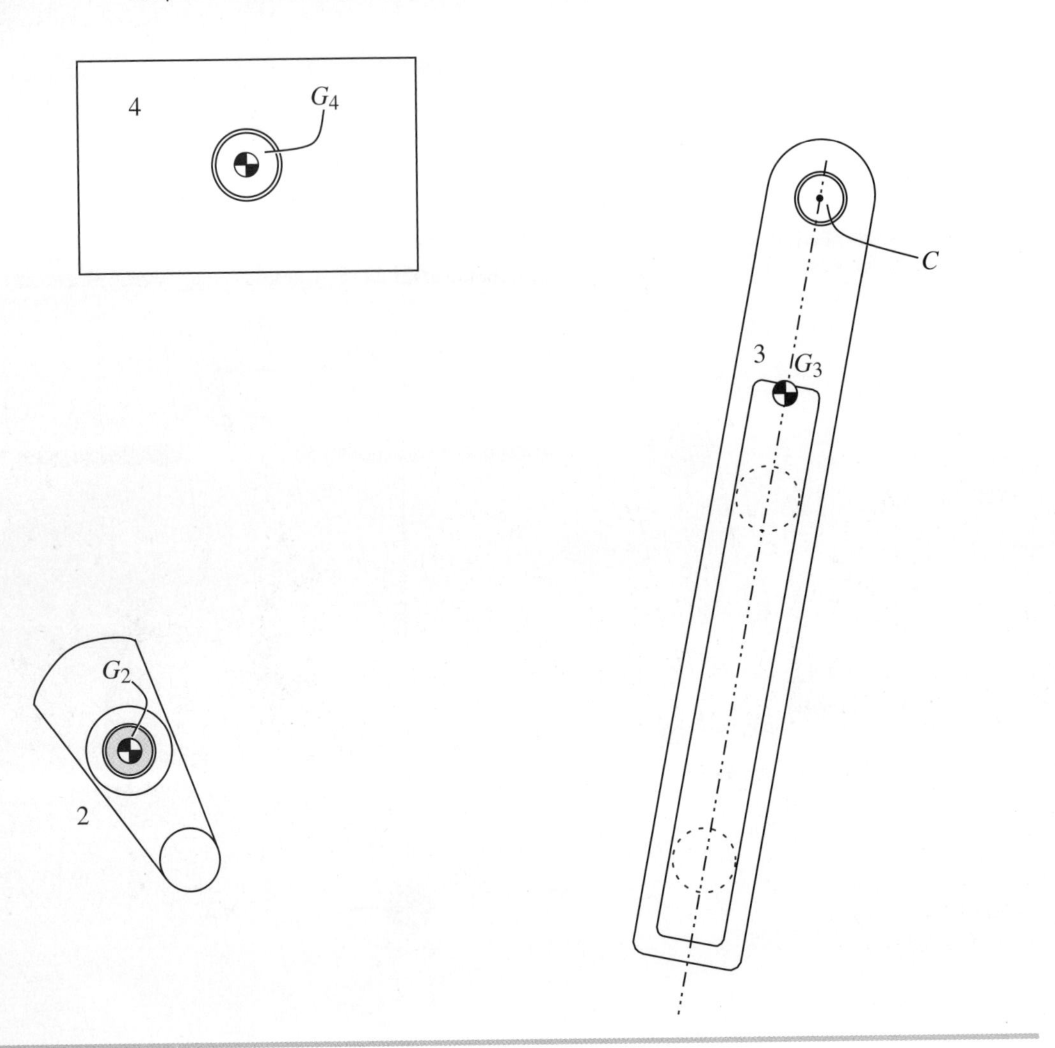

FIGURE 6.22 Free body diagrams

Problem 6.6

Repeat Problem 5.10 and include friction in all the joints. The machine is redrawn in Figure 6.23. Figure 6.24 shows the vector loop. Outline all necessary equations, and show a flowchart for a computer program that calculates the result. Use Figure 6.25 for your free body diagrams.

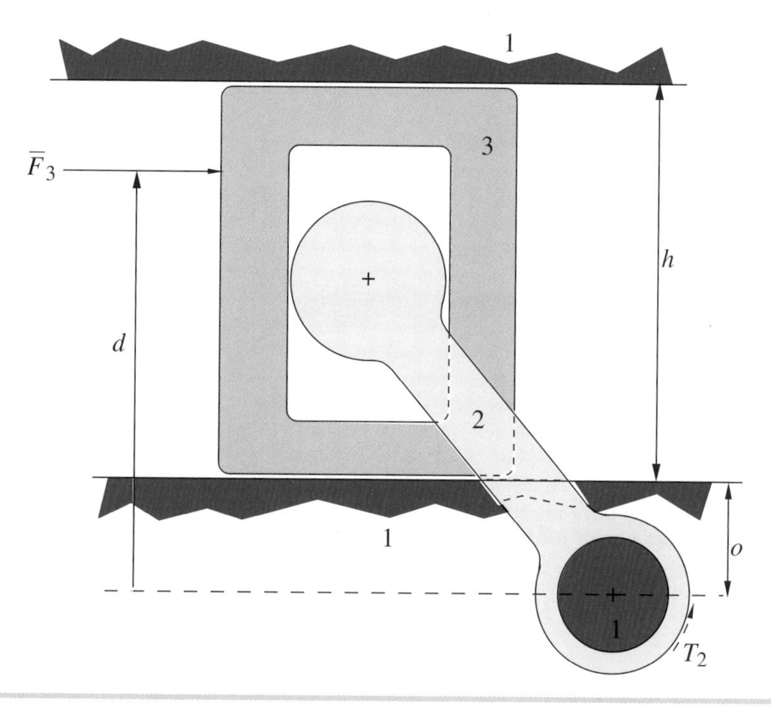

FIGURE 6.23 A machine to be analyzed (Problem 6.6)

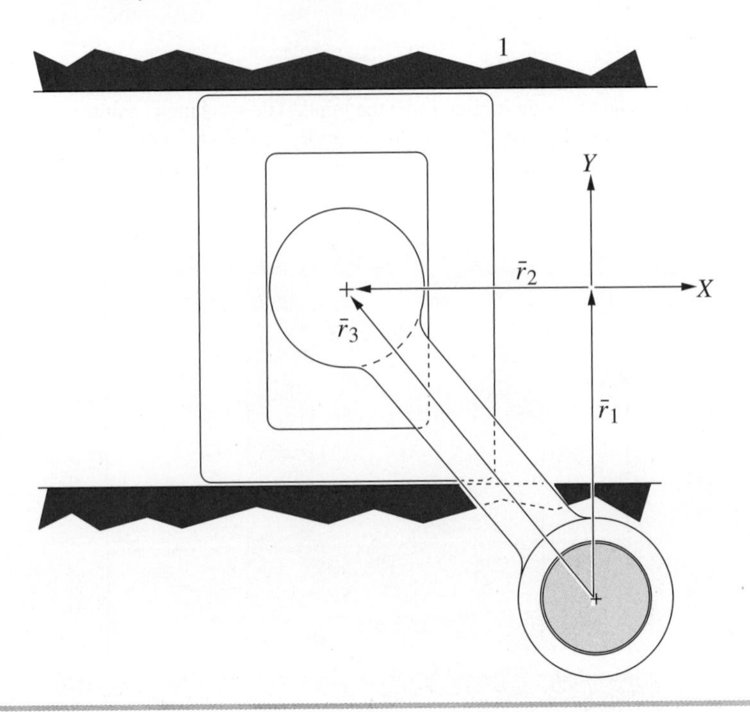

FIGURE 6.24 The vector loop

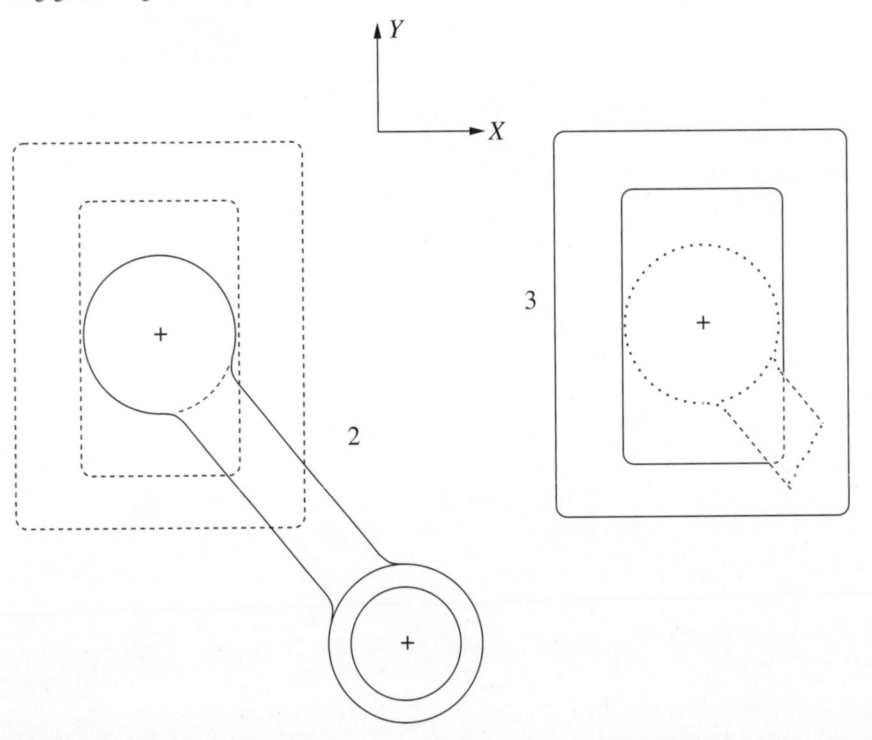

FIGURE 6.25 Free body diagrams

Problem 6.7

Repeat Problem 5.11 and include friction in all the joints. The vector loop is shown in Problem 4.5. Outline all necessary equations, and show a flowchart for a computer program that will calculate the driving force $\overline{F}_2$, which overcomes the load torque $\overline{T}_3$. Use Figure 6.26 for your free body diagrams.

Indicate here the slider block dimensions and the signs of their magnitudes.

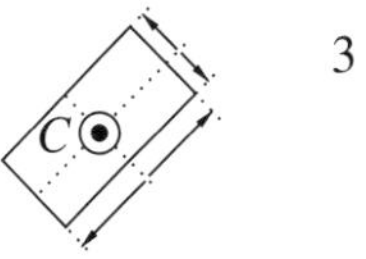

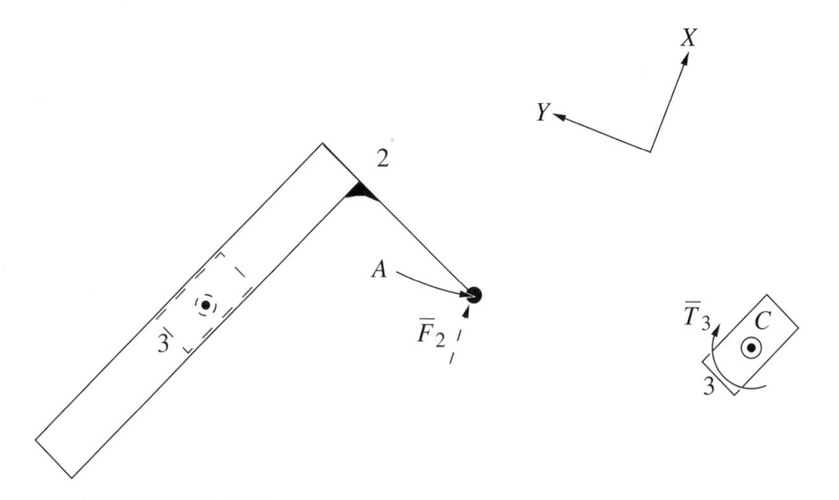

FIGURE 6.26 Free body diagrams

CHAPTER 7

Machine Dynamics Part III: The Power Equation

This chapter examines machine dynamics via an energy-based method known as the power equation. Like other energy methods, the power equation relates the body's motion to its externally applied forces and torques in a quick and straightforward manner. The power equation can be used to solve either the inverse or the forward dynamics problem of a machine. See Chapter 5 if you do not recall the distinction between these two types of dynamics problems. The kinematic coefficients studied in Chapter 4 play a basic role in the power equation. It may very well be that the power equation and the solution to the forward dynamics problem are the most powerful features of the vector loop method and its associated kinematic coefficients.

7.1 DEVELOPMENT OF THE POWER EQUATION

An energy balance on a machine can be written as

$$W = \Delta T + \Delta U + W_f, \tag{7.1}$$

where: W = the net work input to the machine
ΔT = the change in the kinetic energy of the moving parts in the machine
ΔU = the change in the potential energy stored in the machine
W_f = energy dissipated through friction.

In terms of time rates of change, Equation (7.1) becomes the power equation,

$$P = \frac{dT}{dt} + \frac{dU}{dt} + P_f, \tag{7.2}$$

where: P = the net power input to the machine
$\frac{dT}{dt}$ = the rate of change of the kinetic energy of the moving parts in the machine
$\frac{dU}{dt}$ = the rate of change of the potential energy stored in the machine
P_f = power dissipated through friction.

Some terms on the right side of Equation (7.2) may dominate others, allowing the smaller, less significant terms to be neglected. This introduces a slight approximation but yields a great simplification. For example, in high-speed machinery, links tend to have lower inertias, and speeds are very high. In this case the kinetic energy term dominates, and the potential energy term can be neglected. Also, such machines tend to have high-quality bearings, so frictional losses at the bearings can be neglected. In slow-moving, heavy machinery, speeds are low and kinetic energy is small and can be neglected, but since links are massive, potential energy dominates. Such machines may have lower-quality bearings in which case frictional losses at the joints become significant. Many times these massive machines also use springs and counterweights whose potential energy must be taken into account. The distinction between these two types of problems is identical to the distinction between static force analysis and inertia force analysis, discussed in Section 5.3. There may also be other energy-dissipating elements in a machine, such as brakes, clutches, or dampers (shock absorbers). If so, their energy-dissipating effects should be accounted for. In the next subsections, we look at the terms on the right hand side of Equation (7.2) and how they are formulated.

7.1.1 The Rate of Change of Kinetic Energy

A machine consists of many moving elements. Of course some of these are the links of the mechanism, but there may be others, such as the rotating inertia of a motor's windings, the moving inertia of a plunger in a solenoid, the moving mass center of a rapidly stretching spring, or the rotational inertia of a flywheel. Each of these has a kinetic energy that is typically changing in time, and this rate of change is related to kinematic coefficients.

Figure 7.1 shows a moving rigid body that is a link in a machine. In terms of the body's motion and its inertia properties, its kinetic energy T is given by

$$T = \frac{1}{2} m v_g^2 + \frac{1}{2} I_g \omega^2.$$

Let S_i be the input to the mechanism (as defined in Section 4.2), and define the first- and second-order kinematic coefficients of the body's mass center (f_{g_x}, f'_{g_x} and f_{g_y}, f'_{g_y}) as

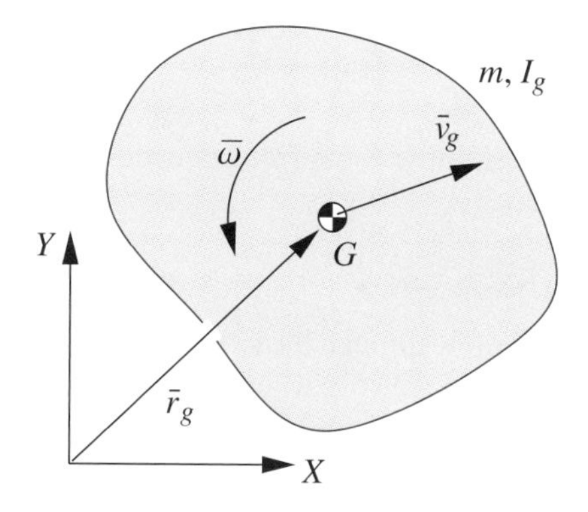

G, body's center of mass
m, body's mass
I_g, body's rotational inertia about g
$\bar{v}_g$, velocity of G
$\bar{\omega}$, angular velocity of the body

FIGURE 7.1 Kinetic energy, T, of a rigid body in planar motion

$$f_{gx} = \frac{dr_{gx}}{dS_i},\ f'_{gx} = \frac{d^2r_{gx}}{dS_i^2} \quad \text{and} \quad f_{gy} = \frac{dr_{gy}}{dS_i},\ f'_{gy} = \frac{d^2r_{gy}}{dS_i^2}, \tag{7.3}$$

where r_{g_x} and r_{g_y} are the components of the vector $\bar{r}_g$, the vector which locates the center of mass in Figure 7.1.

$$\bar{r}_g = \begin{bmatrix} r_{gx} \\ r_{gy} \end{bmatrix}.$$

Also, define the first- and second-order kinematic coefficients h and h' as

$$h = \frac{d\theta}{dS_i} \quad \text{and} \quad h' = \frac{d^2\theta}{dS_i^2},$$

where θ defines the rotation of the body; that is, θ is the direction of a vector that rotates with the body. In terms of these kinematic coefficients, the kinetic energy is given by

$$T = \left[\frac{1}{2}\, m\left(f_{gx}^2 + f_{gy}^2\right) + \frac{1}{2}\, I_g h^2\right] \dot{S}_i^2, \tag{7.4}$$

where $\dot{S}_i$ is the input velocity.

Differentiating Equation (7.4) with respect to time and carefully applying both the product rule and the chain rule for differentiation yields

$$\frac{dT}{dt} = A\ \dot{S}_i\ddot{S}_i + B\ \dot{S}_i^3,$$

where

$$A = \left[m\left(f_{gx}^2 + f_{gy}^2\right) + I_g h^2\right] \quad \text{and} \quad B = \left[m\left(f_{gx}f'_{gx} + f_{gy}f'_{gy}\right) + I_g h h'\right]. \tag{7.5}$$

If one considers the collection of moving parts within a machine, then *for the machine*, the rate of change of kinetic energy is given by

$$\frac{dT}{dt} = (\Sigma A)\ \dot{S}_i\ddot{S}_i + (\Sigma B)\ \dot{S}_i^3, \tag{7.6}$$

where ΣA (a.k.a. the *"sum of the A"s*) and ΣB (a.k.a. the *"sum of the B"s*) are the sums of A and B for all moving parts of the machine.

Equivalent Inertia

Consider the term (ΣA) in Equation (7.6). Substituting for A from Equation (7.5), we find that

$$\Sigma A = \sum \left[m\left(f_{gx}^2 + f_{gy}^2\right) + I_g h^2\right]. \tag{7.7}$$

When the input S_i is a vector length r_i and you follow through the units on the kinematic coefficients, you will see that ΣA has units of mass. Likewise, when the input S_i is a vector direction θ_i and you follow through the units on the kinematic coefficients, you will see that ΣA has units of mass moment of inertia. We can observe from Equation (7.4) that for the machine,

$$T = \frac{1}{2}\,(\Sigma A)\ \dot{S}_i^2.$$

This equation shows that ΣA is an equivalent inertia, which, if attached directly to the input, would have the same kinetic energy as the entire machine. If we let I_e denote the equivalent inertia of the machine, then

$$I_e = \Sigma A.$$

In ΣA the inertia properties are constants, but the kinematic coefficients are generally functions of S_i. So I_e generally changes with the configuration of the machine. Geartrains and transmissions have constant I_e because their kinematic coefficients are constant.

Recall A and B for a single moving part of the machine from Equation (7.5), and note that

$$\frac{dA}{dS_i} = 2\left[m(f_{gx}f'_{gx} + f_{gy}f'_{gy}) + I_g hh'\right] = 2B,$$

or

$$B = \frac{1}{2}\frac{dA}{dS_i}.$$

Thus ΣB for the entire machine is

$$\Sigma B = \Sigma\left(\frac{1}{2}\frac{dA}{dS_i}\right) = \frac{1}{2}\Sigma\left(\frac{dA}{dS_i}\right) = \frac{1}{2}\frac{d(\Sigma A)}{dS_i} = \frac{1}{2}\frac{dI_e}{dS_i},$$

which means that the term ΣB is one-half of the rate of change with respect to S_i of the system's equivalent inertia. For geartrains and transmissions, ΣB is always zero. In terms of the equivalent inertia, Equation (7.6) can be written as

$$\frac{dT}{dt} = I_e\dot{S}_i\ddot{S}_i + \frac{1}{2}\frac{dI_e}{dS_i}\dot{S}_i^3.$$

In machines, potential energy is typically associated with elevation, U_e, and with springs, U_s. The following two subsections look at each of these and how their rate of change is related to kinematic coefficients.

7.1.2 The Rate of Change of Potential Energy Due to Elevation

Figure 7.2 shows a rigid body in a gravity field, $\overline{g}$. Its potential energy, U_e, is a function of its mass center's elevation, H, where H is measured from an arbitrary reference and elevation is understood to be in the opposite direction from the acceleration of gravity, $\overline{g}$. The location of this reference is arbitrary since in the end, we are concerned with *changes* in potential energy. Recall from your study of physics that

$$U_e = mgH,$$

from which

$$\frac{dU_e}{dt} = \frac{d}{dt}(mgH) = mg\frac{dH}{dt}, \tag{7.8}$$

where $\frac{dH}{dt}$ is the rate of change of the mass center's elevation. In Figure 7.2 we define a unit vector $\hat{u}_e$, in the direction of elevation. The direction of elevation is opposite the direction of gravity, which may not necessarily be aligned with one of the axes of

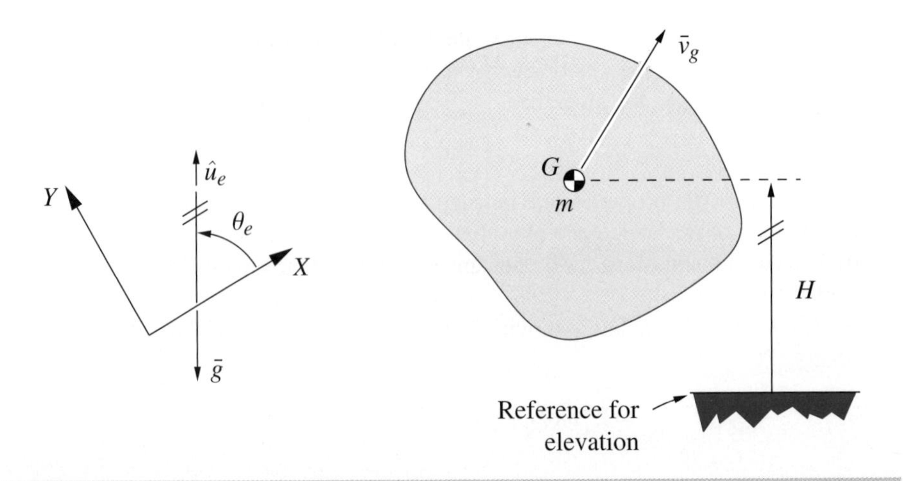

FIGURE 7.2 Potential energy due to elevation

the fixed frame. $\frac{dH}{dt}$ is the component of $\bar{v}_g$ in the direction of elevation (the vertical direction), which can be found from the dot product

$$\frac{dH}{dt} = \bar{v}_g \cdot \hat{u}_e.$$

In terms of its kinematic coefficients, $\bar{v}_g$ is given by

$$\bar{v}_g = \begin{bmatrix} v_{gx} \\ v_{gy} \end{bmatrix} = \begin{bmatrix} f_{gx} \\ f_{gy} \end{bmatrix} \dot{S}_i,$$

so $\frac{dH}{dt}$ is given by

$$\frac{dH}{dt} = \bar{v}_g \cdot \hat{u}_e = \begin{bmatrix} f_{gx} \\ f_{gy} \end{bmatrix} \dot{S}_i \cdot \begin{bmatrix} \cos\theta_e \\ \sin\theta_e \end{bmatrix} = \left(f_{g_x}\cos\theta_e + f_{g_y}\sin\theta_e\right)\dot{S}_i.$$

Defining the kinematic coefficient of elevation, f_e, as

$$f_e = f_{g_x}\cos\theta_e + f_{g_y}\sin\theta_e, \tag{7.9}$$

we have

$$\frac{dH}{dt} = f_e\dot{S}_i$$

and, from (7.8),

$$\frac{dU_e}{dt} = mgf_e\dot{S}_i, \tag{7.10}$$

where f_e is given by Equation (7.9).

The kinematic coefficient of elevation can be determined after the basic kinematic coefficients are known; from this the kinematic coefficients of the mass center are determined, and Equation (7.9) is applied to find f_e. The following example illustrates this for the coupler of a four bar mechanism.

▶ EXAMPLE 7.1

Figure 7.3 shows a four bar linkage whose coupler has a center of mass at G_3. Find the kinematic coefficient of elevation, f_e, for body 3.

Solution

We should probably denote this as f_{e_3} since it is likely that the potential energies of bodies 2 and 4 will also have to be accounted for, and we should distinguish their individual kinematic coefficients of elevation.

The first step is a standard kinematic analysis using the vector loop equation

$$\bar{r}_2 + \bar{r}_3 - \bar{r}_4 - \bar{r}_1 = \bar{0}.$$

If we take θ_2 as S_i, then θ_3 and θ_4 are output variables. The position problem is solved for the output variables, given the link lengths and a value for θ_2. Then the first-order kinematic coefficients h_3 and h_4 are found by differentiating the scalar components of the vector loop equation with respect to θ_2. This procedure should be second nature to you at this point (see Chapters 2 and 4).

The first-order kinematic coefficients of G_3 are found using Equations (4.39) and (4.40) from Section 4.5, where the point C in that discussion is the point G_3 here:

$$f_{g3x} = -r_2 \sin\theta_2 - r_3' \sin(\theta_3 + \phi_3) h_3 \tag{7.11}$$

$$f_{g3y} = r_2 \cos\theta_2 + r_3' \cos(\theta_3 + \phi_3) h_3. \tag{7.12}$$

The kinematic coefficient of elevation for link 3 is now found using Equation (7.9):

$$f_{e_3} = f_{g3x} \cos\theta_e + f_{g3y} \sin\theta_e$$

where θ_e is indicated in Figure 7.3.

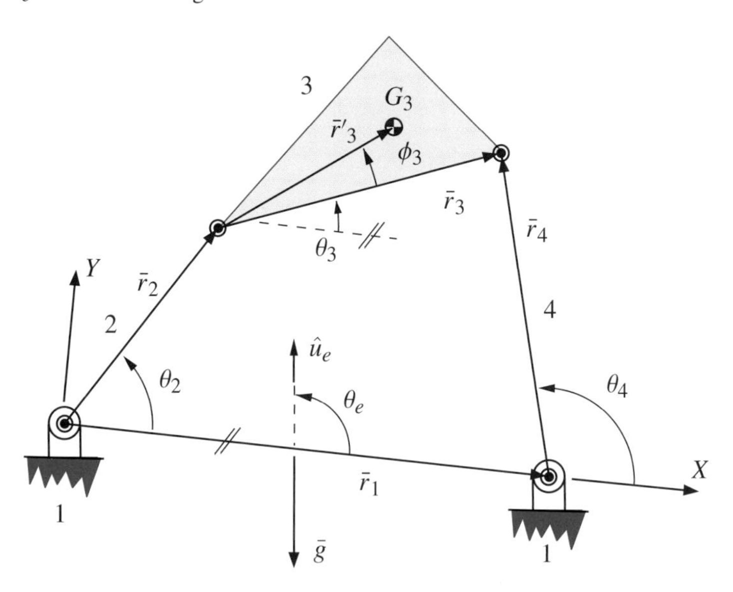

FIGURE 7.3 Finding f_e for the center of mass of the coupler of a four bar mechanism

7.1.3 The Rate of Change of Potential Energy in a Spring

Figure 7.4 shows a linear spring. The magnitude of the spring force, F_s, is linearly proportional to the change in length of the spring,

$$F_s = F_o + k(r_s - r_{s_o}), \tag{7.13}$$

where: F_s = the spring force
F_o = the preload in the spring (the force required to cause the spring to begin to deform)
k = the spring stiffness
r_s = the spring length
r_{s_o} = the length of the spring when $F_s = F_o$.

This is shown in Figure 7.5. The potential energy stored in the spring, U_s, is the area under this force–deflection curve, which is the sum of areas A_1 and A_2, the rectangular and triangular areas, respectively:

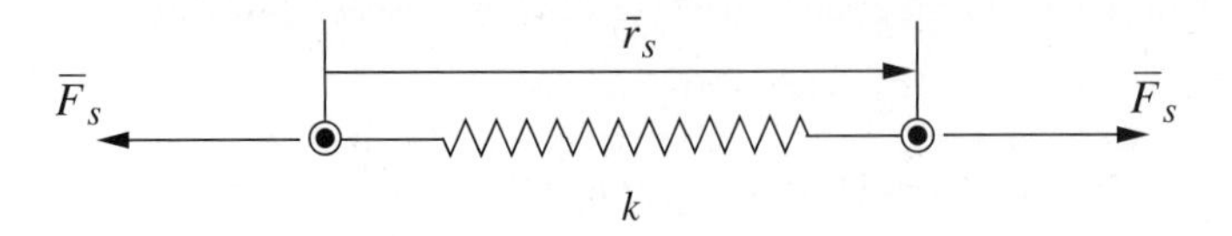

FIGURE 7.4 A linear spring

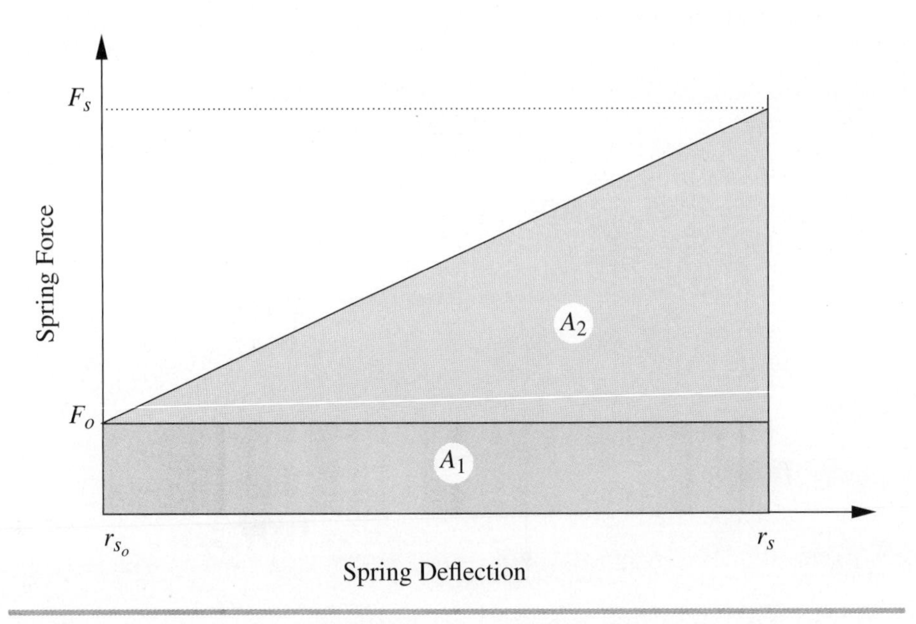

FIGURE 7.5 Spring force–deflection curve for a linear spring

$$U_s = A_1 + A_2 = F_o(r_s - r_{s_o}) + \frac{1}{2}(r_s - r_{s_o})\underbrace{(F_s - F_{s_o})}_{=k(r_s - r_{s_o})} = F_o(r_s - r_{s_o}) + \frac{1}{2}k(r_s - r_{s_o})^2. \quad (7.14)$$

Differentiating Equation (7.14) with respect to time gives our result,

$$\frac{dU_s}{dt} = F_s f_s \dot{S}_i, \quad (7.15)$$

where the kinematic coefficient of the spring, f_s, is defined as

$$f_s = \frac{dr_s}{dS_i}. \quad (7.16)$$

The kinematic coefficient of the spring can be found from the basic kinematic coefficients, using an additional vector loop. The following examples illustrate this.

▶ EXAMPLE 7.2

Many times springs are incorporated into mechanisms and machines. For example, look at Figure 1.13, where a spring is incorporated into a six bar linkage that guides a hood. The spring exchanges potential energy with the hood as the hood is raised and lowered. In Figure 7.6 there is a spring connected between a moving point on body 3 and a fixed point on body 1.

Find the kinematic coefficient of the spring for a given value of the input angle, θ_2.

Solution

Before finding this kinematic coefficient, we must solve the position problem for θ_3 and θ_4 and then solve for the basic first-order kinematic coefficients h_3 and h_4. We learned how to do this in Chapters 2 and 4.

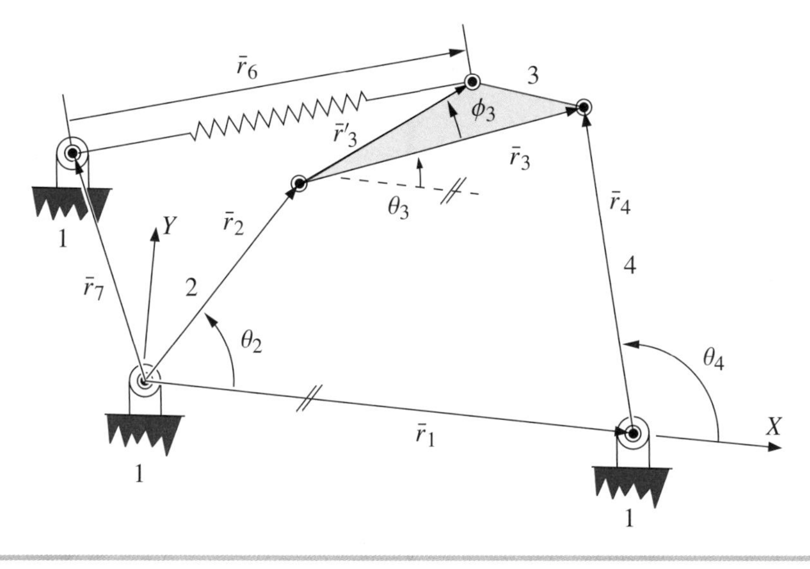

FIGURE 7.6 A spring with one fixed end

To find the kinematic coefficient of the spring, we need an additional vector loop that includes a vector that spans the spring. This second loop is seen in Figure 7.6, and the vector $\bar{r}_6$ spans the spring. The kinematic coefficient of the spring is f_6.

The vector loop equation for this second loop is

$$\bar{r}_7 + \bar{r}_6 - \bar{r}'_3 - \bar{r}_2 = \bar{0},$$

which has the X and Y components

$$r_7\cos\theta_7 + r_6\cos\theta_6 - r'_3\cos(\theta_3 + \phi_3) - r_2\cos\theta_2 = 0 \tag{7.17}$$

$$r_7\sin\theta_7 + r_6\sin\theta_6 - r'_3\sin(\theta_3 + \phi_3) - r_2\sin\theta_2 = 0. \tag{7.18}$$

Equations (7.17) and (7.18) can be solved for the two unknowns r_6 and θ_6 using Newton's Method or in closed form if desired. After r_6 and θ_6 are found, Equations (7.17) and (7.18) can be be differentiated with respect to θ_2, which gives two linear equations in unknowns f_6 and h_6:

$$\begin{bmatrix} \cos\theta_6 & -r_6\sin\theta_6 \\ \sin\theta_6 & r_6\cos\theta_6 \end{bmatrix} \begin{bmatrix} f_6 \\ h_6 \end{bmatrix} = \begin{bmatrix} -r'_3\sin(\theta_3 + \phi_3)h_3 - r_2\sin\theta_2 \\ r'_3\cos(\theta_3 + \phi_3)h_3 + r_2\cos\theta_2 \end{bmatrix}.$$

This equation can be solved for the kinematic coefficient of the spring, f_6, using Cramer's Rule.

The next example considers a more algebraically difficult problem that arises when neither end of a spring is fixed.

▶ EXAMPLE 7.3

Figure 7.7 shows a spring that has one end connected to a moving point on body 3 and the other end connected to a moving point on body 4. Find the kinematic coefficient of the spring for a given value of the input angle, θ_2.

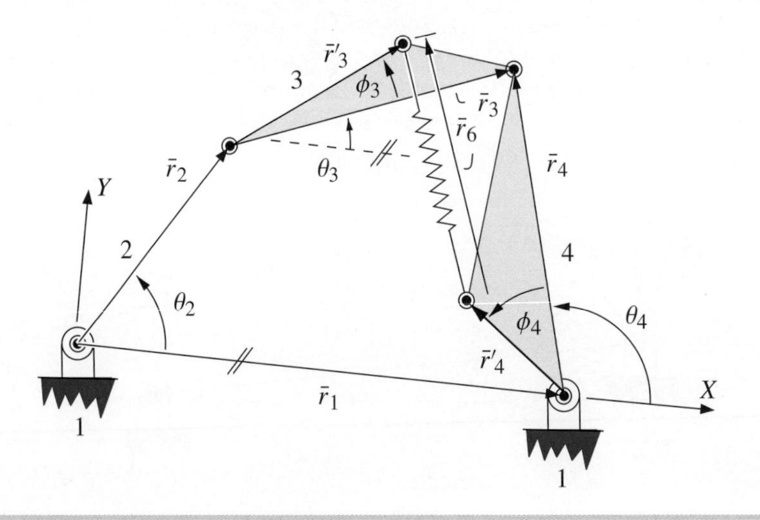

FIGURE 7.7 A spring with no fixed ends

Solution

Before finding this kinematic coefficient, we must solve the position problem for θ_3 and θ_4 and then solve for the basic first-order kinematic coefficients h_3 and h_4. We learned how to do this in Chapters 2 and 4.

To find the kinematic coefficient of the spring, we need an additional vector loop that includes a vector that spans the spring. This second loop is seen in Figure 7.7, and the vector $\bar{r}_6$ spans the spring. The kinematic coefficient of the spring is f_6.

The additional vector loop equation is $\bar{r}_2+\bar{r}'_3-\bar{r}_6-\bar{r}'_4-\bar{r}_1=\bar{0}$. Its X and Y components are

$$r_2\cos\theta_2 + r'_3\cos(\theta_3+\phi_3) - r_6\cos\theta_6 - r'_4\cos(\theta_4+\phi_4) - r_1 = 0$$
$$r_2\sin\theta_2 + r'_3\sin(\theta_3+\phi_3) - r_6\sin\theta_6 - r'_4\sin(\theta_4+\phi_4) = 0.$$

There are two unknowns in these two equations, r_6 and θ_6. They must be found before f_6 can be found. This can be done in closed form or by Newton's Method. Differentiating the position equations with respect to the input θ_2 gives two linear equations in the unknowns f_6 and h_6:

$$\begin{bmatrix} -\cos\theta_6 & r_6\sin\theta_6 \\ -\sin\theta_6 & -r_6\cos\theta_6 \end{bmatrix}\begin{bmatrix} f_6 \\ h_6 \end{bmatrix} = \begin{bmatrix} r_2\sin\theta_2 + r'_3\sin(\theta_3+\phi_3)h_3 - r'_4\sin(\theta_4+\phi_4)h_4 \\ -r_2\cos\theta_2 - r'_3\cos(\theta_3+\phi_3)h_3 + r'_4\cos(\theta_4+\phi_4)h_4 \end{bmatrix}.$$

This matrix equation can be solved for the kinematic coefficient of the spring, f_6.

7.1.4 Power Dissipated by Viscous Damping

Figure 7.8 shows a viscous damper. As the ends are pulled apart or pushed together, a viscous fluid (oil usually) is transferred across the face of the slider through small orifices. The transfer of this fluid resists the relative motion of the two ends.

Let $\bar{r}_c$ be a vector that spans the length of the damper. The resisting force is given by

$$F_c = c\dot{r}_c, \tag{7.19}$$

where c is known as the damping coefficient. This is called a linear damper because F_c is proportional to $\dot{r}_c$. You have seen these linear dampers in your vibrations or controls course. They are energy dissipaters. They take kinetic energy out of a machine and convert it to heat. In your automobile they are either shock absorbers or McPherson struts. In extreme cases such as high-performance off-road vehicles, the heat that develops in the viscous fluid needs to be removed by a heat exchanger. In an automobile the linear dampers often incorporate a linear spring (now called a "coil over shock").

The power dissipated by the damper is given by

$$P_f = F_c\frac{dr_c}{dt} = F_c\dot{r}_c,$$

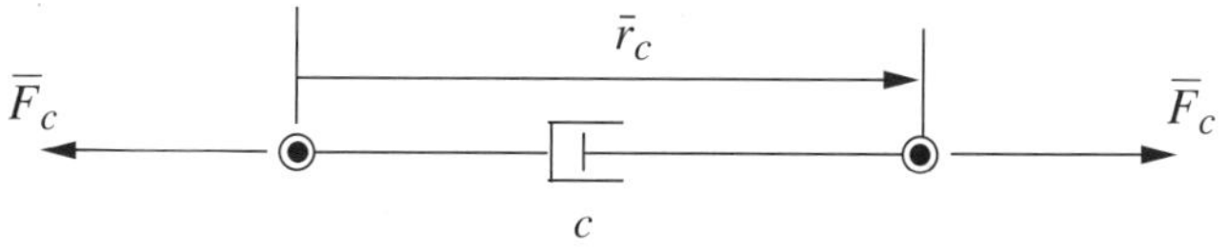

$\bar{r}_c$

$\bar{F}_c$ $\bar{F}_c$

c

FIGURE 7.8 A linear damper

and incorporating Equation (7.19) gives

$$P_f = c\dot{r}_c^2 = cf_c^2\dot{S}_i^2,$$

where

$$f_c = \frac{dr_c}{dS_i}$$

is called the kinematic coefficient of the damper.

In Example 7.2 or 7.3, if the spring were replaced by a damper, you would find f_c in the same manner as you found f_s. Finding f_s and f_c is the same procedure. The procedure begins by defining a vector loop that contains a vector that spans across the spring or the damper.

7.1.5 Power Dissipated by Coloumb Friction

In order to compute the rate of energy dissipation through Coloumb friction, we must know the friction force and rubbing velocity, and/or the friction torque and angular velocity of rubbing. This means that a force analysis, in addition to the kinematic analysis, must be done in order to evaluate this dissipation term in the power equation. The force analysis is an inverse dynamics problem that requires knowledge of the complete state of motion. This can be problematic, and we will discuss it in a later section of this chapter. Let us consider each case of joint friction that we have studied.

The Pin Joint

Recall the discussion of joint friction in Section 6.1, and Figure 6.1. For the pin joint, the power lost due to friction is given by

$$P_f = t_{yx}\omega_{yx}. \tag{7.20}$$

The Pin-in-a-Slot Joint

Recall the discussion of joint friction in Section 6.2, and Figures 6.3 and 6.5. For the pin-in-a-slot joint, the power lost due to friction is given by

$$P_f = f_{yx}v_{c_y/c_x} + t_{yx}\omega_{yx}. \tag{7.21}$$

The Straight Sliding Joint

Recall the discussion of joint friction in Section 6.3, and Figures 6.7 and 6.12. For the straight slider joint, the power lost due to friction is given by

$$P_f = f_{yx}v_{c_y/c_x}. \tag{7.22}$$

The difficulty in including power lost due to friction is that in order to do so, we must know the kinematics, and we must also know the friction forces and/or torques. The friction forces and torques are part of the IDP, which requires knowing the state of motion. If the power equation is being used to solve the FDP, we have a problem in that the state of motion is not completely known, and it is not possible to calculate the friction forces/torques. We will discuss this in Section 7.3.

7.2 THE POWER EQUATION AND THE INVERSE DYNAMICS PROBLEM

Recall the definition of an IDP. In an IDP the state of motion is known, and there is one unknown force or torque (typically the driver). It will be the one unknown in the power Equation (7.2). The following example illustrates how the power equation is applied to the inverse dynamics problem.

▶ EXAMPLE 7.4

Consider the inverted crank-slider mechanism shown in Figure 7.9. The inertia properties of the links are given in Figure 7.9. At this instant, $\theta_2 = 30°$ and ω_2 is a constant of 60 rpm clockwise. So instantaneously,

$$\theta_2 = 30° = \pi/6, \quad \dot{\theta}_2 = -60 \text{ rpm} = -2\pi \text{ rad/s}, \quad \text{and} \quad \ddot{\theta}_2 = 0. \tag{7.23}$$

Suppose a load torque of $T_4 = 150.0$ ft · lb$_f$ in the counterclockwise direction is acting on link 4:

$$T_4 = 150.0 \text{ ft} \cdot \text{lb}_f \text{ ccw}. \tag{7.24}$$

Compute the the required driving torque T_2.

Solution

Let us write the power equation for this machine. To keep the example simple, we will neglect potential energy effects and losses since we expect these to have negligible effects on the solution. In this case the power Equation (7.2) reduces to

$$P = \frac{dT}{dt}. \tag{7.25}$$

The power to the machine is given by

$$P = \overline{T}_2 \cdot \overline{\omega}_2 + \overline{T}_4 \cdot \overline{\omega}_4,$$

where

$$\overline{T}_2 = \begin{bmatrix} 0 \\ 0 \\ 1 \end{bmatrix} T_2, \quad \overline{\omega}_2 = \begin{bmatrix} 0 \\ 0 \\ 1 \end{bmatrix} \dot{\theta}_2, \quad \overline{T}_4 = \begin{bmatrix} 0 \\ 0 \\ 1 \end{bmatrix} T_4, \quad \overline{\omega}_4 = \begin{bmatrix} 0 \\ 0 \\ h_4 \end{bmatrix} \dot{\theta}_2.$$

The signs of T_2 and T_4 are positive if their direction is ccw and negative if their direction is cw. Taking the dot products gives

$$P = (T_2 + T_4\, h_3)\dot{\theta}_2 \quad (h_4 = h_3). \tag{7.26}$$

From Equation (7.6), with $S_i = \theta_2$,

$$\frac{dT}{dt} = (\Sigma A)\dot{\theta}_2\ddot{\theta}_2 + (\Sigma B)\dot{\theta}_2^3. \tag{7.27}$$

Substituting Equations (7.26) and (7.27) into Equation (7.25) and dividing through by $\dot{\theta}_2$ give the power equation for this machine,

$$(T_2 + T_4\, h_3) = (\Sigma A)\ddot{\theta}_2 + (\Sigma B)\dot{\theta}_2^2. \tag{7.28}$$

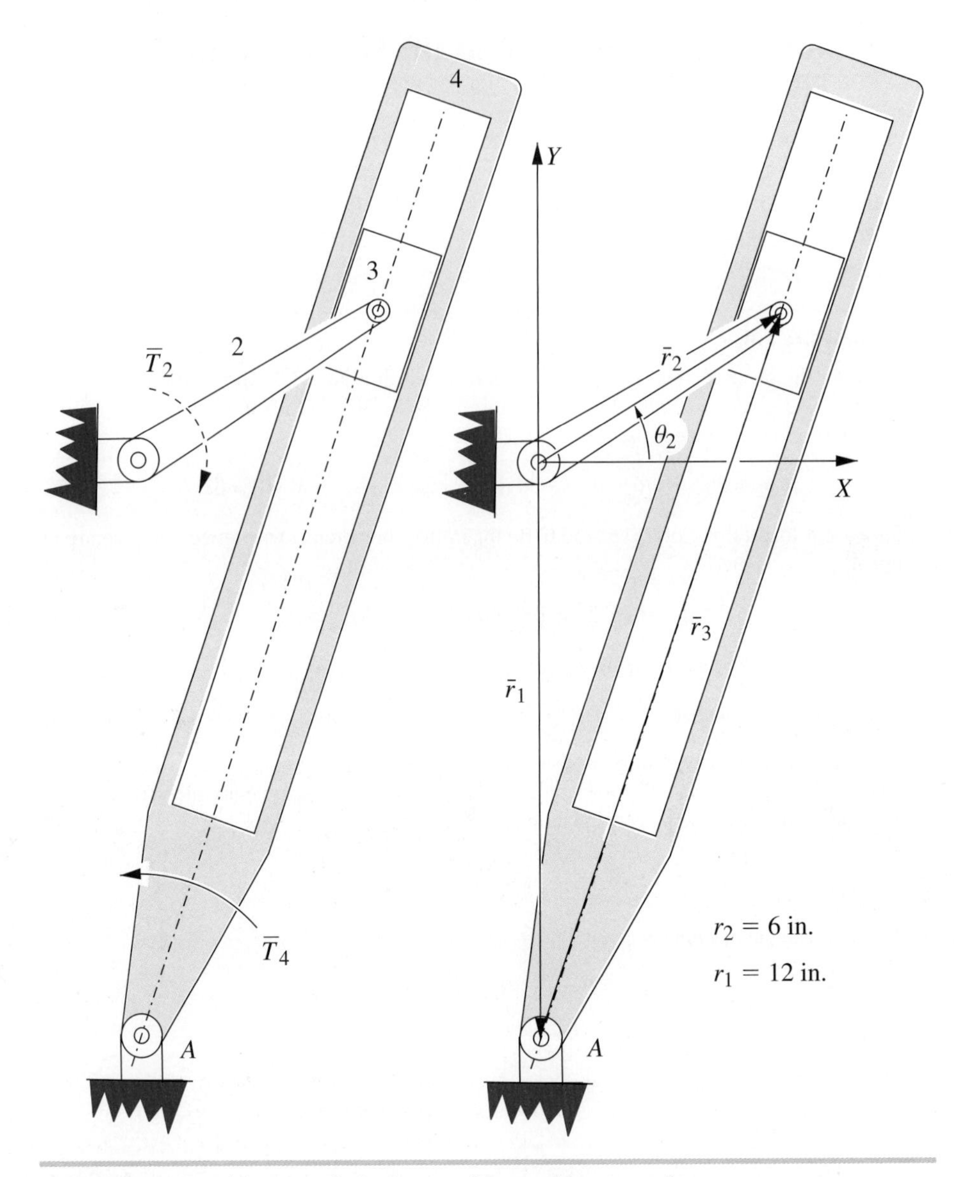

FIGURE 7.9 A quick-return indexing machine and its vector loop

Solving for T_2 gives the solution to the IDP,

$$T_2 = -T_4\, h_3 + (\Sigma A)\ddot{\theta}_2 + (\Sigma B)\dot{\theta}_2^2. \tag{7.29}$$

To compute T_2 from Equation (7.29), we must first compute h_3, (ΣA), and (ΣB) from the kinematics and the inertia properties of the mechanism. The inertia properties are listed in Figure 7.10. In Appendix I of Chapter 5 (Section 5.6), you will find the standard kinematic analysis using the vector loop method that you have learned so far. The vector loop is shown in Figure 7.9. Using the link dimensions given in Figure 7.9 and $\theta_2 = 30° = \frac{\pi}{6}$, the results of the kinematic analysis are the position solution,

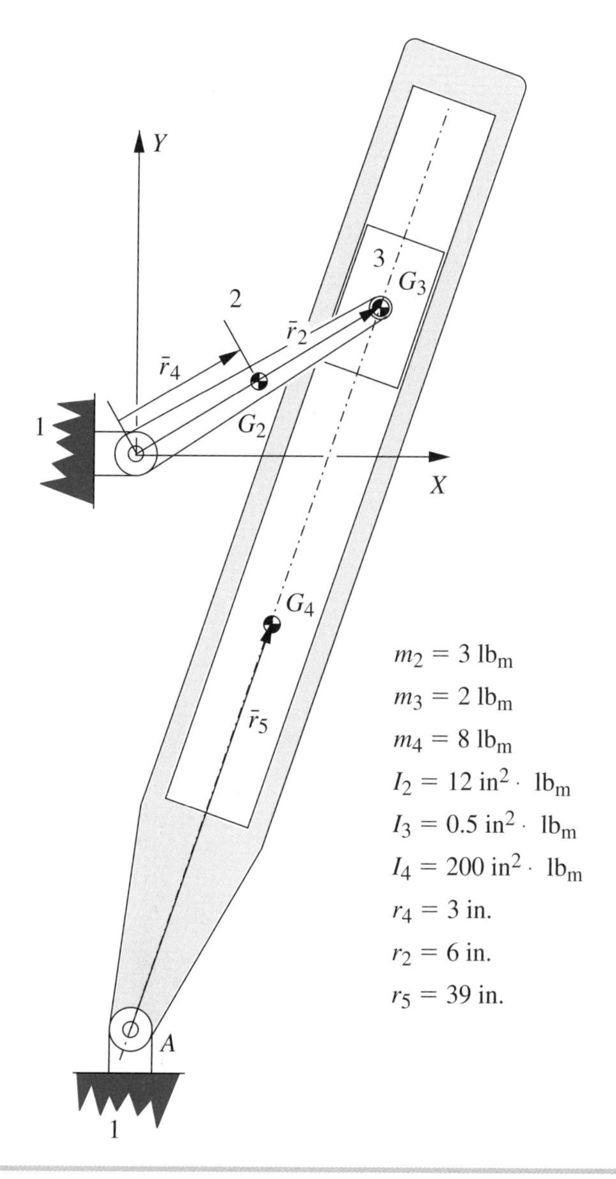

FIGURE 7.10 Inertia properties of the moving links

$$\theta_3 = 70.893°, \quad r_3 = 15.875 \text{ in.},$$

then the basic first- and second-order kinematic coefficients,

$$\begin{aligned} h_3 &= 0.286, & f_3 &= 3.928 \text{ in.} \\ h_3' &= 0.106, & f_3' &= -3.240 \text{ in.}, \end{aligned} \tag{7.30}$$

and then the first- and second-order kinematic coefficients of the mass centers,

$$\begin{aligned} f_{g2_x} &= -1.500 \text{ in.}, & f'_{g2_x} &= -2.598 \text{ in.} \\ f_{g2_y} &= 2.598 \text{ in.}, & f'_{g2_y} &= -1.500 \text{ in.} \end{aligned}$$

$$f_{g3_x} = -3.000 \text{ in.}, \quad f'_{g3_x} = -5.196 \text{ in.}$$
$$f_{g3_y} = 5.196 \text{ in.}, \quad f'_{g3_y} = -3.000 \text{ in.}$$
$$f_{g4_x} = -2.430 \text{ in.}, \quad f'_{g4_x} = -1.142 \text{ in.}$$
$$f_{g4_y} = 0.842 \text{ in.}, \quad f'_{g4_y} = -0.382 \text{ in.}$$

With the inertia properties, we compute A and B for each moving member.

Body 2:

$$A_2 = m_2(f_{g2_x}^2 + f_{g2_y}^2) + I_{g_2}h_2^2 = 39.000 \text{ in}^2 \cdot \text{lb}_\text{m}$$
$$B_2 = m_2(f_{g2_x}f'_{g2_x} + f_{g2_y}f'_{g2_y}) + I_{g_2}h_2h'_2 = 0 \text{ in}^2 \cdot \text{lb}_\text{m}$$

Body 3:

$$A_3 = m_3(f_{g3_x}^2 + f_{g3_y}^2) + I_{g_3}h_3^2 = 108.041 \text{ in}^2 \cdot \text{lb}_\text{m}$$
$$B_3 = m_3(f_{g3_x}f'_{g3_x} + f_{g3_y}f'_{g3_y}) + I_{g_3}h_3h'_3 = 0.015 \text{ in}^2 \cdot \text{lb}_\text{m}$$

Body 4:

$$A_4 = m_4(f_{g4_x}^2 + f_{g4_y}^2) + I_{g_4}h_4^2 = 69.225 \text{ in}^2 \cdot \text{lb}_\text{m}$$
$$B_4 = m_4(f_{g4_x}f'_{g4_x} + f_{g4_y}f'_{g4_y}) + I_{g_4}h_4h'_4 = 25.693 \text{ in}^2 \cdot \text{lb}_\text{m}$$

After this we can compute (ΣA) and (ΣB) for the machine:

$$\Sigma A = A_2 + A_3 + A_4 = 216.266 \text{ in}^2 \cdot \text{lb}_\text{m} = 0.0467 \text{ ft}^2 \cdot \text{slug}$$
$$\Sigma B = B_2 + B_3 + B_4 = 25.708 \text{ in}^2 \cdot \text{lb}_\text{m} = 0.0055 \text{ ft}^2 \cdot \text{slug} \tag{7.31}$$

Substituting h_3 from Equation (7.30), ΣA and ΣB from Equation (7.31), $\dot{\theta}_2$ and $\ddot{\theta}_2$ from Equation (7.23), and T_4 from Equation (7.24) into Equation (7.29) and solving for T_2 gives

$$T_2 = -150.0 \text{ ft} \cdot \text{lb}_\text{f}(0.286) + .047\text{ft}^2 \cdot \text{slug}(0) + .006\text{ft}^2 \cdot \text{slug}(-2\pi \text{ rad/s})^2 \underbrace{\left(\frac{\text{lb}_\text{f} \cdot \text{s}^2}{\text{ft} \cdot \text{slug}}\right)}_{g_c}$$

$$T_2 = -42.64 \text{ ft} \cdot \text{lb}_\text{f} \tag{7.32}$$

The negative sign means that T_2 acts in the cw direction with a magnitude of 42.64 ft $\cdot$ lb_f. This is the solution to the IDP. It agrees with the solution to this IDP that was found independently in Example 5.2, through force analysis and free body diagrams.

7.2.1 The Inverse Dynamics Problem Applied to Motor Selection

The motor selection problem is encountered when you are designing a machine. There are a great many calculations involved in this process, and it can realistically be done only with a computer program. We illustrate this motor selection process by continuing with Example 7.4. In this example, we determine the required power rating of a motor that would drive link 2 in the machine.

▶ EXAMPLE 7.5

Find the power rating of a motor to drive link 2 of the machine shown in Figures 7.9 and 7.10. The machine is a quick-return crank (link 2)–rocker (link 4) mechanism. If 2 is rotated continuously, link 4 oscillates between its limit positions. This was discussed in Example 4.4.

If 2 is driven cw at a constant speed, the ccw rotation of 4 occurs in less time than its cw rotation. During its slower cw rotation, link 4 is performing an indexing operation, and during its quick-return ccw, link 4 is returning to begin another indexing operation. As a result of this intermittent operation, there is a load $T_4 = 150.0\text{ft} \cdot \text{lb}_\text{f}$ that resists the rotation of link 4 only when it is indexing—that is, only when it is rotating cw. On its quick-return ccw rotation, there is no load torque on 4.

The machine should perform 60 indexing operations per minute, which means link 2 is to be driven at a constant speed of $\dot{\theta}_2 = 60\,\text{rpm} = (2\pi\,\text{rad/s})$ cw, while overcoming the directionally dependent load on link 4. Estimate the power required to drive this machine.

Solution

In a computer program, T_4 could be determined with an `if` statement in some code that might look like

```
T4=150
    if t4dot>0
    T4=0
    end
```

where `T4` and `t4dot` are the computer program variables that represent T_4 and $\dot{\theta}_4$, respectively. Let us plot T_2 as a function of θ_2, assuming that $\dot{\theta}_2$ has a constant value of 60 rpm ($\ddot{\theta}_2 = 0$).

For this we write a program that increments θ_2 through one revolution and, at each value of θ_2, computes T_2 from Equation (7.29), just as in Example 7.4. We plot each (θ_2, T_2) pair, generating Figure 7.11. The cusps in the plot are the points where body 4 changes its direction of rotation and the load torque T_4 is discontinuous. Using Simpson's Rule, the plotted function is numerically integrated to find the work done by the driving torque *per revolution of the crank*, W_{rev}:

$$W_{rev} = -157.1\ \frac{\text{ft} \cdot \text{lb}_\text{f}}{\text{rev}}.$$

If the desired speed of the crank is a constant $\dot{\theta}_2 = 60$ rpm cw (that is, –60 rpm), then the power required by the motor providing T_2, P_{req}, is

$$P_{req} = W_{rev}\,\dot{\theta}_2 = \left(-157.1\ \frac{\text{ft} \cdot \text{lb}_\text{f}}{\text{rev}}\right)\left(-60\ \frac{\text{rev}}{\text{min}}\right)\left(\frac{\text{min}}{60\ \text{s}}\right)\left(\frac{\text{hp} \cdot \text{s}}{550\ \text{ft} \cdot \text{lb}_\text{f}}\right) = 0.29\ \text{hp}.$$

So to drive this machine, we are looking at about a 0.3-hp motor that must then incorporate a transmission to adjust the motor's rated speed to match the desired constant speed $\dot{\theta}_2$.

We are ready to go to the manufacturers' catalogs and find a suitable motor—one that has this power rating. The speed at which the motor delivers this power is not likely to match our desired speed of 60 rpm. In that case we will need to incorporate a transmission to bring us the motor's power at the speed we desire. Before proceeding to select a motor, we should briefly discuss prime movers in general.

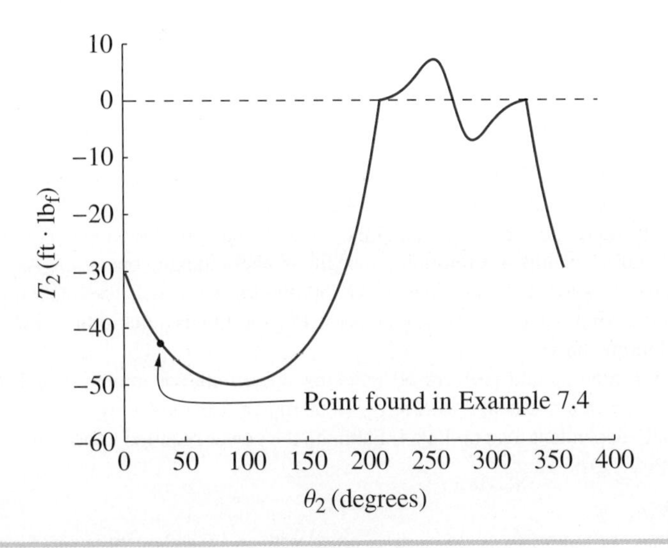

FIGURE 7.11 T_2 vs. θ_2 for $\dot{\theta}_2 = 2\pi$ rad/s and $\ddot{\theta}_2 = 0$

7.2.2 Prime Movers

A prime mover (a.k.a. a driver) is what supplies the power needed to drive the input. These are usually electric motors, but they can also be solenoids (electrical push-pull devices), internal combustion engines, and other power sources. These systems generally do not produce a constant driving torque or force. The driving torque or force usually varies with either the position of the input or the velocity of the input.

In the case of a solenoid, the driver force is dependent on position. As the plunger of a solenoid retracts, the solenoid force decreases quickly, in a nonlinear manner. In the case of an electric motor or internal combustion engine, the output torque is dependent on the output rotational speed. The torque–speed relationship is nonlinear and is described by a torque–speed curve. The torque–speed curve of an engine or motor is found using a piece of equipment known as a dynamometer. First a word about terminology.

Engines convert thermal energy of some form (steam, petrochemical combustion) into mechanical work. An internal combustion (IC) engine produces low torque at low speeds, which is why it is so easy to stall an engine if you let the clutch out too fast in an automobile that is at rest. An IC engine's torque increases with its speed, which is why you slightly rev the engine before letting out the clutch. This increase continues to a limit, after which the torque begins to decrease rapidly. Figure 7.12 shows a typical torque–speed curve for an IC engine.

Motors convert electrical energy (DC or AC) into mechanical work. A motor produces largest torques at low-speed conditions, which is why electric cars accelerate so fast when starting from rest. Motors produce zero torque at their maximum speed; see Figures 7.13 and 7.14 as examples.

It is beyond our scope to look at all the different types of prime movers. We will briefly discuss prime movers that are AC (alternating current) induction motors. These

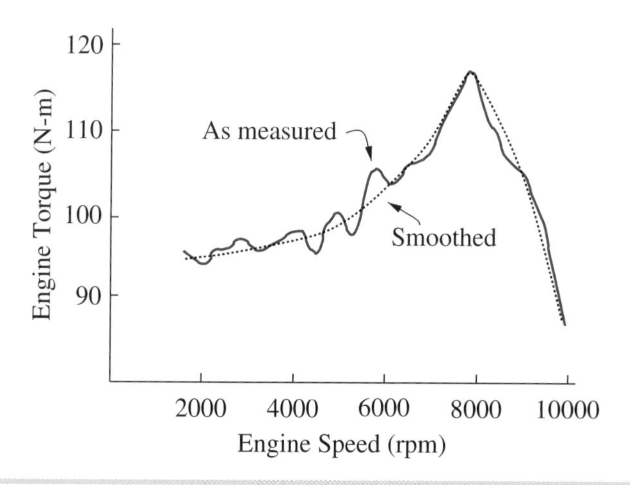

FIGURE 7.12 Torque–speed curve for an internal combustion engine

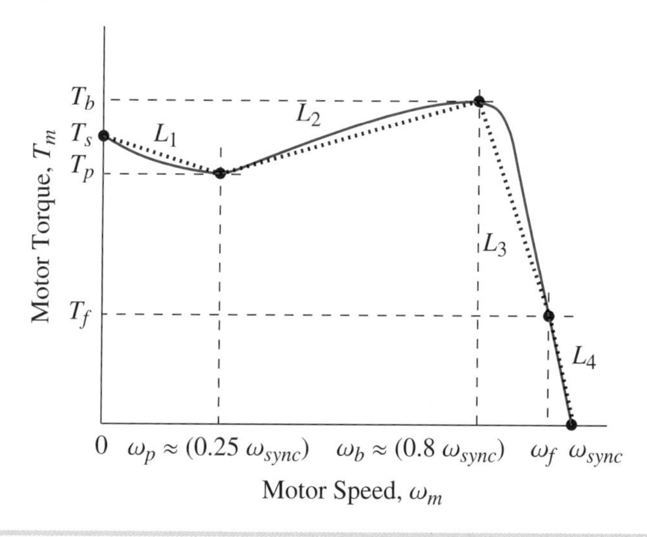

FIGURE 7.13 Torque–speed curve for a NEMA A, B, or C type motor design

are the most common type of prime mover found in machinery today. This is also just one type of electric motor—the most popular type.

7.2.3 AC Induction Motor Torque–Speed Curves

This is not intended to be a thorough discussion of AC (alternating current) induction motors. Standards for electric motors are set by NEMA (National Electrical Manufacturers Association). (For example, NEMA defines standard bolt hole patterns used to mount electric motors to frames.) An induction motor has a synchronous speed that is an integer multiple of the frequency of the electrical power supply. The value of this integer multiple depends on things such as the number of windings in the rotor and the

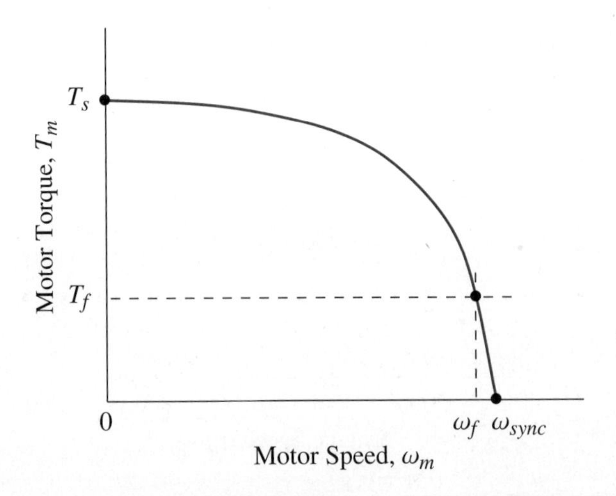

FIGURE 7.14 Torque–speed curve for a NEMA D type motor design

number of poles in the stator, i.e., the design of the motor. This is not so important to the mechanical engineer as simply knowing what this integer multiple is—that is, the synchronous speed of the motor. Line voltage is 60 Hz, so induction motors running off of line frequency have a synchronous speed in revolutions per minute that is an integer multiple of 60. Most often, the synchronous speed is 1800 rpm, but 3600 rpm and 1200 rpm exist. The synchronous speed ω_{sync}, is the no-load motor speed, and it is specified by the manufacturer.

Any amount of load causes the rotor of an induction motor to "slip" and lag behind the synchronous speed. The ideal of a no-load condition never exists, because even if there is no externally applied load, there are always a frictional and a drag load that resist the motor's rotation, however small they may be. So there is always motor slip, and this (among other factors) generates heat, which is typically dissipated by an internal fan driven directly by the rotor. When this heat generation is excessive and the fan cannot cool the motor sufficiently, the motor's temperature rises. It can rise to the point where the insulating materials inside the motor melt or burn, leading to a "burned out" motor.

Motor slip, s, is defined as

$$s = \frac{\omega_{sync} - \omega_m}{\omega_{sync}} = 1 - \frac{\omega_m}{\omega_{sync}}, \tag{7.33}$$

which is the percentage by which the motor speed, ω_m, lags its synchronous speed, ω_{sync}. When the motor is running at its synchronous speed, $s = 0$ (0% lag), and when the motor is not rotating (that is, when the motor is stalled), $s = 1$ (100% lag). (An aside about notation: Many technical documents reserve the symbol ω for angular velocity in rad/s and use the symbol n for angular velocity in revolutions per minute, rpm. Here there is no distinction. The symbol ω is used for angular velocity, and units are not a consideration, but don't be surprised to see this distinction made elsewhere.)

Figure 7.13 shows a representative torque–speed curve for a NEMA Design A, B, or C electric motor (see the NEMA Standards Publication MG10-2001 (R2007)). The torque–speed curve starts at $\omega_m = 0$ ($s = 1$), the stalled condition, and the motor is creating a great deal of torque called the stall torque, T_s. The motor will quickly overheat when stalled because the corresponding 100% slip generates a lot of heat, and since the rotor is stationary, the fan that cools the motor is not turning. As motor speed increases (i.e., s decreases), the motor torque decreases to a local minimum called the pull-up torque, T_p. After this point motor torque increases with motor speed to its maximum value, called the breakdown torque, T_b. After this the motor torque rapidly decreases to zero at the synchronous speed, ω_{sync}. Between T_b and zero motor torque at the synchronous speed, there is the full load operating torque, T_f, which occurs at the full load operating speed, ω_f. T_f and ω_f are always specified by the manufacturer. The motor slip corresponding to ω_f can be computed from Equation (7.33) using $\omega_m = \omega_f$. This is the maximum amount of slip the motor can tolerate while running continuously. The heat being generated by the motor at this amount of slip can be removed by the fan so that at steady state, the motor temperature is safely below the point where the motor burns up. If the fan could remove more heat, then ω_f could be reduced and thus T_f increased. But there is a limit to how much heat the fan can remove.

Thus the shape of the torque–speed curve is defined by the locations on the Y axis of:

T_s, the locked rotor torque, a.k.a. the stall torque
T_p, the pull-up torque
T_b, the breakdown torque
T_f, the full load operating torque

and by the location on the X axis of:

ω_f, the full load operating speed
ω_s, the synchronous speed.

Unfortunately, manufacturers do not specify the motor speed (or slip) at T_p and T_b. They do not expect you to operate in that region. For all practical purposes, the motor speeds at these points may be taken as about (0.25 ω_{sync}) and (0.80 ω_{sync}) respectively. With these assumptions, there are a number of methods to model the torque–speed curve in Figure 7.13.

One method is to fit T_m to a sixth-order polynomial in ω_m,

$$T_m = c_0 + c_1\omega_m + c_2\omega_m^2 + c_3\omega_m^3 + c_4\omega_m^4 + c_5\omega_m^5 + c_6\omega_m^6, \tag{7.34}$$

and solve for the seven constants c_0 to c_6 in order to satisfy the seven conditions.

at $\omega_m = 0$, $T_m = T_s$,
at $\omega_m = 0.25\ \omega_{sync}$, $T_m = T_p$,
at $\omega_m = 0.25\ \omega_{sync}$, $dT_m/ds = 0$,
at $\omega_m = 0.8\ \omega_{sync}$, $T_m = T_b$,
at $\omega_m = 0.8\ \omega_{sync}$, $dT_m/ds = 0$,
at $\omega_m = \omega_f$, $T_m = T_f$, and
at $\omega_m = \omega_{sync}$, $T_m = 0$.

Imposing these seven conditions on Equation (7.34) results in a system of seven linear equations in the seven unknown constants c_0 to c_6. Sometimes this polynomial gives very nice approximations of the torque–speed curve. However, in some cases it may result in unwanted maxima and minima. You will see this in the next example.

A simpler option—and one that always works—is to model the curve as four piecewise linear functions,which are the connected lines L_1 through L_4 in Figure 7.13. The linear segments begin and end at T_s, T_p, T_b, T_f, and at $T_m = 0$.

Figure 7.14 shows the representative curve for the NEMA Design D electric motor. This torque–speed curve can be approximated by

$$T_m = T_s\left(1 - \left(\frac{\omega_m}{\omega_{sync}}\right)^n\right) \quad n = 2, 3, \ldots$$

where larger values of n correspond to more rapid rates of decrease in the motor torque as ω_m approaches ω_{sync}. The value you choose for n will depend on what this decrease looks like for the motor in question. You should use the value of n that best matches the point (T_f, ω_f) for the motor in question.

To see which NEMA design type suits a particular application, refer to NEMA Standards Publication MG10-2001 (R2007). The NEMA Design D is a custom motor and for cost reasons is typically avoided.

The following example continues Example 7.5. There we saw that in order to drive the machine in Example 7.4 at a constant speed of 60 rpm cw and overcome a directionally dependent load torque $T_4 = 150\,\text{ft} \cdot \text{lb}_\text{f}$ that acts on link 4 (only when it is rotating cw), we need an approximately 0.3-hp driver.

▶ EXAMPLE 7.6

Select an AC induction motor to drive the machine in Example 7.4.

Solution

In looking through the online catalogs of various manufacturers of AC induction motors, we come upon the Baldor 1/3-hp motor, VL3501, with the following characteristics, (Google the phrase "`Products: AC Motor: VL3501: Baldor`" and then pick the first site and go to the link "`Performance Data`" and you will find the following numbers.)

$$\omega_s = 1800 \text{ rpm}$$
$$T_f = 1.0 \text{ ft} \cdot \text{lb}_\text{f}$$
$$\omega_f = 1730 \text{ rpm}$$
$$T_s = 3.6 \text{ ft} \cdot \text{lb}_\text{f}$$
$$T_p = 2.55 \text{ ft} \cdot \text{lb}_\text{f}$$
$$T_b = 2.95 \text{ ft} \cdot \text{lb}_\text{f}$$

The motor can be purchased with an attached transmission with a ratio of 30:1. It is the Baldor AC induction motor VL3501, in combination with the Baldor transmission GF3015AGA33. The ratio results in reducing the full load operating speed to $57.\overline{6}$ rpm, which is close enough to the desired 60 rpm. It is also very convenient to purchase the transmission and motor combination as a compatible set.

7.2.4 DC Motor Torque–Speed Curves

DC (direct current) motors are very common in battery-powered mechatronic systems. They have linear torque–speed curves; see Figure 7.15.

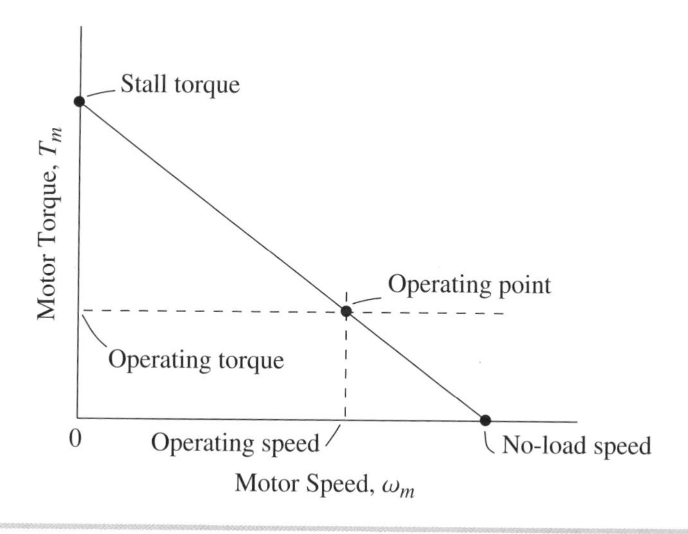

FIGURE 7.15 Torque–speed curve for DC motor

7.3 THE POWER EQUATION AND THE FORWARD DYNAMICS PROBLEM

Recall the power Equation (7.2),

$$P = \frac{dT}{dt} + \frac{dU}{dt} + P_f.$$

Substituting from Equation (7.6) for $\frac{dT}{dt}$ gives

$$P = (\Sigma A)\dot{S}_i\ddot{S}_i + (\Sigma B)\dot{S}_i^3 + \frac{dU}{dt} + P_f. \tag{7.35}$$

Consider now the forward dynamics problem (FDP). In the FDP the driving torque, the load torque, and the position and velocity of the input (S_i and $\dot{S}_i$) are known. The input acceleration ($\ddot{S}_i$) is unknown. In the FDP every term in the power Equation (7.35) is known, except for $\ddot{S}_i$, which can be computed. This is the solution to the FDP.

7.3.1 The Forward Dynamics Problem Applied to Dynamic Simulation

Equation (7.35) is a differential equation of motion. Solving this differential equation would give S_i as a function of time. The differential equation is nonlinear because it involves products of S_i and $\dot{S}_i$—and probably the sines and cosines of S_i as well. The methods you have learned for solving ordinary linear differential equations will not work except in the absolutely simplest cases. However, given a set of initial conditions at some time $t = t_0$, denoted as

$$S_i(t_0) = S_{i_0} \quad \text{and} \quad \dot{S}_i(t_0) = \dot{S}_{i_0},$$

we can solve this FDP and compute $\ddot{S}_i$ at $t = t_0$ from Equation (7.35) so that

$$\ddot{S}_i(t_0) = \ddot{S}_{i_0}$$

is also known. With this information we can generate a numerical solution for $S_i(t)$ using Euler's Method. Euler's Method assumes that during a sufficiently small time step Δt, the acceleration $\ddot{S}_i$ is constant so that during the time step, $\ddot{S}_i = \ddot{S}_{i_0}$. This allows us to calculate the corresponding change in position, ΔS_i, and change in velocity, $\Delta \dot{S}_i$, during the time step using constant acceleration equations. Specifically,

$$\Delta \dot{S}_i = \ddot{S}_{i_0} \Delta t \tag{7.36}$$

$$\Delta S_i = \dot{S}_{i_0} \Delta t + \frac{1}{2} \ddot{S}_{i_0} \Delta t^2, \tag{7.37}$$

from which the position and velocity at $t = t_0 + \Delta t$ are

$$\dot{S}_i = \dot{S}_{i_0} + \Delta \dot{S}_i \tag{7.38}$$

$$S_i = S_{i_0} + \Delta S_i. \tag{7.39}$$

So Euler's Method has generated the initial conditions for the beginning of a second time step. We repeat the process and generate the initial conditions at the start of a third time step... and so on. Euler's Method marches through time in small steps Δt, numerically generating S_i, $\dot{S}_i$, and $\ddot{S}_i$ as functions of time.

What is a reasonable Δt? The smaller Δt is, the more valid the assumption of constant acceleration is, but computation time increases as well. A general rule is that Δt should be small enough that ΔS_i is reasonable. If S_i were an angle θ_i, a reasonable value for $\Delta \theta_i$ would be 5° or less. If this value is exceeded, Δt needs to be reduced. If S_i were a vector length r_i, a reasonable value for Δr_i would depend on the size of the links and is typically no more than 2% of the longest link length in the mechanism.

Instead of checking the value of ΔS_i it is recommended that you set ΔS_i and calculate the corresponding Δt. This way you can be assured that your time step is not too large. If ΔS_i is known, the corresponding Δt comes from Equation (7.37):

$$\Delta t = \frac{-\dot{S}_{i_0} \pm \sqrt{\dot{S}_{i_0}^2 + 2\ddot{S}_{i_0} \Delta S_i}}{\ddot{S}_{i_0}}. \tag{7.40}$$

You should use the + or – sign in Equation (7.40), which will always give you a positive Δt. The value of $\dot{S}_{i_0}$ at the end of the step in S_i would come from Equation (7.36) and Equation (7.38), using the Δt calculated above.

▶ EXAMPLE 7.7

Develop a dynamic simulation of the machine in Example 7.4, which is being driven by the motor selected in Example 7.6. The goal of the dynamic simulation is to predict the crank speed as a function of time. The purpose of the dynamic simulation is threefold:

1. To verify that the selected motor is capable of driving the machine at the desired speed
2. To determine how long it takes the machine to reach steady state operation
3. To determine what the input speed fluctuation is at steady state—that is, to determine how "smoothly" the machine is running at steady state.

Solution

The power equation for this machine was derived in Equation (7.28). We repeat it here:

$$(T_2 + T_4\, h_3) = (\Sigma A)\ddot{\theta}_2 + (\Sigma B)\dot{\theta}_2^2.$$

Solving for $\ddot{\theta}_2$ yields

$$\ddot{\theta}_2 = \frac{T_2 + T_4\, h_3 - (\Sigma B)\dot{\theta}_2^2}{(\Sigma A)}. \tag{7.41}$$

We will use Equation (7.41) to develop a dynamic simulation of this machine using Euler's Method and following the procedure in Section 7.3.1. First we need to develop the torque–speed curve of the VL3501 based on our discussion in Section 7.2.3.

Motor Torque–Speed Curve:

Using the manufacturer's specifications for the motor listed in Example 7.6, and assuming that T_p and T_b occur at $(0.25\ \omega_{sync}) = 450$ rpm and $(0.8\ \omega_{sync}) = 1440$ rpm, respectively, gives a sixth-order polynomial for the torque–speed curve:

$$\begin{aligned} T_m = 3.6 - 0.0222\omega_m + 1.1369 \times 10^{-4}\omega_m^2 - 2.4026 \times 10^{-7}\omega_m^3 \\ +2.346 \times 10^{-10}\omega_m^4 - 1.0532 \times 10^{-13}\omega_m^5 + 1.7543 \times 10^{-17}\omega_m^6, \end{aligned}$$

where T_m has units $\text{ft} \cdot \text{lb}_\text{f}$ and ω_m is in rpm. This is plotted in Figure 7.16. The curve has two questionable minima that are indicated. This is not a good approximation. The piecewise linear approximation is shown in Figure 7.17. The piecewise linear approximation of the torque–speed curve is implemented in a computer program by a series of `if` statements that look like

```
Tm=Ts+wm*((Tp-Ts)/wp);
    if wm>wp
    Tm=Tp+(wm-wp)*((Tb-Tp)/(wb-wp));
end
    if wm>wb
    Tm=Tb+(wm-wb)*((Tf-Tb)/(wf-wb));
    end
        if wm>wf
        Tm=Tf+(wm-wf)*((-Tf)/(wsync-wf));
        end
```

where `Tm`, `Ts`, `Tp`, `Tb`, `Tf`, `wm`, `wp`, `wb`, `wf`, and `wsync` are the variables in the computer program that represent the motor characteristics T_m, T_s, T_p, T_b, T_f, ω_m, ω_p, ω_b, ω_f, and ω_{sync} of the motor selected in Example 7.6.

Having established the motor's torque–speed curve, we can proceed with developing the dynamic simulation. The simulation begins with the initial conditions. At $t = 0$,

$$\theta_2(0) = \theta_{2_0} = 0 \text{ rad} \quad \text{and} \quad \dot{\theta}_2(0) = \dot{\theta}_{2_0} = 0 \text{ rad/s},$$

and we need to compute $\ddot{\theta}_2(0) = \ddot{\theta}_{2_0}$ from Equation (7.41). Consider each term on the right hand side of Equation (7.41).

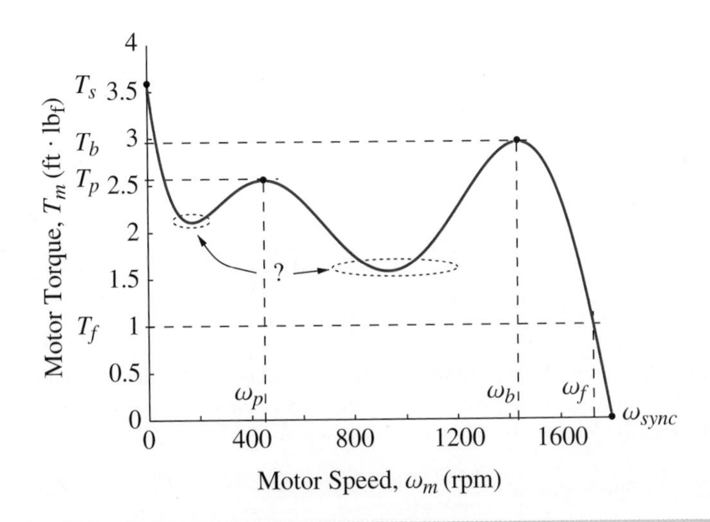

FIGURE 7.16 Sixth-order polynomial approximation of the torque–speed curve for the VL3501 motor

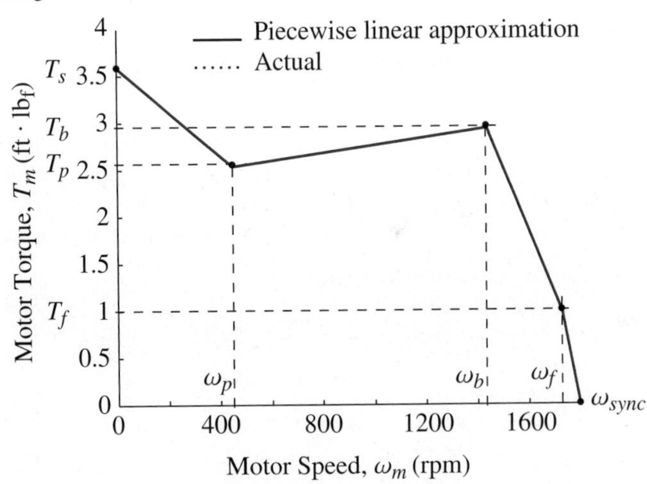

FIGURE 7.17 Piecewise linear approximation of torque–speed curve for the VL3501 motor

T_2

T_2 is coming from the motor. Since the transmission on the motor has a gear ratio of $R = 30$, we know that

$$T_2 = -30T_m, \tag{7.42}$$

where T_m is known from the motor speed ω_m, which is computed from $\dot{\theta}_2$:

$$\omega_m = -30\dot{\theta}_2. \tag{7.43}$$

The negative signs are included in Equation (7.42) and Equation (7.43) because T_2 and ω_2 will be clockwise (negative) and the torque–speed curve does not account for direction of rotation.

A word of warning. In the computer program, $\dot{\theta}_2$ will be in units of rad/s, which means that ω_m will have the same units. The torque–speed curve determined by the computer program logic above is based on ω_m in units of rev/min. It is good programming practice to have two different computer variables for ω_2, one that has units of rad/s (perhaps with a name like `omega2`) and the other that has units of rev/min (perhaps with a name like `w2`, on the previous page).

From the motor's piecewise linear torque–speed curve, Figure 7.17, at $\omega_m = 0$, $T_m = 3.60\ \text{ft}\cdot\text{lb}_\text{f}$. From Equation (7.42), $T_2 = -108.0\ \text{ft}\cdot\text{lb}_\text{f}$.

T_4

T_4 has a value of 150 ft · lb$_\text{f}$ when ω_4 is clockwise and 0 ft · lb$_\text{f}$ when ω_4 is counterclockwise. Example 7.5 describes a computer logic for computing T_4 that depends on the sign of ω_4. According to that logic, at this instant $\omega_4 = 0$ and so $T_4 = 150\ \text{ft}\cdot\text{lb}_\text{f}$.

ΣA and ΣB

ΣA and ΣB are calculated from the kinematics and the inertia properties of all the moving links. The rotational inertia of the motor and its kinematic coefficients must be included. We see from Equation (7.43) that $h_m = -30$ is a constant and so $h'_m = 0$. The motor mass center is stationary and has zero first- and second-order kinematic coefficients. This motor has a rotational inertia of $I_m = 70\ \text{in}^2\cdot\text{lb}_\text{m}$ (found by a phone call to vendor). In many cases, the vendor may not have this information but will be able to supply you with information about the motor's dimensions so that you can make a good estimate of the size of the motor's windings and then approximate I_m by assuming the windings are a solid core of copper.

Kinematic analysis shows that when $\theta_2 = 0$,

$$h_3 = 0.200, \quad (\Sigma A) = 21.4142\ \text{ft}^2\cdot\text{slugs} \ \text{ and } \ (\Sigma B) = 0.0172\ \text{ft}^2\cdot\text{slugs},$$

and we now have values for all the terms on the right hand side of Equation (7.41).

Computing $\ddot{\theta}_2$ at $t = 0$ from Equation (7.41) yields

$$\begin{aligned}\ddot{\theta}_{2_0} &= \frac{T_2 + T_4\,h_3 - (\Sigma B)\dot{\theta}_2^2}{(\Sigma A)} \\ &= \frac{\left(-108.0 + 150(.20)\right)\ \text{ft}\cdot\text{lb}_\text{f}\left(\frac{\text{ft}\cdot\text{slug}}{\text{lb}_\text{f}\cdot\text{s}^2}\right) - 0.0172\ \text{ft}^2\cdot\text{slugs}(0\ \text{rad/s})^2}{21.4142\ \text{ft}^2\cdot\text{slugs}} \\ &= -3.6424\ \text{rad/s}^2. \end{aligned} \tag{7.44}$$

We now use Euler's Method in Equation (7.37) and Equation (7.36) to integrate over a small time step Δt to find $\Delta\dot{\theta}_2$ and $\Delta\theta_2$. Using $\Delta t = 0.1$ s, we compute $\Delta\dot{\theta}_2$ and $\Delta\theta_2$ during the first time step:

$$\Delta\dot{\theta}_2 = \ddot{\theta}_{2_0}(\Delta t) = (-3.6424\ \text{rad/s}^2)(0.1\ \text{s}) = -0.364\ \text{rad/s}$$

$$\begin{aligned}\Delta\theta_2 &= \dot{\theta}_{2_0}(\Delta t) + \frac{1}{2}\ddot{\theta}_{2_0}(\Delta t)^2 = 0(0.1\ \text{s}) + (1/2)(-3.6424\ \text{rad/s}^2)(0.1\ \text{s})^2 \\ &= -0.0182\ \text{rad} = -1.043^\circ.\end{aligned}$$

We see that the change in θ_2 is less than 5° but not excessively small, so Δt is appropriate in size. The values of θ_2 and $\dot{\theta}_2$ at $t = t_0 + \Delta t = (0 + 0.1)\ \text{s} = 0.1\ \text{s}$ (the end of the first time step) are

$$\dot{\theta}_2 = \dot{\theta}_{2_0} + \Delta\dot{\theta}_2 = 0\ \text{rad/s} + (-0.364)\ \text{rad/s} = -0.364\ \text{rad/s}$$
$$\theta_2 = \theta_{2_0} + \Delta\theta_2 = 0\ \text{rad} + (-0.0182)\ \text{rad} = -0.0182\ \text{rad}.$$

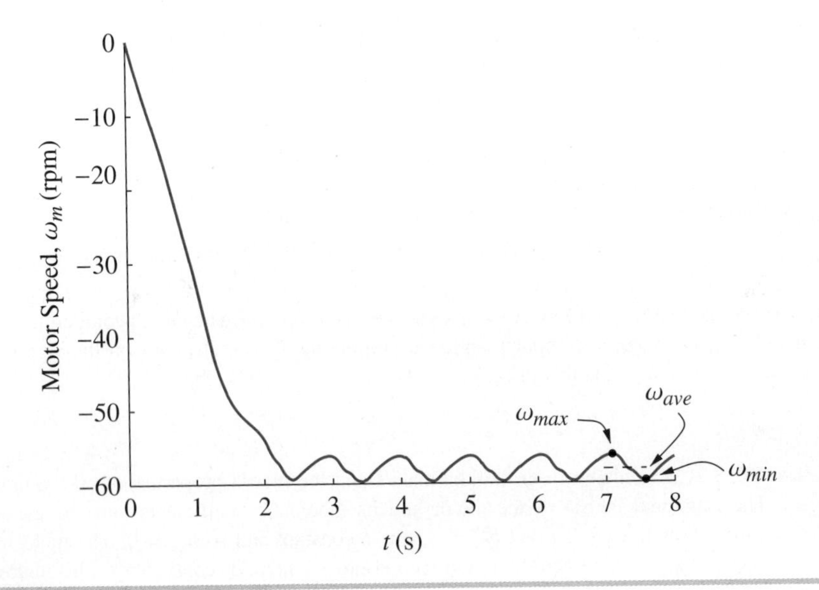

FIGURE 7.18 $\dot{\theta}_2$ as a function of time

We now repeat the process with the new initial conditions that at $t = 0.1$ s:

$$\theta_2(0.1 \text{ s}) = \theta_{2_o} = -0.0182 \text{ rad} \quad \text{and} \quad \dot{\theta}_2(0.1 \text{ s}) = \dot{\theta}_{2_o} = -0.364 \text{ rad/s}.$$

Over 80 time steps, the result shown in Figure 7.18 is developed by a computer program. Decreasing the time step to 0.01 s produces the same result in Figure 7.18. The dynamic simulation shows that steady state is reached at about $t = 2.4$ s and that at steady state the crank speed oscillates between a maximum of 59.85 rpm and a minimum of 56.29 rpm with an average speed of 58.07 rpm (these all correspond to negative crank velocities, because the rotation is clockwise).

Extreme fluctuation in crank speed at steady state is undesirable. It leads to higher bearing forces, higher cyclic stresses on the links, more loading on the motor, and more motor heating due to the repeated motor slip. The speed fluctuation here is not extreme, but we would still like to reduce it a bit to have a smoother steady state operation.

The specifications of the Baldor VL3501 give the full load operating speed as 1730 rpm. Accounting for the ratio 30:1, this corresponds to a crank speed of $57.\overline{6}$ rpm. The average speed from our dynamic simulation is slightly higher than this, as it should be to avoid overheating the motor. We conclude that in Example 7.5 we correctly determined that this machine needs a 0.3-hp motor and that the Baldor VL3501 selected in Example 7.6 was an appropriate choice from the point of view of performance. In reality there would be other factors to consider in making this choice, such as cost, availability, and compatibility with available electric power (single-phase, three-phase, 120 V, 220 V, etc.).

Including Joint Friction in a Dynamic Simulation

To include friction in the power equation, $\dot{S}_i$ must be known so that the rubbing velocities in Equation (7.20)–Equation (7.22) can be computed. This is not problematic because we have values for the initial conditions at the start of each time step, and furthermore, all the kinematic coefficients have been calculated. The friction forces and torques (a.k.a.

the friction terms) must also be known, and this *is* problematic. The friction terms can be found only as part of an IDP and a force analysis, as we saw in Example 6.1. But the dynamic simulation is part of a FDP. The solution to this conundrum is again the Method of Successive Iterations.

For the first time step, the friction terms are assumed to be equal to zero. For all subsequent time steps, the friction terms from the previous time step are used as the assumed values.

Using the assumed friction terms, we solve the FDP for $\ddot{S}_i$ from the power equation. Using this value of $\ddot{S}_i$, we solve the IDP for new values of the friction terms. These updated friction terms are then used in the FDP to solve for $\ddot{S}_i$ again, which is again used in the IDP to update the friction terms, . . . and so on. These iterations repeat until the friction terms and $\ddot{S}_i$ have converged. After convergence, Euler's Method takes a time step and integrates to find the initial value for the next time step. At the start of the next time step, the friction terms are assumed to be those that were converged upon at the end of the last time step, and the Method of Successive Iterations begins again for this new time step.

The end result is that the Method of Successive Iterations is being used in a loop within a loop, because it is used in the IDP to find the friction terms (see Example 6.1), and it is being used again in the FDP (as just described) to compute $\ddot{S}_i$ for the dynamic simulation.

Coefficient of Fluctuation

The steady state speed variation of a machine is quantified by the coefficient of fluctuation, C_f, which is defined as

$$C_f = \frac{\dot{S}_{i_{max}} - \dot{S}_{i_{min}}}{\dot{S}_{i_{ave}}}, \quad \text{where} \quad \dot{S}_{i_{ave}} = \frac{\dot{S}_{i_{max}} + \dot{S}_{i_{min}}}{2}$$

and $\dot{S}_{i_{max}}$, $\dot{S}_{i_{min}}$, and $\dot{S}_{i_{ave}}$ are the maximum, minimum, and average speed of the input at steady state. A general rule of thumb is that C_f should never exceed 10%, and for induction motors that are intended to run continuously, this should probably be closer to a maximum of 5%.

▶ EXAMPLE 7.8

Compute C_f for the machine whose dynamic simulation was developed in Example 7.7.

Solution

For this machine $S_i = \theta_2$, so

$$C_f = \frac{\dot{\theta}_{2_{max}} - \dot{\theta}_{2_{min}}}{\dot{\theta}_{2_{ave}}}.$$

From Figure 7.18, at steady state

$$\dot{\theta}_{2_{max}} = 59.85 \text{ rpm}, \quad \dot{\theta}_{2_{min}} = 56.29 \text{ rpm} \quad \text{and} \quad \dot{\theta}_{2_{ave}} = 58.07 \text{ rpm},$$

so

$$C_f = \frac{59.85 \text{ rpm} - 56.29 \text{ rpm}}{58.07 \text{ rpm}} = 6.13\%,$$

which is slightly high.

Flywheels

A flywheel is a rotating cylindrical mass usually attached to the input crank or directly to the output shaft of the motor, or gearmotor in the case of the Baldor VL 3501. A flywheel adds inertia to a system, and the inertia resists changes in motion. A flywheel also stores and releases kinetic energy. A flywheel is used to smooth out the operation of a machine and to reduce steady state speed fluctuation. A flywheel is balanced so that its mass center is stationary, and the kinematic coefficients of the flywheel's mass center are zero.

▶ EXAMPLE 7.9

Find the dimensions of a flywheel that would be attached to the output shaft of the Baldor VL 3501 of Example 7.7 and would reduce the machine's C_f to less than 3.5%.

Solution

We will use the same dynamic simulation program we developed for Example 7.7. The flywheel rotates directly with link 2, so the flywheel inertia, I_{fly}, is added directly to I_2 when computing A_2 and B_2. If we choose I_{fly} = 120,000 $\text{in}^2 \cdot \text{lb}_\text{m}$, the dynamic simulation produces Figure 7.19, where the maximum, minimum, and average crank speeds at steady state are 59.12, 57.16 and 58.17 rpm, which results in C_f = 3.50%. This value of I_{fly} was found by

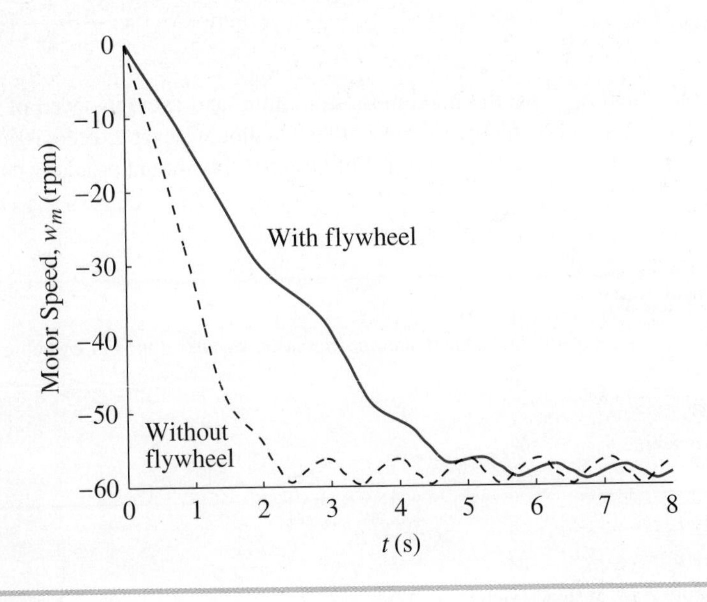

FIGURE 7.19 $\dot{\theta}_2$ as a function of time with and without a flywheel

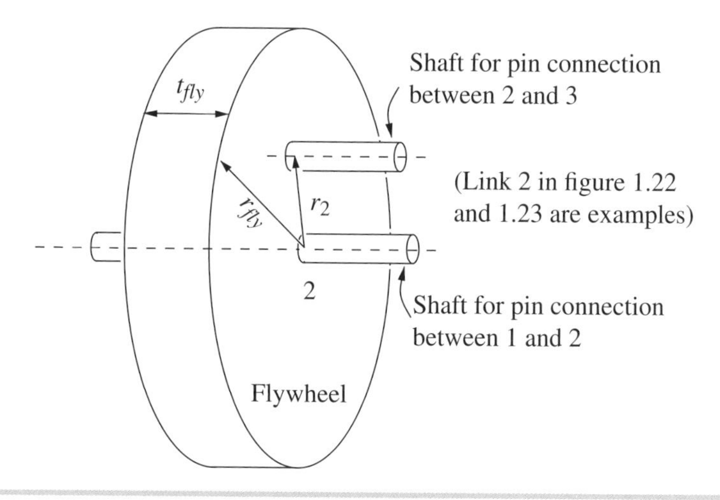

FIGURE 7.20 Link 2, incorporating a flywheel

trial and error. The time to reach steady state has increased to about 6 seconds. More motor heat will be generated during start-up, but this is well compensated for by less heat generation during continuous operation. Many times, the flywheel is incorporated into the driven link so that link 2 becomes a circular disk, and at a radius of $r_2 = 6$ in. there is a shaft for a pin joint connection between 2 and 3, as shown in Figure 7.20. How large is a flywheel with this rotational inertia and how much does it weigh? In terms of the dimensions shown in Figure 7.20 and the density of steel, ρ_{steel}, the rotational inertia is given by

$$I_{fly} = \frac{1}{2}\pi r_{fly}^4 t_{fly} \rho_{steel}.$$

Solving for t_{fly} gives

$$t_{fly} = \frac{2I_{fly}}{\pi r_{fly}^4 \rho_{steel}}. \tag{7.45}$$

$r_{fly} = 12$ in. is a reasonable number. The density of steel is 0.283 lb_m/in^3. Computing t_{fly} from Equation (7.45) results in $t_{fly} = 13.0$ in. The mass of this flywheel would be 833 lb_m and a weight of 833 lb_f. Pretty heavy. If the weight is a concern, you can increase the radius. If r_{fly} were 14 in. the thickness would be 7 in. and the weight would be reduced to 610 lb_f. If the flywheel were running at a higher speed, or if the radius got fairly large, you would be advised to check the stresses in the flywheel due to the centripetal accelerations.

7.4 MECHANICAL ADVANTAGE

In machines, power is supplied by what is known as a *driver*. The driver provides either a torque, T_{driver}, as in a motor, or a force, F_{driver}, as in a solenoid or a hydraulic cylinder. Generically these will all be referred to as the driver, D.

In machines, power is taken by what is known as a *load*. The load may also be in the form of a torque, T_{load}, which resists the rotation of a shaft connected to one of the links, or in the form of a force, F_{load}, which resists the motion of a point on one of the mechanism's links. Generically these will all be referred to as the load, L.

Mechanical advantage, M_a, is defined as

$$M_a = \frac{L}{D}; \quad \text{that is,} \quad L = M_a\, D. \tag{7.46}$$

The right side of this equation shows that M_a is the force or torque amplification achieved by a tool, mechanical device, or machine. M_a is synonymous with leverage and is also an indicator of the effectiveness of the driver in overcoming the load. As M_a increases, the driver has more leverage in overcoming the load. As you will see, infinite M_a corresponds to a limit position in the movement of the link associated with the load, and zero M_a corresponds to a dead position of the link associated with the driver.

M_a can have different units, depending on whether D and L are torques or forces. M_a is a function of the knowns and the unknowns in the position equations. Kinematic aspects of M_a involve the unknowns of the position equations (S_i and all of the S_k), whereas dimensional aspects of M_a involve the knowns from the position equationa, which are all dimensions. In mechanisms that consist entirely of rolling contacts (geartrains), M_a is synonymous with the torque ratio n_t. In geartrains, $M_a = n_t$ and both M_a and n_t are constant and independent of S_i.

When determining M_a, we seek a relationship between D and L so we can solve for their ratio. This can be done by performing a force analysis and deriving a system of equations involving the load, driver, and all forces of interaction. However, it is simpler to use work/energy principles that directly relate the load and driver. In other words, it is simpler to use the power equation Equation (7.2). M_a neglects effects of kinetic or potential energy and losses due to friction, and in this case Equation (7.2) reduces to

$$P = 0, \tag{7.47}$$

where P is the instantaneous power. To formulate P, follow the several preliminary steps listed below. It is wise to follow these steps precisely.

Procedure for Determining the Mechanical Advantage of a Mechanism

Step 1: Solve the position problem, and determine the basic first-order kinematic coefficients of the mechanism.

Step 2: Express the first-order kinematic coefficients of the points of application of any driving or load forces in terms of the basic first-order kinematic coefficients.

Step 3: Express the angular velocity of any body that has an applied driving torque or load torque in terms of the basic first-order kinematic coefficients and the input velocity $\dot{S}_i$,

Step 4: Express the linear velocity of the points of application of any driver or load forces in terms of the first-order kinematic coefficients of their point of application and the input velocity $\dot{S}_i$.

Step 5: Write an expression for the total power, P, and substitute in for the first-order kinematic coefficients.

Step 6: Finally, formulate P, equate it to zero and solve for M_a.

The following examples illustrate this procedure. The first two examples are very simple ones where you probably know the answer. Their purpose is to demonstrate how to apply the VLM and the kinematic coefficients to determining mechanical advantage.

▶ EXAMPLE 7.10

Figure 7.21 shows a wrench on the head of a bolt. The bolt is seated, and the wrench is trying to break it loose (the bolt has a right hand thread). The wrench has a driving force applied to it, and friction in the bolt's threads and between the bolt head and the surface provide a resisting torque load. Apply steps 1–6) in the procedure from Section 7.4 to determine the mechanical advantage. You probably know or sense that $M_a = r_2$.

Solution

Step 1: The VLE is meaningless. It states that

$$\bar{r}_2 - \bar{r}_2 = \bar{0},$$

which is trivial and cannot be counted as an equation. The knowns are r_2 and the unknowns are θ_2. The system has one unknown and no equations, so it has one dof. Take θ_2 as S_i. There are no kinematic coefficients because this system has no S_k.

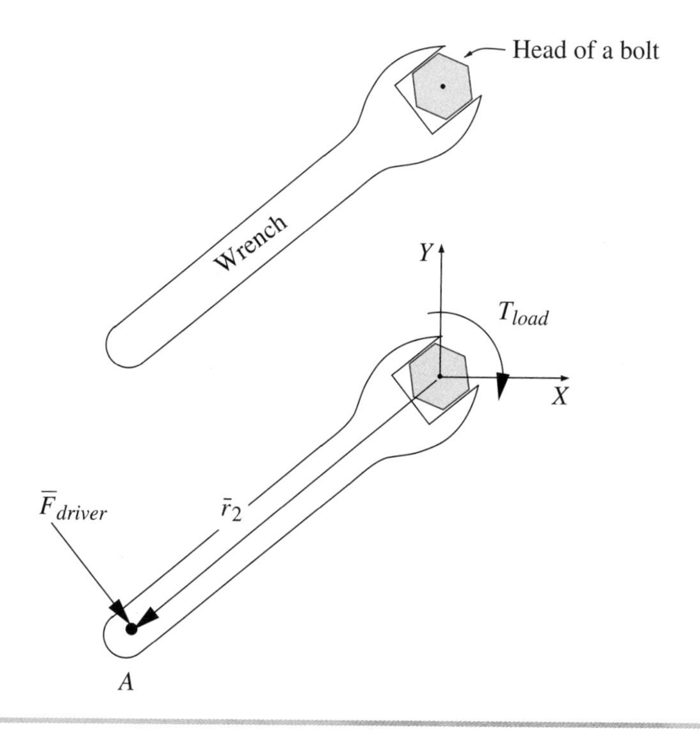

FIGURE 7.21 Mechanical advantage of a wrench

Step 2: The driving force is applied at point A.

$$\bar{r}_a = \bar{r}_2 \quad \rightarrow \quad \begin{array}{l} r_{a_x} = r_2\cos\theta_2 \\ r_{a_y} = r_2\sin\theta_2 \end{array} \quad \rightarrow \quad \begin{array}{l} f_{a_x} = -r_2\sin\theta_2 \\ f_{a_y} = r_2\cos\theta_2 \end{array}. \tag{7.48}$$

Step 3: The load torque is applied to 2, and

$$\overline{\omega}_2 = \begin{bmatrix} 0 \\ 0 \\ 1 \end{bmatrix} \dot{\theta}_2. \tag{7.49}$$

Step 4: The velocity of point A is given by

$$\bar{v}_a = \begin{bmatrix} v_{a_x} \\ v_{a_y} \\ 0 \end{bmatrix} = \begin{bmatrix} f_{a_x} \\ f_{a_y} \\ 0 \end{bmatrix} \dot{\theta}_2. \tag{7.50}$$

Step 5: The power, P, is

$$P = \overline{F}_{driver} \cdot \bar{v}_a + \overline{T}_{load} \cdot \overline{\omega}_2, \tag{7.51}$$

where

$$\overline{F}_{driver} = F_{driver} \begin{bmatrix} \cos(\theta_2 + \frac{\pi}{2}) \\ \sin(\theta_2 + \frac{\pi}{2}) \\ 0 \end{bmatrix} = F_{driver} \begin{bmatrix} -\sin\theta_2 \\ \cos\theta_2 \\ 0 \end{bmatrix} \tag{7.52}$$

and

$$\overline{T}_{load} = T_{load} \begin{bmatrix} 0 \\ 0 \\ -1 \end{bmatrix}. \tag{7.53}$$

Substituting Equations (7.49), (7.50), (7.52), and (7.53) into Equation (7.51) gives

$$P = \left[F_{driver}(-\sin\theta_2 f_{a_x} + \cos\theta_2 f_{a_y}) - T_{load}\right]\dot{\theta}_2.$$

Substituting Equation (7.48) for the kinematic coefficients gives

$$P = (-F_{driver} r_2 + T_{load})\dot{\theta}_2. \tag{7.54}$$

Step 6: Equate Equation (7.54) to zero and solve for M_a:

$$M_a = \frac{T_{load}}{F_{driver}} = r_2.$$

This a long-winded way of getting an obvious answer. However, it demonstrates the procedure and verifies its correctness. M_a is a constant and has units of length. For the wrench there is only a dimensional aspect to M_a, namely r_2. In mechanisms, the kinematic aspects of M_a allow for both zero and/or infinite M_a.

▶ EXAMPLE 7.11

Figure 7.22 shows a simple machine, the lever. Apply steps 1–6 in the procedure from Section 7.4 to determine the mechanical advantage. You probably know that $M_a = (r_2/r_3)$. You studied this simple machine in grade school.

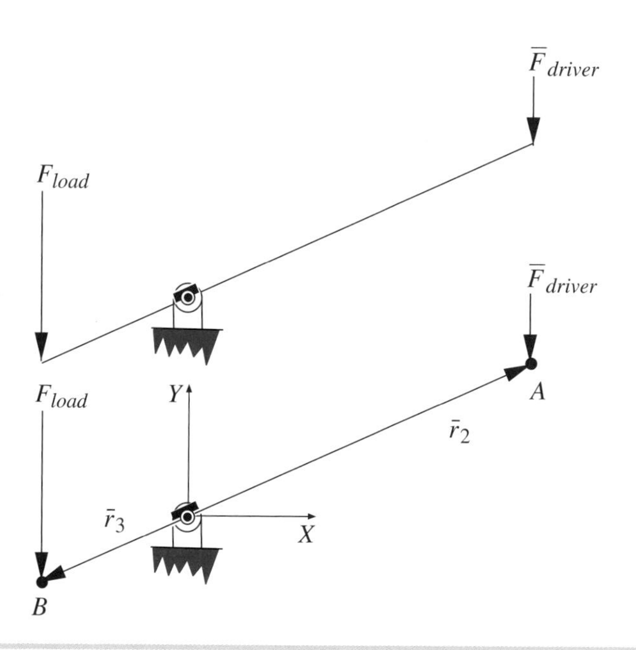

FIGURE 7.22 Mechanical advantage of a lever

Solution

Step 1: The vector loop here is also trivial: $\overline{r}_2 - \overline{r}_2 + \overline{r}_3 - \overline{r}_3 = \overline{0}$. There is, however, a meaningful geometric constraint,

$$\theta_2 - \theta_3 + \pi = 0, \tag{7.55}$$

and it is the only position equation. The knowns are r_2 and r_3. The unknowns are θ_2 and θ_3. There is one equation and there are two unknowns, so this is a one-degree-of-freedom system. Take θ_2 as S_i and differentiate Equation (7.55) to get the kinematic coefficient h_3:

$$1 - h_3 = 0 \quad \rightarrow \quad h_3 = 1. \tag{7.56}$$

Step 2: A driving force is applied at point A; its kinematic coefficients are

$$\overline{r}_a = \overline{r}_2 \quad \rightarrow \quad \begin{matrix} r_{a_x} = r_2 \cos\theta_2 \\ r_{a_y} = r_2 \sin\theta_2 \end{matrix} \quad \rightarrow \quad \begin{matrix} f_{a_x} = -r_2 \sin\theta_2 \\ f_{a_y} = r_2 \cos\theta_2 \end{matrix}. \tag{7.57}$$

The load force is applied at point B; its kinematic coefficients are

$$\overline{r}_b = \overline{r}_3 \quad \rightarrow \quad \begin{matrix} r_{b_x} = r_3 \cos\theta_3 \\ r_{b_y} = r_3 \sin\theta_3 \end{matrix} \quad \rightarrow \quad \begin{matrix} f_{b_x} = -r_3 \sin\theta_3 h_3 \\ f_{b_y} = r_3 \cos\theta_3 h_3 \end{matrix}. \tag{7.58}$$

Step 3: There are no driving or load torques, so this step does not apply here.

Step 4: The velocity of point A is given by

$$\overline{v}_a = \begin{bmatrix} v_{a_x} \\ v_{a_y} \\ 0 \end{bmatrix} = \begin{bmatrix} f_{a_x} \\ f_{a_y} \\ 0 \end{bmatrix} \dot{\theta}_2, \tag{7.59}$$

and the velocity of point B is given by

$$\overline{v}_b = \begin{bmatrix} v_{b_x} \\ v_{b_y} \\ 0 \end{bmatrix} = \begin{bmatrix} f_{b_x} \\ f_{b_y} \\ 0 \end{bmatrix} \dot{\theta}_2. \tag{7.60}$$

Step 5: The total power is given by

$$P = \overline{F}_{driver} \cdot \overline{v}_a + \overline{F}_{load} \cdot \overline{v}_b, \tag{7.61}$$

where

$$\overline{F}_{driver} = F_{driver} \begin{bmatrix} 0 \\ -1 \\ 0 \end{bmatrix} \quad \text{and} \quad \overline{F}_{load} = F_{load} \begin{bmatrix} 0 \\ -1 \\ 0 \end{bmatrix}. \tag{7.62}$$

Substituting Equations (7.59), (7.60), and (7.62) into Equation (7.61) gives

$$P = [-F_{driver} f_{a_y} - F_{load} f_{b_y}]\dot{\theta}_2. \tag{7.63}$$

Substitute Equation (7.56), Equation (7.57), and Equation (7.58) for the kinematic coefficients:

$$P = F_{driver} r_2 \cos\theta_2 + F_{load} r_3 \cos\theta_3$$

Equate Equation (7.63) to zero and solve for M_a:

$$M_a = \frac{F_{load}}{F_{driver}} = -\frac{r_2 \cos\theta_2}{r_3 \cos\theta_3}.$$

Substituting Equation (7.55) for θ_3 above gives

$$M_a = -\frac{r_2 \cos\theta_2}{r_3 \cos(\theta_2 + \pi)} = \frac{r_2}{r_3},$$

as expected. M_a has only dimensional aspects r_2 and r_3 and is constant and dimensionless.

► EXAMPLE 7.12

Figure 7.23 shows an inline crank-slider and a vector loop for the mechanism. There is a driving torque applied to link 2 and a load force acting at point C on the slider. The vector loop equation is $\overline{r}_2 + \overline{r}_3 - \overline{r}_4 = \overline{0}$. The simplified X and Y components are

$$r_2 \cos\theta_2 + r_3 \cos\theta_3 - r_4 = 0$$
$$r_2 \sin\theta_2 + r_3 \sin\theta_3 = 0.$$

There are no geometric constraints or rolling contact equations. The unknowns in these position equations are θ_2, θ_3, and r_4, so there is one degree of freedom. Take θ_2 as the input variable S_i, although for the purpose of finding M_a, any of the variable unknowns could have been chosen as S_i. Prove this to yourself by later reworking this problem, using r_4 as S_i. Apply the procedure outlined in Section 7.4 to do the following:

1. Derive an expression for the mechanical advantage of the mechanism.
2. Determine when the mechanical advantage is either inifinite or at a maximum.
3. Determine when the mechanical advantage is either zero or at a minimum.

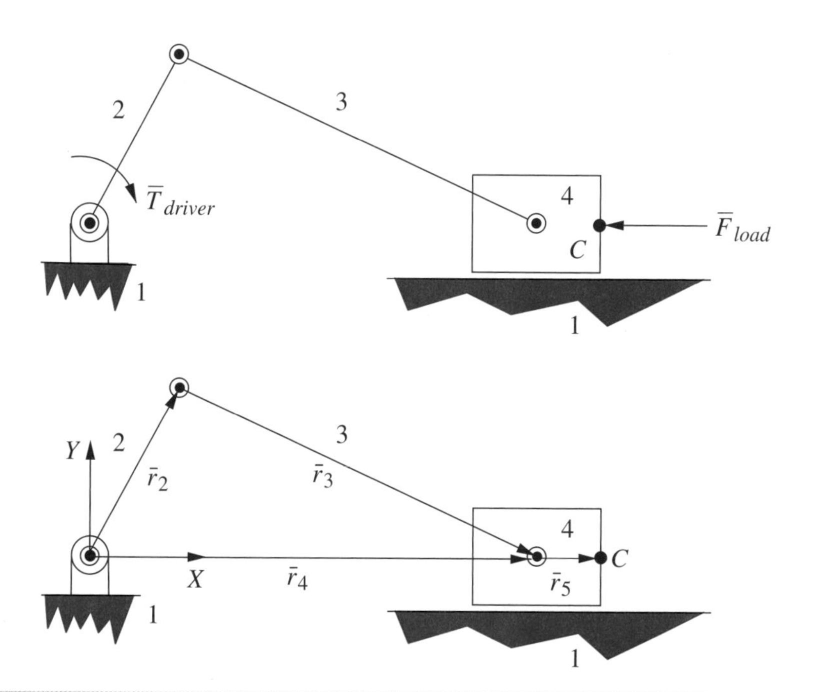

FIGURE 7.23 Mechanical advantage of a crank-slider

Solution

Step 1: Differentiating the position equations with respect to θ_2 and solving for the basic kinematic coefficients gives

$$h_3 = \frac{-r_2\cos\theta_2}{r_3\cos\theta_3} \quad \text{and} \quad f_4 = \frac{r_2\sin(\theta_3 - \theta_2)}{\cos\theta_3}. \tag{7.64}$$

Step 2: Point C is the point of application of the load force. It is located by the vector loop,

$$\bar{r}_c = \bar{r}_4 + \bar{r}_5 \;\rightarrow\; \begin{matrix} r_{c_x} = r_4 + r_5 \\ r_{c_y} = 0 \end{matrix} \;\rightarrow\; \begin{matrix} f_{c_x} = f_4 \\ f_{c_y} = 0 \end{matrix}. \tag{7.65}$$

Step 3: Body 2 has the driving torque applied to it. Its angular velocity is

$$\overline{\omega}_2 = \begin{bmatrix} 0 \\ 0 \\ 1 \end{bmatrix} \dot{\theta}_2. \tag{7.66}$$

Step 4: The velocity of C is

$$\bar{v}_c = \begin{bmatrix} f_{c_x} \\ f_{c_y} \\ o \end{bmatrix} \dot{\theta}_2. \tag{7.67}$$

Step 5: The total power, P, is

$$P = \overline{T}_{driver} \cdot \overline{\omega}_2 + \overline{F}_{load} \cdot \overline{v}_c = 0, \tag{7.68}$$

where

$$\overline{T}_{driver} = T_{driver} \begin{bmatrix} 0 \\ 0 \\ -1 \end{bmatrix} \text{ and } \overline{F}_{load} = F_{load} \begin{bmatrix} -1 \\ 0 \\ 0 \end{bmatrix}. \tag{7.69}$$

Substituting Equations (7.64), (7.65), (7.66), (7.67), and (7.69) into Equation (7.68) gives

$$P = \left(-T_{driver} - F_{load} \frac{r_2 \sin(\theta_3 - \theta_2)}{\cos\theta_3} \right) \dot{\theta}_2.$$

Step 6: Equating P to zero and solving for M_a gives

$$M_a = \frac{F_{load}}{F_{driver}} = -\frac{\cos\theta_3}{r_2 \sin(\theta_3 - \theta_2)}. \tag{7.70}$$

The dimensional aspect of M_a is the term $1/r_2$. To maximize M_a you want r_2 as small as possible. There will be a limit to how small a value of r_2 you can accept, because reducing r_2 reduces the stroke of the slider. The kinematic aspect of M_a is the term $\cos\theta_3/r_2\sin(\theta_3 - \theta_2)$. Due to this kinematc aspect, the mechanical advantage of this mechanism changes with its configuration. M_a is maximum and equal to infinity when the denominator of Equation (7.70) vanishes, i.e., when $\theta_3 - \theta_2 = (0, \pi)$, which means that links 2 and 3 are inline. These are the the limit positions of 4 shown in Figure 7.24. Mechanisms have infinite mechanical advantage in the limit positions of the output link. This is the basis of clamping mechanisms.

M_a is zero when the numerator of Equation (7.70) is zero—that is, when $\theta_3 = \pm\frac{\pi}{2}$. Mechanisms have zero M_a at the dead positions of the input. In this case the dead positions of the crank are shown in Figure 7.25 for $\theta_3 = \frac{\pi}{2}$.

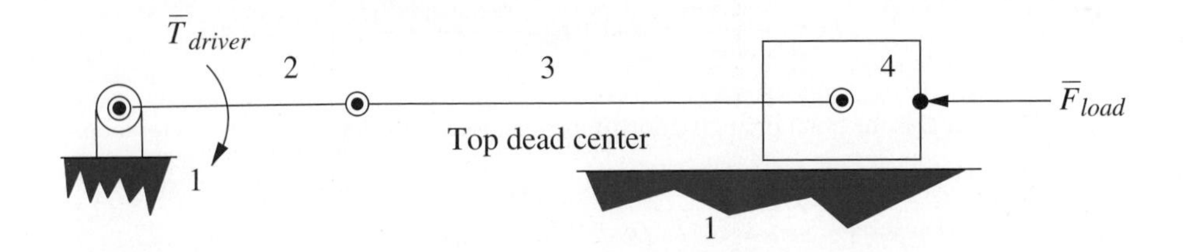

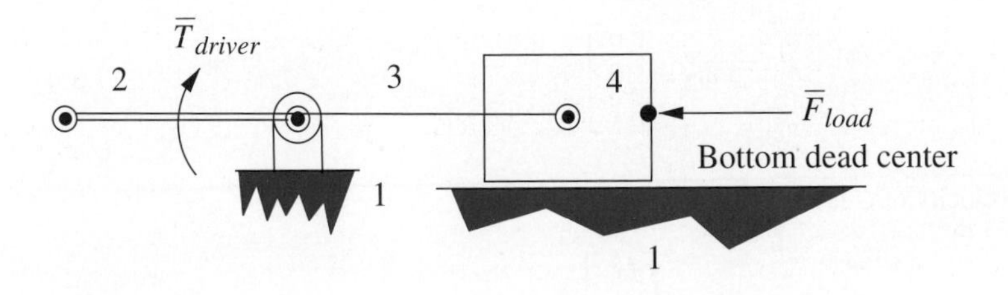

FIGURE 7.24 A crank-slider with infinite mechanical advantage

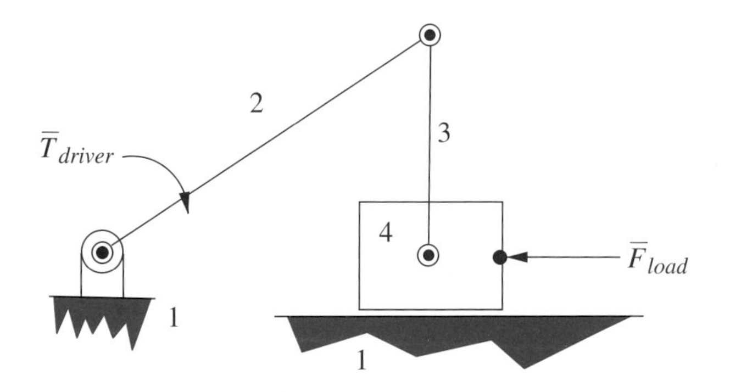

FIGURE 7.25 A crank-slider with zero mechanical advantage

▶ EXAMPLE 7.13

Figure 7.26 shows the side view of a mechanism that might be found on many garage walls near the family recycling station. Its purpose is to crush aluminum beverage cans. Body 2 has a handle on its end at point *D*. The handle is pulled down by a person, and this causes an aluminum can placed under 3 to be crushed. Body 3 slides vertically in a square slot in the housing of the tool, 1. The housing is fastened to the garage wall. Body 2 is pin-jointed to 3 and has a slot at its end that engages a pin on 1.

An expression for the mechanical advantage of the system should be derived, and from this expression the configuration of the mechanism at which $M_a = \infty$, or is a maximum, should be determined. The system should be designed so that this configuration corresponds to the onset of can crushing, since the greatest M_a is needed to initiate buckling of the can.

The crushing force is F_{load} and the force applied by the person is F_{driver}. Both are assumed vertical. Figure 7.27 shows an appropriate vector loop. The fixed coordinate system

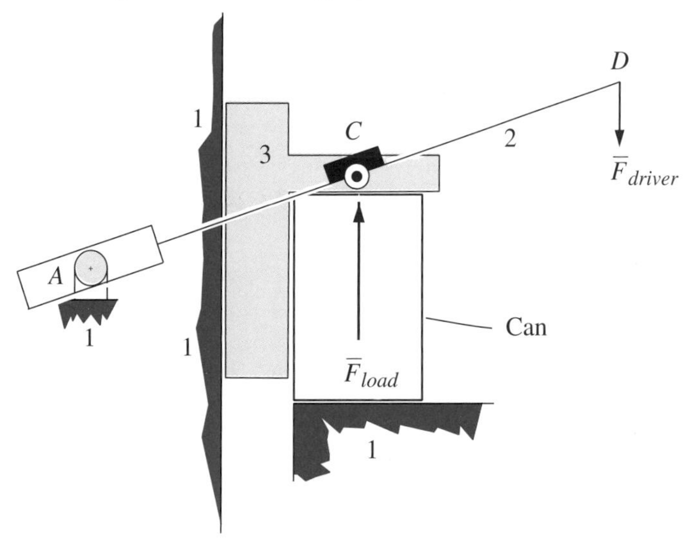

FIGURE 7.26 A can-crushing mechanism

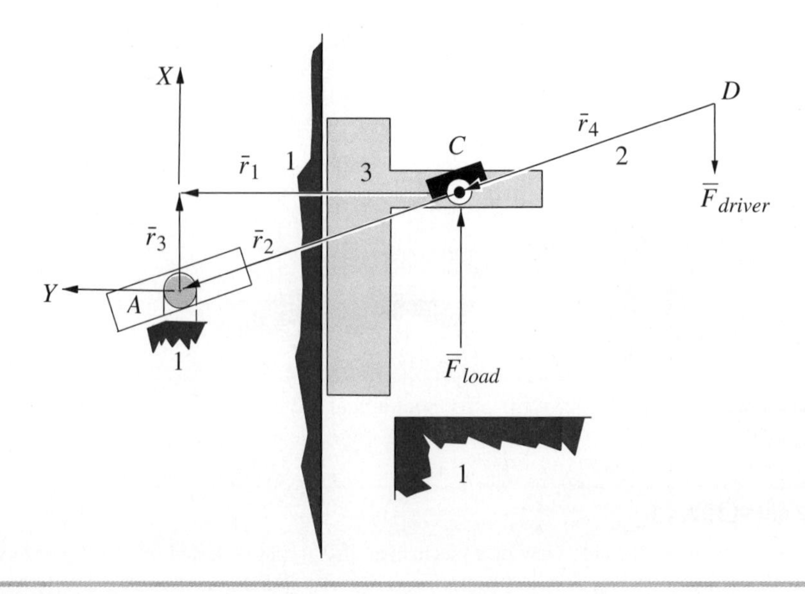

FIGURE 7.27 Vector loop for the can-crushing mechanism

has its origin at A. Noting that $\theta_1 = \frac{\pi}{2}$ and $\theta_3 = 0$, the vector loop equation, $\bar{r}_1 - \bar{r}_3 - \bar{r}_2 = \bar{0}$, has the X and Y components

$$-r_3 - r_2\cos\theta_2 = 0 \tag{7.71}$$

$$r_1 - r_2\sin\theta_2 = 0. \tag{7.72}$$

The unknowns are r_3, r_2, and θ_2, of which the input variable is taken to be r_2 ($S_i = r_2$) and the output variables are taken to be r_3 and θ_2. Follow the procedure outlined in Section 7.4 and:

1. Derive an expression for the mechanical advantage of the mechanism.
2. Determine when $M_a = \infty$ or is a maximum.

Solution

Step 1: Differentiate the position equations, and solve for the basic first-order kinematic coefficients,

$$f_3 = \frac{-1}{\cos\theta_2} \quad \text{and} \quad h_2 = \frac{-\sin\theta_2}{r_2\cos\theta_2}. \tag{7.73}$$

Step 2: Find the kinematic coefficients of points C and D since they are points of application for the load and driver forces, respectively:

$$\begin{aligned}
&\bar{r}_c = -\bar{r}_2 \rightarrow \begin{matrix} r_{c_x} = -r_2\cos\theta_2 \\ r_{c_y} = -r_2\sin\theta_2 \end{matrix} \rightarrow \begin{matrix} f_{c_x} = -\cos\theta_2 + r_2\sin\theta_2 h_2 \\ f_{c_y} = -\sin\theta_2 - r_2\cos\theta_2 h_2 \end{matrix} \\
&\bar{r}_d = -\bar{r}_2 - \bar{r}_4 \rightarrow \begin{matrix} r_{d_x} = -(r_2 + r_4)\cos\theta_2 \\ r_{d_y} = -(r_2 + r_4)\sin\theta_2 \end{matrix} \rightarrow \begin{matrix} f_{d_x} = -\cos\theta_2 + (r_2 + r_4)\sin\theta_2 h_2 \\ f_{d_y} = -\sin\theta_2 - (r_2 + r_4)\cos\theta_2 h_2 \end{matrix}.
\end{aligned} \tag{7.74}$$

Step 3: Since there are no load or driver torques, this step does not apply.

Step 4: The velocity of C and D are given by

$$\bar{v}_c = \begin{bmatrix} f_{c_x} \\ f_{c_y} \\ 0 \end{bmatrix} \dot{r}_2, \quad \bar{v}_d = \begin{bmatrix} f_{d_x} \\ f_{d_y} \\ 0 \end{bmatrix} \dot{r}_2. \tag{7.75}$$

Step 5: The total power is

$$P = \bar{F}_{driver} \cdot \bar{v}_d + \bar{F}_{load} \cdot \bar{v}_c, \tag{7.76}$$

where

$$\bar{F}_{driver} = F_{driver} \begin{bmatrix} -1 \\ 0 \\ 0 \end{bmatrix}, \quad \bar{F}_{load} = F_{load} \begin{bmatrix} 1 \\ 0 \\ 0 \end{bmatrix}. \tag{7.77}$$

Substituting Equations (7.73), (7.74), (7.75), and (7.77) into Equation (7.76) gives

$$P = \left[-(r_2 + r_4 \sin^2\theta_2)\, F_{driver} - r_2\, F_{load} \right] \left(\frac{\dot{r}_2}{r_2 \cos\theta_2} \right). \tag{7.78}$$

Step 6: Equating P to zero and solving for M_a yields

$$M_a = \frac{F_{load}}{F_{driver}} = 1 + \frac{r_4}{r_2} \sin^2\theta_2. \tag{7.79}$$

M_a is a function of configuration since M_a depends on r_2 and θ_2. Sometimes, when the mechanism is complex, it is not possible to obtain an expression for M_a in terms of only one S_k or in terms of S_i alone. In those cases you need to plot M_a as a function of S_i and note the maximum and minimum values from the plot.

In this example, however, it is possible. Using the Y component of the VLE in Equation (7.72), $\sin\theta_2$ in Equation (7.79) can be replaced by r_1/r_2, which gives M_a as a function of the input variable only:

$$M_a = 1 + \frac{r_4 r_1^2}{r_2^3}. \tag{7.80}$$

Equation (7.80) shows that M_a is maximum when r_2 is minimum, which occurs when the handle is horizontal.

Alternatively, using the Y component of the VLE in Equation (7.72), r_2 in Equation (7.79) can be replaced by $(r_1/\sin\theta_2)$, which yields

$$M_a = 1 + \frac{r_4}{r_1} \sin^3\theta_2, \tag{7.81}$$

giving M_a as a function of θ_2 alone. From Equation (7.81) you see that M_a is maximum when $\theta_2 = \pm\frac{\pi}{2}$ (i.e., the handle is horizontal and r_2 is minimum, which is the same conclusion drawn from Equation (7.80).

There is a dimensional aspect to this M_a. To maximize M_a, the ratio of dimensions r_4/r_1 should be maximized. This is more evident from Equation (7.81), but the same conclusion can be drawn from Equation (7.80) if it is rewritten as

$$M_a = 1 + \frac{r_4}{r_1} \frac{r_1^3}{r_2},$$

which indicates that not only does one want to maximize the ratio r_4/r_1 but one should maximize r_1 as well (which just means making the thing bigger). There is a limit to how much either of these can be maximized, because the device would protrude too far from the wall.

The conclusion is that the can crusher should be designed so that the handle, link 2, is horizontal when the can is put into place. This will result in the greatest mechanical advantage at the point where the greatest force is needed in order to initiate buckling of the can. So the mechanism in Figure 7.26 needs to have point A moved up so that the handle is horizontal.

7.4.1 Mechanical Advantage of a Geartrain

M_a and the torque ratio n_t are equivalent. In the case of geartrains, both are a measure of how much the input torque is amplified into the output torque. For a geartrain, the load and driver are each a torque: T_{out} and T_{in}, respectively. From the definition of mechanical advantage,

$$M_a = \frac{T_{out}}{T_{in}} = n_t. \tag{7.82}$$

▶ EXAMPLE 7.14

In the geartrain of Figure 3.23, suppose a driving torque is applied in the direction of rotation of the input gear 6, and a load torque is applied to gear 2. Prove to yourself that $M_a = \frac{-1}{h_2}$—that is, that $M_a = -R$.

7.5 EFFICIENCY AND MECHANICAL ADVANTAGE

The mechanical efficiency of a machine, η, is defined as the ratio of the actual output power, $P_{out_{actual}}$, to the ideal output power, $P_{out_{ideal}}$. $P_{out_{ideal}}$ is the frictionless case.

$$\eta = \frac{P_{out_{actual}}}{P_{out_{ideal}}}$$
$$P_{out_{actual}} = \eta P_{out_{ideal}} \tag{7.83}$$

Let $\dot{S}_{load}$ denote the linear or angular velocity associated with computing $P_{out_{ideal}}$ in the mechanical advantage. Then

$$P_{out_{ideal}} = L_{ideal}\dot{S}_{load},$$

where L_{ideal} is the load that the driver could equilibrate for the ideal case of perfect efficiency. Substituting $L_{ideal} = M_a D$ from Equation (7.46) gives

$$P_{out_{ideal}} = M_a D\dot{S}_{load}. \tag{7.84}$$

$P_{out_{actual}}$ is given by

$$P_{out_{actual}} = L_{actual}\dot{S}_{load}, \tag{7.85}$$

where L_{actual} is the actual load torque the machine can equilibrate. Substituting Equations (7.84) and (7.85) into Equation (7.83) gives

$$
\begin{aligned}
L_{actual}\dot{\S}_{load} &= \eta M_a D\dot{\S}_{load} \\
L_{actual} &= \eta M_a D.
\end{aligned}
\tag{7.86}
$$

Equation (7.86) shows that any inefficiency in a machine results in a reduction in mechanical advantage. Speed is not lost. What is lost is the ability of a driver to overcome a load.

Let M_a, as defined in Section 7.4, be denoted as the ideal mechanical advantage, $M_{a_{ideal}}$. From Equation (7.86) and the definition of mechanical advantage, the actual mechanical advantage, $M_{a_{actual}}$ for a less than perfectly efficient machine would be

$$M_{a_{actual}} = \eta M_{a_{ideal}}.$$

If $M_{a_{actual}}$ were determined by experimentally measuring L_{actual} for a given D, and $M_{a_{ideal}}$ were determined by the methods in Section 7.4, the mechanical efficiency of the machine could be computed by

$$\eta = \frac{M_{a_{actual}}}{M_{a_{ideal}}}.$$

7.6 PROBLEMS

Problem 7.1

In Problem 3.13 you computed the gear ratio R for the transmission shown in Figure 3.32. In this problem you should compute the transmission's mechanical advantage. How does this compare to the torque ratio n_t and gear ratio R? Give your answer in terms of tooth numbers. Gear 2 is fixed and does not rotate. A driving torque acts on 1, and a load torque acts on 5.

Problem 7.2

Compute the mechanical advantage (a.k.a torque ratio) for high and reverse speeds in the Model T transmission shown in Figure 3.24.

Problem 7.3

Compute the mechanical advantage (a.k.a torque ratio) for the reverse speed condition in the two speed automatic transmission shown in Figure 3.25.

Problem 7.4

What type of mechanism/machine would have a *constant* mechanical advantage?

Problem 7.5

In the elliptical trammel mechanism shown in Figure 7.28 the driving force is applied to point *A*, and the load force is applied to point *B*.

1. Derive an expression for the mechanical advantage of the system. If it is necessary to add vectors in order to find the kinematic coefficients of points of interest, then do so.
2. When would this mechanical advantage be infinite? Give a geometric interpretation, and provide a sketch showing this condition.
3. When would this mechanical advantage be zero? Give a geometric interpretation and provide a sketch showing this condition.
4. Identify the dimensional aspect of M_a and the kinematic aspect of M_a.

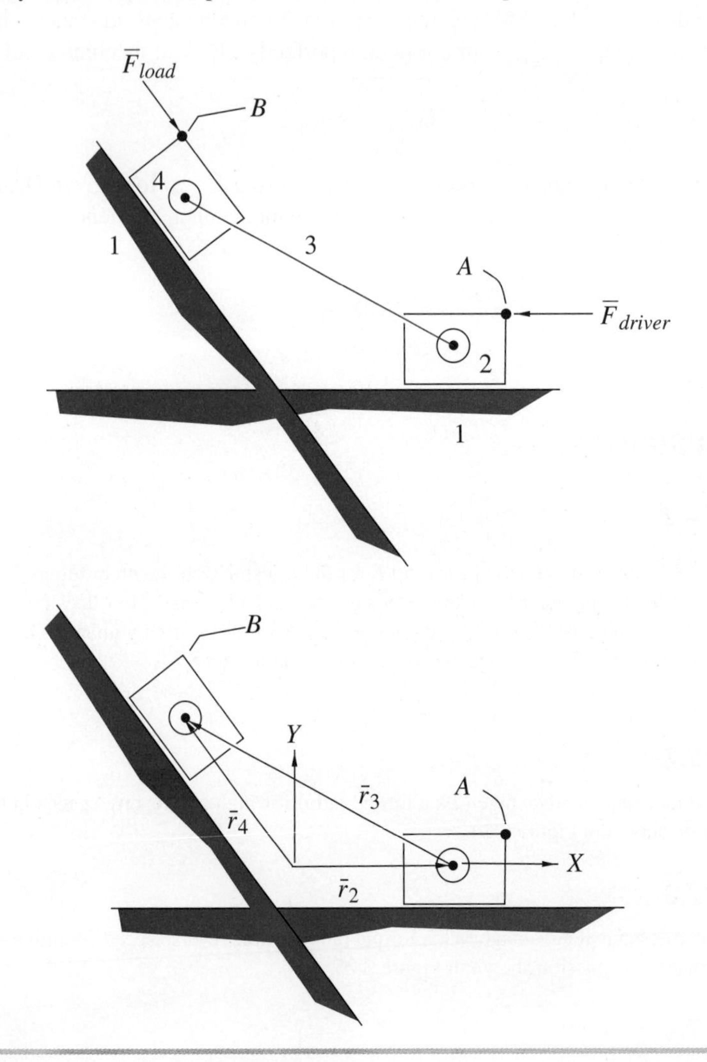

FIGURE 7.28 An elliptical trammel

Problem 7.6

In the inverted crank slider mechanism shown in Figure 7.29 a driving torque is applied to 4, and a load torque is applied to 2.

1. Derive an expression for the mechanical advantage of the system.
2. Identify the dimensional aspect of M_a and the kinematic aspect of M_a.
3. When would this mechanical advantage be infinite? Give a geometric interpretation of this condition.
4. When would this mechanical advantage be zero? Give a geometric interpretation of this condition.

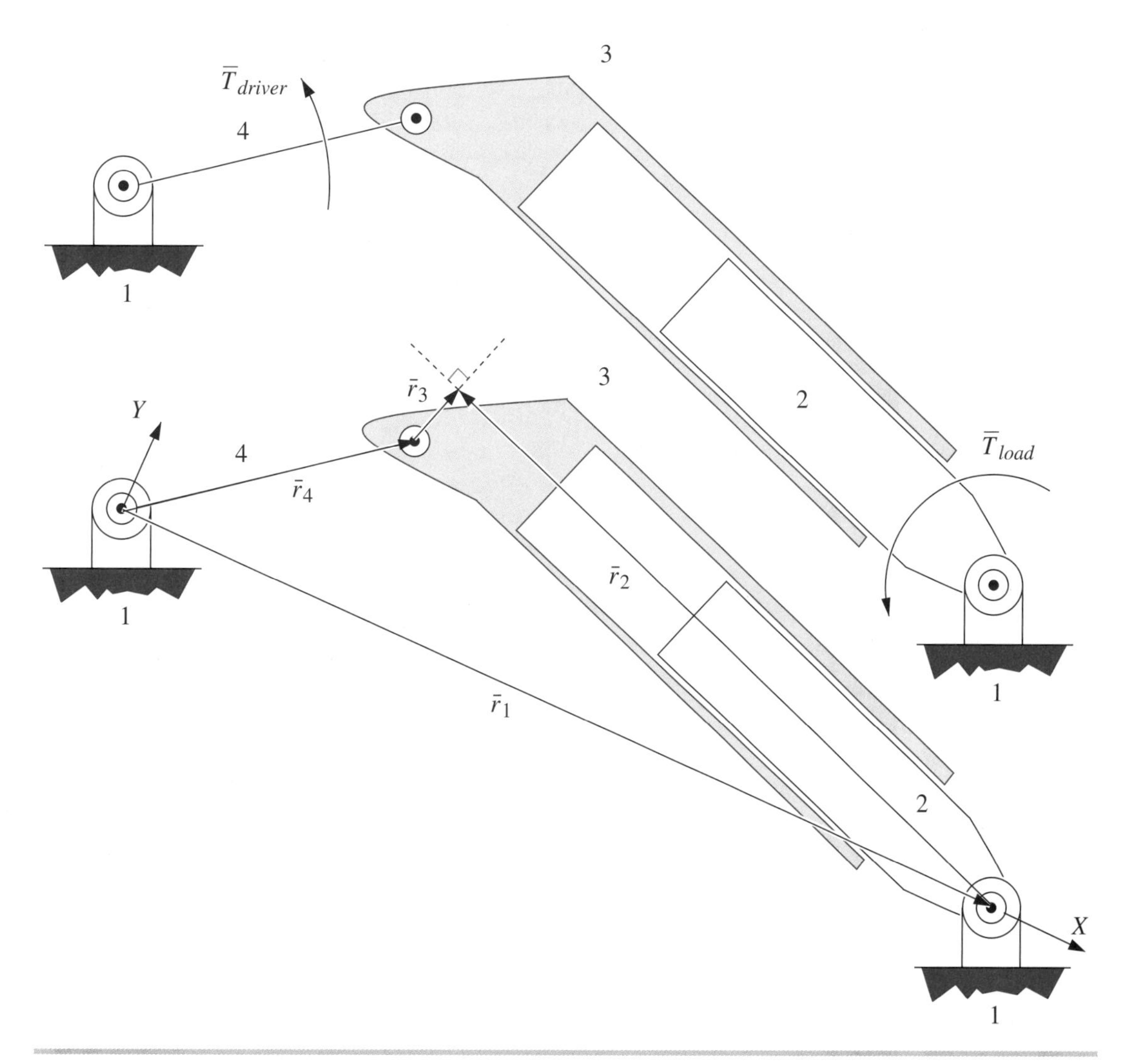

FIGURE 7.29 An inverted crank slider

Problem 7.7

Figure 7.30 shows a tool called a vise grip. (It is not called a *vice* grip. Gripping vices is not recommended!) It is the same tool you saw in Figure 1.25, except that the vise grips there were for clamping onto flat surfaces. These are better for clamping onto curved surfaces. Other than the shape of their jaws, they are effectively identical.

To the right is shown a four bar mechanism superimposed on the tool. On the bottom left is the mechanism with its vector loop. The bottom right shows the driving force $\overline{F}_{driver}$ applied to the handle 3, and the load force $\overline{F}_{load}$ applied to jaw 4. The dimensions of the vise grips are $r_1 = 5$ inches, $r_2 = 3.5$ inches, $r_3 = 1.5$ inches, $r_4 = 1.7$ inches, $r_5 = 4.7$ inches, and $r_6 = 1.2$ inches. The range of motion of θ, corresponding to when the jaws go from closed to open, is $20° \leq \theta_2 \leq 40°$.

Step 1 in the procedure for determining the M_a is completed. S_i was taken as θ_2, and the kinematic coefficients h_3 and h_4 were found to be

$$h_3 = \frac{r_2}{r_3}\frac{\sin(\theta_4 - \theta_2)}{\sin(\theta_3 - \theta_4)} \quad \text{and} \quad h_4 = \frac{r_2}{r_4}\frac{\sin(\theta_3 - \theta_2)}{\sin(\theta_3 - \theta_4)}.$$

1. Complete the remaining five steps in the procedure for computing M_a, and find a symbolic solution for M_a in terms of the link dimensions and the joint variables.
2. Identify the dimensional aspect and the kinematic aspect of the mechanical advantage.
3. Give the geometric condition that corresponds to $M_a = \infty$. If M_a is never infinite, then when is it maximum? Plot M_a for the prescribed range of θ_2 if necessary.
4. Give the geometric condition that corresponds to $M_a = 0$. If M_a is never zero, then when is it minimum? Plot M_a for the prescribed range of θ_2 if necessary.

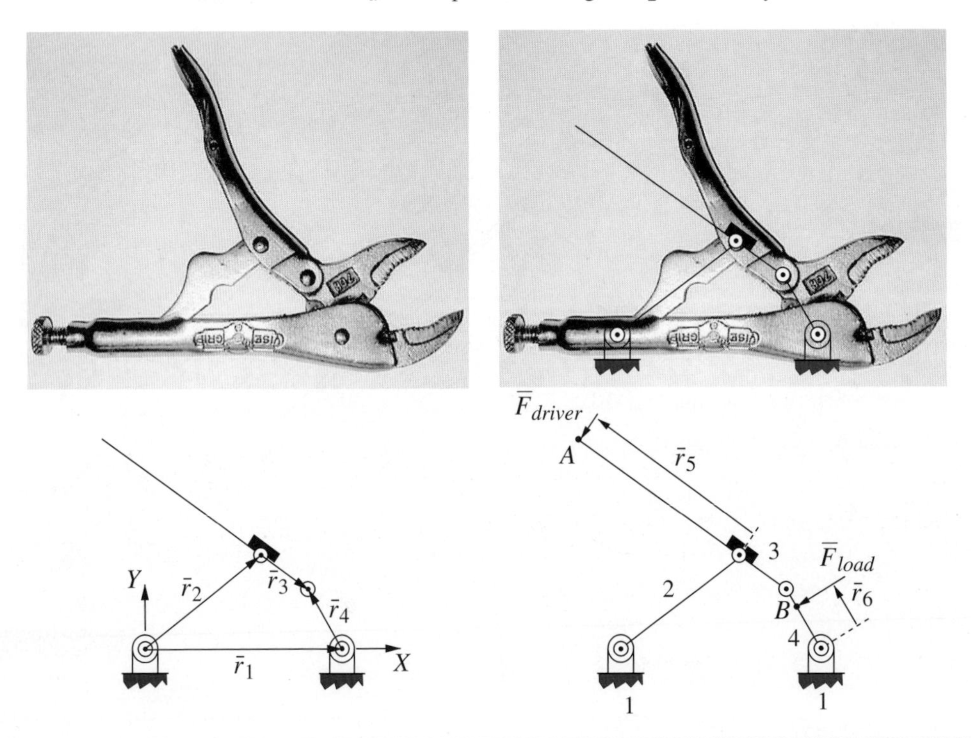

FIGURE 7.30 A visegrip wrench

Problem 7.8

This problem continues with Problem 4.5 and 5.11. Refer to Problem 4.5 for the vector loop and Problem 5.11 for the load and driver. The vector loop equation and kinematic coefficients from Problem 4.5 are repeated below. The position equations are

$$r_1 + r_3\cos\theta_3 + r_2\cos\theta_2 = 0$$
$$r_3\sin\theta_3 + r_2\sin\theta_2 - r_4 = 0$$
$$\theta_3 + (\pi/2) - \theta_2 = 0$$

These three position equations have

scalar knowns: $r_4, \theta_4 = -\pi/2, \theta_1 = 0, r_3$

and

scalar unknowns: $r_1, \theta_3, r_2, \theta_2$.

Using r_1 as the input, the first-order kinematic coefficients are

$$h_2 = h_3 = \frac{\sin\theta_2}{r_2} \quad \text{and} \quad f_2 = -\frac{r_3}{r_2}\cos\theta_3 - \cos\theta_2.$$

Find the mechanical advantage of the mechanism, following each step in Section 7.4. Substitute the given expressions for the basic kinematic coefficients into your expression for mechanical advantage. Identify the geometric conditions that correspond to maximum and minimum mechanical advantage.

7.7 PROGRAMMING PROBLEMS

In the following programming problems, you may use any programming language or script of your choice.

Programming Problem 5

This problem continues with Programming Problem 1 in Section 2.7 and Problem 2.17. The mechanism in Figure 2.26 is now a machine that has a driving torque applied to link 2 and a resisting load force acting at point B on 5 as shown in Figure 7.31. Recall the vector loop from Problem 2.7. Recall the two vector loop equations and the geometric constraint, Equations (2.69),

$$\bar{r}_2 - \bar{r}_3 - \bar{r}_1 = \bar{0}$$
$$\bar{r}_6 + \bar{r}_5 - \bar{r}_4 - \bar{r}_1 = \bar{0}$$
$$\theta_4 - \theta_3 = 0,$$

and the resulting system of five scalar position Equations (2.70),

$$r_2\cos\theta_2 - r_3\cos\theta_3 + r_1 = 0$$
$$r_2\sin\theta_2 - r_3\sin\theta_3 = 0$$
$$r_6 - r_4\cos\theta_4 + r_1 = 0$$
$$-r_5 - r_4\sin\theta_4 = 0$$
$$\theta_4 - \theta_3 = 0,$$

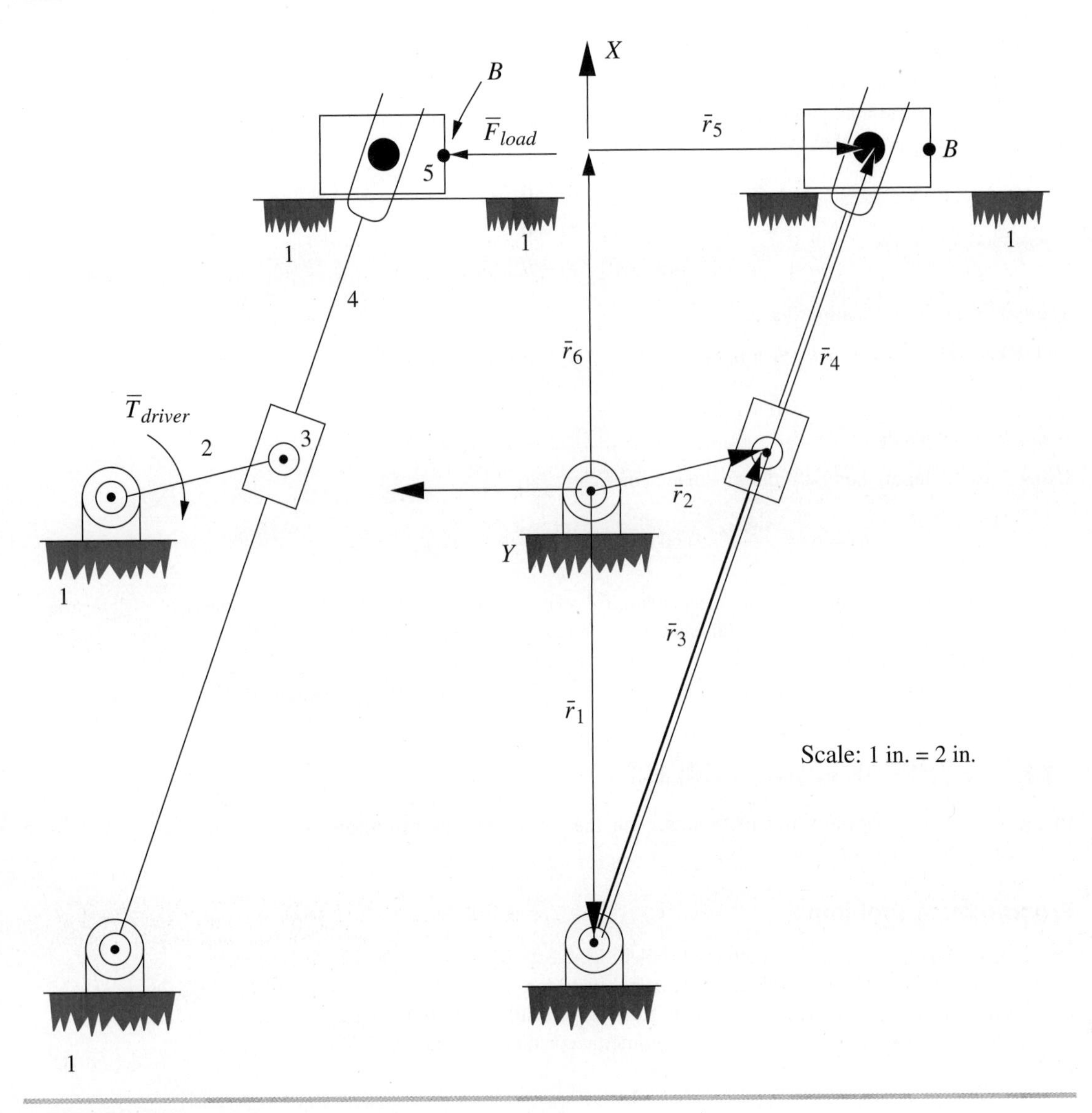

FIGURE 7.31 Shaper mechanism with a load and driver

with

scalar knowns: $r_2, r_1, \theta_1 = -\pi, r_6, \theta_6 = 0, \theta_5 = -\frac{\pi}{2}$

and

scalar unknowns: $\theta_2, \theta_3, r_3, \theta_4, r_4$ and r_5.

Of the six unknowns, take θ_2 as S_i and plot the mechanical advantage of the machine as a function of θ_2. Use your program from Programming Problem 1.

1. When does $M_a = \infty$? Or, if this never occurs, then when is M_a maximum? What are the corresponding configurations of the machine?
2. When does $M_a = 0$? Or, if this never occurs, then when is M_a minimum? What are the corresponding configurations of the machine?

Programming Problem 6

Figure 7.32 shows a tool which heavy-duty branch lopper used to cut through tree branches up to 2 inches thick. Cheaper branch loppers are simple scissor-like devices with a blade at the end of each handle and a single hinge connecting the handles. The cheaper loppers can cut only small branches. They operate exactly like the lever we looked at in Example 7.11. This heavy-duty lopper has a mechanism within it to amplify the strength of the user (that is, to produce a high mechanical advantage), making it possible to cut through thicker branches.

When using the tool to cut, the operator holds the lower handle steady and rotates the upper handle 90° ccw. This causes the cutting blade to rotate ccw and opens the jaws. The open jaws are then placed across a branch, and the upper handle is rotated 90° cw (pulled toward the lower handle). This causes the cutting blade to rotate cw, cutting through the branch with the cutting blade. The device is shown in a configuration where it has just cut all the way through the branch.

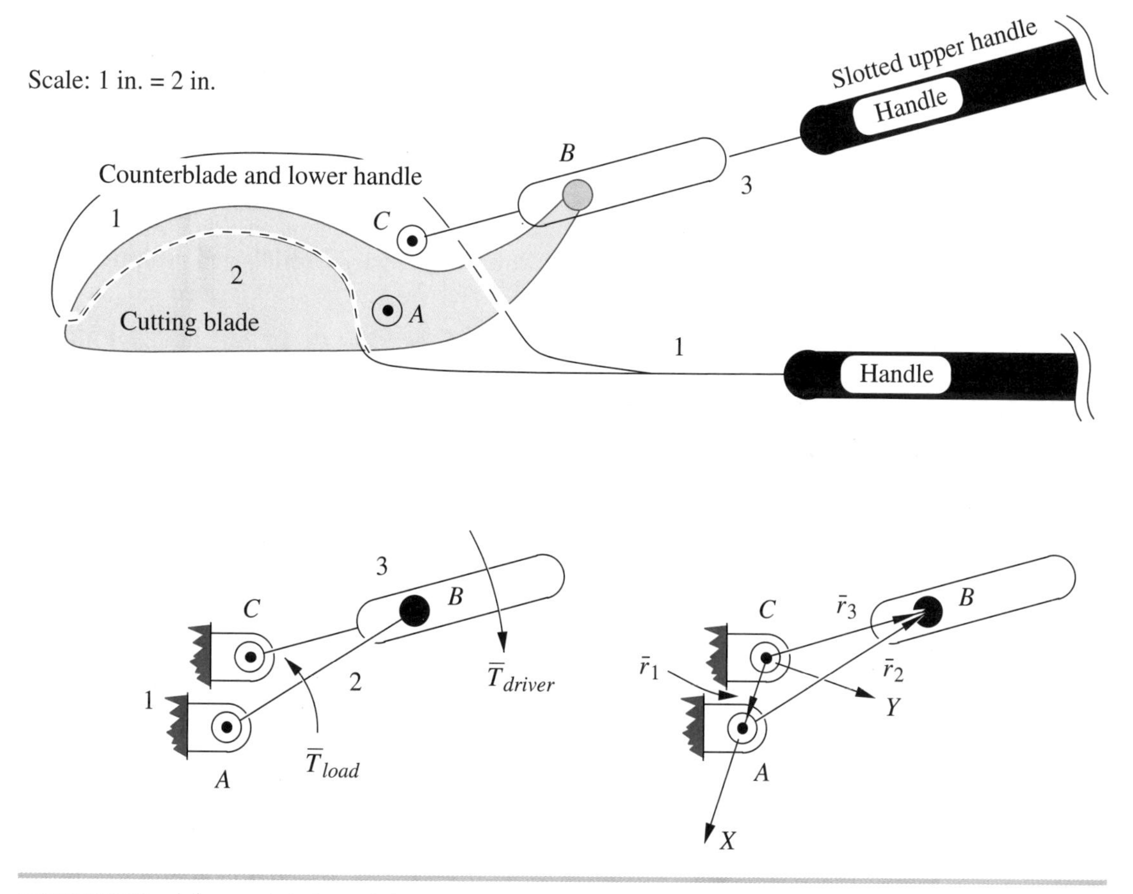

FIGURE 7.32 A heavy duty branch lopper

Below the branch lopper is a skeleton diagram of the mechanism. It considers that 1 (the lower handle) is fixed, and a driving torque is applied to the upper handle (3). A load torque acts against the cutting blade (2).

In this branch lopper, the counterblade is formed onto a part of the lower handle (1). The cutting blade (2) is pinned to the counterblade and lower handle at point *A* and has a pin-in-a-slot connection to the upper handle (3) at point *B*. The upper handle has a pin connection to the lower handle at point *C*.

Next to the skeleton diagram is a drawing of the mechanism with an appropriate vector loop for the system's kinematic analysis. When the jaws are fully open, $\theta_3 = 220°$. When the jaws are fully closed and the branch has been cut, $\theta_3 = 130°$.

Plot the mechanical advantage of the system as θ_3 goes from 220° to 130°. For dimensions use $r_1 = 1.1$inches, and $r_2 = 3.0$ inches.

1. When does $M_a = \infty$? Or, if this never occurs, then when is M_a maximum? What are the corresponding configurations of the machine?
2. When does $M_a = 0$? Or, if this never occurs, then when is M_a minimum? What are the corresponding configurations of the machine?

7.8 PROGRAMMING PROBLEMS—DESIGNING THE DRIVE SYSTEM OF AN AIR COMPRESSOR

The following series of programming problems concerns the design of a drive system for an air compressor. Each programming problem will result in a computer code that will then serve as a subroutine (if using Fortran) or function (if using Matlab) in the programming problems that follow it. This way you will be building up a rather complex code one piece at a time, debugging each piece as you go. This is good programming practice.

The single-stage air compressor in question is shown in Figure 7.33. The machine uses an inline crank-slider mechanism to compress the air. The air compressor intakes atmospheric air and delivers compressed air at 120 psig (134.7 psia). The ultimate goal of this project is to select the motor and pulley to drive this machine so that it delivers 3 cfm of the compressed air.

There has been some preliminary design of the machine, and the following information is known.

Dimensions

Crank: 1.5 in. from the center of the main bearing to center of the crank bearing.

Connecting rod: 4.0 in. from the center of the crank bearing to the center of the wrist pin.

Piston: diameter, d_p, is 3.2 in.

Clearance: clearance, *C*, is 7/64 in.

Inertia Properties

Crank: G_2 is located 0.50 in. below the center of the main bearing in Figure 7.33. The crank has a mass of 4.5 lb_m, and the rotational inertia about g_2 is 16.0 $\text{in}^2\text{lb}_\text{m}$.

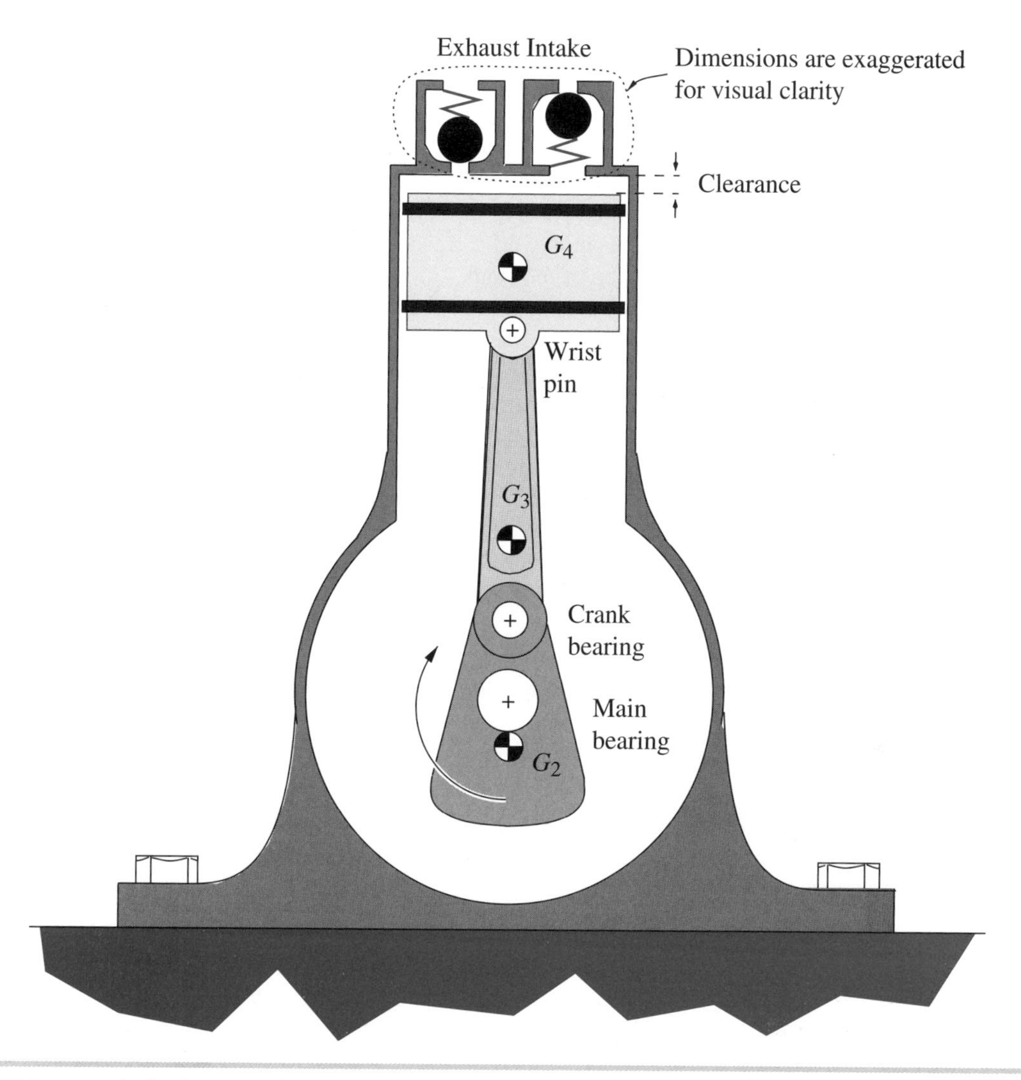

FIGURE 7.33 A single-stage air compressor

Source: Based on Runton Engineering, Inc., BRUIT FORCE SM, Industrial Solenoids, http://www.runton.com/AC-Datasheet25.htm

Connecting Rod: G_3 is located 1.50 in. above the center of the crank bearing in Figure 7.33. The connecting rod has a mass of 2.5 lb_m, and the rotational inertia about g_3 is 4.0 $in^2 lb_m$.

Piston: The piston has a mass of 1.4 lb_m.

The rotational inertias of the motor and the other transmission elements are unknown. You will select these components, and this will set their inertias.

7.8.1 Pump Operation

The pump uses check valves for both the intake and the exhaust. The check valves use a ball and spring to control when the valves open and close. The spring pushes against one side of the ball, causing the ball to push against the orifice, keeping the valve closed. When the pressure acting against the other side of the ball is sufficiently large, the ball is pushed against the spring, causing the spring to compress, which allows the valve to open. To explain how the pump works, we will walk through one working cycle, which corresponds to one rotation of the crank.

Consider that the piston is at bottom dead center (BDC) and the pump is beginning a compression stroke. The piston rises toward top dead center (TDC), and in the process both valves are closed and the pressure of the air trapped in the cylinder is increasing. When the pressure in the cylinder reaches 120 psig (134.7 psia), the spring in the exhaust check valve begins to compress, and high-pressure air is then pushed out of the valve at constant pressure and into a manifold that directs it to a storage tank.

After the piston reaches TDC, most of the high-pressure air has been moved out. Only the high-pressure air in the clearance volume remains. When the piston begins to retract and move back toward BDC, the high-pressure air in the clearance volume expands, and at this point the exhaust valve closes. The air left in the clearance volume will continue to expand as the piston moves toward BDC, and when the cylinder pressure drops below 0 psig (14.7 psia), the spring in the intake valve compresses and the cylinder draws in constant-pressure ambient air for the remainder of its intake stroke. The piston then reaches BDC and begins the compression stroke again. The intake valve closes almost immediately, and the cycle repeats itself.

Programming Problem 7

Use the vector loop shown in Figure 7.34 for your kinematic analysis, and use the mass center coordinates given in the inertia properties.

1. Develop a program that computes θ_3 and θ_4 for a given value of θ_2. Check your results graphically.
2. Write a program that computes the basic first- and second-order kinematic coefficients. The previous program should be a subroutine or function in this program. Check your results graphically.
3. Write a program that computes the first- and second-order kinematic coefficients of the mass centers. The previous program should be a subroutine or function in this program.
4. Write a program that plots ΣA and ΣB for one revolution of θ_2. In all of these programs, you should consider what units you will be computing with.

The crank of the compressor rotates clockwise while θ_2 is measured counterclockwise positive. We would like to define the crank angle of the compressor, ϕ as the angle measured clockwise from BDC, as shown in Figure 7.34, so

$$\phi = \pi - \theta_2 = \pi - \theta_2 \qquad (\phi = \phi - 2\pi) \tag{7.87}$$

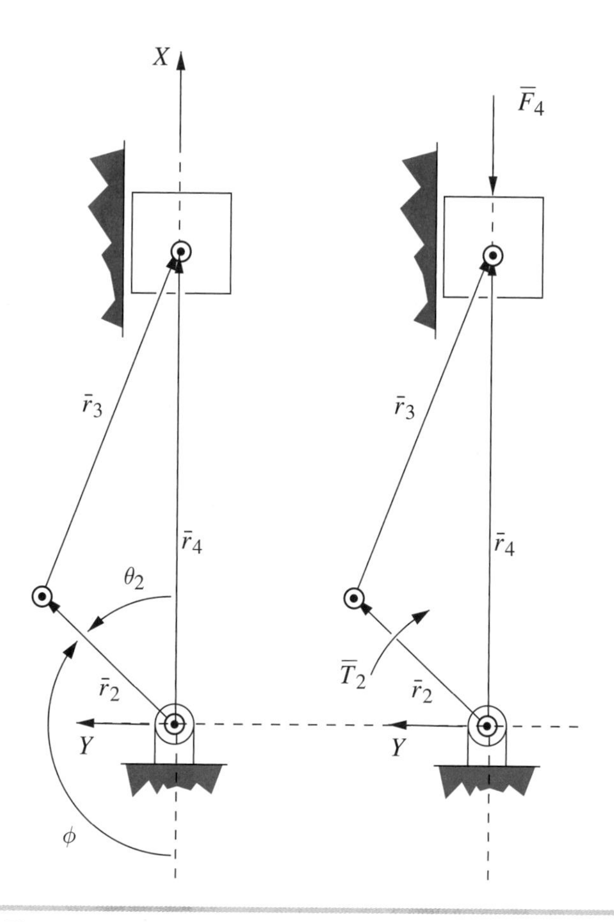

FIGURE 7.34 The crank-slider mechanism in an air compressor

Programming Problem 8

The goal of this programming problem is to plot the torque on link 2, T_2, which is required to compress the air, for one cycle of motion. To do so we need to be able to predict the gas pressure in the cylinder as a function of the crank angle. Knowing this pressure and the piston area, we know the load force, and from that we can use the power equation to compute T_2.

In the following discussion we present a thermodynamic model for the gas being compressed in the cylinder in order to determine the gas pressure as a function of the crank angle. Figure 7.35 shows the in-line crank-slider mechanism within the air compressor. The crank-slider starts at BDC, as shown on the left. At this moment the cylinder is full of ambient air, and both check valves are closed. Even though θ_2 is measured ccw positive, the crank is required to rotate cw. The figure in the center shows the vector loop. The link lengths are r_2 and r_3, and the position of the slider is defined by r_4.

We will assume adiabatic (no heat transfer), isentropic (perfectly efficient) compression and expansion of the air trapped in the cylinder. That is,

$$pv^k = \text{constant}, \tag{7.88}$$

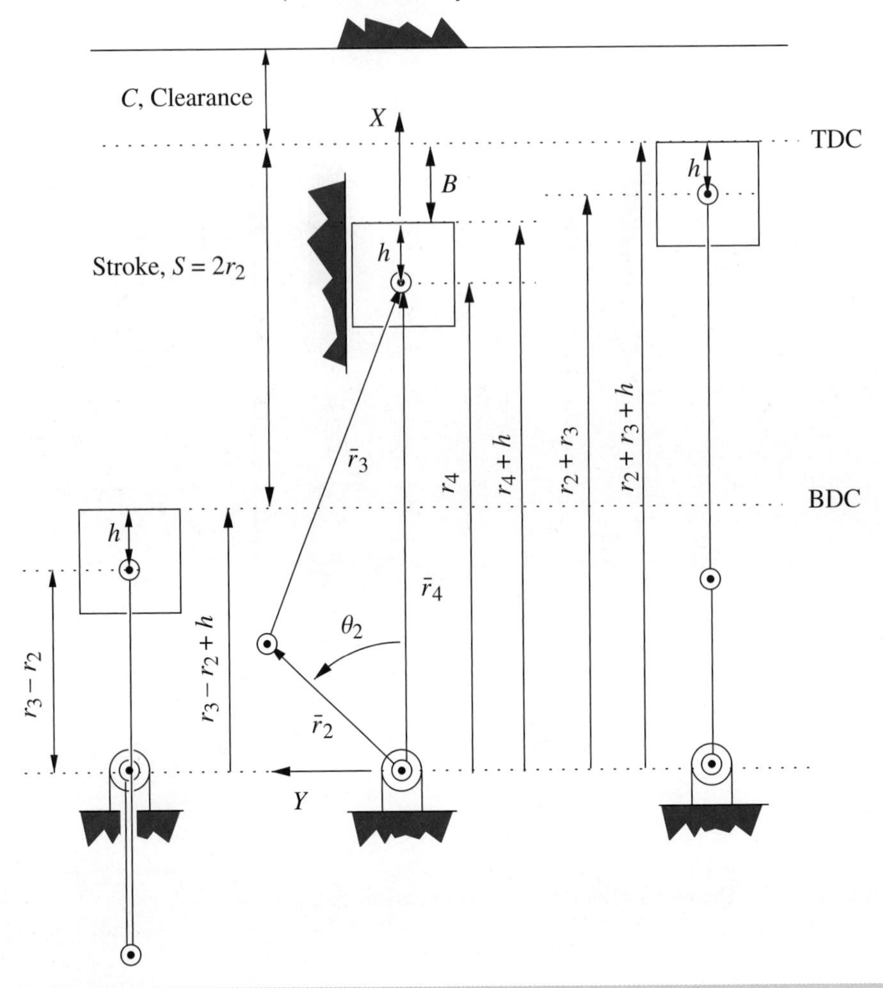

FIGURE 7.35 The crank-slider mechanism in an air compressor and a model for gas pressure

where: p = the gas pressure
v = the specific volume
k = 1.4 for air.

While the air in the cylinder either is being compressed or is expanding, it is in a closed system (the cylinder), and the mass of the air is constant. Therefore, the specific volume can be replaced by the volume of the cylinder, V. Thus during expansion or compression,

$$pV^k = \text{constant}, \tag{7.89}$$

and V is given by

$$V = (B + C)A_p, \tag{7.90}$$

where: B = the displacement of the top of the piston from its TDC location
C = the clearance between the top of the piston at TDC and the bottom of the cylinder head
A_p = the area of the top of the piston; that is, $A_p = \frac{\pi d_p^2}{4}$.

For a value of θ_2, B is known from the position solution. Referring to Figure 7.35,

$$B = (r_2 + r_3 + h) - (r_4 + h) = r_2 + r_3 - r_4, \tag{7.91}$$

so

$$V = (r_2 + r_3 - r_4 + C)A_p. \tag{7.92}$$

Adiabatic Compression and Constant-Pressure Exhaust

The compression stroke begins at BDC. At BDC,

$\theta_2 = 180°$, $p = p_o = 14.7$ psia (atmospheric pressure), and $V = V_o = (S + C)A_p = (2r_2 + C)A_p$,

which enables us to calculate the constant on the r.h.s. of Equation (7.89) for the compression portion of the exhaust stroke. Specifically, during the compression,

$$\text{constant} = p_o V_o^k, \tag{7.93}$$

so during the compression,

$$pV^k = p_o V_o^k. \tag{7.94}$$

That is,

$$\begin{aligned} p &= p_o\left(\frac{V_o}{V}\right)^k \\ &= 14.7 \text{ psia}\left(\frac{(2r_2 + C)A_p}{(r_2 + r_3 - r_4 + C)A_p}\right)^k \\ &= 14.7 \text{ psia}\left(\frac{(2r_2 + C)}{(r_2 + r_3 - r_4 + C)}\right)^k. \end{aligned} \tag{7.95}$$

Equation (7.95) models the pressure in the cylinder during the compression portion of the exhaust stroke. When this pressure exceeds 134.7 psia (120 psig), the exhaust valve is opened, and the gas pressure for the remainder of the exhaust stroke is a constant 120 psig while the gas is being exhausted to the manifold directing it to the storage tank.

Adiabatic Expansion and Constant-Pressure Intake

The intake stroke begins at TDC. At TDC,

$\theta_2 = 0°$, $p = p_{max} = 120$ psig $= 134.7$ psia, and $V = V_o = CA_p$,

which enables us to calculate the constant on the r.h.s. of Equation (7.89) for the expansion portion of the intake stroke. Specifically, during the expansion,

$$\text{constant} = p_o V_o^k, \tag{7.96}$$

so during the expansion,

$$pV^k = p_{max} V_o^k. \tag{7.97}$$

That is,

$$\begin{aligned} p &= p_{max}\left(\frac{V_o}{V}\right)^k \\ &= 134.7 \text{ psia}\left(\frac{CA_p}{(r_2 + r_3 - r_4 + C)A_p}\right)^k \\ &= 134.7 \text{ psia}\left(\frac{C}{(r_2 + r_3 - r_4 + C)}\right)^k. \end{aligned} \tag{7.98}$$

Equation (7.98) models the pressure in the cylinder during the expansion portion of the intake stroke. When this pressure falls to 14.7 psia, the intake valve is opened, and the gas pressure for the remainder of the intake stroke is a constant 14.7 psia while a fresh charge of atmospheric air is being drawn into the cylinder.

1. Incorporate this model of gas pressure into a computer program that uses the previous programs as subroutines or functions, and plot T_2 vs. θ_2.
2. Plot T_2 vs. r_4.
3. Integrate T_2 vs. θ_2 to determine the amount of work done per revolution when compressing the air, W_{rev}.
4. From the plot of T_2 vs. r_4, compute the amount of pressurized air delivered each revolution, V_{rev}.
5. Knowing that the compressor must deliver 3 cfm of pressurized air and V_{rev}, compute the required operating speed of the compressor, $\omega_{2_{oper}}$.
6. Knowing $\omega_{2_{oper}}$ and W_{rev}, compute the required horsepower of a motor that would drive link 2.
7. Select an appropriate AC induction motor from the Baldor Corporation website, and give its operating characteristics, T_s, T_p, T_b, T_f, ω_f, and ω_s.
8. Select appropriate pulleys for the motor and link 2, so that the motor drives link 2 at the desired operating speed.

Programming Problem 9

Plot the torque–speed curve for your selected motor.

Programming Problem 10

As a "test case", let $\theta_2 = 80°$ and $\dot{\theta}_2 = -50$ rad/s and compute $\ddot{\theta}_2$ using the power equation and the motor we have selected.

To do this we need to include the rotational inertia of the motor in the ΣA and ΣB. Only the rotational inertia is needed because the mass center of the motor does not move. One could use the dimensions of the motor and approximate the size of the cylindrical ferrous core, then compute the rotational inertia based on that. In this case, a call to Baldor Corp. tech support revealed that the motor had a rotational inertia, $I_m = 0.319\ \text{lb}_\text{m}\ \text{ft}^2$ which should be converted into the consistent units

$$I_m = \frac{0.319}{32.2}\text{slug ft}^2$$

and in ΣA and ΣB we should include A_m and B_m,

$$A_m = I_m h_m^2 \quad \text{and} \quad B_m = I_m h_m h_m' = 0\ (h_m' = 0).$$

Programming Problem 11

Develop a dynamic simulation of the air compressor. For now, use the initial conditions $\theta_2 = 0$ and $\dot{\theta}_2 = 0$.

1. Determine how long it takes to reach steady state.
2. Determine C_f.
3. In your differential equations course, you should have learned about the relationship of initial conditions to steady state response. What is it? If you do not recall, experiment with your initial conditions to find out.

7.9 DESIGNING THE DRIVE SYSTEM OF A FAIL-SAFE QUICK VALVE SHUT-OFF SYSTEM

In the event of a power loss, it is necessary to prevent any backflow through a pipe in a plant. For this purpose a fail-safe quick valve shut-off system has been designed. The system and its dimensions are shown in Figure 7.36. The system incorporates a four bar mechanism. The valve in question rotates with link 2. Figure 7.36 shows the position of link 2 when the valve is fully open and fully closed. In going from fully open to fully closed, link 2 rotates 148° counterclockwise.

The principle of the fail-safe operation is as follows: The plunger of a solenoid pulls a cable that is wrapped around a pulley attached to link 4. When there is power to the plant and the solenoid is energized, an electromagnetic force pulls the plunger to the right, where it seats itself on the base of the solenoid. While energized, the solenoid is holding the valve in its fully opened position. When power to the plant is lost, the solenoid is de-energized, the electromagnetic force disappears, and a pair of tension springs connected between S_1 and S_2 rotate 4 counterclockwise until the valve is fully closed. (Figure 7.36 shows only one spring, but there are actually two of them, one on each side of the four bar mechanism. The tension springs have been selected so that they will close the valve in the required amount of time. The shut-off mechanism is fail-safe because the springs guarantee that the valve will be closed if power to the plant is lost. The springs have a combined stiffness of 1.2 lb_f/in., a free length of 15.81 in., and a combined preload of 1.5 lb_f (these are close-wound springs.)

When power to the plant is restored, product once again begins to flow through the pipe, and it is essential that the valve be reopened by the solenoid within 0.4 second. Otherwise, the product will back up and overflow in upstream equipment. While the valve is being opened, there is a torque that resists its opening, T_valve. This has been measured as

$$T_\text{valve} = 2\left(\frac{\text{in.}\cdot\text{lb}_\text{f}}{\text{degree}}\right)\phi_{2_d} + 2\ \text{in.}\cdot\text{lb}_\text{f}\,, \quad \text{where } 0° \leq \phi_{2_d} \leq 148°.$$

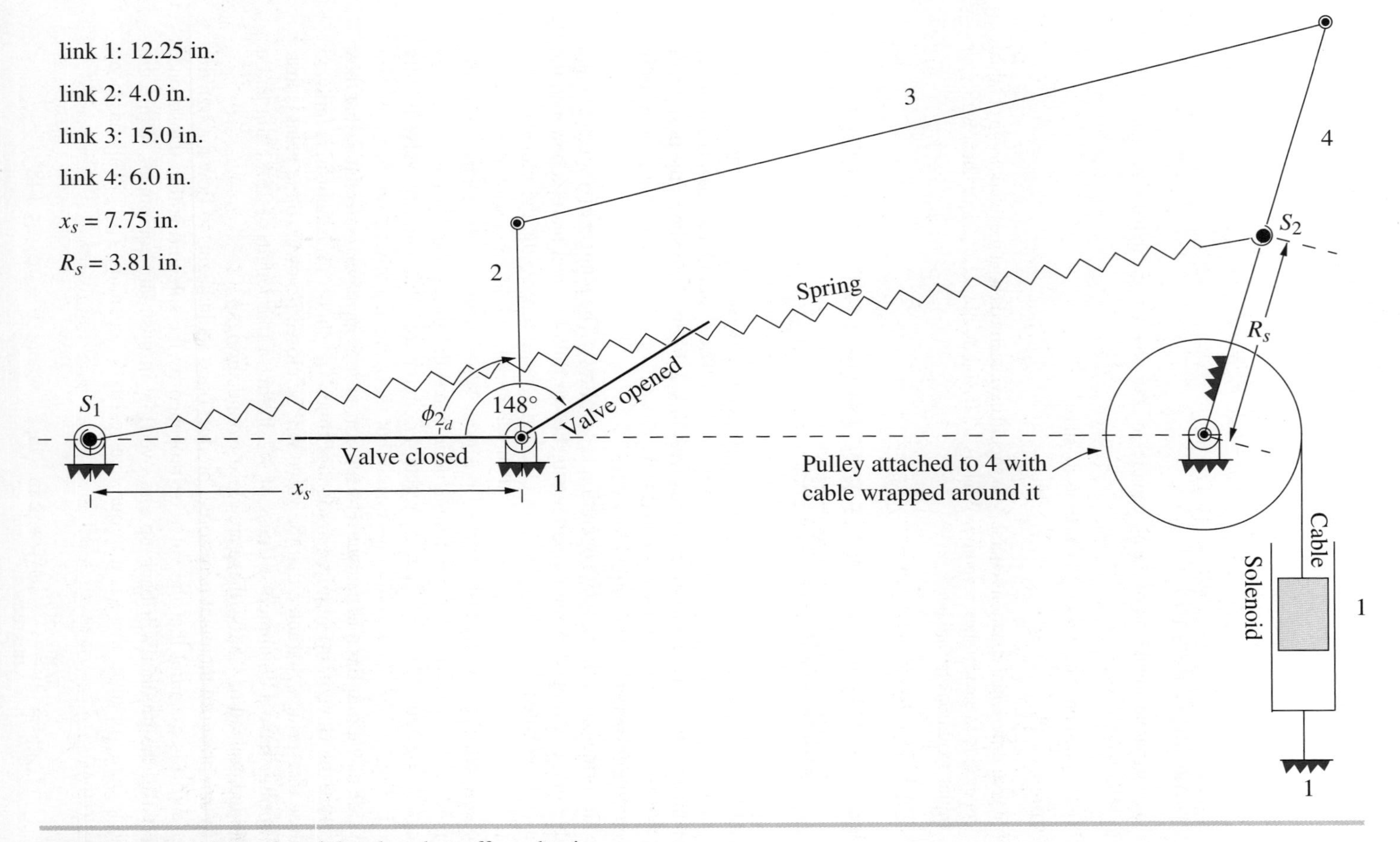

FIGURE 7.36 A fail-safe quick valve shut-off mechanism

Your task as the engineer assigned to this project is to select the appropriate electric solenoid so that the required reopening time of 0.4 second is met. From the catalog of the solenoid vendor, we find the graphs in Figure 7.38, which shows the standard pull curves for the vendor's F-300 series of AC solenoids. You should select the solenoid from this F-300 series.

The moving components of the system include links 2, 3, and 4, the rotating valve and pulley, the springs, and the plunger of the solenoid. We must account for their inertia properties. The inertia properties of the moving components follow.

Link 2:
$m_2 = 1.37\ \text{lb}_\text{m}, I_{g_2} = 3.03\ \text{lb}_\text{m}\text{in}^2, r_2' = 2.00$ in.

Valve:
$I_{g_\text{valve}} = 72.01\ \text{lb}_\text{m}\text{in}^2$

Link 3:
$m_3 = 2.78\ \text{lb}_\text{m}, I_{g_3} = 56.36\ \text{lb}_\text{m}\text{in}^2, r_3' = 7.50$ in.

Link 4:
$m_4 = 2.67\ \text{lb}_\text{m}, I_{g_4} = 12.40\ \text{lb}_\text{m}\text{in}^2, r_4' = 4.50$ in.

Pulley:
The pulley is a steel plate 0.5 in. thick. To determine I_{g_pulley} we need to have its radius. The radius depends on the stroke of the solenoid.

Solenoid plunger:
Figure 7.39 is a copy of a page from the vendor's catalog, which gives the plunger weight for each solenoid.

Springs:
The combined mass of both springs is 1.94 lb_m.

Figure 7.37 shows the locations of the mass centers of the links. The mass center of the spring is at its center.

To complete your task, solve the following programming problems.

Programming Problem 12

1. Adapt your Program from Programming Problem 2 to calculate the pulley radius required by each F-300 series solenoid.
2. Calculate the amount of work needed to open the valve, and on the basis of this, determine which of the F-300 series solenoids are suitable.

Programming Problem 13

Write a computer program that computes the pull force of the solenoid as a function of θ_2. Plot this for several of the suitable solenoids.

Programming Problem 14

Write a computer program that predicts the amount of time required to open the valve, and determine which of the suitable solenoids meets the required reopening time.

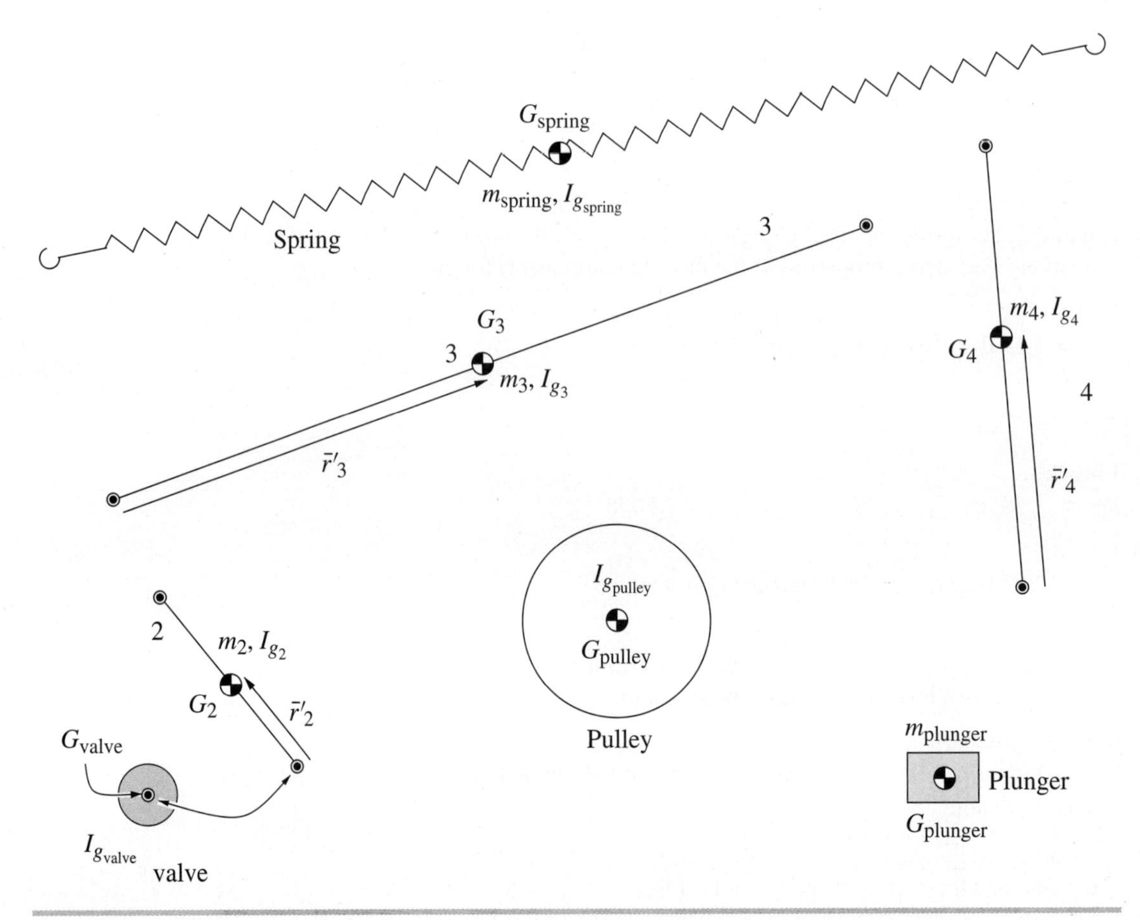

FIGURE 7.37 Inertia properties

Programming Problem 15

As an independent check on the design of the shut-off mechanism, your project engineer wants you to predict the amount of time it takes for the springs to close the valve in the event of a power failure. Do so.

7.10 DESIGN PROBLEMS

Design Problem 1

Use the dynamic simulation you developed in Programming Problem 11 to select a flywheel for the air compressor. The purpose of the flywheel is to reduce the level of the steady state speed variation in the machine.

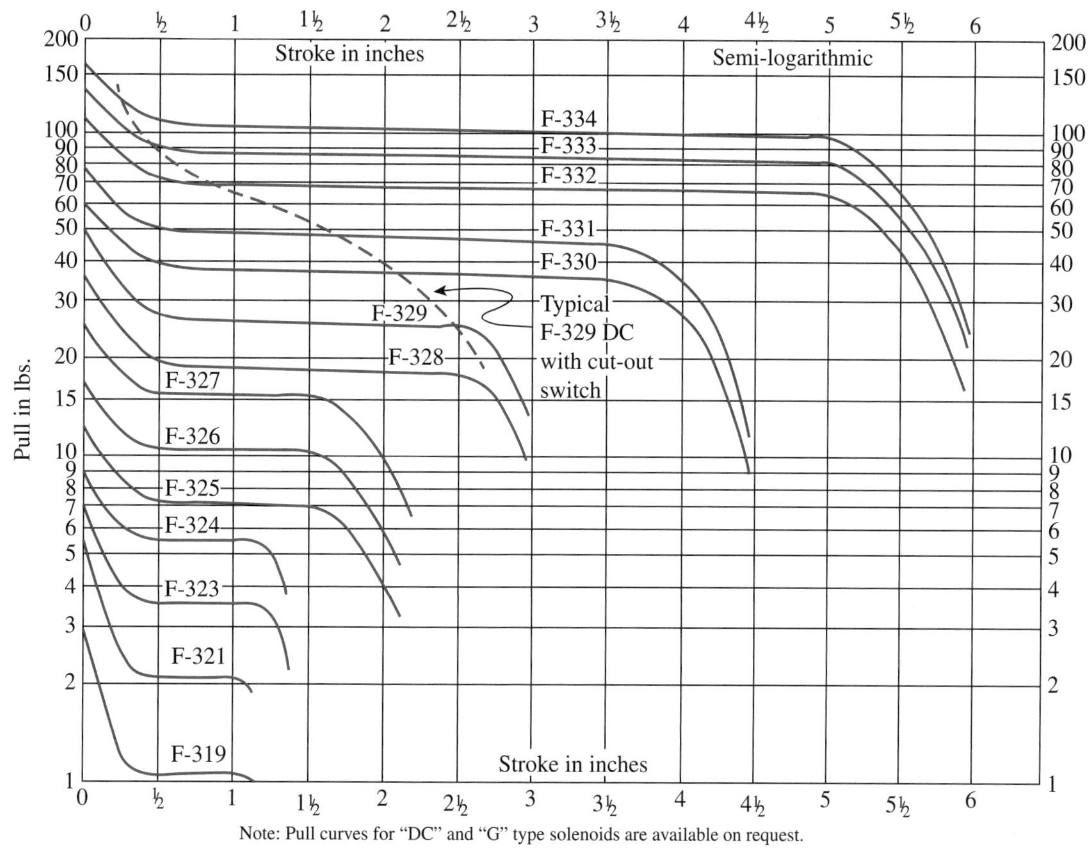

FIGURE 7.38 Page from solenoid vendor catalog showing standard pull curves

F-300 & G-200
SELECTION & ENGINEERING DATA

Model G-200*	Model F-300	Max. Stroke in inches	Pull in lb. Continuous Duty at Max. 85%	Pull in lb. Continuous Duty Stroke 100%	AC Current 220 Volt 60 Hz at Max. Stroke	AC Current 220 Volt 60 Hz Seated	DC Current 220 Volt Inrush (Unseated)	DC Current 220 Volt ** Seated	Plunger Weight	Solanoid Shipping Weight
—	F-319	1.00	1.0	1.4	1.0	0.2	—	—	0.13	1.0
G-207	—	1.00	0.7	0.9	0.7	0.1	—	—	0.13	1.0
—	F-321	1.00	2.0	2.8	1.8	0.3	—	—	0.19	1.3
G-208	—	1.00	0.9	1.3	1.3	0.2	—	—	0.19	1.3
G-209	F-323	1.25	3.0	4.2	3.2	0.5	1.7	0.02	0.25	2.3
G-209M	F-324	1.25	5.0	7.0	5.2	0.5	3.0	0.02	0.44	3.0
G-210	F-325	1.50	7.0	9.8	8.2	1.1	3.0	0.03	0.56	4.5
G-211	F-326	1.50	10.0	14.0	11.2	1.2	3.2	0.03	0.88	6.6
—	F-327	1.50	16.0	22.0	16.2	1.4	3.4	0.03	1.30	8.0
—	F-328	2.50	18.0	25.2	33.5	2.3	3.6	0.04	2.19	14.0
G-206	—	2.00	22.7	31.6	30.0	2.3	3.6	0.04	2.30	13.5
—	F-329	2.50	25.0	35.0	39.0	2.0	4.0	0.04	3.19	17.0
—	F-330	3.50	35.0	49.0	68.0	4.2	4.3	0.05	4.50	31.0
G-212	—	3.00	21.7	30.2	37.0	3.2	3.7	0.04	2.80	18.3
—	F-331	3.50	45.0	63.0	94.0	5.3	5.5	0.06	6.00	33.0
G-213	—	3.00	42.0	57.0	70.0	5.5	4.2	0.04	5.30	26.5
—	F-332	5.00	65.0	91.0	187.0	5.5	6.0	0.10	10.00	52.0
G-214	—	3.50	51.0	73.0	110.0	6.0	5.5	0.06	6.80	41.0
—	F-333	5.00	80.0	112.0	226.0	7.0	7.0	0.10	12.80	70.0
G-215	—	3.50	70.0	97.0	140.0	8.0	6.2	0.10	9.50	56.5
—	F-334	5.00	100.0	140.0	280.0	14.0	10.0	0.15	17.00	73.0

* Type G-200 solenoids serve as direct replacements for General Electric series CR-9503 solenoids and utilize similar identification numbers to simplify ordering procedures. For example: the Trombetta G-212 replaces the GE CR-9503-**212**.

** Seated current values for DC solenoids are achieved with a two-section coil and cut-out switch for continuous duty operation.

FIGURE 7.39 Page from solenoid vendor catalog giving plunger weights

Source: Based on data from Runton Engineering, Inc., BRUIT FORCE SM, Industrial Solenoids, http://www.runton.com/AC-Datasheet25.htm

CHAPTER 8

Mechanism Synthesis Part I: Freudenstein's Equation

In this chapter you will learn how to design four bar and crank-slider mechanisms for function generation. This is done by using Freudenstein's Equation. Freudenstein's Equation is named for its discoverer, Ferdinand Freudenstein, the originator of mechanism design in the United States. His equation is one of the most useful and easy-to-use tools for designing function-generating four bar mechanisms and crank-slider mechanisms.

8.1 FREUDENSTEIN'S EQUATION FOR THE FOUR BAR MECHANISM

DEFINITION:

Function Generation in Four Bar Mechanisms *A design (synthesis) problem where the link lengths of a four bar mechanism must be determined so that the rotations of the two levers within the mechanism, ϕ and ψ, are functionally related.*

The desired relation is represented by $f(\phi, \psi) = 0$. It is illustrated in Figure 8.1.

Figure 8.2 shows the four bar mechanism and the vector loop necessary for the mechanism's analysis. The vector loop equation is

$$\bar{r}_2 + \bar{r}_3 - \bar{r}_4 - \bar{r}_1 = \bar{0}, \tag{8.1}$$

which can be decomposed into the scalar X and Y components:

$$r_2\cos\theta_2 + r_3\cos\theta_3 - r_4\cos\theta_4 - r_1 = 0 \tag{8.2}$$

$$r_2\sin\theta_2 + r_3\sin\theta_3 - r_4\sin\theta_4 = 0. \tag{8.3}$$

Equations (8.2) and (8.3) can be reduced to a single equation relating θ_2, θ_4, and the four link lengths by eliminating θ_3. Equations (8.2) and (8.3) are nonlinear in the

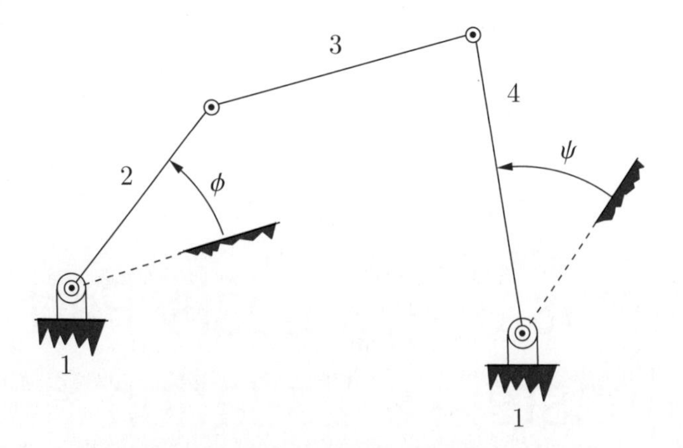

FIGURE 8.1 A function-generating four bar mechanism developing $f(\phi, \psi) = 0$

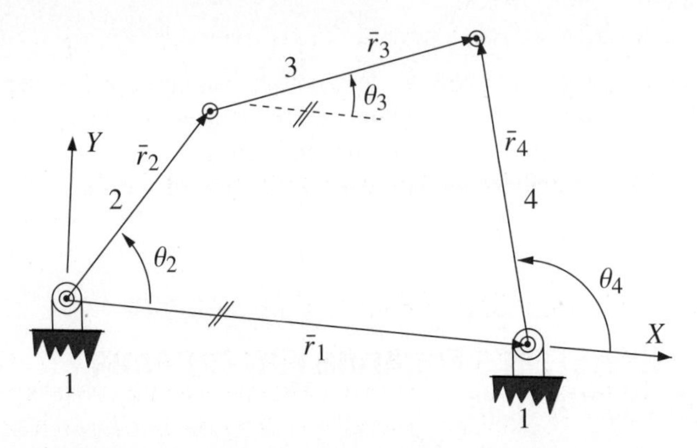

FIGURE 8.2 The four bar mechanism and its vector loop

angles, so elimination of θ_3 is not by simple back substitution. To eliminate θ_3, rewrite Equation (8.2) and Equation (8.3) as

$$r_3\cos\theta_3 = -r_2\cos\theta_2 + r_4\cos\theta_4 + r_1$$

$$r_3\sin\theta_3 = -r_2\sin\theta_2 + r_4\sin\theta_4,$$

and then square both sides of each equation and add the results together:

$$r_3^2 = r_2^2 + r_4^2 + r_1^2 - 2r_2r_4\cos\theta_2\cos\theta_4 + 2r_1r_4\cos\theta_4 - 2r_2r_1\cos\theta_2 - 2r_2r_4\sin\theta_2\sin\theta_4.$$

Dividing both sides by $2r_2r_4$ gives the result

$$\frac{r_2^2 - r_3^2 + r_4^2 + r_1^2}{2r_2r_4} + \frac{r_1}{r_2}\cos\theta_4 - \frac{r_1}{r_4}\cos\theta_2 = \cos\theta_2\cos\theta_4 + \sin\theta_2\sin\theta_4.$$

Define the three parameters:

$$R_1 = \frac{r_1}{r_4}, \quad R_2 = \frac{r_1}{r_2}, \quad \text{and} \quad R_3 = \frac{r_2^2 - r_3^2 + r_4^2 + r_1^2}{2r_2r_4}. \tag{8.4}$$

Then the previous equation can be written as

$$R_3 + R_2\cos\theta_4 - R_1\cos\theta_2 = \cos\theta_2\cos\theta_4 + \sin\theta_2\sin\theta_4.$$

Recognizing the double angle on the right hand side simplifies the result to

$$\boxed{R_3 + R_2\cos\theta_4 - R_1\cos\theta_2 = \cos(\theta_2 - \theta_4).} \tag{8.5}$$

Equation (8.5) is **Freudenstein's Equation**. It is the relationship between input rotation θ_2 and output rotation θ_4 as determined by the link lengths r_1 through r_4. In function generation via Freudenstein's Equation, the idea is to use Equation (8.5) to determine a set of link lengths that will result in a θ_2-θ_4 relationship that matches a desired function.

Two four bar mechanisms that are scaled versions of one another will have the same θ_2-θ_4 relationship since the scaling factor will be eliminated in the ratios R_1, R_2, and R_3, so function-generating four bar mechanisms are scale invariant. Function generators are also rotation invariant, which means they can be rotated into any orientation and the function generation is not affected.

Function generation is similar to curve fitting. There are two basic methods:

1. Point-matching method, matching a set of θ_2-θ_4 pairs in the desired range of the function.

2. Derivative-matching method, matching the function and its derivatives at one point in the desired range of the function.

8.1.1 Function Generation in Four Bar Mechanisms

As seen in Figure 8.2, to develop Freudenstein's Equation from a vector loop, the angular positions of the input and output are indicated by θ_2 and θ_4, respectively. Function generation in four bar mechanisms is a relation between an input rotation, ϕ, and an output rotation, ψ, which may not necessarily be referenced from the positive direction of the X axis, as the angles θ_2 and θ_4 are. This is shown in Figure 8.3. From Figure 8.3, these pairs of rotations are related by

$$\phi = \theta_2 - \theta_{2_i}, \tag{8.6}$$

where θ_{2_i} is a constant that is the offset between the reference for θ_2 and the reference for ϕ. Likewise,

$$\psi = \theta_4 - \theta_{4_i}, \tag{8.7}$$

where θ_{4_i} is a constant that is the offset between the reference for θ_4 and the reference for ψ. Equations (8.6) and (8.7) can be differentiated to show that

$$d\phi = d\theta_2 \quad \text{and} \quad d\psi = d\theta_4. \tag{8.8}$$

In other words, the changes in the angles are equal, but the angles themselves are offset by a constant value. Likewise for θ_4 and ψ. You will see that in the point-matching method you are free to choose both θ_{2_i} and θ_{4_i}. In the derivative-matching method you indirectly choose θ_{4_i} after which θ_{2_i} is defined.

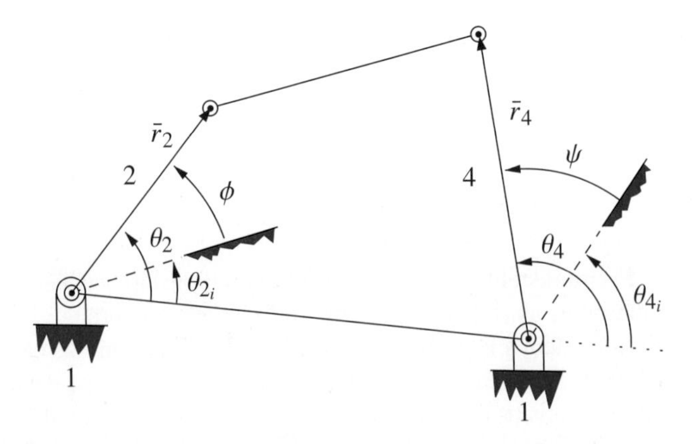

FIGURE 8.3 Relationship among θ_2, θ_4, ϕ, and ψ

8.1.2 Point-Matching Method with Freudenstein's Equation

The point-matching method with Freudenstein's Equation is a method by which a four bar mechanism is designed to match three desired pairs of $\{\phi, \psi\}$ values that satisfy the desired function generation $f(\phi, \psi) = 0$. These three pairs of values are known as the *precision points* and will be denoted as

$$\{(\phi)_1, (\psi)_1\}, \quad \{(\phi)_2, (\psi)_2\}, \quad \{(\phi)_3, (\psi)_3\}.$$

These three precision points are free choices of the designer. The designer then chooses values for θ_{2_i} and θ_{4_i}. These are also free choices. Clearly, there is a lot of room for optimization through varying these five free choices.

Using Equations (8.6) and (8.7), the three $\{\phi, \psi\}$ precision points are converted into three $\{\theta_2, \theta_4\}$ precision points denoted as

$$\{(\theta_2)_1, (\theta_4)_1\}, \quad \{(\theta_2)_2, (\theta_4)_2\}, \quad \{(\theta_2)_3, (\theta_4)_3\}.$$

These three $\{\theta_2, \theta_4\}$ precision points can be substituted into Equation (8.5) one pair at a time to generate three linear equations in three unknowns R_1, R_2 and R_3:

$$\begin{bmatrix} -\cos(\theta_2)_1 & \cos(\theta_4)_1 & 1 \\ -\cos(\theta_2)_2 & \cos(\theta_4)_2 & 1 \\ -\cos(\theta_2)_3 & \cos(\theta_4)_3 & 1 \end{bmatrix} \begin{bmatrix} R_1 \\ R_2 \\ R_3 \end{bmatrix} = \begin{bmatrix} \cos\left((\theta_2)_1 - (\theta_4)_1\right) \\ \cos\left((\theta_2)_2 - (\theta_4)_2\right) \\ \cos\left((\theta_2)_3 - (\theta_4)_3\right) \end{bmatrix}. \tag{8.9}$$

After solving Equation (8.9) for R_1, R_2, and R_3, a value is assumed for one of the four link lengths, and Equations (8.4) are solved for the remaining three. The assumed link length affects the scale of the mechanism.

A word about precision points: The choice of precision points is a design variable that can be varied to optimize matching the desired function. Precision points at the extreme ends of the range are generally not good. This results in larger errors in the middle of the range and no error at the ends. Usually the idea is to minimize the overall error. However, in some instances it is necessary to exactly match end conditions. One example is when the mechanism's output is opening and closing a valve, and the open and closed positions of the valve must exactly correspond to certain values of the input. Exact determination of the optimal precision points is not possible. Several individuals have proposed techniques for tuning in to the optimal precision points, the most popular being Chebyshev Spacing. Research it if you are interested.

In applications where high levels of torque must be transmitted from the input to the output and the rotations are oscillatory (not continuous), a four bar mechanism is more appropriate than a pair of gears. Mechanisms in general have several orders of magnitude higher "torque density" than gears. Torque density is the ratio of the torque capacity of a mechanism to its weight. Gears are used when a constant speed ratio is required for continuous rotations of the input and output as in a transmission. In that case only gears will work. The following example illustrates the use of Freudenstein's Equation to design a four bar that generates a linear function that can replace a pair of gears with limited rotations.

▶ EXAMPLE 8.1

An example of the usefulness of a four bar mechanism's high torque density occurs in the raising of an urban bridge to allow for the passing of a boat/ship where the bridge needs to be rotated ("lifted") a finite amount and the torque required to do this is large even if partially counterbalanced.

Consider designing a four bar mechanism to replace a pair of gears with a 2:1 gear ratio ($R = 2$). Consider ϕ as the input rotation and ψ as the output rotation. In such a gear pair, the function generation would be linear and given by

$$\psi = \frac{1}{2}\phi \quad \rightarrow \quad f(\phi, \psi) = \psi - \frac{1}{2}\phi = 0, \tag{8.10}$$

where

$$0^\circ \leq \phi \leq 180^\circ \quad \text{and, correspondingly,} \quad 0^\circ \leq \psi \leq 90^\circ.$$

Solution

For precision points take the three $\{\phi, \psi\}$ pairs in Table 8.1, which are evenly spaced. To apply Freudenstein's Equation, it is necessary to establish three precision point $\{\theta_2, \theta_4\}$ pairs that correspond to these three precision point $\{\phi, \psi\}$ pairs. This requires establishing values of θ_{2_i} and θ_{4_i} so that Equations (8.6) and (8.7) can be used to do this. θ_{2_i} and θ_{4_i} are free choices. Take $\theta_{2_i} = 20^\circ$ and $\theta_{4_i} = 40^\circ$. Equations (8.6) and (8.7) give the three precision point $\{\theta_2, \theta_4\}$ pairs in Table 8.2.

Substituting the values in each row of Table 8.2 into Equation (8.9) generates three linear equations in the three unknowns R_1, R_2, and R_3:

$$\begin{bmatrix} -0.423 & 0.462 & 1 \\ 0.342 & 0.087 & 1 \\ 0.906 & -0.301 & 1 \end{bmatrix} \begin{bmatrix} R_1 \\ R_2 \\ R_3 \end{bmatrix} = \begin{bmatrix} 0.999 \\ 0.906 \\ 0.676 \end{bmatrix}. \tag{8.11}$$

ϕ (degrees)	ψ (degrees)
45°	22.5°
90°	45°
135°	67.5°

TABLE 8.1 The three assumed $\{\phi, \psi\}$ precision points

θ_2 (degrees)	θ_4 (degrees)
65°	62.5°
110°	85°
155°	107.5°

TABLE 8.2 The three $\{\theta_2, \theta_4\}$ precision points

Solving Equation (8.11) gives

$$R_1 = 0.5880, \quad R_2 = 1.4475, \text{ and } R_3 = 0.5790.$$

Take $r_1 = 10$ ft., and solving Equations (8.4) for r_2, r_3, and r_4 gives

$$r_2 = 6.866 \text{ ft.}, \quad r_4 = 16.886 \text{ ft., and} \quad r_3 = 17.278 \text{ ft.},$$

which completes the design of the function generator. Although it is not the case here, it could happen that either r_4 or r_2 computes as a negative number. This means the vector would be pointing opposite the assumed direction of θ_4 or θ_2, respectively.

Figure 8.4 shows the designed four bar mechanism that realizes the three (θ_2, θ_4) pairs for which it was designed. Figure 8.5 shows the performance of the mechanism. It appears to approximately generate the desired linear relationship for $40° < \theta_2 < 200°$, which corresponds to $55° < \theta_4 < 125°$. This corresponds to $20° < \phi < 180°$, which does not quite make the desired range of ϕ.

To improve the design, several things can be tried:

1. Try spreading the precision points farther apart.
2. Try choosing different values of θ_{2_i} and θ_{4_i}.

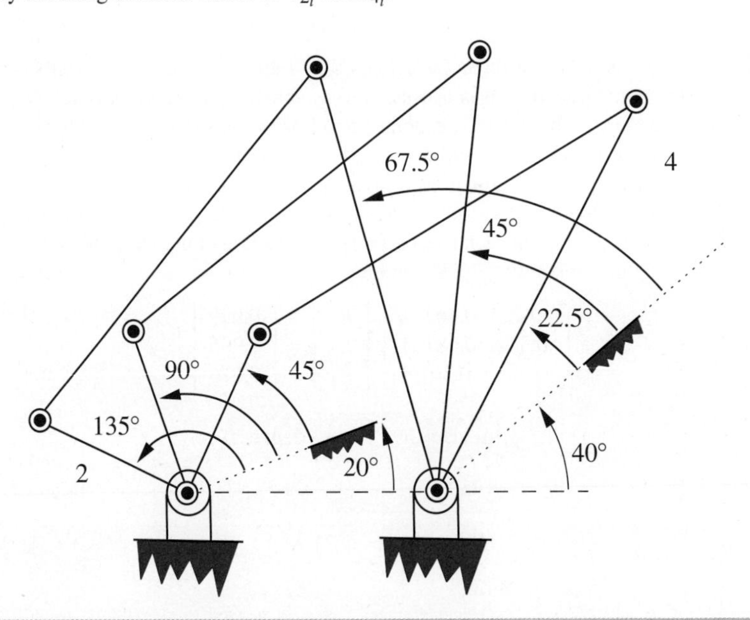

FIGURE 8.4 A four bar mechanism approximating a 2:1 ratio gear pair

There are an infinite number of choices for each precision point and for each value of θ_{2_i} and θ_{4_i}. Hence there are an infinity to the fifth power of choices for the set of all five parameters (three for each precision point and one each for θ_{2_i} and θ_{4_i}). There could definitely be some good optimization from that. In order to judge which design is better than another, the performance plot in Figure 8.5 has to be compared between competing designs. Other performance characteristics, such as the force/torque transmission capability (mechanical advantage), also need to be considered. Figure 8.6 shows the four bar mechanism being implemented in a bridge-lifting system.

8.1.3 Derivative-Matching Method with Freudenstein's Equation

In the discussion here, Freudenstein's Equation is used to design a four bar mechanism to match a value of θ_4 and its first three derivatives with respect to θ_2. The application of this to function generation will be explained later. But for now, we will focus on the theory.

Define the first-, second-, and third-order kinematic coefficients h_4, h_4', and h_4'' as the first, second, and third derivatives of θ_4 with respect to θ_2:

$$h_4 = \frac{d\theta_4}{d\theta_2}, \quad h_4' = \frac{d^2\theta_4}{d\theta_2^2}, \text{ and } \quad h_4'' = \frac{d^3\theta_4}{d\theta_2^3}. \tag{8.12}$$

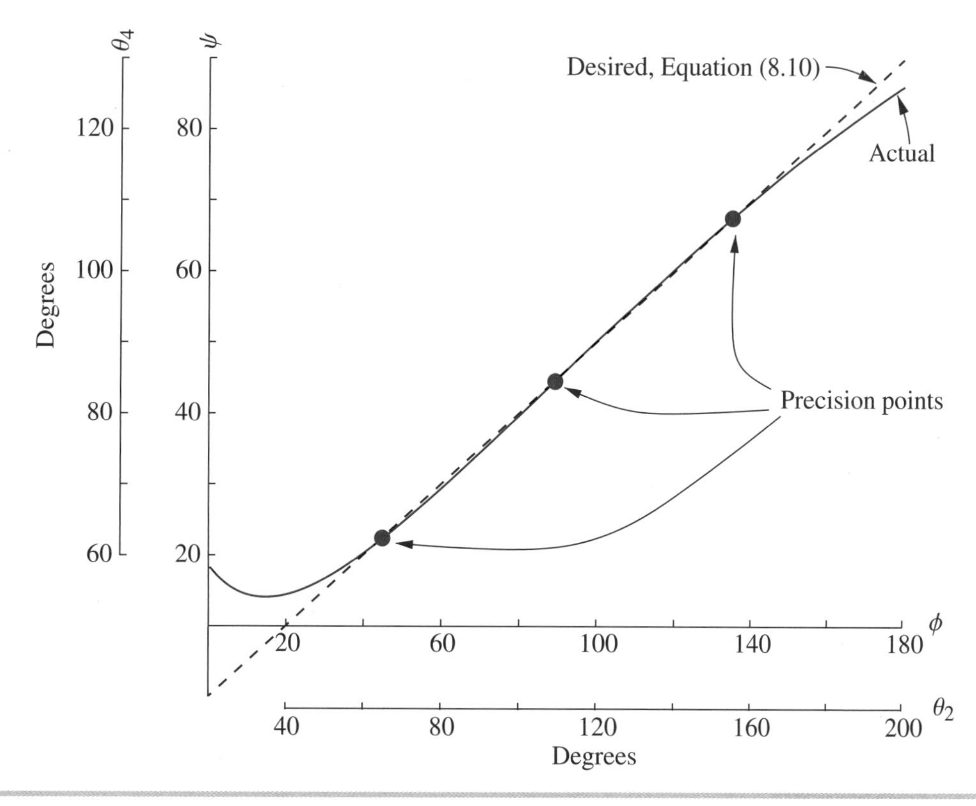

FIGURE 8.5 Performance of the four bar mechanism approximating a 2:1 ratio gear pair
© Cengage Learning.

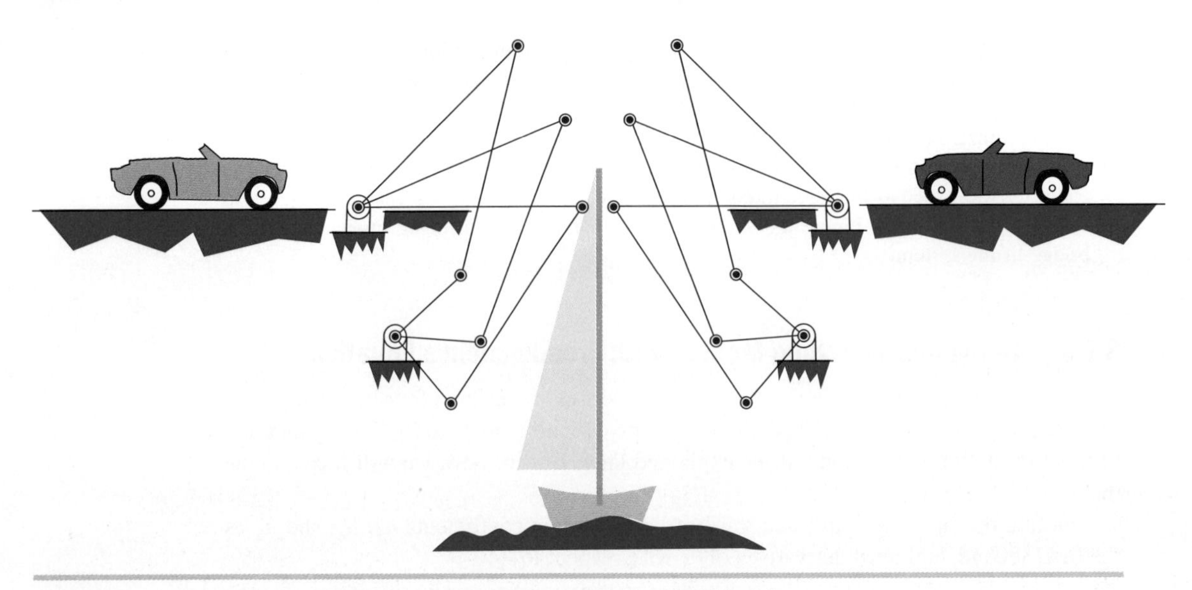

FIGURE 8.6 The four bar mechanism being implemented in a bridge lifter

Differentiating Equation (8.5) with respect to θ_2 yields an equation in the first-order kinematic coefficient,

$$R_1 \sin\theta_2 - R_2 \sin\theta_4 h_4 = -\sin(\theta_2 - \theta_4)(1 - h_4). \tag{8.13}$$

Differentiating Equation (8.13) again with respect to θ_2 yields an equation in the first-order and second-order kinematic coefficients,

$$R_1 \cos\theta_2 - R_2 \cos\theta_4 h_4^2 - R_2 \sin\theta_4 h_4' = -\sin(\theta_2 - \theta_4)(-h_4') - \cos(\theta_2 - \theta_4)(1 - h_4)^2,$$

which is rewritten as

$$R_1 \cos\theta_2 - R_2 \left(h_4' \sin\theta_4 + h_4^2 \cos\theta_4\right) = \sin(\theta_2 - \theta_4)h_4' - \cos(\theta_2 - \theta_4)(1 - h_4)^2. \tag{8.14}$$

Differentiating Equation (8.14) again with respect to θ_2 yields an equation in terms of the first-, second-, and third-order kinematic coefficients,

$$\begin{aligned} -R_1 \sin\theta_2 - R_2 \left(h_4' \cos\theta_4 h_4 + h_4'' \sin\theta_4 - h_4^2 \sin\theta_4 h_4 + 2h_4 h_4' \cos\theta_4\right) = \\ + \cos(\theta_2 - \theta_4)h_4'(1 - h_4) + \sin(\theta_2 - \theta_4)h_4'' + \sin(\theta_2 - \theta_4)(1 - h_4)^2(1 - h_4) \\ - \cos(\theta_2 - \theta_4)\left(2(1 - h_4)(-h_4')\right). \end{aligned}$$

In this equation, collect the terms on the left hand side that multiply R_2 with $\sin\theta_4$ and $\cos\theta_4$, and collect terms on the right hand side that multiply $\sin(\theta_2 - \theta_4)$ and $\cos(\theta_2 - \theta_4)$,

$$\begin{aligned} -R_1 \sin\theta_2 - R_2\left((h_4'' - h_4^3)\sin\theta_4 + 3h_4 h_4' \cos\theta_4\right) = \\ \sin(\theta_2 - \theta_4)\left(h_4'' + (1 - h_4)^3\right) + \cos(\theta_2 - \theta_4)\left(3h_4'(1 - h_4)\right). \end{aligned} \tag{8.15}$$

i	a_i	b_i	c_i	d_i
1	$-h_4$	0	$-(1-h_4)$	0
2	$-h'_4$	$-h_4^2$	h'_4	$-(1-h_4)^2$
3	$h''_4-h_4^3$	$3h_4h'_4$	$-h''_4-(1-h_4)^3$	$-3h'_4(1-h_4)$

TABLE 8.3 Definitions simplifying Equations (8.13), (8.14), and (8.15)

Equations (8.13), (8.14), and (8.15) have a common form. All three can be written as

$$R_1\begin{Bmatrix}\sin\theta_2\\ \cos\theta_2\end{Bmatrix}+R_2(a_i\sin\theta_4+b_i\cos\theta_4)=$$
$$c_i\sin(\theta_2-\theta_4)+d_i\cos(\theta_2-\theta_4)\quad i=1,2,3,\quad (8.16)$$

where i represents the number of differentiations of Equation (8.5). So $i = 1, 2, 3$ corresponds to Equations (8.13), (8.14), and (8.15), respectively. The first term in Equation (8.16) is $R_1\sin\theta_2$ for odd values of i and $R_1\cos\theta_2$ for even values of i, and the constants a_i, b_i, c_i, and d_i are known in terms of the kinematic coefficients from Equations (8.13), (8.14), and (8.15) and are given in Table 8.3.

Equations (8.5), (8.13), (8.14), and (8.15) are a system of four equations in four unknowns R_1, R_2, R_3, and θ_2. They are linear in R_1, R_2, and R_3 but nonlinear in θ_2. A subset of these four equations, namely Equations (8.13), (8.14), and (8.15), are a system of three equations in three unknowns R_1, R_2, and θ_2. From this subset of equations a closed-form solution for θ_2 can be found. Here we go.

Expand the double angles on the right hand side of Equation (8.16),

$$R_1\begin{Bmatrix}\sin\theta_2\\ \cos\theta_2\end{Bmatrix}+R_2(a_i\sin\theta_4+b_i\cos\theta_4)=$$
$$c_i(\sin\theta_2\cos\theta_4-\cos\theta_2\sin\theta_4)+d_i(\cos\theta_2\cos\theta_4+\sin\theta_2\sin\theta_4)\quad i=1,2,3,$$

and group the coefficients of $\sin\theta_2$ and $\cos\theta_2$ on the right hand side:

$$R_1\begin{Bmatrix}\sin\theta_2\\ \cos\theta_2\end{Bmatrix}+R_2(a_i\sin\theta_4+b_i\cos\theta_4)=$$
$$(c_i\cos\theta_4+d_i\sin\theta_4)\sin\theta_2+(-c_i\sin\theta_4+d_i\cos\theta_4)\cos\theta_2\quad i=1,2,3.\quad (8.17)$$

In Equation (8.17), define another set of constants:

$$E_i=a_i\sin\theta_4+b_i\cos\theta_4,\quad D_i=c_i\cos\theta_4+d_i\sin\theta_4,\quad\text{and}\quad F_i=-c_i\sin\theta_4+d_i\cos\theta_4.\quad (8.18)$$

E_i, D_i, and F_i are known from a_i, b_i, c_i, and d_i in Table 8.3, and θ_4 has an assumed value known as the design value, $\theta_4=\theta_{4_{des}}$.

With these definitions, Equation (8.17) becomes

$$R_1\begin{Bmatrix}\sin\theta_2\\ \cos\theta_2\end{Bmatrix}+R_2E_i=D_i\sin\theta_2+F_i\cos\theta_2\quad i=1,2,3.\quad (8.19)$$

R_1 and R_2 will now be eliminated from Equations (8.19) to leave one equation in one unknown, θ_2. To do so, write Equation (8.19) for $i = 1$ and $i = 2$,

$$R_1\sin\theta_2 + R_2E_1 = D_1\sin\theta_2 + F_1\cos\theta_2 \quad (i = 1)$$
$$R_1\cos\theta_2 + R_2E_2 = D_2\sin\theta_2 + F_2\cos\theta_2 \quad (i = 2),$$

which is written in matrix form as

$$\begin{bmatrix} \sin\theta_2 & E_1 \\ \cos\theta_2 & E_2 \end{bmatrix}\begin{bmatrix} R_1 \\ R_2 \end{bmatrix} = \begin{bmatrix} D_1\sin\theta_2 + F_1\cos\theta_2 \\ D_2\sin\theta_2 + F_2\cos\theta_2 \end{bmatrix}$$

and is solved for R_1 and R_2 using Cramer's Rule,

$$R_1 = \frac{G_1}{G} \quad \text{and} \quad R_2 = \frac{G_2}{G}, \tag{8.20}$$

where

$$G = \begin{vmatrix} \sin\theta_2 & E_1 \\ \cos\theta_2 & E_2 \end{vmatrix} = E_2\sin\theta_2 - E_1\cos\theta_2 \tag{8.21}$$

$$\begin{aligned} G_1 &= \begin{vmatrix} D_1\sin\theta_2 + F_1\cos\theta_2 & E_1 \\ D_2\sin\theta_2 + F_2\cos\theta_2 & E_2 \end{vmatrix} \\ &= (D_1\sin\theta_2 + F_1\cos\theta_2)E_2 - (D_2\sin\theta_2 + F_2\cos\theta_2)E_1 \end{aligned} \tag{8.22}$$

$$\begin{aligned} G_2 &= \begin{vmatrix} \sin\theta_2 & D_1\sin\theta_2 + F_1\cos\theta_2 \\ \cos\theta_2 & D_2\sin\theta_2 + F_2\cos\theta_2 \end{vmatrix} \\ &= (D_2\sin\theta_2 + F_2\cos\theta_2)\sin\theta_2 - (D_1\sin\theta_2 + F_1\cos\theta_2)\cos\theta_2. \end{aligned} \tag{8.23}$$

Now write Equation (8.19) for $i = 3$,

$$R_1\sin\theta_2 + R_2E_3 = D_3\sin\theta_2 + F_3\cos\theta_2, \tag{8.24}$$

substitute Equations (8.20) into Equation (8.24), and multiply through by G to obtain

$$G_1\sin\theta_2 + G_2E_3 = G(D_3\sin\theta_2 + F_3\cos\theta_2), \tag{8.25}$$

which is an equation with only one unknown θ_2. It can be rearranged into a more easily solved form. Substitute Equations (8.21), (8.22), and (8.23) into Equation (8.25) and expand the result to obtain

$$\begin{aligned} &[(D_1\sin\theta_2 + F_1\cos\theta_2)E_2 - (D_2\sin\theta_2 + F_2\cos\theta_2)E_1]\sin\theta_2 \\ &+ [(D_2\sin\theta_2 + F_2\cos\theta_2)\sin\theta_2 - (D_1\sin\theta_2 + F_1\cos\theta_2)\cos\theta_2]\,E_3 \\ &\qquad = [(E_2\sin\theta_2 - E_1\cos\theta_2)(D_3\sin\theta_2 + F_3\cos\theta_2)]. \end{aligned}$$

Further expansion yields

$$\begin{aligned} &D_1E_2\sin^2\theta_2 + F_1E_2\cos\theta_2\sin\theta_2 - D_2E_1\sin^2\theta_2 - F_2E_1\cos\theta_2\sin\theta_2 \\ &+D_2E_3\sin^2\theta_2 + F_2E_3\cos\theta_2\sin\theta_2 - D_1E_3\sin\theta_2\cos\theta_2 - F_1E_3\cos^2\theta_2 \\ &\quad = E_2D_3\sin^2\theta_2 - E_1D_3\cos\theta_2\sin\theta_2 + E_2F_3\cos\theta_2\sin\theta_2 - E_1F_3\cos^2\theta_2. \end{aligned}$$

Group the terms that multiply $\sin^2\theta_2$, the terms that multiply $\sin\theta_2\cos\theta_2$, and the terms that multiply $\cos^2\theta_2$ to obtain

$$\sin^2\theta_2\lambda_1 + \sin\theta_2\cos\theta_2\lambda_2 + \cos^2\theta_2\lambda_3 = 0,$$

that is,

$$\tan^2\theta_2\lambda_1 + \tan\theta_2\lambda_2 + \lambda_3 = 0, \tag{8.26}$$

where

$$\begin{aligned}\lambda_1 &= D_2(E_3 - E_1) + E_2(D_1 - D_3)\\ \lambda_2 &= E_2(F_1 - F_3) - D_1E_3 + E_1D_3 + F_2(E_3 - E_1)\\ \lambda_3 &= E_1F_3 - F_1E_3.\end{aligned} \tag{8.27}$$

Equation (8.26) can be solved for $\tan\theta_2$ with λ_1, λ_2, and λ_3 given by Equations (8.27).

8.1.4 Design Procedure

The design procedure is as follows:

1. Choose a value for ϕ and ψ which satisfy the desired function generation $f(\phi, \psi) = 0$, at which three derivatives of ψ with respect to ϕ will be matched. These values of ϕ and ψ are called the *design position* (or *precision point*) and are denoted as ϕ_{des} and ψ_{des} respectively.
2. According to Equation (8.8) we can compute the desired values of h_4, h_4' and h_4'' at the design position via,

$$h_4 = \frac{d\theta_4}{d\theta_2} = \left.\frac{d\psi}{d\phi}\right|_{\substack{\phi=\phi_{des}\\ \psi=\psi_{des}}} \quad h_4' = \frac{d^2\theta_4}{d\theta_2^2} = \left.\frac{d^2\psi}{d\phi^2}\right|_{\substack{\phi=\phi_{des}\\ \psi=\psi_{des}}} \quad \text{and } h_4'' = \frac{d^3\theta_4}{d\theta_2^3} = \left.\frac{d^3\psi}{d\phi^3}\right|_{\substack{\phi=\phi_{des}\\ \psi=\psi_{des}}}.$$

3. Knowing h_4, h_4' and h_4'' at the design position, the values of a_i, b_i, c_i and d_i in Table 8.3 are computed.
4. A value of θ_4 in the design position, denoted as $\theta_{4_{des}}$, is assumed and the values of D_i, E_i and F_i $(i = 1, 2, 3)$ in Equations (8.18) are computed.
5. The values of λ_1, λ_2 and λ_3 in Equations (8.27) are computed.
6. Solve Equation (8.26), a quadratic in $\tan\theta_2$. It gives two solutions for $\tan\theta_2$.
7. Each value of $\tan\theta_2$ gives a pair of solutions for θ_2 so there are a total of four values of θ_2. Choose one of these as the value of θ_2 in the design position, denoted as $\theta_{2_{des}}$.
8. G, G_1 and G_2 are then computed from Equations (8.21), (8.22), and (8.23).
9. R_1 and R_2 are computed from Equations (8.20) and R_3 from Equation (8.5) using $\theta_2 = \theta_{2_{des}}$ and $\theta_4 = \theta_{4_{des}}$.
10. One link length is assumed and the remaining three are computed from equations (8.4).
11. We can now compute the values of θ_{2_i} and θ_{4_i} from Equations (8.6) and (8.7) since we know that ϕ_{des} corresponds to $\theta_{2_{des}}$ and ψ_{des} corresponds to $\theta_{4_{des}}$, so

$$\theta_{2_i} = \theta_{2_{des}} - \phi_{des} \quad \text{and} \quad \theta_{4_i} = \theta_{4_{des}} - \psi_{des}.$$

This completes the design of a four bar mechanism which matches three derivatives of ψ with respect to ϕ at the chosen design postion ϕ_{des} and ψ_{des}.

It is possible that for the value of $\theta_{4_{des}}$ chosen in step 3, the quadratic in $\tan\theta_2$ in step 6 gives imaginary solutions. In that case the assumed value of $\theta_{4_{des}}$ is unacceptable. It is advisable to search through values of $\theta_{4_{des}}$ beforehand to see which range of $\theta_{4_{des}}$ values gives real solutions for the $\tan\theta_2$.

Note that a value of $\theta_{4_{des}}$ and a value of $(\theta_{4_{des}}+\pi)$ will give opposite values of E_i, D_i, and F_i in Equations (8.18). The values of λ_1, λ_2, and λ_3 in Equation (8.26) are products of these, so the double negative causes λ_1, λ_2, and λ_3 to be the same for these two values of θ_4. As a result, $\theta_{4_{des}}$ and $(\theta_{4_{des}}+\pi)$ yield the same quadratic for $\tan\theta_2$ and the same values of θ_2 (that is, $\theta_{2_{des}}$). Then the values of G and G_2 will be opposite, and the values of G_1 will be the same, in Equations (8.21), (8.22), and (8.23). This results in the same values for R_2 and opposite values of R_1 in Equations (8.20). So $\theta_{4_{des}}$ and $(\theta_{4_{des}}+\pi)$ will give opposite values of r_4 from Equation (8.4). But since the two values of $\theta_{4_{des}}$ differ by π, they produce the same vector $\bar{r}_4$, and thus $\theta_{4_{des}}$ and $(\theta_{4_{des}}+\pi)$ develop the same four bar mechanism. Hence it is only necessary to check for admissible values of $\theta_{4_{des}}$ in the range $0 \le \theta_{4_{des}} \le \pi$.

8.1.5 Number of Solutions

A word about the number of solutions. In step 5 there were two solutions for $\tan\theta_2$, and in step 6 each value of $\tan\theta_2$ generated a pair of values for $\theta_{2_{des}}$ that differ by π. Consider one of these pairs of $\theta_{2_{des}}$ and $(\theta_{2_{des}}+\pi)$ values. The sines and cosines of $\theta_{2_{des}}$ and $(\theta_{2_{des}}+\pi)$ will be opposite, which means that in step 7 the values of G and G_1 will be opposite, but the values of G_2 will be the same. Then, in step 8, the values of R_1 will be the same, but the values of R_2 will be opposite. Then, in step 9, if a length for r_1 is assumed (for example), the same values will be computed for r_4, and opposite values will be computed for r_2. But since the two values of $\theta_{2_{des}}$ differ by π, they produce the same vector $\bar{r}_2$, and thus $\theta_{2_{des}}$ and $(\theta_{2_{des}}+\pi)$ develop the same four bar mechanism. So in step 6 only one of $\theta_{2_{des}}$ or $(\theta_{2_{des}}+\pi)$ needs to be retained for each value of $\tan\theta_2$, and the problem has only two solutions, not four.

This is exactly the same as for the two values of θ_4, and this should be expected since θ_2 and θ_4 can be interchanged. In other words, who is to say which is the input and which is the output? The following example illustrates using the derivative-matching method to design a four bar linkage that generates a linear function.

▶ EXAMPLE 8.2

Design a four bar mechanism so that an output rotation ψ is related to an input rotation ϕ by the equation

$$\psi = -\phi + 10^\circ \quad (0^\circ < \psi < 110^\circ) \ (-100^\circ < \phi < 10^\circ) \tag{8.28}$$

over the indicated range of output and input. This function corresponds to a pair of externally rolling gears with equal pitch radii i.e. $R = -10$.

Solution

The process begins by choosing a "design position." This is the designer's free choice. The design position can be varied to try to optimize the mechanism's performance. Call these design values ψ_{des} and ϕ_{des}. Take $\psi_{des} = 55° = 0.960$ radian, in which case $\phi_{des} = -45°$ $= -0.440$ radian. At the design position, compute the first-, second-, and third-order kinematic coefficients:

$$h_4 = \frac{d\theta_4}{d\theta_2} = -1, \quad h_4' = \frac{d^2\theta_4}{d\theta_2^2} = 0, \text{ and } \quad h_4'' = \frac{d^3\theta_4^3}{d\theta_2^3} = 0.$$

This completes step 1. Now compute the elements in Table 8.3.

i	a_i	b_i	c_i	d_i
1	1	0	–2	0
2	0	–1	0	–4
3	1	0	–8	0

This completes step 2. Now arbitrarily choose $\theta_{4_{des}} = 135°$. From Equations (8.18), compute

$$\begin{aligned} &E_1 = 0.7071, \ E_2 = 0.7071, \ E_3 = 0.7071, \\ &D_1 = 1.4141, \ D_2 = -2.828, \ D_3 = 5.656, \\ &F_1 = 1.4141, \ \ F_2 = 2.828, \ \ F_3 = 5.656. \end{aligned}$$

This completes step 3, and from Equation (8.27),

$$\lambda_1 = -3, \qquad \lambda_2 = 0, \text{ and } \qquad \lambda_3 = 3.$$

This completes step 4, and from Equation (8.26),

$$-3\tan^2\theta_2 + 3 = 0,$$

from which

$$\tan^2\theta_2 = 1 \quad \Longrightarrow \quad \begin{aligned} &\tan\theta_2 = 1 \ \ \Longrightarrow \theta_2 = 45° \text{ or } \not{2}25° \\ &\tan\theta_2 = -1 \Longrightarrow \theta_2 = -45° \text{ or } 1\not{3}5°. \end{aligned}$$

The second value of θ_2 coming from each value of $\tan\theta_2$ has been "stricken out" because, as has been shown, it leads to the same four bar mechanism as the first value of θ_2. Either of the two remaining values of θ_2 is acceptable. Use $\theta_{2_{des}} = -45°$. This completes steps 5 and 6. From Equations (8.21), (8.22), and (8.23),

$$G = -1, \ \ G_1 = -2.828, \text{ and } \ \ G_2 = -2.828,$$

which completes step 7. R_1 and R_2 are computed from Equations (8.20) and R_3 from Equation (8.5):

$$R_1 = 2.828 \quad R_2 = 2.828 \quad R_3 = 3.0,$$

which completes step 8. Take $r_1 = 10$ inches. Then, from Equation (8.4),

$$r_4 = \frac{r_1}{R_1} = 3.535 \text{ inches}$$

$$r_2 = \frac{r_1}{R_2} = 3.535 \text{ inches}$$

$$r_3 = \sqrt{r_1^2 + r_2^2 + r_4^2 - 2R_3 r_2 r_4} = 7.072 \text{ inches},$$

which completes step 9, and the four bar mechanism is known. Figure 8.7 shows a performance plot. In the design position, the slope of the curve, which is $d\theta_4/d\theta_2 = h_4$, is −1. It remains nearly so over a large range because the second and third derivatives are both zero in the design position.

The following example illustrates the use of the derivative-matching method to design a four bar linkage that generates a nonlinear function. You should note that this is no more or less general than a mechanism that generates a linear function. Both require matching the first, second, and third derivatives. In the case of the linear function, the

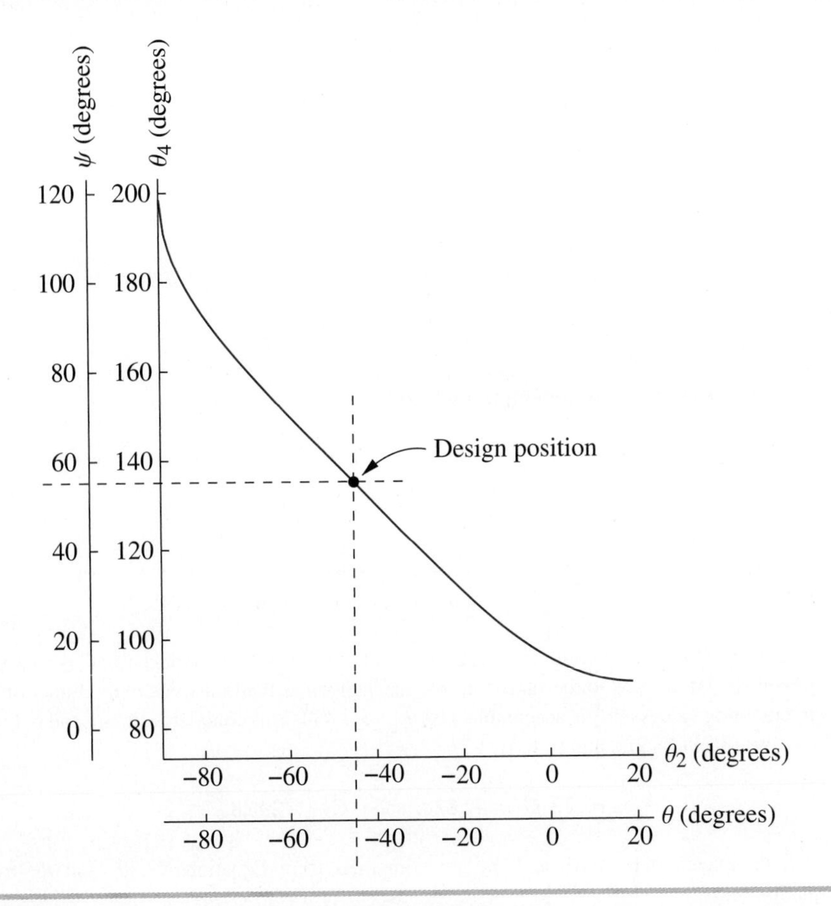

FIGURE 8.7 θ_4 versus θ_2 for the developed design

second and third derivatives are zero. It is no easier to match zero-value derivatives than nonzero-valued derivatives.

▶ EXAMPLE 8.3

Design a four bar mechanism so that the output rotation, ψ_d (in degrees), is related to the input rotation, ϕ_d (in degrees), by the equation

$$\psi_d = C_1\phi_d^2 - C_2\phi_d^3 \quad (0° < \phi_d < 90°) \text{ and } (0° < \psi_d < -155.78°), \tag{8.29}$$

where

$$C_1 = \frac{1}{100°} \quad \text{and} \quad C_2 = \frac{1}{(54.77°)^2}.$$

Figure 8.8 shows the desired function generation.

Solution

Equation (8.29) must be converted into a relationship between ψ and ϕ where each rotation is in units of radians; otherwise, the units on the derivatives h_4, h_4', and h_4'' will be incorrect. Making the substitution,

$$\phi_d = \phi\frac{180°}{\pi} \quad \text{and} \quad \psi_d = \psi\frac{180°}{\pi},$$

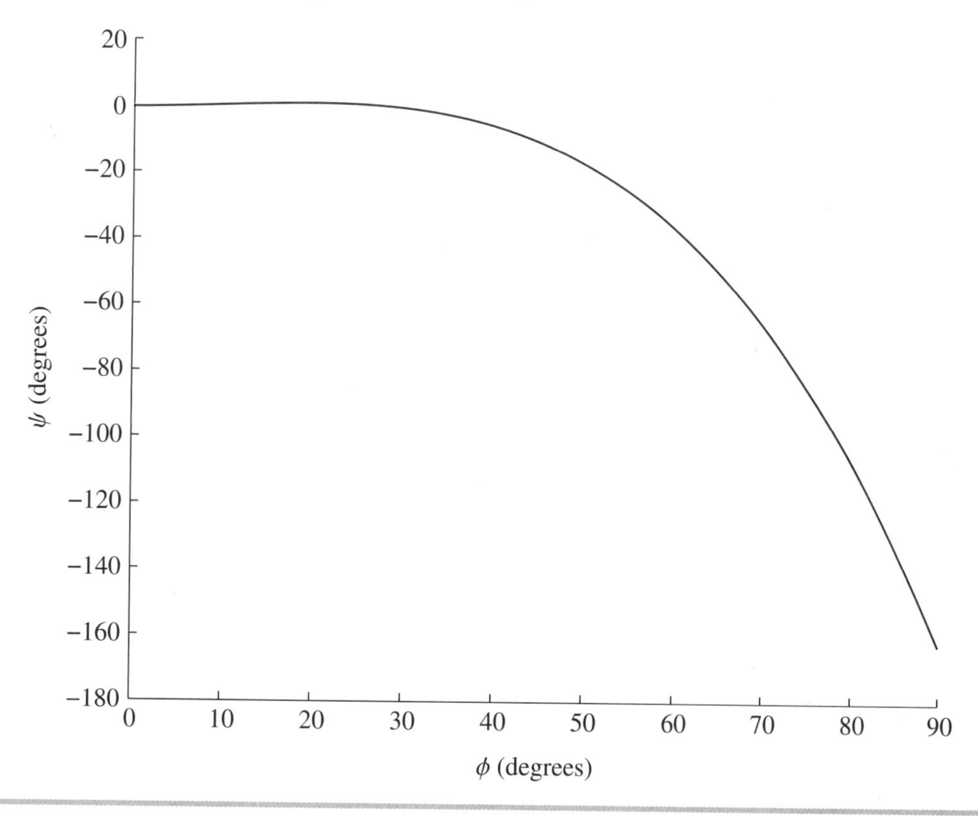

FIGURE 8.8 Desired function generation

gives

$$\psi = 0.573\phi^2 - 1.094\phi^3. \tag{8.30}$$

Equation (8.30) is differentiated three times, which, recalling Equation (8.8), gives

$$h_4 = \frac{d\theta_4}{d\theta_2} = \frac{d\psi}{d\phi} = 1.146\phi - 3.282\phi^2 \tag{8.31}$$

$$h_4' = \frac{d^2\theta_4}{d\theta_2^2} = \frac{d^2\psi}{d\phi^2} = 1.146 - 6.564\phi \tag{8.32}$$

$$h_4'' = \frac{d^3\theta_4}{d\theta_2^3} = \frac{d^3\psi}{d\phi^3} = -6.564 \tag{8.33}$$

$$h_4''' = \frac{d^4\theta_4}{d\theta_2^4} = \frac{d^4\psi}{d\phi^4} = 0. \tag{8.34}$$

Note these are dimensionless, as they should be. We will consider the fourth derivative when we try to optimize the design.

A design position ϕ_{des} and ψ_{des} must be selected, at which point three derivatives of ψ with respect to ϕ will be matched. Looking at Figure 8.8, it seems reasonable to choose a value of ϕ for the design position as $\phi_{des} = 55°$. From Equation (8.29) the corresponding value of ψ in the design position is $\psi_{des} = -25.21°$. In step 1, the values of the derivatives to the third order are computed from Equations (8.31), (8.32), and (8.33) in the design position:

$$h_4 = -1.93, \quad h_4' = -5.16, \text{ and } \quad h_4'' = -6.57.$$

Step 2 computes the values of a_i, b_i, c_i, and d_i $(i = 1, 2, 3)$ in Table 8.3.

i	a_i	b_i	c_i	d_i
1	1.93	0	−2.93	0
2	5.16	−3.71	−5.16	−8.56
3	0.57	29.78	−18.46	45.25

Choose a value of θ_4 in the design position as $\theta_{4_{des}} = 155°$; then, in step 3, the values of E_i, D_i, and F_i $(i = 1, 2, 3)$ are computed from Equations (8.18):

$$\begin{array}{lll} E_1 = 0.81 & E_2 = 5.53 & E_3 = -26.75 \\ D_1 = 2.65 & D_2 = 1.06 & D_3 = 35.85 \\ F_1 = 1.24 & F_2 = 9.93 & F_3 = -33.21. \end{array}$$

In step 4 the values of λ_1, λ_2, and λ_3 are computed from Equation (8.27):

$$\lambda_1 = -213.02, \quad \lambda_2 = 17.03, \text{ and } \quad \lambda_3 = 6.05.$$

Using these values in Equation (8.26) and solving the quadratic give two values for $\tan\theta_2$ in step 5. In step 6 each of these values for $\tan\theta_2$ gives two values of θ_2, only one of which is retained from each $\tan\theta_2$. The two retained values are $\theta_2 = (172.41°, 12.03°)$; the first of these will be used, so $\theta_{2_{des}} = 172.41°$. In step 7 the values of G, G_1, and G_2 are computed. In step 8 the values of R_1, R_2, and R_3 are computed. In step 9 the value of r_1 is assumed to be 10 inches, and the remaining link lengths are computed, giving the four bar mechanism,

$$\theta_{2_{des}} = 172.41°, \quad r_1 = 10 \text{ inches}, \quad r_2 = -7.15 \text{ inches},$$
$$r_3 = 3.50 \text{ inches, and} \quad r_4 = 5.04 \text{ inches.}$$

Notice the negative value of r_2. To accommodate a positive value of r_2, $\theta_{2_{des}}$ is changed to $\theta_{2_{des}} = 172.41° - 180° = -7.59°$. The mechanism is shown in the design position as the solid image in Figure 8.9. The initial values of θ_2 and θ_4 are given by

$$\theta_{2_{initial}} = \theta_{2_{des}} - \phi_{des} = -7.59° - 55° = -62.79°$$
$$\theta_{4_{initial}} = \theta_{4_{des}} - \psi_{des} = 155° - (-25.21°) = 180.21°.$$

The dashed image shows the mechanism as near to $\phi = 0°$ as it can come. The input is about 6° short of reaching $\phi = 0°$ (that is, of reaching $\theta_{2_i} = \theta_{2_{des}} - \phi_{des} = -62.59°$). The performance curve is shown in Figure 8.10. Away from the design position, the four bar deviates from the desired function fairly quickly. A check on the fourth derivative in the design position reveals $h_4''' = 266.38$, which is considerably far from the desired value of zero in Equation (8.34). So an improved design should be possible.

Some numerical experimentation showed that as the value of $\theta_{4_{des}}$ was reduced, the value of the fourth-order kinematic coefficient got smaller and closer to the desired value of zero. There came a limiting point, however, since links 3 and 4 were becoming 20 times as long as link 2. A value of $\theta_{4_{des}} = 85°$ was settled on, giving a four bar mechanism,

$$\theta_{2_{des}} = 158.28°, \quad r_1 = 10 \text{ inches}, \quad r_2 = -6.67 \text{ inches},$$
$$r_3 = 53.53 \text{ inches, and} \quad r_4 = 50.63 \text{ inches,}$$

whose performance curve is shown in Figure 8.11.

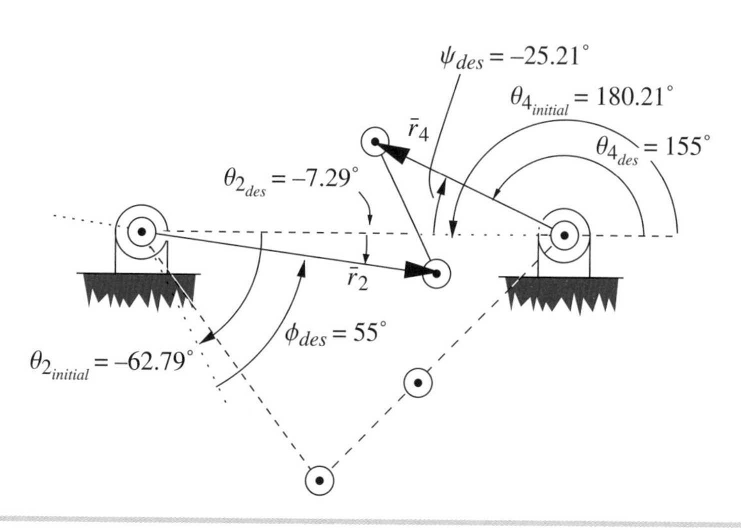

FIGURE 8.9 Design with $\phi_{des} = 55°$ and $\theta_{4_{des}} = 155°$

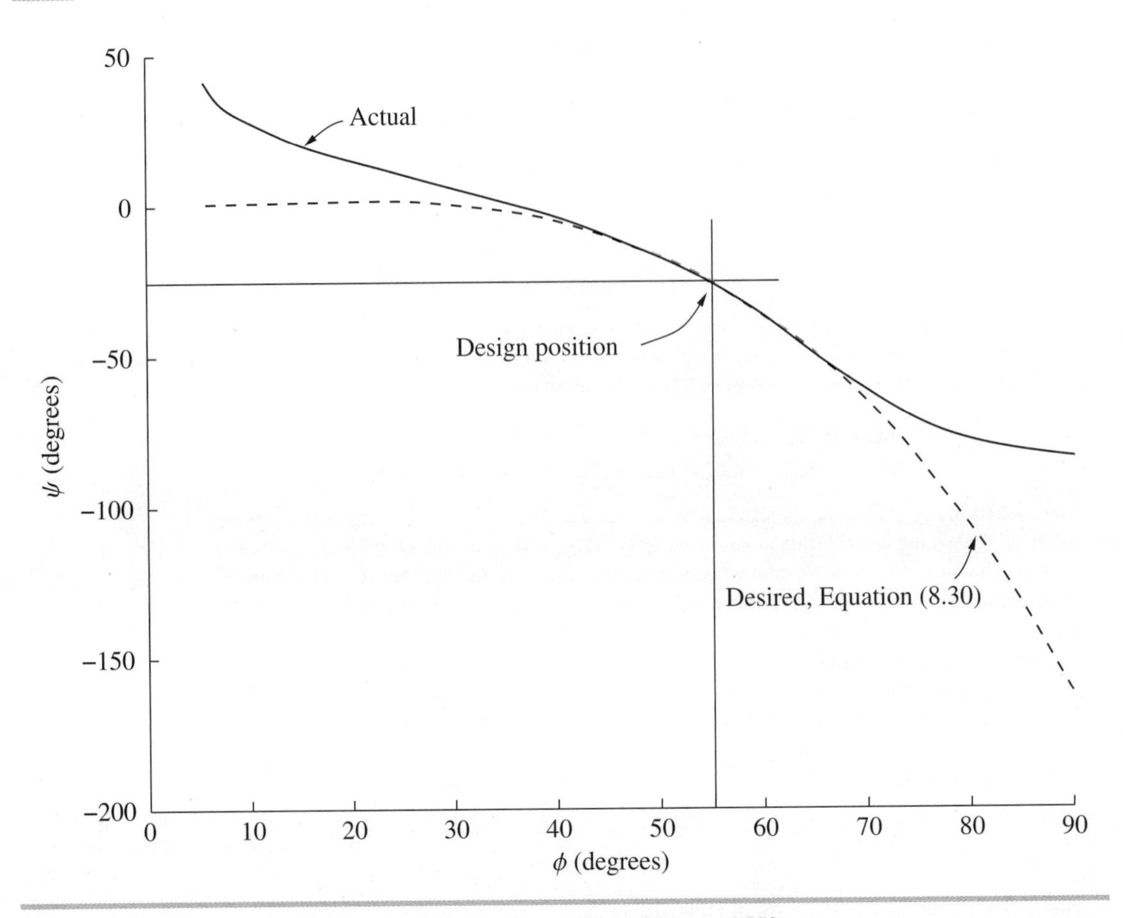

FIGURE 8.10 Design with performance curve $\phi_{des} = 55°$ and $\theta_{4_{des}} = 155°$

Once again the mechanism is short of reaching the initial position where $\phi = 0°$, but it does considerably better at matching the desired function because it has a fourth-order kinematic coefficient $h_4''' = 80.15$. The error is worse toward the end of the input range, so it was thought to move the design position a small amount in that direction by choosing $\phi_{des} = 58°$. Looking at the fourth-order kinematic coefficient for different values of $\theta_{4_{des}}$ showed that $\theta_{4_{des}} = 230°$ had reasonable link lengths and a good fourth-order kinematic coefficient $h_4''' = 12.23$. This four bar mechanism was given by

$$\theta_{2_{des}} = 161.91°, \quad r_1 = 10 \text{ inches}, \quad r_2 = -6.04 \text{ inches},$$
$$r_3 = 13.68 \text{ inches, and} \quad r_4 = 17.70 \text{ inches}.$$

The performance curve is shown in Figure 8.12. At first glance it looks quite good. It is able to start at $\phi = 0°$, but it comes far short of the maximum value of $\phi = 90°$, so its performance on that end is not as good as the previous mechanism. Overall, however, it does seem better. It would be beneficial to consider the designs that would develop from using the alternative values of $\theta_{2_{des}}$ that come out of step 6, as well as other values of ϕ_{des} and $\theta_{4_{des}}$. A mechanism that realizes the entire input range and closely matches the desired function generation is very likely to exist.

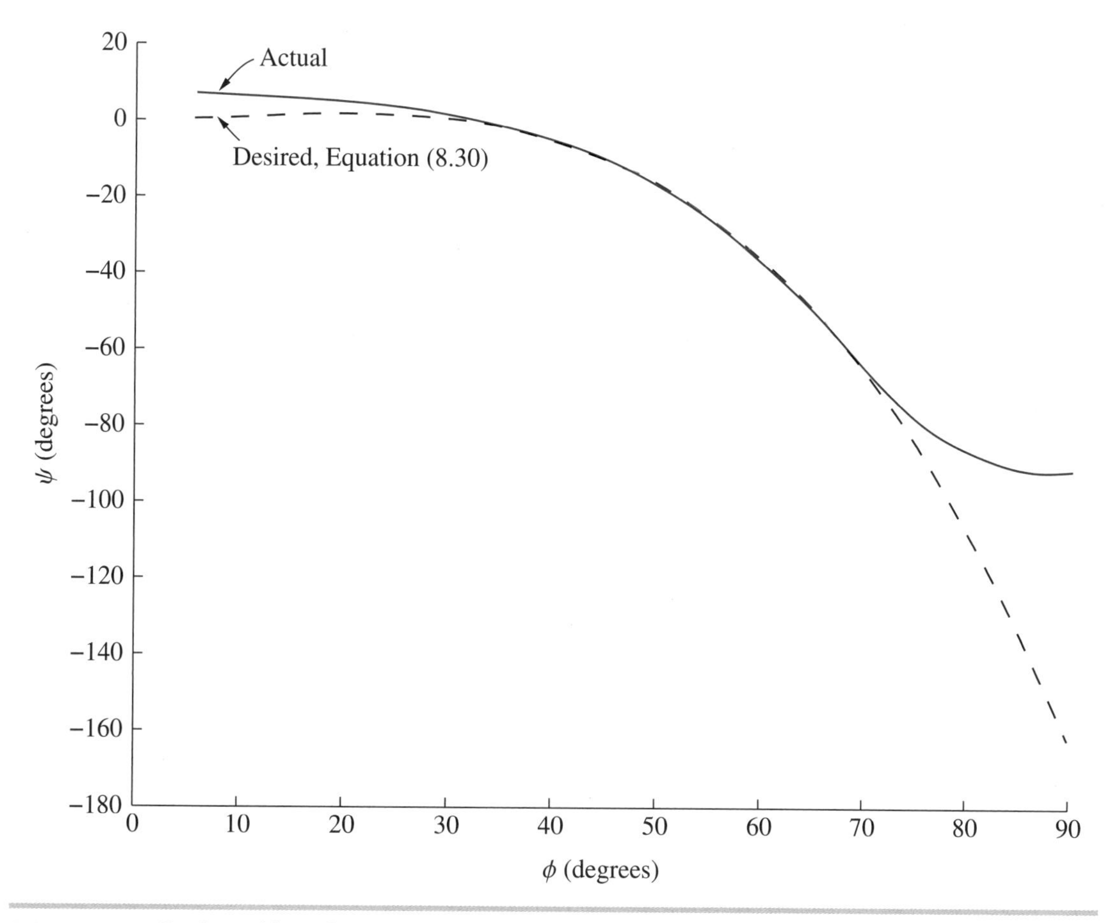

FIGURE 8.11 Design with performance curve $\phi_{des} = 55°$ and $\theta_{4_{des}} = 85°$

8.2 FREUDENSTEIN'S EQUATION FOR THE CRANK-SLIDER MECHANISM

DEFINITION:

Function Generation in Crank-Slider Mechanisms *A design (synthesis) problem where the link lengths of a crank-slider mechanism must be determined so that the translation r and the rotation ϕ are functionally related.*

The desired relation is represented by $f(\phi, r) = 0$. This is illustrated in Figure 8.13. Figure 8.14 shows a crank-slider mechanism and its vector loop. The vector loop equation is

$$\overline{r}_2 + \overline{r}_3 - \overline{r}_4 - \overline{r}_1 = \overline{0},$$

with X and Y components

$$r_2\cos\theta_2 + r_3\cos\theta_3 - r_4 = 0$$
$$r_2\sin\theta_2 + r_3\sin\theta_3 - r_1 = 0.$$

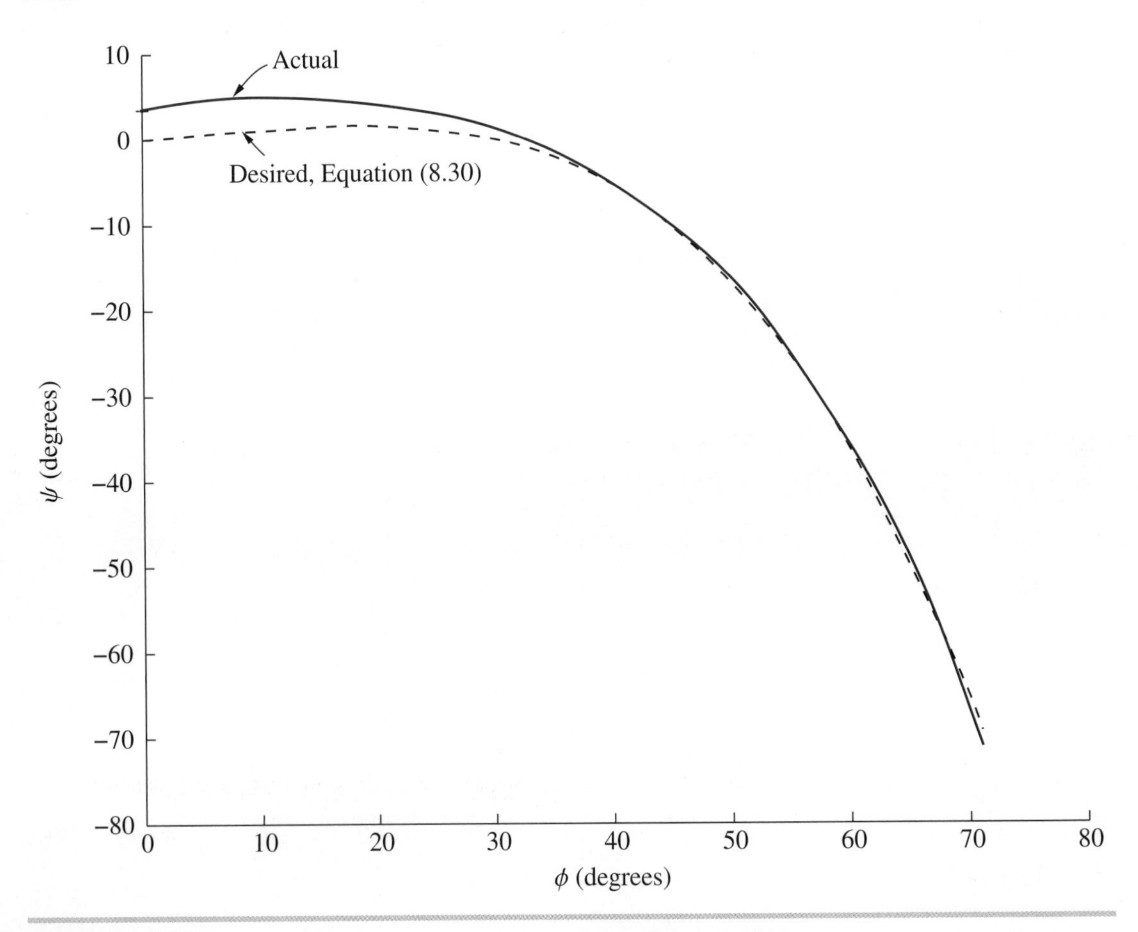

FIGURE 8.12 Design with performance curve $\phi_{des} = 58°$ and $\theta_{4_{des}} = 230°$

Rewriting yields

$$r_3 \cos\theta_3 = -r_2 \cos\theta_2 + r_4$$
$$r_3 \sin\theta_3 = -r_2 \sin\theta_2 + r_1.$$

Then, squaring both sides, adding the equations together, and rearranging give **Freudenstein's Equation for the crank-slider mechanism**:

$$\boxed{S_1 r_4 \cos\theta_2 + S_2 \sin\theta_2 + S_3 = r_4^2} \tag{8.35}$$

where

$$S_1 = 2r_2, \quad S_2 = 2r_1 r_2, \quad \text{and} \quad S_3 = r_3^2 - r_2^2 - r_1^2. \tag{8.36}$$

Unlike the four bar mechanism, function-generating crank-slider mechanisms are not scale invariant because, unlike R_1, R_2 and R_3 in Equation (8.4), S_1, S_2 and S_3 in Equation (8.36) are not dimensionless ratios.

8.2.1 Function Generation in Crank-Slider Mechanisms

As seen in Figures 8.13 and 8.14, the rotation θ_2 and displacement r_4 are referenced differently than rotation ϕ and displacement r in the function generation. The references for these displacements are related, and this can be seen in Figure 8.15.

The rotations are related by

$$\phi = \theta_2 - \theta_{2_i}, \tag{8.37}$$

and the displacements are related by

$$r = r_4 - r_{4_i}, \tag{8.38}$$

where θ_{2_i} and r_{4_i} are constants. Differentiating Equation (8.37) and Equation (8.38) shows that, in a manner similar to the four bar mechanism,

$$d\phi = d\theta_2 \quad \text{and} \quad dr = dr_4.$$

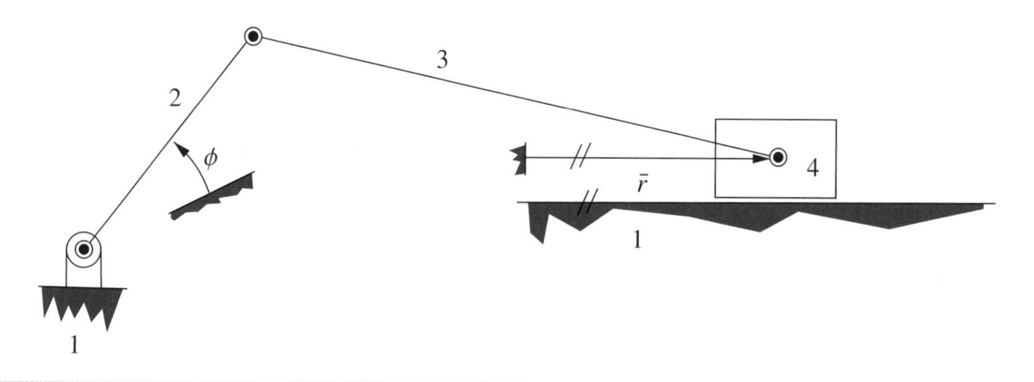

FIGURE 8.13 A function-generating crank-slider mechanism developing $f(\phi, r) = 0$

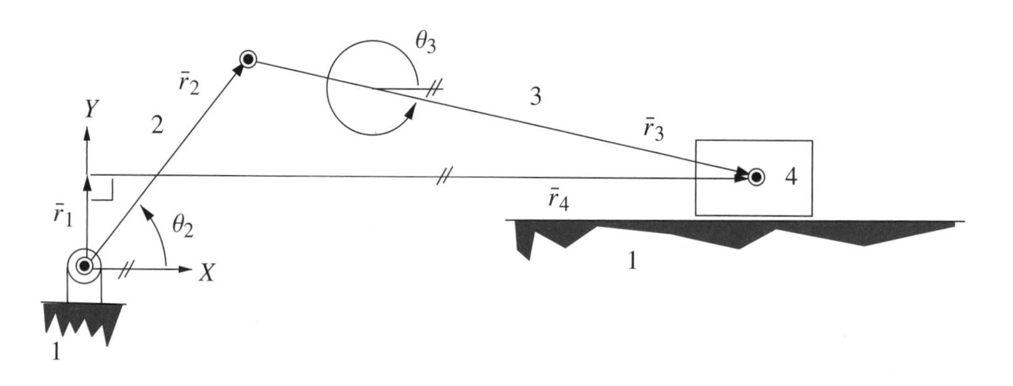

FIGURE 8.14 The vector loop for the crank-slider mechanism

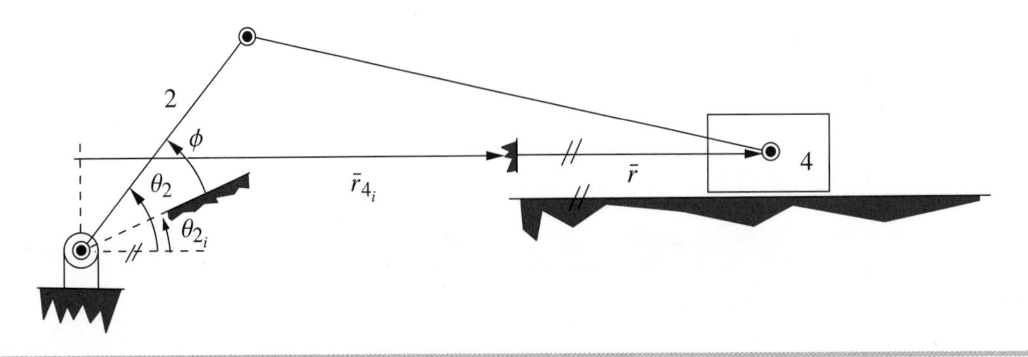

FIGURE 8.15 Relationship among θ_2, r_4, ϕ, and r

8.2.2 Point-Matching Method with Freudenstein's Equation

The point-matching method with Freudenstein's Equation is a method by which a crank slider mechanism is designed to match three desired pairs of $\{\phi, r\}$ values that satisfy the desired function generation $f(\phi, r) = 0$. These three pairs of values are known as the *precision points* and will be denoted as

$$\{(\phi)_1, (r)_1\}, \quad \{(\phi)_2, (r)_2\}, \quad \{(\phi)_3, (r)_3\}.$$

These three precision points are free choices of the designer. The designer then chooses values for θ_{2_i} and r_{4_i}. These are also free choices and again, there is a lot of room for optimization through varying these five free choices.

Using Equations (8.37) and (8.38), the three $\{\phi, r\}$ precision points are converted into three $\{\theta_2, r_4\}$ precision points denoted as

$$\{(\theta_2)_1, (r_4)_1\}, \quad \{(\theta_2)_2, (r_4)_2\}, \quad \{(\theta_2)_3, (r_4)_3\}.$$

Substituting each $\{\theta_2, r_4\}$ pair into Equation (8.35) yields a system of three equations that are linear in the three unknowns S_1, S_2 and S_3,

$$\begin{bmatrix} (r_4)_1\cos(\theta_2)_1 & \sin(\theta_2)_1 & 1 \\ (r_4)_2\cos(\theta_2)_2 & \sin(\theta_2)_2 & 1 \\ (r_4)_3\cos(\theta_2)_3 & \sin(\theta_2)_3 & 1 \end{bmatrix} \begin{bmatrix} S_1 \\ S_2 \\ S_3 \end{bmatrix} = \begin{bmatrix} (r_4^2)_1 \\ (r_4^2)_2 \\ (r_4^2)_3 \end{bmatrix}. \tag{8.39}$$

Solve Equation (8.39) for S_1, S_2, and S_3, then solve Equations (8.36) for the dimensions r_2, r_3, and r_1. The following example illustrates the design of a flowmeter using the point-matching method of Freudenstein's Equation for a crank-slider mechanism.

▶ EXAMPLE 8.4

Figure 8.16 shows a mechanical flowmeter. The flow of an incompressible liquid in a pipe passes through an orifice. The orifice causes a pressure drop so that the fluid pressure before the orifice, P_{high}, is greater than the fluid pressure after the orifice, P_{low}. The pressure drop, ΔP, is given by $\Delta P = P_{high} - P_{low}$. ΔP is related to the volumetric flow rate $\dot{q}$ by the equation

$$\Delta P = C_1\dot{q}^2, \tag{8.40}$$

where C_1 is determined experimentally. (You will learn this in fluid mechanics.)

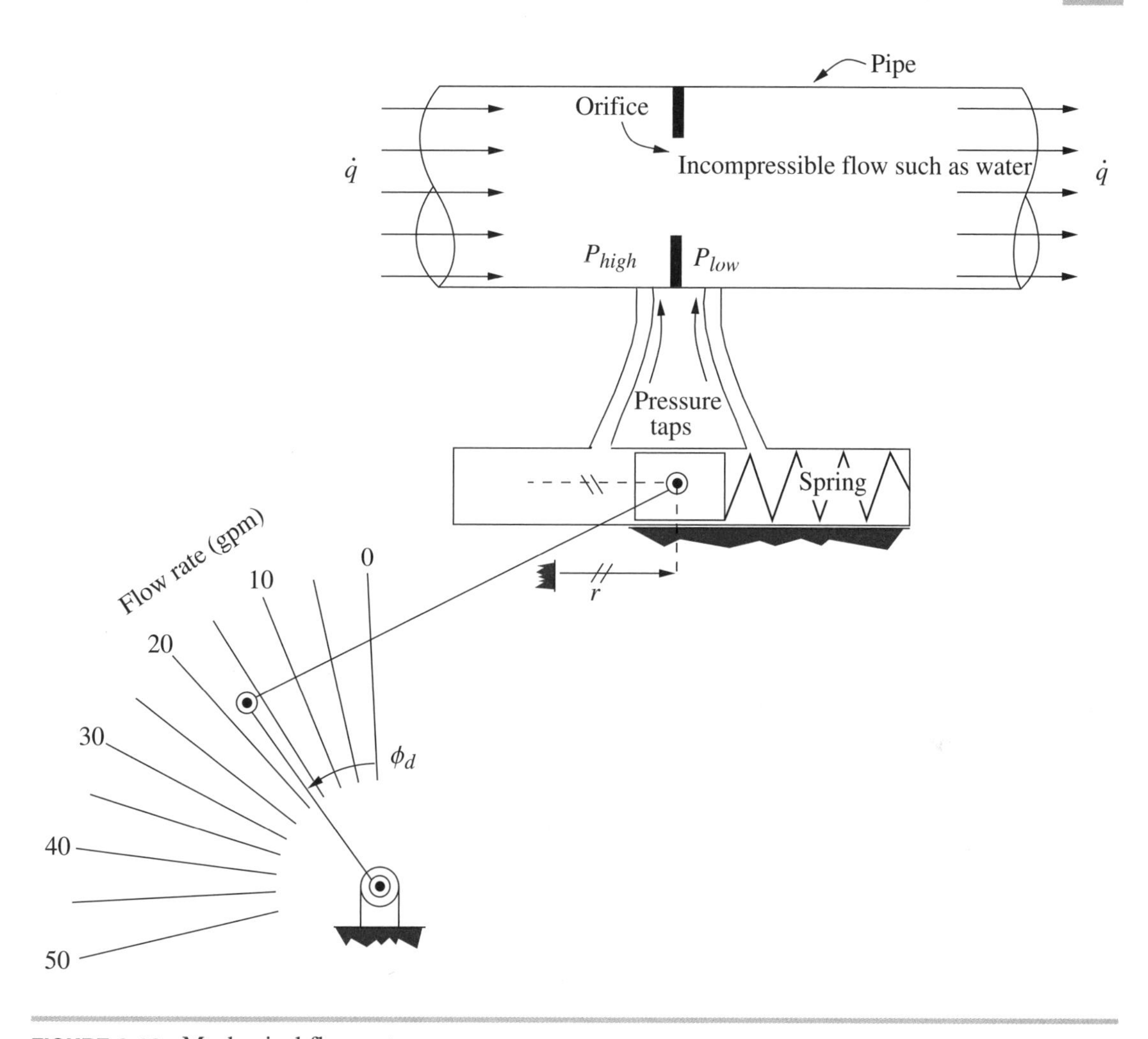

FIGURE 8.16 Mechanical flowmeter

Pressure taps on opposite sides of the orifice feed some of the fluid into a cylinder, where each pressure exerts a force on one face of a slider. The pressure drop ΔP causes the slider to move to the right, until it is statically equilibrated by a compression spring.

ΔP is linearly related to slider displacement, r:

$$\Delta P = C_2 r, \tag{8.41}$$

where C_2 is the ratio of spring stiffness to cross-sectional area of the piston (slider), and r is referenced from the position of the pin joint on the slider, which corresponds to the unstretched spring length. Incorporating Equation (8.41) in Equation (8.40) relates the flow velocity to the slider displacement:

$$C_1 \dot{q}^2 = C_2 r \quad \longrightarrow \quad r = \frac{C_1}{C_2} \dot{q}^2.$$

In this system, when the flow is measured in gallons per minute (gpm), the ratio C_1/C_2 has a value of $0.00240 \frac{\text{in.}}{(\text{gpm})^2}$,

$$r = 0.00240\dot{q}^2. \tag{8.42}$$

Equation (8.42) relates the flow rate and the slider displacement. The relationship is nonlinear. The movement of the slider could be used to indicate flow rate, but the scale indicating the flow rate would also be nonlinear, and this is undesirable in a meter.

To develop a linear flowmeter, connect the slider to a crank as shown in Figure 8.16, and design the crank-slider mechanism so that the flow rate can be read from a linear scale on the crank rotation ϕ, as shown. To do this, choose a sensitivity C_3 for this scale. We choose to have a 2° change in ϕ correspond to a change in flow of 1 gpm, which should give us a very readable scale, making C_3 equal to $0.5 \frac{(\text{gpm})}{\text{degree}}$.

$$\dot{q} = C_3\phi_d \quad (\phi_d \text{ is } \phi \text{ in degrees}) \tag{8.43}$$

$$\dot{q} = 0.5\phi_d. \tag{8.44}$$

Substituting Equation (8.44) into Equation (8.42) gives the required function generation:

$$r = C_4\phi_d^2, \quad \text{where } C_4 = 0.00060 \frac{\text{in.}}{\text{degree}^2}, \tag{8.45}$$

where ϕ_d is in degrees and r_4 is in inches. This relationship will cause the scale of the flowmeter to read increasing flow rates to the left, as Figure 8.16 shows, which might be awkward. If you want the scale to read increasing to the right, then make C_4 negative.

Solution

Choose the precision points $\{\phi, r_4\}$ shown in Table 8.4, which can be adjusted later to improve performance.

In order to use Freudenstein's Equation, you need to convert these into $\{\theta_2, r_4\}$ pairs by using Equations (8.37) and (8.38). The values of θ_{2_i} and r_{4_i} in Equations (8.37) and (8.38) are free choices. You can also use them to optimize your design. Use $r_{4_i} = 2$ in. and $\theta_{2_i} = 30°$. These were arbitrary choices.

Substituting the values in each row of Table 8.5 into Equation (8.39) gives

$$\begin{bmatrix} 1.4398 & 0.7660 & 1.0000 \\ 0.0000 & 1.0000 & 1.0000 \\ -3.4300 & 0.8660 & 1.0000 \end{bmatrix} \begin{bmatrix} S_1 \\ S_2 \\ S_3 \end{bmatrix} = \begin{bmatrix} 5.0176 \\ 17.3056 \\ 47.0598 \end{bmatrix}.$$

Solving gives

$$S_1 = -8.6474, \quad S_2 = -0.6965, \text{ and } S_3 = 18.0021,$$

ϕ (degrees)	r (inches)
20°	0.24
60°	2.16
90°	4.86

TABLE 8.4 The three assumed $\{\phi, r\}$ precision points

θ_2 (degrees)	r_4 (inches)
50°	2.24
90°	4.16
120°	6.86

TABLE 8.5 The three $\{\theta_2, r_4\}$ precision points

θ_2 (degrees)	r_4 (inches)
230°	2.24
270°	4.16
300°	6.86

TABLE 8.6 $\{\theta_2, r_4\}$ Precision points adjusted to obtain positive values of r_2

and from Equation (8.36),

$$r_2 = \frac{S_1}{2} = -4.3237 \text{ inches}$$

$$r_1 = \frac{S_2}{2r_2} = 0.0805 \text{ inches}$$

$$r_3 = (S_3^2 + r_2^2 + r_1^2)^{\frac{1}{2}} = 6.0583 \text{ inches.}$$

The vector $\bar{r}_2$ has a negative magnitude. This means the vector is pointing in the opposite direction from θ_2, which is $\theta_2 \pm 180°$ (not $-\theta_2$!). Thus Table 8.5 becomes and the value of θ_{2_i} goes from 30° to 210°.

Figure 8.17 shows the performance of the system. The results are quite impressive. To improve performance, the first precision point could be moved down 10° or 15°. One might even consider moving the first and third precision points to $\phi = 0°$ and $\phi = 180°$.

8.2.3 Derivative-Matching Method with Freudenstein's Equation

Here Freudenstein's Equation will be used to design a crank-slider mechanism so that it matches a value of r_4 and three derivatives of r_4 with respect to θ_2, denoted as

$$f_4 = \frac{dr_4}{d\theta_2}, \quad f_4' = \frac{d^2 r_4}{d\theta_2^2}, \text{ and } \quad f_4'' = \frac{d^3 r_4}{d\theta_2^3}, \tag{8.46}$$

where f_4, f_4', and f_4'', are first-, second-, and third-order kinematic coefficients of the crank-slider mechanism. Recall Freudenstein's Equation for the the crank-slider mechanism, Equation (8.35). Differentiating it three times with respect to θ_2 yields

$$S_1(E_1\cos\theta_2 + D_1\sin\theta_2) + S_2\cos\theta_2 = F_1 \tag{8.47}$$

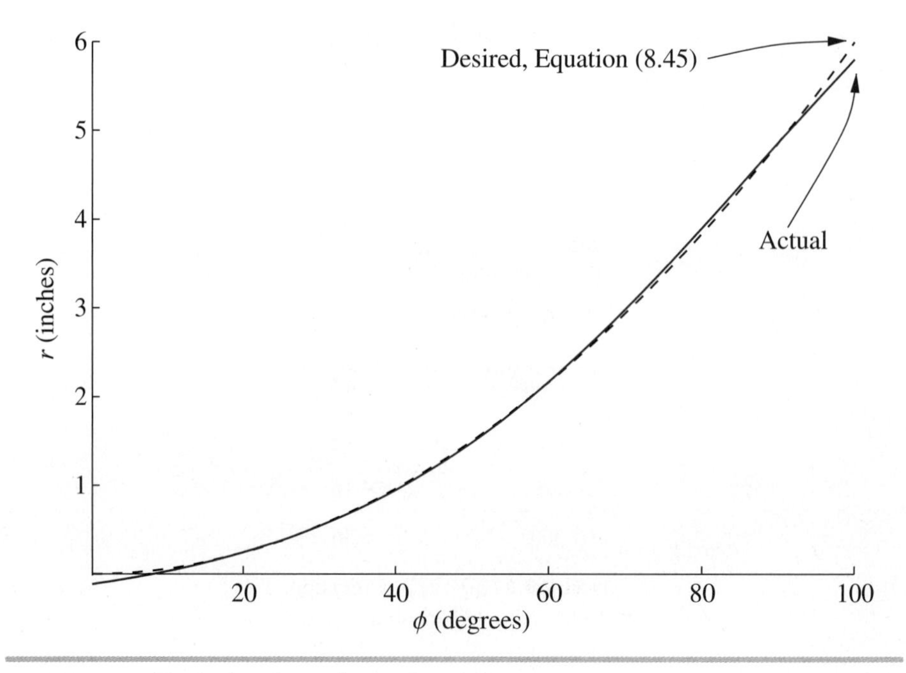

FIGURE 8.17 Desired and actual r vs. ϕ

$$S_1(E_2\cos\theta_2 + D_2\sin\theta_2) + S_2\sin\theta_2 = F_2 \tag{8.48}$$

$$S_1(E_3\cos\theta_2 + D_3\sin\theta_2) + S_2\cos\theta_2 = F_3, \tag{8.49}$$

where

$$\begin{array}{lll} E_1 = f_4 & D_1 = -r_4 & F_1 = 2r_4 f_4 \\ E_2 = r_4 - f_4' & D_2 = 2f_4 & F_2 = -2(f_4^2 + r_4 f_4') \\ E_3 = 3f_4 - f_4'' & D_3 = 3f_4' - r_4 & F_3 = -2(3f_4 f_4' + r_4 f_4''). \end{array} \tag{8.50}$$

Equation (8.35) and Equations (8.47) through (8.49) are a system of four equations in four unknowns: S_1, S_2, S_3, and θ_2. Equations (8.47) through (8.49) are a subset of three equations in three unknowns: S_1, S_2, and θ_2. Solve these three equations, and then return to Equation (8.35) to find S_3.

Write the first two Equations (8.47) and (8.48) in matrix form, and use them to solve for S_1 and S_2 in terms of θ_2,

$$\begin{bmatrix} (E_1\cos\theta_2 + D_1\sin\theta_2) & \cos\theta_2 \\ (E_2\cos\theta_2 + D_2\sin\theta_2) & \sin\theta_2 \end{bmatrix} \begin{bmatrix} S_1 \\ S_2 \end{bmatrix} = \begin{bmatrix} F_1 \\ F_2 \end{bmatrix}.$$

Using Cramer's Rule yields

$$S_1 = \frac{G_1}{G} \quad \text{and} \quad S_2 = \frac{G_2}{G}, \tag{8.51}$$

where

$$\begin{aligned} G_1 &= F_1 \sin\theta_2 - F_2 \cos\theta_2 \\ G_2 &= (E_1 \cos\theta_2 + D_1 \sin\theta_2)F_2 - (E_2 \cos\theta_2 + D_2 \sin\theta_2)F_1 \\ G &= (E_1 \cos\theta_2 + D_1 \sin\theta_2)\sin\theta_2 - (E_2 \cos\theta_2 + D_2 \sin\theta_2)\cos\theta_2. \end{aligned} \tag{8.52}$$

Substituting Equations (8.51) for S_1 and S_2 into Equation (8.49) gives one equation in the one unknown θ_2:

$$G_1(E_3 \cos\theta_2 + D_3 \sin\theta_2) + G_2 \cos\theta_2 = GF_3.$$

Expanding and collecting the coefficients of $\sin^2\theta_2, \sin\theta_2\cos\theta_2$, and $\cos^2\theta_2$ gives

$$\lambda_1 \sin^2\theta_2 + \lambda_2 \sin\theta_2 \cos\theta_2 + \lambda_3 \cos^2\theta_2 = 0,$$

that is,

$$\lambda_1 \tan^2\theta_2 + \lambda_2 \tan\theta_2 + \lambda_3 = 0, \tag{8.53}$$

where

$$\begin{aligned} \lambda_1 &= F_1 D_3 - F_3 D_1 \\ \lambda_2 &= F_1 E_3 - F_3 E_1 + D_1 F_2 - D_2 F_1 + F_3 D_2 - F_2 D_3 \\ \lambda_3 &= E_1 F_2 - E_2 F_1 + F_3 E_2 - F_2 E_3. \end{aligned} \tag{8.54}$$

Equation (8.53) can be solved for two values of $\tan\theta_2$.

8.2.4 Design Procedure

The design procedure is as follows:

1. Choose a value for ϕ and r that satisfy the desired function generation $f(\phi, r) = 0$, at which three derivatives of r with respect of ϕ will be matched. These values of ϕ and r are called the *design position* (or *precision point*) and are denoted as ϕ_{des} and r_{des} respectively.
2. From Equations (8.37) and (8.38) we see that $d\theta_2 = d\phi$ and $dr_4 = dr$ so the desired values of f_4, f_4' and f_4'' at the design position can be computed via,

$$f_4 = \frac{dr_4}{d\theta_2} = \left.\frac{dr}{d\phi}\right|_{\substack{\phi=\phi_{des}\\ r=r_{des}}} \quad f_4' = \frac{d^2r_4}{d\theta_2^2} = \left.\frac{d^2r}{d\phi^2}\right|_{\substack{\phi=\phi_{des}\\ r=r_{des}}} \quad \text{and } f_4'' = \frac{d^3r_4}{d\theta_2^3} = \left.\frac{d^3r}{d\phi^3}\right|_{\substack{\phi=\phi_{des}\\ r=r_{des}}}.$$

3. A value for $r_4 = r_{4_{des}}$ is assumed.
4. Compute E_i, D_i and F_i $(i = 1, 2, 3)$ in Equation (8.50).
5. Compute λ_1, λ_2 and λ_3 in Equation (8.54).
6. Solve Equation (8.53) for two values of $\tan\theta_2$, and keep either value. Either is legitimate.
7. For the chosen value of $\tan\theta_2$, compute two values of θ_2. These are in fact the same solution. This sets the value of $\theta_{2_{des}}$.
8. Compute G_1, G_2, G then S_1 and S_2 from Equations (8.52) and (8.51).
9. Compute S_3 from Equation (8.35) using $\theta_2 = \theta_{2_{des}}$ and $r_4 = r_{4_{des}}$.

10. Compute the links lengths from Equation (8.36).
11. We can now compute the values of θ_{2_i} and r_{4_i} from Equations (8.37) and (8.38) since we know that ϕ_{des} corresponds to $\theta_{2_{des}}$ and r_{des} corresponds to $r_{4_{des}}$, so

$$\theta_{2_i} = \theta_{2_{des}} - \phi_{des} \quad \text{and} \quad r_{4_i} = r_{4_{des}} - r_{des}.$$

8.2.5 Number of Solutions

In step 7, you will compute two values of θ_2 that differ by π and have opposite signed sines and cosines. Working this sign change through Equations (8.52), (8.51), and finally Equation (8.36) shows that the value of r_2 is opposite signed. This means the two solutions for θ_2 give the same design, so only one of them needs to be considered.

▶ EXAMPLE 8.5

Here we will consider the flowmeter design in Example 8.4, but this time we will use the derivative-matching method. You should see that although this method is more complicated, it produces better designs.

Solution

Recall Equation (8.45), the required function generation. It relates the slider position r to the angle ϕ_d, which has units of degrees. Unlike the point-matching method, units are very important here because derivatives have units. If we leave Equation (8.45) as it is, the units on f_4, then f_4' and f_4'' will be (length/degree), (length)/(degree)^2, and (length)/(degree)^3, respectively, whereas all three should have units of length. You can see this by considering the units in Equations (8.50) through (8.54). To convert Equation (8.45) into one where ϕ has units of radians, make the substitution

$$\phi_d = \phi \frac{180°}{\pi},$$

which yields

$$r = C_5\phi^2, \quad \text{where } C_5 = 1.9697 \text{ inches.} \tag{8.55}$$

Choose the design position

$$\phi_{des} = 45° = 0.7854 \text{ radians,} \tag{8.56}$$

and from Equation (8.55),

$$r_{des} = 1.215 \text{ in.} \tag{8.57}$$

Differentiate Equation (8.55) three times with respect to ϕ, and evaluate at ϕ_{des} and r_{des}:

$$f_4 = 3.094 \text{ inches}, \quad f_4' = 3.9394 \text{ inches, and} \quad f_4'' = 0.0 \text{ inches.} \tag{8.58}$$

Choose $r_{4_{des}} = 10$ in. and substitute it and the derivatives in Equation (8.58) into Equation (8.50) and compute

$$\begin{array}{lll} E_1 = 3.094 & D_1 = -10.000 & F_1 = 61.88 \\ E_2 = 6.061 & D_2 = 6.188 & F_2 = -97.93 \\ E_3 = 9.282 & D_3 = 1.818 & F_3 = -73.13. \end{array} \tag{8.59}$$

Substitute these values into Equation (8.54) and compute

$$\lambda_1 = -618.8 \text{ in}^3, \quad \lambda_2 = 1122. \text{ in}^3, \text{ and } \lambda_3 = -212.2 \text{ in}^3.$$

Substitute these values into Equation (8.53), compute two values for $\tan\theta_2$, and find the corresponding values of θ_2 in the design position, $\theta_{2_{des}}$:

$$\theta_{2_{des}} = 12.10° \quad \text{or} \quad \theta_{2_{des}} = 57.99°.$$

Use $\theta_{2_{des}} = 12.10°$, and from Equation (8.52),

$$G_1 = 108.7 \text{ in}^2, \quad G_2 = -537.9 \text{ in}^3, \text{ and } \quad G = -6.868 \text{ in}.$$

Substituting these into Equations (8.51) gives

$$S_1 = -15.83 \text{ in.} \quad \text{and} \quad S_2 = 78.33 \text{ in}^2,$$

and from Equation (8.35),

$$S_3 = 238.4 \text{ in}^2,$$

and the link lengths are computed from Equation (8.36):

$$r_1 = -4.95 \text{ in}, \quad r_2 = -7.92 \text{ in, and} \quad r_3 = 18.04 \text{ in}.$$

The reference for displacement r, r_{4_i}, is

$$r_{4_i} = r_{4_{des}} - r_{des} = 10 \text{ in.} - 1.215 \text{ in.} = 8.785 \text{ in.} \tag{8.60}$$

To make the value of r_2 positive, add 180° to the value of θ_2 in the design position, so $\theta_{2_{des}} = 12.10° + 180.0° = 192.1°$. The reference for ϕ, θ_{2_i}, is given by

$$\theta_{2_i} = \theta_{2_{des}} - \phi_{des} = 192.10° - 45° = 147.10°.$$

The function-generating crank-slider mechanism is now fully defined and is shown in Figure 8.18. Figure 8.19 shows the performance of the mechanism, which is nearly a perfect match to the desired function.

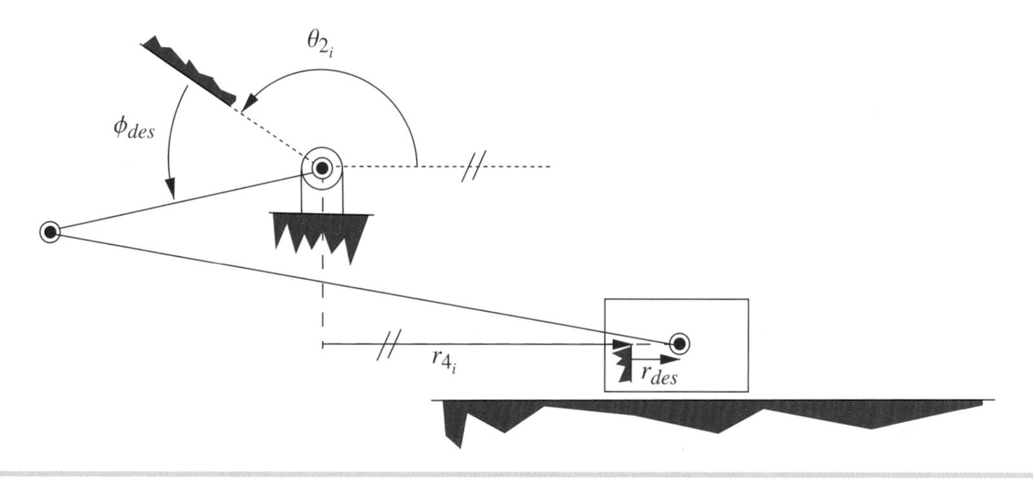

FIGURE 8.18 Crank-slider mechanism in the design position

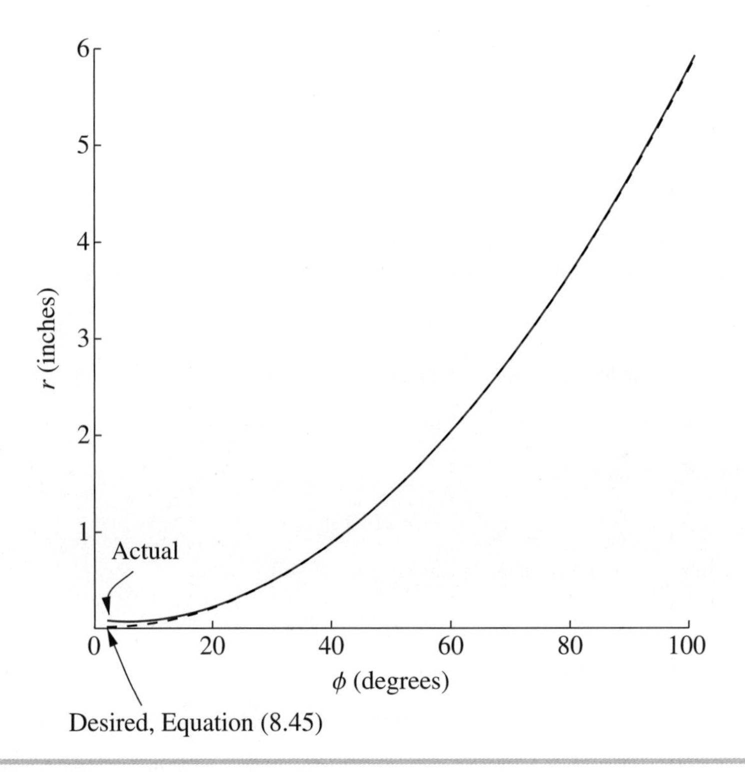

FIGURE 8.19 Performance of the function generator

8.3 DESIGN PROBLEMS

The following problems are examples of function generators.

Design Problem 2

The pipe flow of a liquid material in a plant is regulated by the butterfly valve shown in Figure 8.20. When $\psi_d = 0°$ the valve is fully closed. When $\psi_d = 90°$ the valve is fully open.

The mass flow rate of the product, $\dot{m}$, is related to the angular position of the valve, ψ, in a nonlinear fashion. The flow tends to surge when the valve is near its closed position and is relatively unaffected by the valve position near its open position. Measurements have been conducted in the plant relating the flow rate to the valve position. The data points are shown in Figure 8.21.

A function, $\dot{m} = f(\psi_d)$, must be developed to relate the mass flow to the valve position. Some requirements of this function follow.

- At $f(\psi_d = 0°)$, it is equal to 0.
- At $f(\psi_d = 90°)$, it is equal to 50.
- At $\frac{df}{d\psi}(\psi_d = 90°)$, it is equal to 0.
- $\frac{df}{d\psi} > 0$ for $0 \leq \psi_d \leq 90°$.

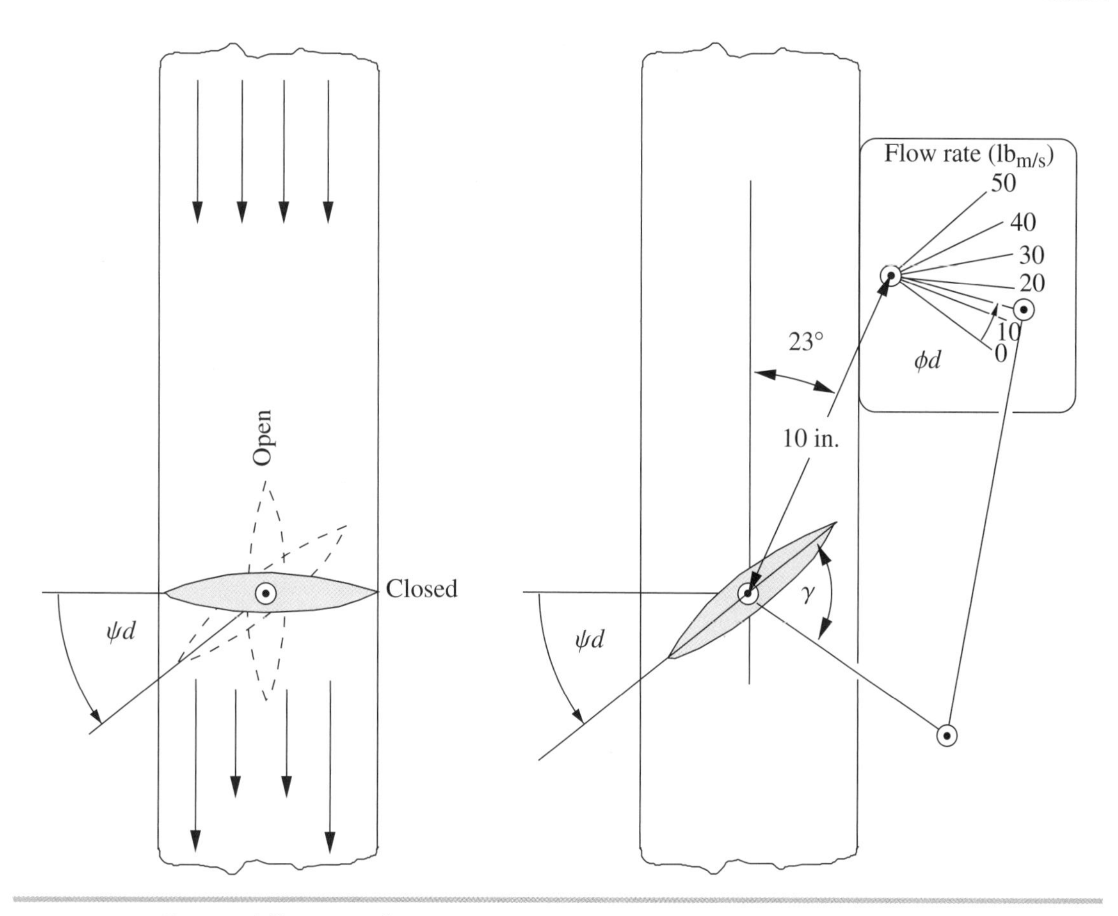

FIGURE 8.20 Proposed flow control system

The experimental data look more polynomial than exponential or sinusoidal, and the one-parameter family of curves,

$$\dot{m} = (50\ \text{lb}_\text{m}/\text{s})\left(1 - \left(1 - \frac{\psi_d}{90°}\right)^n\right), \quad \psi_d \text{ in degrees,} \tag{8.61}$$

seems to be a good model. A nonlinear regression to this model (courtesy of *Mathematica*) yielded a least-squares fit value of $n = 4.83000$.

The nonlinear nature of this relationship is making it difficult for an operator or an automatic control system to regulate the product flow near the closed position, due to the sensitivity of the flow rate to the valve opening.

It is desired to linearize this relationship so as to simplify (linearize) the flow control. The system shown in Figure 8.20 has been proposed to linearize the control of the flow rate. It incorporates a four bar mechanism whose function is to convert the nonlinear relationship into a linear one. To do this, we would like to have the angular position of the input, ϕ_d, be proportional to the mass flow rate. The proportionality is the designer's free choice. Let's begin by choosing

$$\dot{m} = C\phi_d, \quad \text{where} \quad C = 1\ \text{lb}_\text{m}/\text{deg}\cdot\text{s and } \phi_d \text{ is in degrees.} \tag{8.62}$$

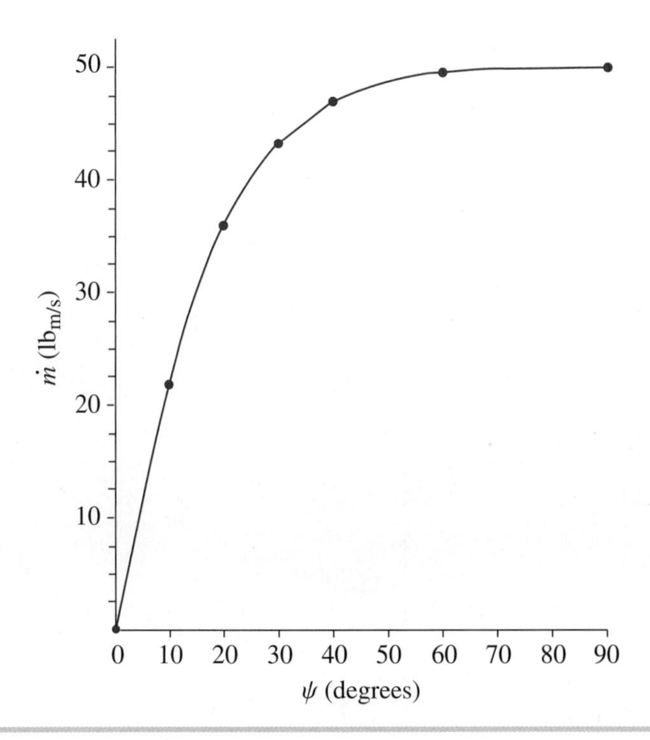

FIGURE 8.21 Data relating mass flow rate to valve position

Equating the right sides of Equations (8.61) and (8.62) gives the desired function generation:

$$(1\ \mathrm{lb_m/deg \cdot s})\phi_d = (50\ \mathrm{lb_m/s})\left(1 - \left(1 - \frac{\psi_d}{90°}\right)^n\right)$$

$$\phi_d = (50°)\left(1 - \left(1 - \frac{\psi_d}{90°}\right)^n\right). \tag{8.63}$$

It is necessary to convert this relationship to one where the angles ϕ and ψ are in radians so that the derivatives are dimensionless. To do this, substitute

$$\phi_d = \phi\ \frac{180°}{\pi} \quad \text{and} \quad \psi_d = \psi\ \frac{180°}{\pi}$$

into Equation (8.63), which yields

$$\phi = (5\pi/18)\left(1 - \left(1 - \frac{\psi}{\pi/2}\right)^n\right).$$

Start the design by using $\psi_{des} = 10°$, and determine the corresponding range of acceptable values of $\theta_{4_{des}}$. Then, in the design procedure, use $\theta_{4_{des}} = 20°$ (it is in the range of acceptable values). The whole class should do this case and should report the following numbers (After that, you may change ψ_{des} and $\theta_{4_{des}}$ on your own, and optimize.)

- The value of ϕ_{des}
- *Algebraic* expressions for the three derivatives: $\frac{d\psi}{d\phi}$, $\frac{d^2\psi}{d\phi^2}$, and $\frac{d^3\psi}{d\phi^3}$
- The values of these derivatives evaluated at $\psi = \psi_{des}$

- The values of a_i, b_i, c_i, and d_i, $i = 1, 2, 3$
- The values of E_i, D_i, and F_i, $i = 1, 2, 3$
- The values of λ_1, λ_2, and λ_3
- The two values of $\tan\theta_2$
- The two values of $\theta_{2_{des}}$, of which you will use the one of your choice
- The values of G, G_1, and G_2
- The values of R_1, R_2, and R_3
- Your chosen value of r_1 and the corresponding values of r_2, r_3, and r_4.
- The values of θ_{2_i} and θ_{4_i}
- The angle γ in Figure 8.20, which defines the attachment angle of link 4 to the valve

In your results, you should have plots of $\dot{m}$ vs. ϕ_d from the desired function generation, Equation (8.62), and the actual function generation coming from your four bar mechanism. You should also plot the error vs. ϕ, where the error is the desired value of $\dot{m}$ minus the actual value of $\dot{m}$.

If you wish, you may try to optimize the design. If you do this, be certain first to report the numbers for the case above so your report can be graded, and then give the numbers for your optimal result and perhaps a few words about the logic you applied in finding it.

Design Problem 3

Design a four bar mechanism to replace a pair of gears, where the ratio of input speed to output speed to is 3:1, i.e. $n_s = 3$ and $R = \frac{1}{3}$.

Design Problem 4

Figure 8.22 shows a vehicle making a left turn of radius R. The track is the lateral distance between the centers of the wheels, t, and the wheelbase is the distance between the front and rear wheels, w. The rear wheels are fixed in direction and rotate independently. The front wheels are pivoted on a kingpin axis perpendicular to the plane of motion so as to steer the vehicle around the turn. In order that the front wheels roll without slipping, the velocity of their wheel centers must be in the directions shown, which are perpendicular to the line from the center of the turn to the respective kingpin axes.

From the two right triangles in Figure 8.22 we see that,

$$\tan\phi = \frac{\omega}{R - \frac{t}{2}}, \qquad \tan\psi = \frac{\omega}{R + \frac{t}{2}}. \tag{8.64}$$

Positive R corresponds to turning left and negative R corresponds to turning right. Eliminating R gives the relationship between ϕ and ψ required to turn a corner of any radius without slipping at the front wheels,

$$\frac{\omega}{\tan\phi} + t - \frac{\omega}{\tan\psi} = 0. \tag{8.65}$$

This relationship is called the Ackerman Effect.

The purpose of this design problem is to synthesize a four bar steering mechanism such as that shown in Figure 8.23, which connects the front wheels of the vehicle and produces the desired Ackerman Effect described in Equation (8.65).

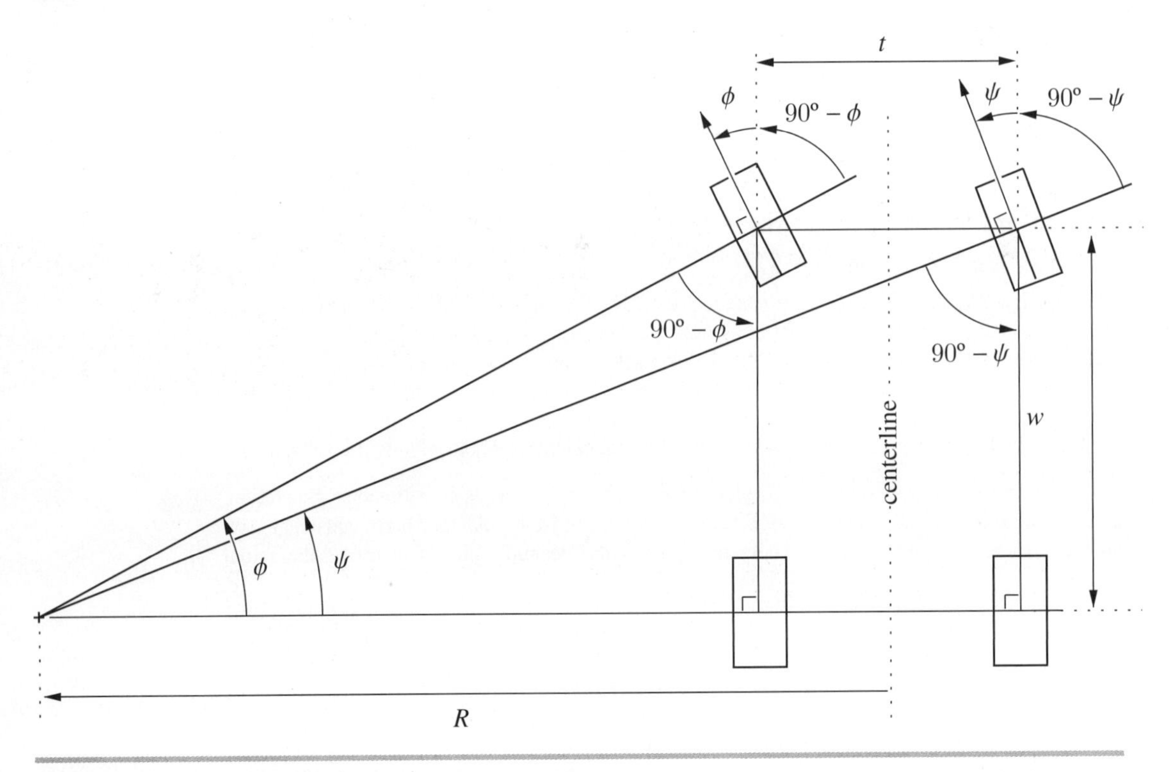

FIGURE 8.22 Ackerman Steering

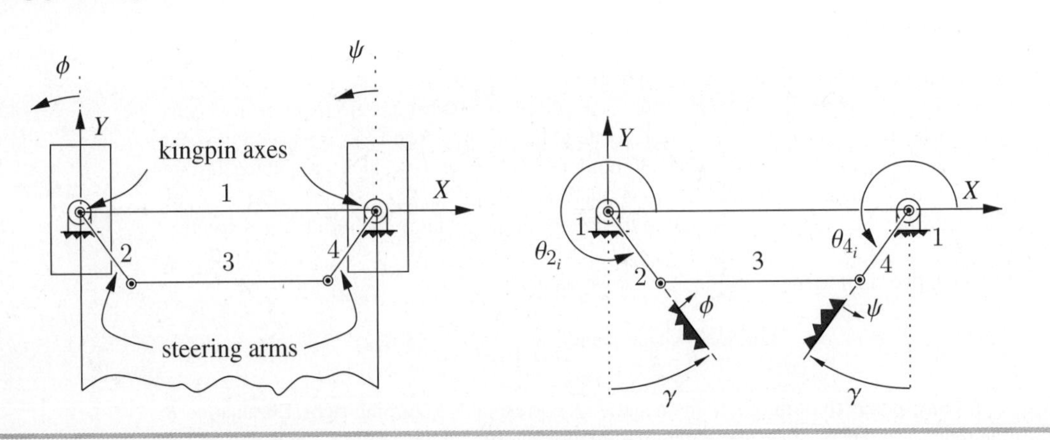

FIGURE 8.23 Four bar mechanism developing Ackerman Steering

The two levers of the four bar mechanism, 2 and 4, are called the steering arms. The steering arms directly rotate their respective tires about their kingpin axes. In Figure 8.23 the vehicle is moving on a straight line so $R = \infty$ and from Equation (8.64), ϕ and ψ are both zero. The right side of Figure 8.23 shows the references for measuring ϕ and ψ in this function generating four bar mechanism, defined by angles θ_{2_i} and θ_{4_i}.

For your vehicle use

$$t = 4 \text{ feet} \quad \text{and} \quad \omega = 5.5 \text{ feet}$$

Part A: Design the four bar steering mechanism using the Point Matching Method.

We might expect that when the mechanism is turning a radius $-R$ (to the right) that it would be a reflection of itself turning an equal but opposite radius R (to the left). So our intuition might tell us that the best four bar mechanism for this design would be symmetric.

Since driving straight is most common and critical we choose one precision point as,

1st Precision Point: $R = \infty \longrightarrow$ from Equation (8.64) $(\phi)_1 = 0°$ and $(\psi)_1 = 0°$, and to get a symmetric design lets choose the two following symmetric precision points,
2nd Precision Point: $R = 20$ feet $\longrightarrow$ from Equation (8.64) $(\phi)_2 = 16.99°$ and $(\psi)_2 = 14.03°$,
3rd Precision Point: $R = -20$ feet $\longrightarrow$ from Equation (8.64) $(\phi)_3 = -14.03°$ and $(\psi)_3 = -16.99°$,

To get a symmetric design such as seen in the bottom of Figure 8.23 we also need symmetric references for ϕ and ψ, which requires equal angles γ,

$$\theta_{2_i} - 270° = 270° - \theta_{4_i} \quad \longrightarrow \quad \theta_{4_i} = 180° - \theta_{2_i} \tag{8.66}$$

For purposes of comparison and verifying correctness, complete the design process using Freudenstein's equation to match the three given precisions points, using $\theta_{2_i} = 300°$, then from Equation (8.66) $\theta_{4_i} = 240°$. After your result for this design checks against your Instructor's and your peers', then you can vary the precision points and the references for ϕ and ψ (θ_{2_i} and θ_{4_i}) to try and improve performance. It is suggested you continue to seek symmetric designs, i.e. enforce Equation (8.66) on θ_{2_i} and θ_{4_i}.

Part B: Design the four bar steering mechanism using the Derivative Matching Method.

The derivative matching method precisely matches the desired function at only one point, which is the design position. If the vehicle is typically driving on a straight path, it would seem logical to choose $\phi_{des} = \psi_{des} = 0$.

We need to compute values of the three derivatives,

$$h_4 = \left.\frac{d\psi}{d\phi}\right|_{\substack{\psi = \psi_{des} \\ \phi = \phi_{des}}}, \quad h_4' = \left.\frac{d^2\psi}{d\phi^2}\right|_{\substack{\psi = \psi_{des} \\ \phi = \phi_{des}}}, \quad h_4'' = \left.\frac{d^2\psi}{d\phi^2}\right|_{\substack{\psi = \psi_{des} \\ \phi = \phi_{des}}}$$

To get determinate results, we should rewrite the Ackerman Function, Equation (8.65) as

$$\tan\psi = \frac{\tan\phi}{1 + a\,\tan\phi} \qquad \text{where: } a = \frac{t}{w}. \tag{8.67}$$

Differentiating Equation (8.67) three times with respect to ϕ and evalating at $\phi = \phi_{des} = 0$ and at $\psi = \psi_{des} = 0$ gives

$$h_4 = 1, \quad h_4' = -2a, \quad h_4'' = 6a^2$$

where $a = \frac{t}{w} = \frac{4\text{ft}}{5.5\text{ft}} = 0.\overline{72}$ is given. You should be able to reproduce these derivatives. With values for these three derivatives we are ready to apply the procedure outlined in section 8.1.4. To get results that we can compare to the results of the point matching method, use $\theta_{4_{des}} = 240°$. Several interesting things happen in this problem.

1. All values of $\theta_{4_{des}}$ are acceptable. This is rarely the case.
2. For each value of $\theta_{4_{des}}$, one of the two corresponding values of $\theta_{2_{des}}$ gives link lengths of zero and infinity, which is not possible.
3. The remaining value of $\theta_{2_{des}}$ always gives a symmetric four bar mechanism!

CHAPTER 9

Mechanism Synthesis Part II: Rigid Body Guidance

This chapter introduces Burmester Theory, which is named for Ludwig Burmester. In the late 1800s Burmester made many of the foundational discoveries in mechanism synthesis. Burmester Theory is a method of designing one-degree-of-freedom planar mechanisms so that a coupler link belonging to the mechanism passes through a number of specified positions. For four bar mechanisms, up to five arbitrary positions can be specified. Burmester developed his theory geometrically. Our development will be more algebraic. Each approach has its benefits. The geometric development is more difficult, but it clearly reveals the kinematic geometry and is easier to apply. The algebraic development is simpler, but it reveals the kinematic geometry less directly and is more tedious to apply.

DEFINITIONS:

Rigid Body Guidance *A mechanism design (synthesis) problem where the link lengths of a mechanism need to be determined so that a moving body is guided through a set of finitely separated positions.*

This is illustrated for three positions in Figure 9.1.

Circling Point *A point on the moving body whose locations in the desired positions of the body all lie on a circle.*

Center Point *The center of the circular path of a circling point.*

Figure 9.1 illustrates a rigid body guidance problem in which the goal is to move the body through the three positions shown at the top. The four bar mechanism at the bottom achieves this goal. Notice that the moving hinges are circling points and their respective center points are the fixed hinges. In rigid body guidance, given a set of desired positions of the body, the idea is to locate circling points on the guided body and then to connect them to their respective center points with links. The resulting four bar mechanism will then realize the desired set of finitely separated positions.

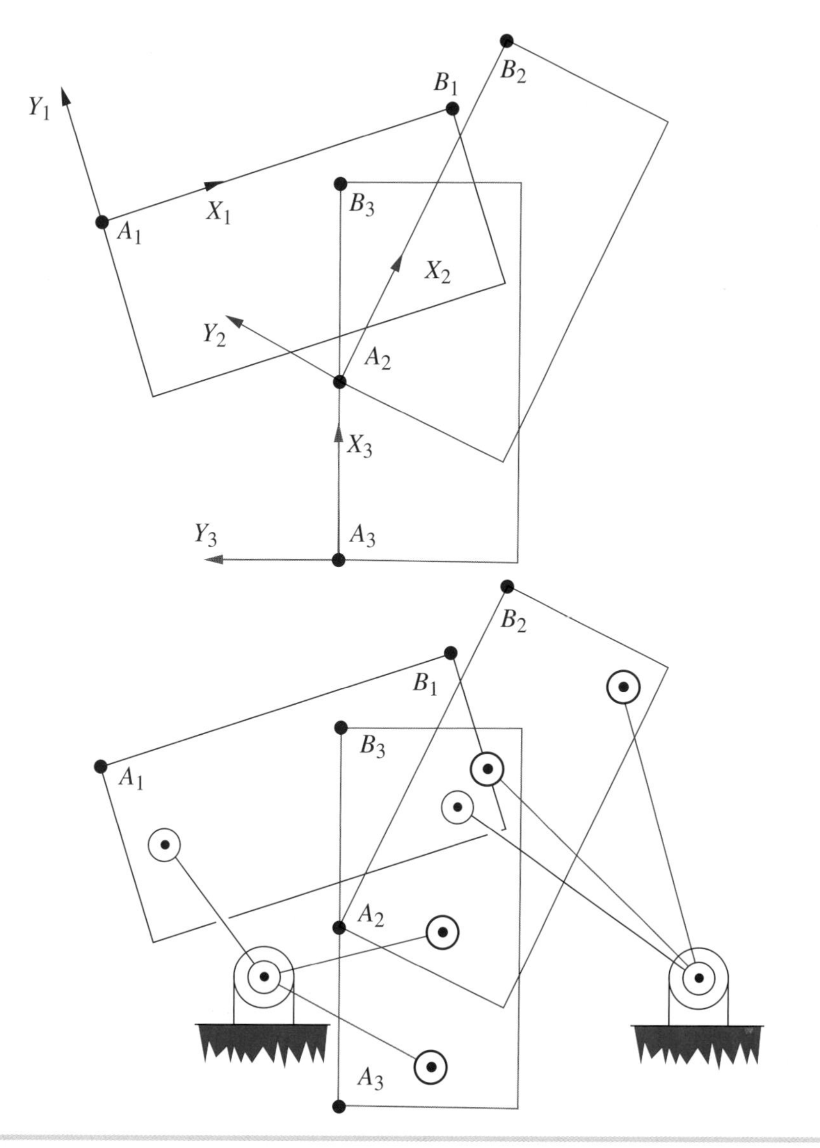

FIGURE 9.1 Circling points and center points for a three-position guidance problem

9.1 MATHEMATICAL MODEL OF A PLANAR RIGID BODY DISPLACEMENT

Before studying the three-position problem algebraically, we must develop a mathematical model of a planar rigid body displacement. The l.h.s. of Figure 9.2 shows a body in some initial position called position 1. The r.h.s. of the figure shows that same body after it has gone through some planar displacement into a position i. The body has attached to it an XY coordinate system that is located at X_1Y_1 in position 1 and at X_iY_i in position i. The coordinate system represents the body.

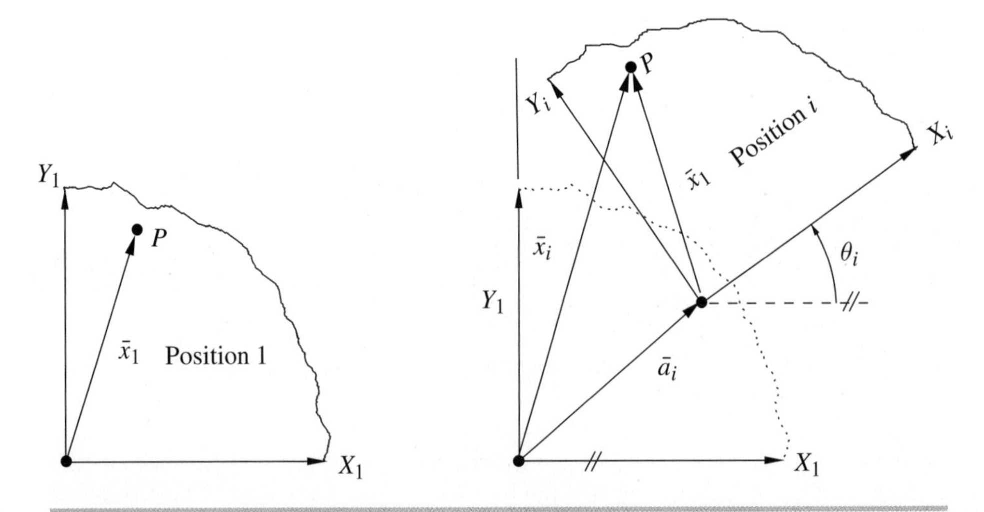

FIGURE 9.2 Mathematical description of a planar rigid body displacement

Use the coordinate system in position 1 as the reference for the displacement to position i. In displacing from position 1 to position i, the body (the attached coordinate system) has undergone a rotationθ_i and a translation $\bar{a}_i = [a_{ix}, a_{iy}]^T$.

Consider now an arbitrary point P belonging to the body. In position 1, it occupies a location relative to X_1Y_1 defined by the vector $\bar{x}_1 = [x_1, y_1]^T$. After the displacement, P occupies a position relative to X_1Y_1 defined by the vector $\bar{x}_i = [x_i, y_i]^T$. The relationship between $\bar{x}_i$ and $\bar{x}_1$ is through $\bar{a}_i$ andθ_i: $\bar{x}_i = \bar{a}_i + \bar{x}_1$. To find this relationship, work through the geometry on the r.h.s. of Figure 9.2 to obtain

$$x_i = a_{ix} + x_1\cos\theta_i + y_1\cos(\theta_i + 90°)$$

$$y_i = a_{iy} + x_1\sin\theta_i + y_1\sin(\theta_i + 90°).$$

Using the double angle formulas,

$$\cos(\theta_i + 90°) = -\sin\theta_i \quad \text{and} \quad \sin(\theta_i + 90°) = \cos\theta_i,$$

yields

$$x_i = a_{ix} + x_1\cos\theta_i - y_1\sin\theta_i$$
$$y_i = a_{iy} + x_1\sin\theta_i + y_1\cos\theta_i.$$

These two equations can be written in matrix form as

$$\begin{bmatrix} x_i \\ y_i \end{bmatrix} = \begin{bmatrix} \cos\theta_i & -\sin\theta_i \\ \sin\theta_i & \cos\theta_i \end{bmatrix} \begin{bmatrix} x_1 \\ y_1 \end{bmatrix} + \begin{bmatrix} a_{ix} \\ a_{iy} \end{bmatrix},$$

that is,

$$\bar{x}_i = A_i\bar{x}_1 + \bar{a}_i, \tag{9.1}$$

where $\bar{x}_i$ and $\bar{x}_1$ have been defined and

$$A_i = \begin{bmatrix} \cos\theta_i & -\sin\theta_i \\ \sin\theta_i & \cos\theta_i \end{bmatrix} \quad \text{and} \quad \bar{a}_i = \begin{bmatrix} a_{ix} \\ a_{iy} \end{bmatrix}. \tag{9.2}$$

Equation (9.1) is a mathematical model of a planar displacement. Matrix A_i is an orthogonal matrix known as the rotation matrix. Together, A_i and $\bar{a}_i$ define the ith position of the rigid body relative to position 1.

9.2 THE THREE-POSITION PROBLEM

To specify a set of n finitely spaced positions of a rigid body for a rigid body guidance problem, a set of matrices A_i and corresponding vectors $\bar{a}_i$ in Equation (9.2) are specified for $i = 1, 2, \ldots, n$, where $A_1 = I_2$ and $\bar{a}_1 = \bar{0}$. The case when $n = 3$ is known as the three-position problem. The goal is to design a mechanism so that the coupler link is guided through the three positions. The goal can be achieved by locating two circling points on the moving body and connecting them to ground at their respective center points with links.

Any three points will define a circle, and this makes the three-position problem a very simple one. In the three-position problem, *any point on the moving body is a circling point* because it will have a known location in each of the body's three positions, and its corresponding center point can be found. Therefore, the three-position problem has a number of four bar mechanism solutions equal to ∞^4.

When two circling points and their corresponding center points are found, the rigid body guidance problem is solved by connecting the moving body to ground with a link between each circling point and its center point. The link's length is the radius of the circle.

Suppose that $\bar{x}_i = [x_i, y_i]^T$ are the coordinates of a circling point, $\bar{x}^* = [x^*, y^*]^T$ are the coordinates of the corresponding center point, and R is the radius of the circle, as shown in Figure 9.3. Three equations of the form

$$R^2 = (\bar{x}_i - \bar{x}^*) \cdot (\bar{x}_i - \bar{x}^*), \qquad i = 1, 2, 3 \tag{9.3}$$

must be satisfied in order for the three positions of $\bar{x}_i$ ($i = 1, 2, 3$) to lie on a circle. For the case $i = 1$, Equation (9.3) can be solved for R,

$$R^2 = (\bar{x}_1 - \bar{x}^*) \cdot (\bar{x}_1 - \bar{x}^*), \tag{9.4}$$

and used to eliminate R^2 from Equation (9.3) for the cases $i = 2, 3$, thereby reducing Equation (9.3) to a system of two equations:

$$(\bar{x}_1 - \bar{x}^*) \cdot (\bar{x}_1 - \bar{x}^*) = (\bar{x}_i - \bar{x}^*) \cdot (\bar{x}_i - \bar{x}^*), \qquad i = 2, 3. \tag{9.5}$$

In terms of the specified displacements to positions 2 and 3, Equation (9.5) is rewritten as

$$(\bar{x}_1 - \bar{x}^*) \cdot (\bar{x}_1 - \bar{x}^*) = (A_i\bar{x}_1 + \bar{a}_i - \bar{x}^*) \cdot (A_i\bar{x}_1 + \bar{a}_i - \bar{x}^*), \qquad i = 2, 3.$$

Substituting for A_i and $\bar{a}_i$ gives

$$\begin{bmatrix} x_1 - x^* \\ y_1 - y^* \end{bmatrix} \cdot \begin{bmatrix} x_1 - x^* \\ y_1 - y^* \end{bmatrix} = \begin{bmatrix} x_1\cos\theta_i - y_1\sin\theta_i + a_{ix} - x^* \\ x_1\sin\theta_i + y_1\cos\theta_i + a_{iy} - y^* \end{bmatrix} \cdot \begin{bmatrix} x_1\cos\theta_i - y_1\sin\theta_i + a_{ix} - x^* \\ x_1\sin\theta_i + y_1\cos\theta_i + a_{iy} - y^* \end{bmatrix}.$$

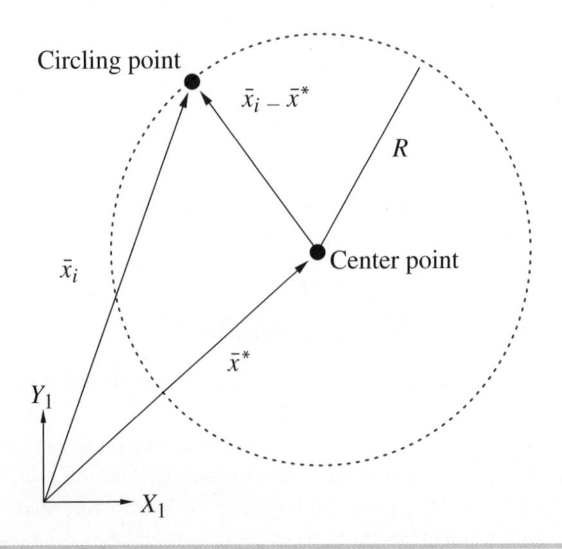

FIGURE 9.3 Equation of a circle

Taking the dot products yields

$$(x_1 - x^*)^2 + (y_1 - y^*)^2 = (x_1\cos\theta_i - y_1\sin\theta_i + a_{ix} - x^*)^2 + (x_1\sin\theta_i + y_1\cos\theta_i + a_{iy} - y^*)^2,$$

and expanding both sides gives

$$\begin{aligned}
&\cancel{x_1^2} - 2x_1x^* + \cancel{x^{*2}} + \cancel{y_1^2} - 2y_1y^* + \cancel{y^{*2}} = \\
&\quad x_1^2\cos^2\theta_i + y_1^2\sin^2\theta_i + a_{ix}^2 + \cancel{x^{*2}} + x_1^2\sin^2\theta_i + y_1^2\cos^2\theta_i + a_{iy}^2 + \cancel{y^{*2}} \\
&\quad \cancel{-2x_1y_1\cos\theta_i\sin\theta_i} + 2x_1a_{ix}\cos\theta_i - 2x_1x^*\cos\theta_i - 2y_1a_{ix}\sin\theta_i \\
&\quad +2y_1x^*\sin\theta_i - 2a_{ix}x^* + \cancel{2x_1y_1\sin\theta_i\cos\theta_i} + 2x_1a_{iy}\sin\theta_i \\
&\quad -2x_1y^*\sin\theta_i + 2y_1a_{iy}\cos\theta_i - 2y_1y^*\cos\theta_i - 2a_{iy}y^*, \qquad i = 2, 3.
\end{aligned} \tag{9.6}$$

Given a point on the body in position 1 and the displacements to positions 2 and 3, Equation (9.6) can be solved for the corresponding center point. This is developed in the next section.

9.2.1 Given the Circling Point and Determining the Center Point

Collecting the coefficients x^* and y^* in Equation (9.6) yields

$$U_ix^* + V_iy^* = W_i, \quad i = 2, 3, \tag{9.7}$$

where

$$U_i = x_1(\cos\theta_i - 1) - y_1\sin\theta_i + a_{ix}$$
$$V_i = x_1\sin\theta_i + y_1(\cos\theta_i - 1) + a_{iy} \tag{9.8}$$
$$W_i = \frac{a_{ix}^2 + a_{iy}^2}{2} + x_1(a_{ix}\cos\theta_i + a_{iy}\sin\theta_i) + y_1(a_{iy}\cos\theta_i - a_{ix}\sin\theta_i).$$

With Cramer's Rule, Equation (9.7) can then be solved for the coordinates of the center point:

$$\bar{x}^* = \begin{bmatrix} x^* \\ y^* \end{bmatrix},$$

where

$$x^* = \frac{\begin{vmatrix} W_2 & V_2 \\ W_3 & V_3 \end{vmatrix}}{\begin{vmatrix} U_2 & V_2 \\ U_3 & V_3 \end{vmatrix}} \quad \text{and} \quad y^* = \frac{\begin{vmatrix} U_2 & W_2 \\ U_3 & W_3 \end{vmatrix}}{\begin{vmatrix} U_2 & V_2 \\ U_3 & V_3 \end{vmatrix}}. \tag{9.9}$$

9.2.2 Given the Center Point and Determining the Circling Point

Given a desired center point and the displacement to positions 2 and 3, Equation (9.6) can be solved for the location of the corresponding circling point in position 1. Collecting the coefficients of x_1 and y_1 in Equation (9.6) gives

$$B_i x_1 + C_i y_1 = D_i, \quad i = 2, 3, \tag{9.10}$$

where

$$B_i = x^*(\cos\theta_i - 1) + y^*\sin\theta_i - a_{ix}\cos\theta_i - a_{iy}\sin\theta_i$$
$$C_i = -x^*\sin\theta_i + y^*(\cos\theta_i - 1) + a_{ix}\sin\theta_i - a_{iy}\cos\theta_i \tag{9.11}$$
$$D_i = \frac{a_{ix}^2 + a_{iy}^2}{2} - a_{ix}x^* - a_{iy}y^*.$$

With Cramer's Rule, Equation (9.10) can then be solved for the coordinates of the circling point in position 1:

$$\bar{x}_1 = \begin{bmatrix} x_1 \\ y_1 \end{bmatrix},$$

where

$$x_1 = \frac{\begin{vmatrix} D_2 & C_2 \\ D_3 & C_3 \end{vmatrix}}{\begin{vmatrix} B_2 & C_2 \\ B_3 & C_3 \end{vmatrix}} \quad \text{and} \quad y_1 = \frac{\begin{vmatrix} B_2 & D_2 \\ B_3 & D_3 \end{vmatrix}}{\begin{vmatrix} B_2 & C_2 \\ B_3 & C_3 \end{vmatrix}}. \tag{9.12}$$

9.2.3 Crank-Slider Design

One can think of a crank-slider mechanism as a four bar mechanism in which the path of one circling point is a circle of infinite radius, which is simply a straight line. The center of this path is located at infinity along a line perpendicular to the straight-line path of sliding. Then the length of the corresponding four bar link becomes infinite. Therefore, crank-slider mechanisms can be designed to solve the three-position rigid body guidance problem by finding a circling point whose corresponding center point is located at infinity. Consider Equation (9.9). The center point is located at infinity when the determinant—that is, the common denominator of x^* and y^*—is equal to zero. Expanding this determinant and simplifying yield

$$E(x_1^2 + y_1^2) + Fx_1 + Gy_1 + H = 0,$$

where

$$\begin{aligned}
E &= \sin\theta_3(\cos\theta_2 - 1) - \sin\theta_2(\cos\theta_3 - 1)\\
F &= a_{2x}\sin\theta_3 - a_{2y}(\cos\theta_3 - 1) - a_{3x}\sin\theta_2 + a_{3y}(\cos\theta_2 - 1)\\
G &= a_{2x}(\cos\theta_3 - 1) + a_{2y}\sin\theta_3 - a_{3x}(\cos\theta_2 - 1) - a_{3y}\sin\theta_2\\
H &= a_{2x}a_{3y} - a_{2y}a_{3x}.
\end{aligned}$$

This is the equation of a circle having a radius of $\frac{\sqrt{F^2+G^2-4EH}}{2E}$ and a center located at $[-\frac{F}{2E}, -\frac{G}{2E}]^T$. It is known as the circle of sliders in position 1. (Had all of the equations in this chapter been derived relative to position 2 instead of relative to position 1, the circle of sliders in position 2 would have been generated, and likewise for position 3.) For every point on this circle, taken as a point in the moving body in position 1 and termed the slider point, the locations of that point in positions 1, 2, and 3 are collinear, which is to say that the path of that point in the moving body can be a straight line between the three positions. The direction of the straight-line path between the positions is different for each point on the circle of sliders and is typically called the "path of sliding." The direction can be easily determined once the point $[x_1, y_1]^T$ has been selected, since the other points on the line can be directly calculated using Equation (9.1), and any two points define the equation of a line. Of course, the locations of the slider point in positions 2 and 3 will not lie on the circle of sliders.

Note that there are fewer crank-slider designs to solve the three-position problem than there are general four bar designs. Not every point in the moving body can be chosen as a slider point because three arbitrary points are not collinear. Therefore, instead of an ∞^2 number of choices for the circle point, there is a single infinity of choices for the slider point; it must lie on the circle of sliders rather than lying anywhere in the plane. Combining this with the ∞^2 number of choices for the other circling/center point, the number of crank-slider designs that solve the three-position problem is ∞^3. One can intuit this result by considering the design parameters of a four bar mechanism. Returning to the concept of a crank-slider mechanism as a four bar mechanism with one infinitely long link, the *a priori* specification of the link length of one member reduces the design space by adding one constraint. The same would be true regardless of the specified length. In this case, it happens to be infinite rather than, say, 2 inches, or any other value. If other constraints were added, such as specifying the direction of the path of sliding, the solution space would similarly be reduced.

9.2.4 Inverted Crank-Slider Design

In a crank-slider mechanism, the moving body is the coupler between the crank and the slider that slides on the ground. If this mechanism is inverted, the moving body slides on one of the cranks. In light of the crank-slider synthesis approach described above, it is not surprising that designing inverted crank-slider mechanisms to solve the three-position problem involves locating a center point for which the corresponding circling point is at infinity. Following the same approach, the determinant in the denominator of Equation (9.12) is equated to zero, yielding

$$K(x^{*2} + y^{*2}) + Lx^* + My^* + N = 0,$$

where

$$K = \sin\theta_2(\cos\theta_3 - 1) - \sin\theta_3(\cos\theta_2 - 1)$$
$$L = a_{2x}[1 + \sin(\theta_3 - \theta_2)] + a_{2y}[\cos(\theta_3 - \theta_2) - 1] + a_{3x}[\sin(\theta_3 - \theta_2) - 1] + a_{3y}[1 - \cos(\theta_3 - \theta_2)]$$
$$M = a_{2x}[1 - \cos(\theta_3 - \theta_2)] + a_{2y}[1 + \sin(\theta_3 - \theta_2)] + a_{3x}[\cos(\theta_3 - \theta_2) - 1] + a_{3y}[\sin(\theta_3 - \theta_2) - 1]$$
$$N = (a_{2x}a_{3y} - a_{2y}a_{3x})\cos(\theta_3 - \theta_2) - (a_{2x}a_{3x} + a_{2y}a_{3y})\sin(\theta_3 - \theta_2).$$

As anticipated, this is again the equation of a circle having a radius of $\frac{\sqrt{L^2 + M^2 - 4KN}}{2K}$ and a center located at $[-\frac{L}{2K}, -\frac{M}{2K}]^T$. Every point on this circle is a candidate center point for which the corresponding circling point is located at infinity. Thus any point on this circle can be used to design an inverted crank-slider mechanism that can at least be assembled with its moving body in each of the three prescribed positions.

▶ EXAMPLE 9.1

Design of a Conveyor Transporter A four bar mechanism should be designed to carry a box from the lower conveyor to the upper conveyor, while passing through the chosen intermediate position as shown in Figure 9.4. In many cases, only the initial and final positions of the body are specified, which leaves the intermediate position as a design choice. Relative to position 1, the second and third positions are specified as follows.

Position 2:

$$\bar{a}_2 = \begin{bmatrix} -2 \\ 6 \end{bmatrix} \text{ ft and } \theta_2 = 40^\circ$$

Position 3:

$$\bar{a}_3 = \begin{bmatrix} -10 \\ 8 \end{bmatrix} \text{ ft and } \theta_3 = 90^\circ$$

Solution

The top left corner of the box in position 1 has its other two positions on a circle whose center visually appears to be well located. For this reason, a moving pivot might be tried near that point. Choose the moving pivot

$$\bar{x}_1 = \begin{bmatrix} -3 \\ -1.5 \end{bmatrix} \text{ ft.}$$

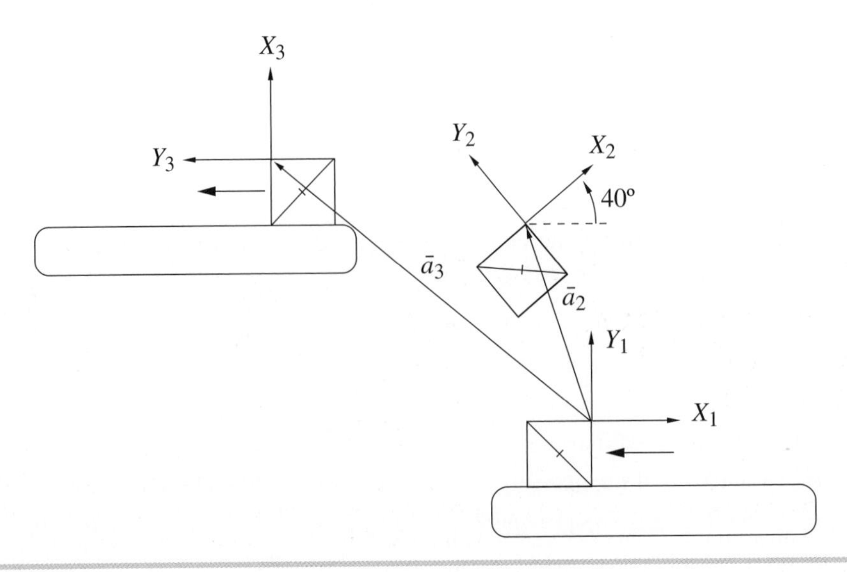

FIGURE 9.4 Design specification for transporter between conveyors

Then, from Equation (9.8),

$$U_2 = -0.33 \quad V_2 = 4.42 \quad W_2 = 4.20$$

$$U_3 = -5.5 \quad V_3 = 6.5 \quad W_3 = 43.0,$$

and from Equation (9.9), the fixed pivot is at

$$\bar{x}^* = \begin{bmatrix} -7.35 \\ 0.40 \end{bmatrix} \text{ ft},$$

and from Equation (9.4), the link length is $R = 4.74$ ft. Choosing another moving pivot at

$$\bar{x}_1 = \begin{bmatrix} -6 \\ 6 \end{bmatrix} \text{ ft}$$

and following the same procedure result in a fixed pivot at

$$\bar{x}^* = \begin{bmatrix} -9.32 \\ -0.20 \end{bmatrix} \text{ ft},$$

with a link length of $R = 7.04$ ft. Figure 9.5 shows the four bar mechanism in the design position, position 1.

Figure 9.6 shows an animation of the four bar mechanism guiding the package between the conveyors and passing through the three designated design positions.

9.3 THE FOUR-POSITION AND FIVE-POSITION PROBLEMS

The three-position problem is the simplest rigid body guidance problem of any significance, because any point on the moving body will have its three positions on a circle, so there are ∞^2 circling points to choose from for each of the moving hinges, and hence ∞^4 four bar mechanisms solve the three-position problem.

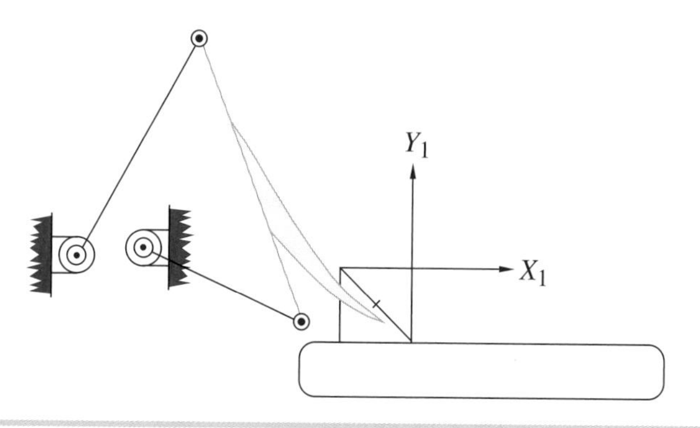

FIGURE 9.5 Design of transporter in position 1

The four-position problem is over-constrained, because in general, the four positions of an arbitrary point on the moving body will not lie on a circle. However, it can be shown that there exists an infinity of points on the moving body that *do* have their four positions on a circle, and these points all lie on a cubic curve known as the circling point curve. So for the four-position problem, there are an infinite number of choices for each moving hinge, and thus there are ∞^2 four bar mechanisms that solve a four-position problem.

The five-position problem is cast as two four-position problems, where each four-position problem considers a different set of four of the five desired positions. Each of the four-position problems has its associated circling point curve. The intersection of these two cubics yields a finite set of points whose five positions lie on a circle. These are called the Burmester points. The four- and five-position problems are beyond the scope of this undergraduate text. You are encouraged to study them in a graduate course that you may be able to take as a technical elective in your course of study.

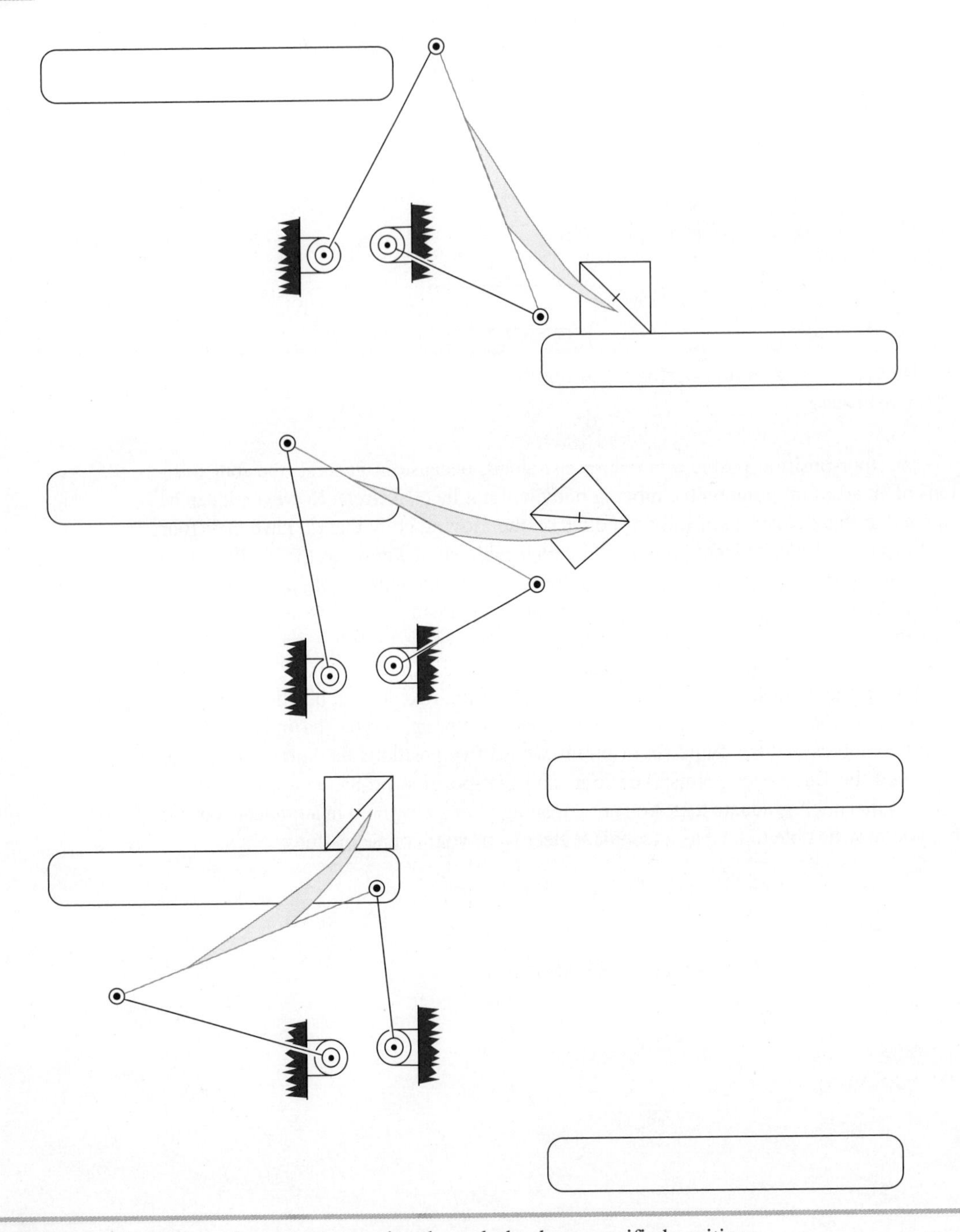

FIGURE 9.6 Animation of transporter passing through the three specified positions

9.4 DESIGN PROBLEMS

Design Problem 5

Design a four bar mechanism to guide a bicycle and the bicycle rack to which it is attached from the position shown to the overhead position. Ideally, the ground pivots would be located on the footboard. The links cannot collide with the wall or floor as the bicycle is lifted. The constraint on the location of the fixed hinges may be compromised if necessary. Not allowing link collisions with the wall and the floor is a hard constraint.

In your results, show an animation of your design that verifies the rigid body guidance.

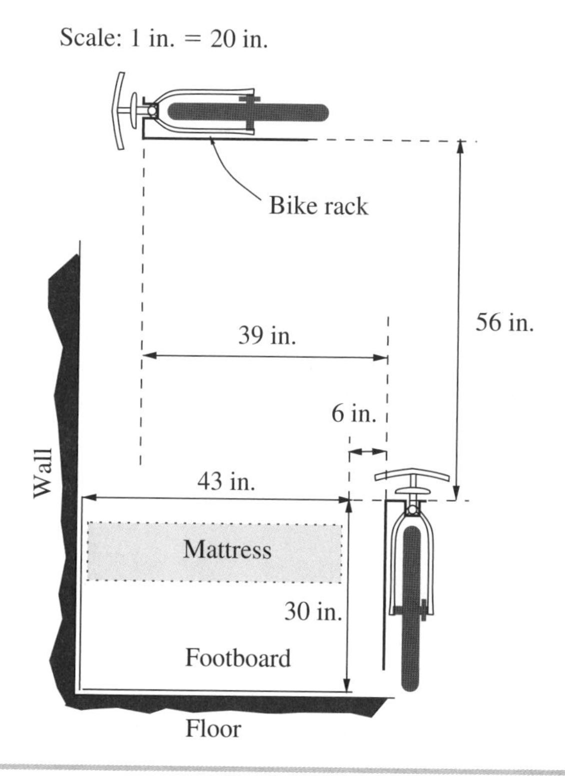

FIGURE 9.7 A dormitory bicycle storage mechanism

Index

D

E

F

G

L

M

R

T

U

V

W